Tourism, Tradition and Culture: A Reflection on their Role in Development

Tourism, Tradition and Culture: A Reflection on their Role in Development

By

David Harrison

Middlesex University

CABI

CABI is a trading name of CAB International

CABI
Nosworthy Way
Wallingford
Oxfordshire OX10 8DE
UK

Tel: +44 (0)1491 832111
Fax: +44 (0)1491 833508
E-mail: info@cabi.org
Website: www.cabi.org

CABI
WeWork
One Lincoln St
24th Floor
Boston, MA 02111
USA

T: +1 (617)682-9015
E-mail: cabi-nao@cabi.org

A catalogue record for this book is available from the British Library, London, UK.

Library of Congress Cataloging-in-Publication Data

Names: Harrison, David, 1941- author. | C.A.B. International.
Title: Tourism, tradition and culture : a reflection on their role in development / David Harrison, Middlesex University.
Description: First Edition. | Boston : CAB International, 2020. | Includes bibliographical references and index. | Summary: "David Harrison reflects on the role of tourism in development and its conflict with culture, tradition and the need for sustainability. The book deliberates David's research from the past 30 years, which looks at the impacts of tourism on societies emerging from Western colonialism and societies previously part of the Soviet system"-- Provided by publisher.
Identifiers: LCCN 2020026154 (print) | LCCN 2020026155 (eBook) | ISBN 9781789245899 (hardback) | ISBN 9781789245905 (eBook) | ISBN 9781789245912 (ePub)
Subjects: LCSH: Heritage tourism. | Culture and tourism.
Classification: LCC G156.5.H47 H369 2020 (print) | LCC G156.5.H47 (eBook) | DDC 338.4/791--dc23
LC record available at https://lccn.loc.gov/2020026154
LC eBook record available at https://lccn.loc.gov/2020026155

References to Internet websites (URLs) were accurate at the time of writing.

ISBN-13: 9781789245899 (hardback)
9781789245905 (ePDF)
9781789245912 (ePub)

Commissioning Editor: Claire Parfitt
Editorial Assistant: Lauren Davies
Production Editor: Marta Patiño

Typeset by Exeter Premedia Services Pvt Ltd, Chennai, India
Printed and bound in the UK by Severn, Gloucester

Contents

Acknowledgements

The author and publisher wish to thank the following who have kindly given permission for the use of copyright materials:

Marie Francois Lanfant, for Harrison, D. (1992) Tourism, capitalism and tradition. In: Lanfant, M.F. (ed.) *International Tourism Between Tradition and Modernity*, Actes du Colloque International, Nice, 19–21 Novembre, 1992, UREST-CNRS (Paris) pp. 109–122.

John Wiley and Sons Ltd., for Harrison, D. (1992) Tradition, modernity and tourism in Swaziland. In: Harrison, D. (ed.) *Tourism and the Less Developed Countries*, Belhaven Press, London, pp. 148–162.

Elsevier, for Harrison, D. (1994) Tourism and prostitution: sleeping with the enemy? The case of Swaziland. *Tourism Management* 15(6), 435–443.

John Wiley and Sons Ltd., for Harrison, D. (1992) Tourism to less developed countries: the social consequences. In: Harrison, D. (ed.) *Tourism and the Less Developed Countries*, Belhaven Press, London, pp. 19–34.

Bloomsbury Academic and Professional, for Harrison, D. (1996) Sustainability and tourism: reflections from a muddy pool. In: Briguglio, L., Archer, B., Jafari, J. and Wall, G. (eds) *Sustainable Tourism in Islands and Small States: Issues and Policies*, Pinter, London, pp. 69–89.

Taylor and Francis Group, for Harrison, D. (1996) Learning from the Old South by the New South? The case of tourism. *Third World Quarterly* 15(4), 707–721.

Cognizant Communication Corporation, for Harrison, D. (1998) The world comes to Fiji: who communicates what, and to whom? *Tourism, Culture and Communication* 1, 29–138.

Taylor and Francis Group, for Harrison, D. (2001) Islands, image and tourism. *Tourism Recreation Research* 26(3), 9–14.

Arvind Patel, of the University of the South Pacific, for Harrison, D. (2004) Tourism in Pacific islands. *Journal of Pacific Studies* 26(1–2), 1–28.

Taylor and Francis Group, for Harrison, D. (2004) Contested narratives in the domain of world heritage. *Current Issues of World Tourism* 7(4–5), 281–290.

Taylor and Francis Group, for Harrison, D. and Schipani, S. (2007) Lao tourism and poverty alleviation: community-based tourism and the private sector. *Current Issues in Tourism* 10(2–3), 194–230.

Taylor and Francis Group, for Harrison, D. (2008) Pro-poor tourism: a critique. *Third World Quarterly* 29(5), 851–868.

Elsevier, for Harrison, D. (2013) Cocoa, conservation and tourism: Grande Riviere, Trinidad. *Annals of Tourism Research* 34(3), 269–279.

Taylor and Francis Group, for Harrison, D. and Lugosi, P. (2013) Tourism culture(s): the hospitality dimension. *Tourism Recreation Research* 38(3), 269–279.

Taylor and Francis Group, for Harrison, D. (2007) Towards developing a framework for analysing tourism phenomena: a discussion. *Current Issues in Tourism* 10(1), 61–86.

for Harrison, D. (2014) Tourism and development: from development theory to globalisation. In: Lew, A., Michael Hall, C. and Williams, Allan M. (eds) *The Wiley Blackwell Companion to Tourism*, Wiley Blackwell, Chichester, pp. 143–154.

Ploysri Porananond, of Asian Journal of Tourism Research, for Harrison, D. (2016) Looking East but learning from the West: mass tourism and emerging nations. *Asian Journal of Tourism Research* 1(2), 1–16.

CAB International, for (2017) Introduction: mass tourism in a small world (with Richard Sharpley). In: Harrison, D. and Sharpley, R. (eds) *Mass Tourism in a Small World*, CAB International, Wallingford, pp. 1-14.

Elsevier, for Harrison, D. (2017) Tourism, mobilities and paradigms. *Tourism Management* 63, 329–337.

Taylor and Francis Group, for Harrison, D. (2010) Anthropologists, development and tourism: networks, encounters and shadows of a colonial past. *Tourism Recreation Research* 35(2), 109–118.

Personal Acknowledgements

At a more personal level, I first want to thank Senimili Kamikamica for being around and for just being herself, especially during lockdown. She has put up with me for more than 10 years, through thick and thin (and the last 2 years, in particular, have been unexpectedly and particularly thick). Others who helped along the way were Graham and Valerie Walters, Pete Saunders and Anita Emery, Greta Bowman and Ian, along with members of Kew Gardens Rotary Club, and I really appreciated the support of Claire Parfitt and Lauren Davies, of CAB International, who were helpful from start to finish.

My thanks to you all, and I hereby state that the many faults in the book are mine and mine alone.

1

Essaying Tourism: Reflections on Three Decades of International Tourism

For much of my life, I have seldom been far from the sea. A Londoner, but born in wartime Blackpool, England, with annual holidays spent largely on the Kent coast at my grandparents' home and, later, a doctorate based on fieldwork in a coastal village in Trinidad, I worked for more than two decades at the University of Sussex, Brighton, and latterly spent nine years in Fiji. It was perhaps no accident, then, that I should be attracted to international tourism, especially as it impacted on islands, and numerous events in my private life, some quite tragic, served to reinforce the tendency. Such influences are not recorded in any published output but, in reality, they have been at least as influential as any (quite often contrived) academic justification for any particular strand of research.

While these individual events and challenges were (more or less randomly) influencing my fledgling academic career, the wider world was also changing. Tourists had been travelling in increasingly large numbers since the mid-19th century, but only after the Second World War, as the middle class of the West flexed its financial muscle, and entrepreneurs began to utilize aircraft made surplus to post-war military requirements, did they begin to travel *overseas*. In 1950, a mere 25.3 million international tourist trips took place. By 1960, this had increased to 69.3 million; in 1970 it was 165.8 million and by 1980 it was 278.1 million (Harrison and Sharpley, 2017, pp. 1–3). And so on and so on, until in 2018 it had reached 1.4 billion (UNWTO (United Nations World Tourism Organization), 2019). International tourism was big business and, it seemed, then, its growth was unstoppable.

Tourism's social and economic benefits impacts increasingly permeated 'the Third World' and, later, the 'emerging economies' of the former Soviet bloc and, most recently, China, the once-sleeping giant of the East, until such regional distinctions became blurred in the global travel network. And as overseas resorts gained in tourist numbers, the share of the market taken by long-established resorts in the West correspondingly declined. The reasons, though complex, were succinctly summarized in a postcard written in 1980 by a disgruntled visitor to Blackpool: 'Weather foul. And there is nowhere worse than this place if its raining!'

This book contains a selection of my writing, mainly on international tourism, over three decades, a juggernaut which, despite concerns about overtourism (Dodds and Butler, 2019), showed no signs of abating until 2020. Because of the reflective approach I have adopted, this introduction is deliberately personal, and readings are sometimes contextualized by reference to positions I held at the time.

Chapters 2 and 3 of this book focus on tradition and its relationship to tourism. In Chapter 2, *Tourism, Capitalism and Tradition* (1992), tradition is portrayed in 'Not in front of the Tourists,' a (very partial) British TV documentary on Turkey, as highly vulnerable to international capital, a position also taken by such tourism academics as MacCannell and Graburn. 'Authentic' traditional life is replaced by commerce and commoditization, much of what was once given freely is now for sale, and local capital (by implication favoured by both film maker and academic) is swallowed up by the juggernaut of international tourism.

It is noteworthy that Greenwood's angry condemnation of the 'commoditisation' of the *alarde* of Fuentarrabia is in similar vein, only for it to be later retracted (Greenwood, 1989, pp. 171–185). Indeed, the 1990s formation of the Jaizkibel Company, a procession that included *women*, continues to offer a radical and much-disputed alternative to 'tradition' (https://www.youtube.com/watch?v=tg9dXIncMec, accessed 4 December, 2019) thus supporting my general argument that tradition, far from being a delicate flower that withers at the first exposure of social change, is a highly malleable set of institutions that can be harnessed to support a range of social groups anxious to strengthen their access to power.

The case study of tourism in Swaziland in Chapter 3, *Tradition, Modernity and Tourism in Swaziland* (1992), continues themes raised earlier. In the period prior to independence in 1968, King Sobhuza II and his allies revived and interpreted Swazi tradition, leaning heavily on the support of Hilda Kuper, a social anthropologist, firmly ensuring it bolstered their own position. At independence in 1968, this highly partisan package of Swazi culture, along with a revival of Swazi ceremonies and the promotion of the Swazi language, became official. The position of the king was legitimated and the monarchy became both the guardian and the epitome of Swazi 'culture'. 'Tradition' was also central to the marketing of tourism, in which the monarchy was also heavily involved. Here, the Swazi were (and are) presented as an exotic and picturesque people. However, this marketing ploy gave rise to contradictions: tradition, with all its paraphernalia, was a justification of the present by an appeal to the past, and where this was questioned, for example, by those who queried the authority or behaviour of the king, the status quo was quickly reinforced. Hopes that Swaziland would become a modern state proved unfounded as the language of tradition bolstered the absolute power of the monarchy and ensured its pre-eminence over modernity.

The fourth reading, *Tourism and Prostitution: Sleeping with the Enemy?* (1994), was specifically written to counter assertions that prostitution was increasingly providing the basis of tourism in Swaziland (Crush and Wellings, 1983, 1987), a position which seemed to me to be somewhat sensationalist. The overall findings were that prostitution clearly predated tourism, was often linked to migrant labourers returning from the Republic of South Africa, and that where tourists (often male business tourists – a minority of tourists visiting the country) were involved, their relationships with Swazi women varied considerably in intensity, ranging from casual sex arranged at a discotheque to stable relationships across the colour line that were not possible under apartheid in the Republic.

One prominent critic of the above paper complained that it lacked theory, thus implying that such case studies were an inadequate form of research. That is not the position taken here. There is a place for empirical investigation that subjects popular views to scrutiny, and in the Swazi case it was simply untrue to suggest that tourism was based on prostitution. However, reflecting on these early contributions to the tourism literature, I recognize they could be seen as a somewhat disjointed defence of international tourism, a charge to which I must plead guilty. It was a reaction to the way both sensation-seeking journalists and serious academics were demonizing international tourists when, in fact, with all their differences, and despite our relatively unsuccessful attempts to categorize them, they are similar, if not identical, to the populations from which they are drawn. Put differently, they are us! To focus only on 'deviant behaviour' (as did many social scientists) may make for good copy, but does international tourists, and by extension the tourist 'industry', a disservice, and has led to frequent accusations concerning the tourist 'industry', which has now become reified to the position of an important but largely impersonal actor.

In fact, in *Tourism and the Less Developed Countries: the Social Consequences* (1992) (reading five), I attempted a more theoretical approach to tourism, in which I discussed tourism's possible social consequences and related these to social problems, as perceived by key actors in the continued tension between tradition and modernity. The approach taken here was quite structural, focusing as I did on economic and social *institutions* which were exposed, changed or which newly emerged as a result of the appearance of tourism. Social change was not inevitable. However, where tourism involved increased commoditization, for example, there was also change in the social structure and patterns of ownership and entrepreneurship. The occupational and family structure were also likely to be impacted, possibly with pressure put on traditional patterns of behaviour and beliefs, with new social institutions and new patterns of social interaction.

None of these changes can be taken for granted. Society is comprised of economic, social and cultural institutions, which will be more or less impacted by tourism, but just how *much* they change will depend, *inter alia*, on their resilience and, importantly, any changes that *do* occur are not, in themselves, good or bad, positive or negative. Rather, how they are assessed will depend on the value judgements exercised by critics, including social scientists. This institutional approach to tourism seemed then, and does now, to be productive, allowing as it does an initial focus on institutions, and then an analysis of how – if at all – they change when faced with increased tourism.

For decades, there has been a recognition that 'development', if it is indeed to be considered (as the term suggests) a form of progress, must involve consideration of wider, non-economic features of the environment, namely the physical environment and (less often) the social and cultural aspects of social life. In Chapter 6, *Sustainability and Tourism: Reflections from a Muddy Pool* (1996), an effort is made to explore such concepts and, in particular, some of the ramifications of social and cultural 'sustainability' which, though important, are rarely prominent in the debate over whether or not 'development' is occurring. If they are prioritized, it soon becomes evident that problems arise. Many host societies, for example, are traditionally far from egalitarian, and difficulties inevitably arise in determining who should benefit most from tourism-fostered growth. More particularly, it is difficult to say how far *culture* remains the same when major changes occur over time, as in the transition from fishing villages to centres of tourism. Indeed, the longer the timescale involved, the greater the changes that may occur in social and cultural practices.

In Chapter 7, *Learning from the Old South by the New South? The Case of Tourism* (1994), focus is on the implosion of the Soviet system in 1989, disintegration of the 'Second World', and the subsequent 'emerging economies', which were thus distinguished from the 'new states' of the 1950s and 1960s. Clearly, the 'emerging' economies had more developed industrial sectors, but in many other respects they had much in common with the 'old south' There was intense pressure to reform Soviet-oriented economies and replace them with market economies, the old ruling class was challenged by new entrepreneurs (sometimes with links to the former regimes), ethnicity and nationalism were on the rise, and aid was sought from the West, primarily the European Union.

It was suggested that these 'new states' should not abandon planning, which was necessary if tourism development were to occur and that, similarly, the state should monitor and set limits to such development. At the same time, it should be recognized that tourism has the *potential* to contribute to social unrest, especially when wide differences in culture and wealth between tourists and members of the receiving society are present.

Much of my writing has been a reaction to positions taken by other commentators of tourism or to important events outside tourism in the wider world – hence the papers on tourism and prostitution and on changes in the world order. My 1996 move to Fiji led to a more pronounced focus on tourism's role in the development of small island states. In reading eight, *The World Comes to Fiji: Who Communicates What, and to Whom?* (1998), I revive the notion of the 'plural society', a somewhat discredited approach when applied to small Caribbean societies but much more applicable in the Fiji context, and examine the portrayal of ethnic Fijians and Indians in the tourist brochures. Utilizing a quite basic form of content analysis, I examine the content of tourism brochures for Fiji and the potential they bring for cultural change. While the literature commonly considers tourism's role in changing attitudes and/or behaviour of members of the *receiving* society (especially that of young people and women), the so-called 'demonstration effects' or inward acculturation - it tends to ignore possible changes among tourists - which I term 'outgoing acculturation. The paper finds that the people portrayed in the brochures are almost entirely ethnic Fijian. East Indians, while employed in the tourism sector, are usually backroom staff and virtually absent from the brochures. At the same time, the culture of ethnic Fijians is largely protected from *inward* acculturation by their continued residence in traditional villages, where the position of the chiefs and elders continues to dominate (though access to alternative world views is provided by the Internet. Tourists are not the only exemplars of modernity).

A more historical approach is taken in Chapter 9, *Islands, Image and Tourism* (2004). Recognizing the current attraction of tropical island destinations, in particular, for Western tourists, I examine how they have been portrayed as utopian escapes from Western reality or dystopian harbours of dark and unknown human and non-human forces, sometimes used to isolate the criminal, rebel, or dissident in prisons and detention centres or – even today – place the sick in quarantine (Kingdon, 2015). Such tropical islands have retained their fascination through the ages, morphing from travellers' tales and Tudor exploration to adventures of Victorian children and their rousing stories for the nation's children, through to arenas of bloody battles of the Second World War and, in peacetime, 'paradise' for the global tourist, if not for the residents of these islands, many of whom seek residence in the very same countries tourists want to leave.

Chapter 10, *Tourism in Pacific Islands*, is a review of tourism throughout the region. Bringing together themes raised in earlier papers, it situates tourism in the colonial experience, and shows that its development has often been constrained by political upheaval and unrest and, sometimes, 'traditional' land tenure. In some countries, however, linked to the USA, mass tourism has been able to develop. At the same time, sustainable tourism development is the subject of much rhetoric, which regrettably is rarely matched in practice.

Chapter 11, *Contested Narratives in the Domain of World Heritage* (2004) focuses primarily on how 'heritage' is interpreted by various social actors, either individually or collectively. What is 'natural' is contested; heritage is a performance, and debates occur as to what is genuine, authentic or fake. Memories, too, are selective: victors in power struggles ignore or downplay shameful historical episodes in favour of those that show them in a favourable light, and World Heritage status is usually accorded only to sites that have gone through a rigorous negotiation process to determine *which* of them is of *universal* value (and reflects most favourably on the dominant government and interest groups of the time).

Throughout this paper, and in many of the others reprinted here, my emphasis is on social realities that are the result of *choice* and human action. Nothing is inscribed on tablets of stone. This is not to deny the difficulties involved in challenging traditions and the established ways of doing things. Those who support the *status quo* more often than not have control of the big battalions. However, today's innovation can become tomorrow's tradition, and the young Turks of today may become tomorrow's elite (which applies *inter alia* to academia, as discussed in Chapter 20 of this book).

In the early 2000s I was a member of a team of consultants for South-east Asia for the Asian Development Bank (ADB) – perhaps the most enjoyable project in which I have ever participated. This involved extensive travel in South-east Asia, visiting areas proposed by governments to be developed as tourist attractions. The research skills I had accumulated as an academic were highly appropriate to the more 'quick and dirty' research carried out by consultants, who often relied upon publications such as the *Lonely Planet* guides for most of their information. In the course of this consultancy, I came across the Nam Ha Ecotourism Project in Lao PDR, then a flagship project of the ADB, and recipient of considerable funding (some million US dollars) from a variety of overseas agencies (overseen by my co-author, Stephen Schipani). I *also* researched private sector tourism development of small hotels and guest houses on the island of Don Det in southern Laos. It soon became evident that the quite different basis of their development – private capital, often raised through the sale of cattle or buffalo – would repay careful research. The key result of the research, reported in Chapter 12, was that the development of small hotels and guest houses by the private sector, a little studied aspect of pro-poor tourism, was at least as efficient as, and certainly more cost effective than, the international project.

While my 2007 paper recommended greater involvement by the private sector in the Nam Ha project, such suggestions need to be tempered to circumstance. Indeed, since the paper was written, private operators in the Luang Nam Ha project have multiplied, but at the cost of the integrity of the experience. A more recent report (Cunha, 2014) catalogues a variety of problems, including inadequate reporting of fees, lack of business experience and inconsistent service (Cunha, 2014). Clearly, monitoring their behaviour across a wide geographical area, with few trained human resources, has represented a major challenge to the Lao authorities.

Small, community-based tourism enterprises (such as the Nam Ha project) have often been promoted by advocates of pro-poor tourism as alternatives to mass tourism for poverty reduction, often giving the impression that such pro-poor tourism (PPT) enterprises were the future of the tourism sector. In Chapter 13, *Propoor Tourism: a Critique* (2008), I argue that the perception of PPT as a distinct (and small scale) variant of tourism involves conceptual and substantive problems. There is, indeed, a place for such tourism but it is not and will never be the *alternative* to mass tourism, which will continue as the majority pursuit, and which must incorporate both sustainability and equity if genuine PPT is to occur.

Chapter 14, *Cocoa, Conservation and Tourism* (2007), re-visits the small Trinidadian village of Grande Riviere, which was the subject of my PhD thesis in the early 1970s (Harrison, 1975). It was then an isolated settlement, 'behind God's back', suspicious of outsiders and extremely insular. However, in the early 2000s, I learned it was attracting tourists, who came to experience the vast numbers of leatherback turtles which had begun (for reasons that remain unclear) to nest on the beach, thus making Grande Riviere one of the most important turtle breeding areas in the world. The opportunity was thus presented of studying a village which had moved from a reliance on cocoa as the primary source of income to tourism, and where the nature of village life both before and after tourism was known. The results were clear: although tourism had barely been able to halt the decline of the village, villagers enthusiastically welcomed opportunities to gain income from tourism, they were more open to the world outside, they welcomed outsiders and embraced conservation of the turtle population (which had previously been primarily a source of food). In Grande Riviere, conservation and tourism combined to bring new opportunities to the village and to offer a tourist product that – albeit relatively small scale – was genuinely pro-poor.

In 2008, when I returned to Fiji as Head of the School of Tourism and Hospitality Management at the University of the South Pacific, I became increasingly aware of the need to consider the hospitality component of our programme more closely. The degrees we offered included some fairly basic hospitality skills, which were required learning for our students. However, in exposing them to these *practical* subjects, we were also making a major incursion into indigenous culture. The layout of the dinner table, Western cuisine, the importance of alcohol and, indeed, the ethics of *service*, for example, involved the student in a tourist culture that was often in stark contrast to their indigenous Fijian culture. We were, in fact, teaching them a different way of life.

In Chapter 15, *Tourism Culture(s): The Hospitality Dimension* (2013), Peter Lugosi and I consider three quite different perspectives that assist in understanding how tourism cultures, especially those found in the hotel sector, may be related to those brought *into* hotels by Fijians, to whom the hotel sector and the service culture they incorporate may be entirely foreign. In the paper, we suggest three possible approaches: first, that such hotel cultures represent an uncertain, negotiated, third space which is hybrid in nature; second, they are part of a sociotechnical system, where tourism is the interface between the formal rule and procedures of the organization (the hotels) and the culture taken into the hotel context by hospitality providers; and, finally, hotels can be seen as examples of total institutions, separated from the world outside by quite distinct cultural practices and maintained by clear methods of social control. Tourism and hospitality education, rather than being a neutral array of skills without cultural significance, represented a cultural imperative to which entrants to the hotel sector must commit if they are to succeed. Put differently, they must Westernize.

There is a vast array of scholarly writing about tourism which, because of its very bulk and variety, seems at first sight to lack any central focus. In Chapter 16, *Towards Developing a Framework for Analysing Tourism Phenomena: A Discussion* (2007), I address this issue from a sociological perspective. I contend that tourism phenomena can be situated in any one of three levels of analysis: the most basic is the level of social interaction, where focus is on role-following individuals. Their clusters of roles are features of institutions, the basic building blocks of social structure which make up the intermediate level, and such structures, in turn, make up wider social systems, the third level. The features of these different levels of analyses are discussed in some detail, and prominent examples of research are provided. Recognizing these levels, and their inter-relation, would facilitate a more coherent understanding of tourism's role and importance in a wide range of societies, and also situates empirical research and theoretical analysis in a wider framework.

In Chapter 17, *Tourism and Development: From Development Theory to Globalisation (2014)*, I continue to explore the relationship of tourism to development. In the 1970s, it seemed to me that the role of tourism in 'developing' societies (and, later, 'emergent' societies) could be separated from mainstream development and analysed as a distinct phenomenon. Hence the focus on 'traditional' societies and their response to international visitors. Indeed, in Chapter 17, I try to outline the components of the tourism system, again making a distinction between the three *levels* at which it can be analysed. Such an approach may be too comprehensive, but it does have the merit of allowing us to situate the work of (say) MacCannell (on the nature of tourism in late capitalism) with that of Greenwood (whose focus is on the impacts of a town's tourism on tradition). More recently, though, my attention turned to the complex *relationship* between tourist-sending and tourist-receiving societies. As I indicate in this chapter, 'what goes on in Greece, Cyprus or Italy, Egypt and Tunisia, or the UK and France is often inter-related, and the processes are comparable.' It is no longer possible to limit studies of tourism development to developing societies, for they have much in common with the First World, including reliance on China as both a major recipient of tourists and a provider *of* tourists. And the 2020 onset of Coronavirus and the global lockdown, with the subsequent collapse of global *and* domestic tourism, must surely be the defining indicator that tourism is truly a global phenomenon.

In Chapter 18, *Looking East but Learning from the West? Mass Tourism and Emerging Nations* (2016), I reprise the growth of tourism in the West before focusing on its emergence in Japan and then, for the rest of the chapter, China. Some have argued that outbound tourism from China (and other 'emergent' nations) has been sufficiently distinct to require a different 'paradigm' (Winter, 2009; Cohen and Cohen, 2012), but so much of the experience of early international tourists from Europe has been repeated by the Chinese, and the rapid growth of studies of Chinese tourism, in Chinese and English, which are quite compatible with studies of Western tourism, suggests that such a 'paradigm shift' (if it can be so described) is unnecessary (an issue taken up in much greater detail in Chapter 20). This is not to deny the shift in the centre of gravity of international tourism from the West to China, on which many South-east Asian economies and Europe and North America increasingly depend for their prosperity, whether from international tourism or on other economic sectors. Here, perhaps, is indicated the continued need for commentary on international tourist movement. The movement patterns may change; players move from the side stage to the centre, but for much of the time the *processes* remain the same.

In Chapter 19, *Introduction: Mass Tourism in a Small World* (2017), co-authored by Richard Sharpley, we focus more specifically on mass tourism, which has long seemed to me to have been largely neglected by tourism academics. And yet it remains (or will, after we recover from the global lockdown caused by the Coronavirus) by far the dominant mode of travel for most international tourists. In the concluding sector of the chapter, an outline of the key features of an ideal type of mass tourism is presented, in the hope that it can be used in comparative studies of tourism phenomena across the global divide. As we have again learned, albeit painfully, what goes on in (say) a Chinese wet market can have major global ramifications. When someone sneezes in China, not only are the plans of intending travellers from the UK, France or the USA, for example, put on hold, but also human social interaction at its most basic, the operation of major social and economic institutions, and even of international government and non-government institutions, in and beyond tourism, can be radically changed.

Over the years, the notion of paradigm, and its applicability to tourism studies, has appeared and re-appeared, but in about 2014 – following attendance at a tourism conference in which several eminent tourism scholars referred to the need for a new paradigm to enable proper consideration of the increasingly dominant role of China in international tourism – I finally decided to tackle the issue head-on. The result was Chapter 20, *Tourists, Mobilities and Paradigms* (2017), a paper on which I probably spent more time and effort than any other piece of work. As a reward, the paper was rejected outright by one major tourism journal (Annals of Tourism Research), only to be accepted, with barely a changed word, by Tourism Management! This says much about the nature of peer reviews, refereeing, and the process by which reviewers are selected, not to mention their responsibility to be objective (rather than partisan) in their assessments. However, in the paper, I establish (I think conclusively) that the term 'paradigm' is inappropriately applied to tourism perspectives, which – despite claims to the contrary – have not been dominated by 'the discourse of authenticity'; rather, concepts initially developed by Western scholars studying Western tourism have increasingly been taken up, and used successfully, by Western and Chinese scholars studying *Chinese* tourism. There is no distinct Western social science, but social science which is applicable across the board, using concepts which are more or less applicable to tourists and tourism across a range of societies.

My final chapter, *Anthropologists, Development and Tourism Networks: Encounters and Shadows of a Colonial Past* (2010), is a reflection on the role of anthropologists when measured against the contribution of other stakeholders involved in tourism development projects. Anthropologists were greatly assisted by the colonial system, and for some time their work was tarnished by association. Indeed, the distinction between pure and applied anthropology remains, and employment of anthropologists by government continues to be problematic. Many tourism academics continue to be, at best, ambivalent towards the tourists they study, but few have successfully combined consultancy work with their academic positions and there is a continued divide between tourism academics and consultants. For academics in developing societies, relatively generous remuneration by international organizations may be considered a way of enhancing an otherwise poorly funded position, whereas (the relatively few) Western academics who become consultants may consider their involvement more of an extension of their academic interests. Such differences in approach should be explored during development projects, and the ensuing reflexivity can lead to genuine cooperation across cultural divides. More often than not, however, Western consultants continue to apply their preferred templates to new projects, and their local counterparts continue to pick up the quite substantial crumbs from the project table, with little attempt at reflexivity on either side.

On reflecting on my work on tourism over the last three decades, it would be good to report a coherent set of themes that had been followed up over the years. Regretfully, though, that has not been the case. True, I have written on the role tradition has played in tourism development, on the links between tourism and prostitution, and the similarities of the 'new nations' of the 1970s to the 'emerging nations' of the late 1980s. I have also explored the place of islands in the Western psyche, and tried to develop more coherent approaches to international tourism, considering at the same time that all such efforts are situated *within* a scientific approach that, rather than being a 'Western' invention, is of universal significance and application. And in such halting steps, I have realised that, once again, international tourism must be seen in the round, where events in one part of the world are intimately connected with events that occur elsewhere. This has been evident for years, but any doubts should surely be dispelled by the sudden and catastrophic onset of COVID-19 early in 2020, which is discussed in the final chapter.

In reflecting on my work over three decades, I must conclude it is unfinished business.

However, that is the nature of all science, and in applying its methods to the role of international tourism I have inevitably been highly selective in my choice of topics and subjective but (I hope) clear in my assessments. But 'doing science' is not easy:

> The fate of our times is characterised by rationalisation and intellectualisation and, above all, by the 'disenchantment' of the world ... To the person who cannot bear the fate of times like a man, one must say: may he rather return silently, without the usual publicity build-up of renegades, but simply and plainly. The arms of the old churches are opened widely and compassionately for him. After all, they do not make it hard for him. One way or another, he has to bring his 'intellectual sacrifice' - that is inevitable (Weber, 1948).

But for the committed scientist, the road less-travelled awaits:

> (I)n the lecture rooms of the university no other virtue holds but plain intellectual integrity. We shall set to work and meet 'the demands of the day,' in human relations as well as in our vocation. This, however, is plain and simple, if each finds and obeys the demon who holds the fibres of his very life (Weber, 1948).

References

Cohen, E. and Cohen, S.A. (2012) Current sociological theories and issues in tourism. *Annals of Tourism Research* 39(4), 2177–2202. DOI: 10.1016/j.annals.2012.07.009.

Crush, J. and Wellings, P. (1983) The southern African pleasure periphery, 1966–83. *The Journal of Modern African Studies* 21(4), 673–698. DOI: 10.1017/S0022278X00024265.

Crush, J. and Wellings, P. (1987) Forbidden fruit and the export of vice: tourism in Lesotho and Swaziland. In: Britton, R. and Clarke, W.C. (eds) *Ambiguous Alternative: Tourism in Small Developing Countries*. University of the South Pacific, Suva, pp. 91–112.

Cunha, F.M. (2014) *Nam Ha: Community-based Ecotourism: Current Status and Challenges*. Lao People's Democratic Republic.

Dodds, R. and Butler, R.W. (eds) (2019) *Overtourism: Issues, Realities and Solutions*. De Gruyter, Amsterdam.

Greenwood, D. (1989) Culture by the pound: an anthropological perspective on tourism as cultural commoditization. In: Smith, V.L. (ed.) *Hosts and Guests: The Anthropology of Tourism*, 2nd edn.

Harrison, D. (1975) Social relations in a Trinidadian village. PhD thesis. Department of Social Anthropology, University College London.

Harrison, D. and Sharpley, R. (2017) Introduction: mass tourism in a small world. In: Harrison, D. and Sharpley, R. (eds) *Mass Tourism in a Small World*. CAB International, Wallingford, pp. 1–14.

Kingdon, A. (2015) Five islands we are really sick of. Available at: https://www.hakaimagazine.com/article-short/five-islands-we-are-really-sick (accessed 13 April 2019).

UNWTO (United Nations World Tourism Organization) (2019) UNWTO world tourism barometer and statistical Annex. Available at: www2.unwto.org (accessed 1 December 2019).

Weber, M. (1948) Science as a vocation. In: Gerth, H.H. and Wright Mills, C. (eds) *From Max Weber: Essays in Sociology*. Routledge and Kegan Paul, London, pp. 129–156.

Winter, T. (2009) Asian tourism and the retreat of anglo-western centrism in tourism theory. *Current Issues in Tourism* 12(1), 21–31. DOI: 10.1080/13683500802220695.

2

Tourism, Capitalism and Tradition

Introduction

The growth of international tourism since the 1950s is well known. By 1990, about 425 million international tourist arrivals (as defined by the World Tourism Organization) were recorded, a 16-fold increase since 1950, and it is estimated that by the year 2000 this figure will increase to around 638 million (World Tourism Organization (WTO), 1991a). This immense but temporary migration of people is already the third largest item in world trade, accounting for more than 7% of all world exports and a level of expenditure exceeding the Gross National Product of all countries except the USA. When it is noted that domestic tourism is probably about ten times the extent of international tourism, it should be evident that tourism is big business.

The global movement of tourists is uneven. In 1989, 'developed' countries attracted 65% of all international arrivals and 72% of all tourism receipts, with 'developing countries and territories' (which then excluded European and Asian socialist countries) attracting 21% of international tourists and 26% of all tourism receipts (World Tourism Statistics, 1991b). Europe remains the main destination and sending region, even though its market share has declined over the years: in 1989, it accounted for 64% of all tourist arrivals and provided more than 60% of international tourists. By contrast, East Asia has become increasingly attractive and by the end of the 1980s was visited by 11% of all international arrivals (World Tourism Organization (WTO), 1991a).

In these circumstances, we have witnessed a headlong rush by local, regional and national authorities, both in developed and less developed countries, to join the tourist party. The rationale given is invariably economic: tourism – whether domestic or international – brings with it increased employment, increases in individual and collective incomes, new entrepreneurial opportunities and (in the case of international tourism) foreign exchange. There are, of course, continued debates about multiplier effects and leakages, and inflation has often been an inevitable result of tourism expansion but, in general, the consensus appears to be that, economically, tourism is beneficial.

At the social level, opinions are more varied. Indeed, lists of social 'costs' and 'benefits' are frequently provided, with the negative aspects normally being given more weight. The following example, from the world of television journalism, is fairly typical.

The Invasion of Arcadia

Written and presented by Robert Chesshyre, the documentary '*Not in front of the Tourists*', was produced by the British Broadcasting Corporation

(BBC), 1988. The 40-minute programme focused on tourism development in the Marmaris region of Turkey, especially on its effects on some smaller villages around Marmaris. In essence, the story was of the negative effects of tourism on traditional village life and the role of a local entrepreneur, owner of a small hotel. Throughout the film, tradition is presented as rooted in Arcadia and tourism (and tourists) as the equivalent of the snake, acting for the capitalist devil, in paradise.

At the beginning of the commentary, we are left in no doubt who the 'baddies' are going to be. The flight of the TV crew was delayed at Gatwick Airport by the behaviour of 'yobs', who unfortunately still managed to catch the flight, which was later described as 'a flying cattle truck', Then follows a series of Arcadian scenes, with complementary commentary:

Camera	Commentary
A long shot of peasants in the fields A barefoot man walks in mud behind a horse-drawn plough An elderly peasant woman prepares food Younger women prepare food in a group	Time, like the light, has a different quality here ... In the country, people follow the rhythm of the seasons. ... squatting on the ground like a biblical figure... 'traditionally shy, they still cannot face the camera'.

The rural idyll once established, we discover is in the process of change. In the course of constructing a warehouse, a village man decides, instead, to build a small hotel, which some (educated, non-package) tourists are later shown visiting. Arcadia is under attack.

Camera	Commentary
His young daughter, an English-speaker, serving guests in the restaurant and passing on the orders to her older brother. A crowded beach, with plenty of tourist bodies on display, including a long close-up of a woman sunbathing topless. Another long beach scene, and then a mosque with no worshippers ...	'In the country, where women are subordinate, she "bosses" the men...' 'Mammon, not God, determines what happens on Turkey's beaches. The work ethic, the low crime rate: all are vulnerable to the temptations of Western pleasure-seeking'. 'Attendance at prayers has declined, despite the efforts of a benevolent and tolerant iman'.

In the same manner, we are informed that the translator's commentator, 'a devout muslim', believes

> ...(Y)oung people have been spoiled by the new opportunities. They are keener to hang about in dark glasses with a packet of Marlborough in their pockets than work as their parents once worked.

Furthermore, adds the commentator,

> Tourism has achieved more secularisation in 3 years than Attaturk - the founder of modern Turkey - imposed in twenty. The driving force is now money.

Camera	Commentary
A busy street in Marmaris. Young women trying on (expensive) jewellery. Vendors pointing to their carpets. Motor bikes in narrow streets.	In this environment, we are told, tourism involves 'haggling and leather jackets, necklaces and hassle'.

The scene now shifts to 'another, simple hotel', a small hotel in the hills, where the new entrepreneur is preparing to welcome his first coach party:

Camera	Commentary
A goat is being prepared for the meal. The coach party arrives. A guest is hoping it will not be killed in front of them. The guests are shown eating and enjoying the goat.	'Our dinner was to be dispatched with ceremony and, I suppose, dignity. Or is it sentimental to imagine it is better to die, under the sun on a dusty hill than in an abattoir, reeking with death?' 'The wretched goat, which 2 hours before had been dragged protesting to slaughter'.

The scene shifts again, this time to a larger hotel:

Camera	
Waiters in traditional costume lining up to serve guests. Dancers, including a belly dancer (noted by the commentator to be an import into Turkish culture), entertain the guests.	Commentator: It's rather sad the image of the Red Indian people came to mind – the idea that people are in a reservation and brought forward and asked to perform. There's a loss of dignity involved. 'Foreign hotelier: It all becomes commercial Commentator: Isn't it sad that something once freely (given) has to be paid for?

After an interview with taxi drivers, complaining that they and their villages do not receive the benefits of enclave tourism, the film ends with the entrepreneur receiving money for his efforts.

The above description is not a caricature. Throughout the documentary, the simple, fulfilling and 'natural' life of rural Turkey is favourably contrasted with the commercialized, monetized and uncivilized bustle of tourism. 'Tradition' becomes inauthentic and is replaced by commerce. Spontaneous giving is replaced by the profit motive. What was once given freely now has to be paid for. Good is being replaced by bad.

Tradition and the Sociology of Tourism

In the documentary, tradition is assumed to be good and tourism development is defined as socially and morally disruptive. Whatever the villagers themselves think of the opportunities tourism brings, the commentator implies, in general, their lot is going to change for the worse.

Of course, television commentaries are not formally written sociological texts and it is in the nature of the mass media to simplify and exaggerate. However, in so far as there is a sociology of tourism (a matter to which I return later), is its approach to 'tradition' so different? By way of illustration, I focus on a few examples of what might be described as 'best practice' in the approaches of various sociologists to this topic – not because I want to single anyone out for special treatment but to isolate what might be seen as tendencies which we may all share.

MacCannell is correctly recognized as a key figure in the sociological study of tourism. In his early work (1976), the tourist is an epitome of 'modern man', living in a structurally differentiated world, searching for security and authenticity in a context of increasing mass leisure, where work itself becomes an object of curiosity. Through tourism, the distinction between 'modern' and 'traditional' is broken down and a new and unifying ideology is formed, with authenticity at its heart. This ideology, located in the Western middle class and a central feature of the modernization process, is ultimately unsatisfactory: modernization continues its destructive path of differentiation, destroying only to re-unify and court further destruction, and so on.

Brought into contact with the modern, non-modern culture is artificially preserved and reconstructed *within* modern society (MacCannell, 1976, p. 8). The 'non-modern' is thus despoiled in its contact with modernity. Modernization, in which tourism is a key component, separates cultural objects

> ... from the people and places that made them, breaks up the solidarity of the groups in which they originally figured as cultural elements, and brings the people liberated from traditional attachments into the modern world where, as tourists, they may attempt to discover or reconstruct a cultural heritage or a social identity (MacCannell, 1976, p. 13).

Tradition 'is undermined and it is forced into a new role': embedded in modernity but in a position of servitude, 'it is recalled to satisfy nostalgic whims or provide coloration or perhaps a sense of profundity for a modern theme' (MacCannell, 1976, p. 34). Indeed, more recently MacCannell suggests that traditional societies have disappeared completely. With the disintegration of communist and capitalist ideologies, old forms of analysis, including (what he sees as) the 'problematic of modernization' (1992, pp. 295–296), with its (alleged) polarization of 'traditional' and 'modern' societies, are no longer appropriate. Instead, there is continual social flux among three distinct, inter-related and alternative cultural modes: the ex-primitive, the modern and the postmodern. 'Primitive' cultures have become 'ex-primitive' and, at least in theory, societies may apparently move from 'ex-primitive' status to either 'postmodern' or 'modern' and back again (MacCannell, 1992, pp. 302–303).

At this point, a caveat is necessary: MacCannell is clearly aware, as other passages (and many of his sources) indicate, that 'tradition' is not a set mode of thinking or behaving. To borrow the language of modernization theory, acculturation has been occurring all along the line, both in the written presentation of so-called 'primitive people' and in pasts and presents of their many-faceted and diverse cultures. Put differently, if postmodernism involves new contexts of style and culture, pastiche, variety, parody and so on, it is evident not only in the *styles* adopted by a new generation of anthropologists but also (and not coincidentally) in the cultures they present (or present). That said, the impression received from the passages cited is that traditional cultures either are artificially preserved, in a state of inauthenticity, for the dubious benefit of modern or postmodern man (presumably in the company of his postmodern mate), or they disappear completely. Like the dead parrot in the famous Monty Python sketch, traditional society is no more. It is dead. Deceased. It is gone. It is an ex-primitive.

In for a penny, in for a pound, or a pounding: let me move from MacCannell to Graburn, especially to the latter's discussion of cultural change in 'traditional' societies over the post-colonial period, and the ways they may interact with external cultural forms (Graburn, 1984). Interestingly, Graburn focuses on acculturation, a process closely associated with modernization perspectives, which MacCannell considers should be abandoned (1992, pp. 295–296). More specifically, Graburn charts the movement from the production of 'functional traditional arts', primarily for use in key religious rituals in traditional communities, to the production of 'commercial' replicas of the same items for purchase by tourists, and then to the production of souvenirs and novelties, some of which may later be incorporated into the local culture and form the basis of a revised tradition. At this point, the indigenous culture may respond to outside influences, absorbing, adopting and/or adapting art forms originating from the wider 'world' culture and producing 'assimilated arts' which, although barely distinguishable from art forms produced elsewhere, may attain the status of 'popular arts' within the society if purchased by sufficient members of the indigenous culture.

As Graburn recognizes, acculturation is not restricted to the production of art forms which may, in any case, mirror and even modify the self-perception and behaviour of members of a community. A parallel process can be noted in changing language patterns: like tourist arts, pidgin languages are used cross-culturally, involve simplification and standardization, despite violating traditional rules and practices, and generally remain secondary, despite having considerable commercial viability.

Although Graburn clearly recognizes the articulation of 'traditional' and 'modern' societies, implicit in his account is the idea that as 'traditional' art forms are brought into contact with 'world' culture, they become less central to members of those societies and there is a corresponding decline in the quality of 'traditional' social life. This depiction of cultural change in 'indigenous' art forms is not without its problems. Tourist arts sometimes seem, in effect, to be defined by their market, rather than any intrinsic quality they might possess or through any meanings attributed to them by their producers or their buyers. It is because Eskimo soapstone sculpture, for example, is made for sale to outsiders that it is defined as tourist art; if purchased by the Eskimos themselves, it would apparently be considered an example of 'popular art' (Graburn, 1984). In addition, as I shall mention later, I think there are problems (which we may all experience) with how one is to assess the *quality* of various art forms.

Although Graburn's discussion relates specifically to art forms of ethnic groups, his position has been echoed in other quarters. Some critics of tourism in Western societies, for example, focus on the extent to which the increased market orientation of such institutions as museums and art galleries has led to a deterioration in the very 'product' they are marketing;

> The commodity which is the result of the vast system of cultural production is, I submit, Heritage. Those universal values in the United Nations World Heritage Convention, those values of stewardship, scholarship and cultural identity are now subservient to questions of cash flow and customer orientation. They are threatened by the ideology that asks 'Can the truth be commercially viable?' (Hewison, 1989).

The implication is that once the heritage 'industry' gets its hands on the past, our past, genuine tradition, whatever that might be, is thrown out of the window. That said, it is not clear if critics are actually recommending a return to the Victorian idea of museums as rather gloomy places in which exotic and/or dusty artefacts are crammed into cabinets, decorated with the instruction 'Do not lean on the glass.'

For a final example of a sociological perspective of the influence of tourism on traditional culture, we need go no further that Fuentarrabia, northern Spain. As I have discussed elsewhere (Harrison, 1992a), Greenwood's accounts (1989, first published 1978) of changes in the town's culture throughout the 1950s and 1960s set the tone for similar studies over the next decade. In particular, he focused in the *Alarde*, the major public ritual of the town, which celebrated victory over the French in 1638. According to Greenwood, the ritual (an annual event, coincidentally held during the tourist season) functioned to reproduce solidarity and re-create a sense of community among the townspeople, who saw it as an event primarily for insiders. However, in 1969 the municipal government decreed that, for the benefit of tourists, the *Alarde* should be held twice on the same day:

> In service of simple pecuniary motives, it defined the *Alarde* as a *public show to be performed for outsiders* who, because of their economic importance in the town, had the *right* to see it (Greenwood, 1989).

The result of this decision, as interpreted by Greenwood, was that the 350-year-old ritual lost meaning for the people of Fuentarrabia and, as a consequence, it became difficult for them to participate. Indeed, 'what was a vital and exciting ritual had become an obligation to be avoided.... The ritual has become a performance for money. The meaning is gone' (Greenwood, 1989).

Greenwood's reflections on this article (in the Epilogue to the 1989 edition) make interesting reading and deserve to be quoted in full. However, germane to the present discussion is his conclusion that, contrary to his initial expectation, the *Alarde* continued to be a popular public event,and is imbued now with contemporary political significance as part of the contest over regional political rights in Spain (1989, p. 181). Indeed, he adds that his negative evaluation of the effects of tourism in Fuentarrabia, and on other traditional cultures, was 'an expression of both anger and concern' (1989, p. 181) and 'provides only a partial view of the process' (1989, p. 182). With the benefit of hindsight, it appears that the traditional culture of Fuentarrabia was more durable than he initially imagined.

Retreat from Arcadia?

Clearly, I have been selective in my choice of sociologists (among whom I include anthropologists) and in my choice of quotes. Other parts of their work take up other themes and there are also prominent sociologists I have not mentioned (e.g. Cohen), in whose work 'tradition' is presented very differently. However, to pull the argument together, there appears to be at least some justification for making statements of a very general nature about how the relationship of tourism and tradition has been perceived.

First, there has perhaps been too great a willingness to consider tradition as a fixed entity, enshrined in some distant past and dislodged (even destroyed) by a relatively recent incorporation into the world system. Tradition may be manipulated, managed, enforced, even invented, by groups or classes with a specific interest in maintaining a power base against forces from within or from outside the society (Hobsbawm, 1983; Harrison, 1992b). This position is hardly new: the diffusionist school within anthropology made it clear, sometimes fancifully so, that 'traditional' societies have long been exposed to influences from outside. Even prominent exponents of modernization theory were well aware of the dangers of polarizing tradition and modernity (cf, Bendix, 1967, p. 314; Rudolph and Rudolph, 1967, p. 3–14) and recent studies of tourism indicate traditional peoples are far from helpless when coping with visitors from outside. Strategies may include laughing at the tourists, incorporating them into their ceremonies, and using tourist interest to claim increased class or national status (Evans-Pritchard, 1989; Sweet, 1989; Picard, 1990; Harrison, 1992b). In fact, it may be more useful to see 'tradition' as a resource, to be used to the advantage of specific members of a 'traditional' society - as, for example, in Brittany, where locals struggle to maintain their knowledge of the traditional language and dances, *solely* because they recognize their value as tourist attractions (McDonald, 1987).

Second, it is easy to glorify the past, especially if it is that of someone else. Latter-day Durkheimians and Marxists, we complain bitterly about the degree of anomie or alienation in the modern, postmodern, post-industrial or post-structural (or is it post = paradise) world, but do we seriously seek a return to the days of our grandparents or great-grandparents? In a recent re-showing of a TV documentary on the Amazonian Indians, much was made, and quite properly, of the damage done to different tribes with the expansion of modern Brazil, and an account was provided of how some threatened tribes were brought together in newly-established reservations. At one point, however, a slight hitch in the proceedings was reported: a man in one tribe recognized, in another grouping, his daughter, who had been abducted as part of the spoils of war, many years before. Eventually, the matter was settled – she stayed with the tribe that had captured her – but the incident was a troubling reminder, nevertheless, that not all in the Amazonian garden had necessarily been lovely before the arrival of the Europeans.

At a more personal level, there have been occasions when, carrying out anthropological fieldwork, I have had to grit my teeth and continue taking notes, despite the conviction that what I was witnessing and recording was, in some senses, demeaning, dangerous, even inhuman, for some of the actors involved – often children. More generally, is anyone really going to claim, in all sanity, that they would willingly change places with a 'primitive' or 'ex-primitive' from Papua New Guinea? Moving across cultures, it is highly unlikely that every iman can be described as 'benign and tolerant'. The truth of the matter is that in 'traditional' or 'primitive' society, as in some 'modern' societies, especially in times of war, life was frequently 'nasty, brutish and short'; the daily round of work was backbreakingly hard and although religion may have been the basis of consensus, it could also be regarded as thoroughly oppressive. Undoubtedly, certain groups and categories – including, in many societies, women – led lives which might be regarded, if the luxury of reflection could be afforded, as far from emancipating or fulfilling.

Third, if this exalted view of tradition, especially other peoples' tradition, is indeed found among sociologists of tourism, we need to ask why. The answer is quite simple: many sociologists are ideologically opposed to capitalism and feel guilty, especially when dealing with less developed societies, about their association with it. Unless this is a result of a genetic programme peculiar to sociologists, it is because of our cultural backgrounds, which tend to be middle class, and because of our education and social networks, which are rooted within an academic and 'liberal' establishment. As a consequence, we also disapprove of monetization, commoditization, wage labour and competition among entrepreneurs. And in so far as international tourism encourages such phenomena – *especially* if it involves transporting the Western proletariat to other parts of the globe – we disapprove of that, too.

The accuracy of such assertions is relatively easy to demonstrate, again, from work which is deservedly regarded as seminal. MacCannell, for instance, suggests that once commercial considerations are introduced, local cultures are virtually doomed:

> The dividing line between structure genuine and spurious is the *realm of the commercial*. Spurious social relations and structural elements can be bought, sold, traded and distributed throughout the world. Modern economies are increasingly based on this exchange. The line is the same one between furniture and priceless antiques or between prostitution and 'true love,' which is supposed to be beyond price. It is also the same as the distinction that is commonly made between a gift that has been purchased, which is thought to be inferior, as opposed to one that has been made by the giver especially for the receiver (MacCannell, 1976, pp. 155–156. Emphasis in the original).

Confronted by modernity, placed in a position of 'servitude ... artificially preserved and reconstructed', tradition is taken from the discourse of priceless antiques and true love into that nether world dominated by the cash nexus, of furniture and prostitution.

At the heart of this critique there is an ideal of a new and better community, one which is to occupy the Empty Meeting Grounds:

> A new solidarity, emerging at the community level to the benefit of general humanity, already has assumed discernible forms. It is building a base in the social and economic planning of Euro-socialism; in the anti-war and anti-nuclear movements; in the struggles for ethnic self-determination; in feminist movements and theory, especially non-Western feminism; in new social theory - structuralism, semiotics, Lacanian psychoanalysis, and neo-Marxist; in the environmental movement, and in the non-aligned 'Third World countries' (MacCannell, 1992).

Furthermore:

> The emerging structure, namely the composite community in which culturally different peoples are obliged to work together, is not essentially flawed. Quite the opposite: the composite community, in which there is a place for every human being, even those among us who are bi- and tricultural, sexual and lingual, may be our invention and gift to the future, eventually regarded with the same respect that we give in language, kinship, music, and the other earlier inventions (MacCannell, 1992).

The central focus is no longer tourism as such but the juxtaposition of differences promoted by modernity, of which tourism is a key feature, and how these differences are dealt with. One may query the nature of the 'composite community', and the suggestion that it is already in evidence. However, MacCannell is unequivocal on capitalism: although its role in modernization is unclear, perhaps because of his reliance on Marxist and neo-Marxist commentators (cf, MacCannell, 1992), we are left in no doubt that capitalism, to which he has long been hostile (1992, p. 74), is a bad thing: 'late or decadent ... a sublimated form of the anal-sadistic type of homosexuality' (1992, p. 58), which may reserve' some of its highest rewards for those who are successfully involved in the promotion of rock music groups, export of toxic waste to Third World countries, windfall oil profits from international belligerency, marketing illegal drugs, looting the savings and loan industry and/or the public treasury, and so on (MacCannell, 1992).

At first sight, at least, there seems little chance of mistaking that passage for anything other than a passionate indictment of capitalism.

For Graburn, as for MacCannell, the incorporation of 'traditional' societies in the world system involves a loss of authenticity. By implication, this also involves some decline in the quality of life – at least, in the quality of artistic life. Production for tourists is considered 'less satisfying for local artists' than production for the local community, offering the producer a 'symbiotically incomplete' experience (Graburn, 1984). And although indigenous peoples may still recognize, in tourist arts, some of their history, the process of 'museumification' alienates not only producer and product, but also product and tradition:

> Even though they may derive wealth from continued production, the art forms are no longer totally 'theirs' as control passes into the hands of the dominant commercial and political groups of the pluralistic nations of which they find themselves irreversibly a small part (Graburn, 1984).

Clearly, the indictment here is not as strong as that from MacCannell even though, elsewhere, Graburn comes close to identifying himself with the view that tourism, at least in a metaphorical sense, is akin to prostitution (Graburn, 1983). Again, nevertheless, the spread of capitalism, with its attendant monetization, commoditization and loss of control, is not merely described but evaluated. And it is hard to avoid the impression that this evaluation, which includes an evaluation of art forms themselves, is being made from an apparently privileged position, as when we learn that commercial traditional art depends on 'buyers who know and care enough about the traditional arts to demand "authenticity"', whereas souvenir art requires a 'mass audience which neither knows nor particularly cares for "authentic artefacts": it is simply a "mass ignorant market"' (Graburn, 1984, p. 400).

Such comments, incidentally, raise questions about how sociologists see themselves in relationship to various types of tourist. Others have noted the somewhat uncomfortable links between anthropology and tourism (Crick, 1985; Errington and Gewertz, 1989, pp. 37–38); however, it may be that the cultural background of many sociologists, especially those who place a high value on 'tradition' and a correspondingly low one on capitalism, predisposes them to hold an inherently unfavourable view of working class tourists – package tourists – in particular. This should not be regarded as surprising: after all, it is but an extension of the normal pattern of relationships in metropolitan society. Ironies abound in such circumstances: supporting 'quality' tourism (high spenders) against 'mass tourism' (the working class) and favouring small-scale, locally-owned hotels (local capital – the petty bourgeoisie) over the large-scale conglomerates (international capital), we effectively reject our own proletariat in favour of foreign capitalists. Not for the first time, the question is asked of sociology: 'Whose side are we on?' (Becker, 1967; Gouldner, 1975).

If many sociologists dislike capitalism and desire to distance themselves (at least on holiday, but probably also during the rest of the year) from members of their own working class, this would help explain the prevalent focus on the 'negative' aspects of international tourism. A simple example indicates that the connection is not, at least, entirely spurious. From Greenwood's comments on the effects of tourism in Fuentarrabia, which are quoted earlier, it should be evident that he, too, held capitalism responsible for the destruction of the town's tradition. When culture became a product, it became meaningless:

> Culture is being packaged, priced and sold like building lots, rights-of-way, fast food and room service, as the tourism industry inexorably extends its grasp (Greenwood, 1989).

Holding capitalism to blame, through international tourism, for the disruption of local tradition, it is then easy to attribute to tourism all kinds of negative social effects:

> ... (I)t has become increasingly obvious that the increasing maldistribution of wealth and resultant social stratification are widespread *results* of touristic development.

Various remedies are proposed to counteract these *problems* (Greenwood, 1989, author's emphasis).

The Sociology of Tourism: From Finish to Start?

Given an existing hostility to capitalism on the part of sociologists of tourism, it is easy to see how the onset of capitalism and its accompanying social *effects* come to be seen by them, if not by the subjects of their research, as social *problems*. We have thus moved, almost imperceptibly, from the level of description and analysis to evaluation, where the 'commoditization of culture' becomes *inauthentic*, where 'traditional' people, especially the young, adopt *unhealthy* 'modern' values, where the spread of the cash nexus and market relationships *unfortunately* replace 'traditional' networks and obligations and where there is a *regrettable* increase in social stratification and 'maldistribution' of wealth.

Unable to withstand the onset of modernity – that is capitalism and the monetization and commoditization it necessarily entails – 'tradition' has disappeared. Once again, the 'biblical' old women and the 'rhythm of the seasons', the dignified goat and the 'benign and tolerant iman' have been replaced by young women who give orders to their elder (and better?) menfolk, by tourist hordes, and by a rampant consumerism and commoditization which spell the end of authenticity. And the main agents of this catastrophic state of affairs are the Western proletariat in alliance with international capital.

Where does this leave the sociology of tourism? It has been argued that, with changes occurring in Western society, a new kind of sociology is required, which does not focus on 'work arrangements, class, status, power and related sociological antiquities' (MacCannell, 1976, p. 35). This is debatable, and there is an alternative position, which asserts that the sociology of tourism, far from being in the stages of advanced senility, is 'a rapidly developing specialism ..... characterised by intellectual unevenness' (Urry, 1991). In fact, despite previous attempts to lay down foundations for its development (Lanfant, 1960; Cohen, 1972, 1979, 1984; Jafari, 1987), there is no single sociology of tourism. Instead, there have been several attempts to understand sociologically different aspects of tourism, departing from a number of theoretical perspectives (Dann and Cohen, 1991).

The problem of theoretical fragmentation was also raised by Allcock at a previous meeting of the ISA Tourism Thematic Group, when he argued for a closer association between the sociologies of tourism and leisure (Allcock, 1990, p. 1).

Although work is yet to be carried out in arriving at an understanding of what is involved in the sociology of tourism, the basis for an agreement may already exist. Tourism development involves, but not uniquely, quite specific and well established changes. First, regions once peripheral to the main tourist-sending centres (including peripheral areas within developed countries) have been increasingly incorporated into a world tourist system. This tourist system is but one aspect of the world capitalist system, which (following attempts by some 'bourgeois' theorists) dependency theory and other varieties of underdevelopment theory attempted to describe throughout the 1980s, a perspective also much used in analyses of the role of tourism in specific societies (Harrison, 1992a).

Indeed, Lanfant's article on the 'internationalisation' of tourism (1980) was a notable precursor to such studies. At the 'supply' side, then, the response has been well documented. By contrast, on the 'demand' side, although a considerable (and largely functionalist) literature has emerged on the possible psychological, social psychological and sociological reasons for mass holiday taking, the complex causes of such a vast movement of population have yet to be demonstrated.

Second, at a more micro level, greater integration at the level of economic systems is matched, in some respects, by increased interaction between (at least some) tourist 'guests' and indigenous 'hosts', often mediated by one or more categories of guide. The extent of interaction will depend on the context within which it occurs. However, one result will be increased acculturation, a series of processes not restricted to tourism. Cultures once enjoying (if that is the appropriate term) a relative degree of closure and autonomy have been subjected to new, powerful and largely 'Western' pressures, with a disproportionate effect on the younger members of such societies, who may, as a consequence, increasingly engage in forms of behaviour (the so-called demonstration effect) disapproved of by their 'traditional', and invariably senior, authorities. In such circumstances, issues of individual and collective identity, and a reworking of tradition, may arise.

Third, at the social structural level, tourism expansion has led to the emergence of new forms of entrepreneurship. 'Marginal' groups have frequently taken advantage of the new opportunities and the state, too, has actively promoted tourism development, sometimes in conjunction with (or in opposition to) transnational companies involved in tourism, or some related aspect of it. The role of transnational companies in international tourism is well known and has been noted in numerous studies (Harrison, 1992a). At the same time, new social and political institutions, some of which are cross-national in character, have emerged to promote or oppose tourism, or particular aspects of it.

As Urry has noted (1991, p. 49), the related issues of authenticity and the nature of the tourist 'experience' have been key features of many sociological analyses. However, these in turn rest on my fourth factor: the undeniable process of commoditization, again, a process which is hardly unique to tourism but which has been dramatically highlighted in the incorporation of many 'traditional' communities, in the developed world as well as in developed countries, into the domestic and international tourist circuit. Monetary values and new meanings have been given to buildings and forms of behaviour considered part of national 'heritage'. These include 'traditional' rituals, dances, songs, arts and crafts, perhaps anything that might be defined by a tourist 'client' or indigenous 'entrepreneur' as an addition to local colour.

This list is not exhaustive and contains little of a controversial nature. Indeed, the social phenomena described demonstrate that international tourism is both a symptom and a spur of the continuing involvement of tourist-sending and tourist-receiving areas as inter-linked regions within the international capitalist system. To borrow terminology from a different perspective, they imply production for sale rather than for own use, the sale of labour power, the mediation of exchange through money, external control of the production process and financial decisions and competition among capitalists – in short, the classical definition of capitalism (Bottomore *et al.*, 1983, pp. 64–65).

One need not be a Marxist to regard such changes as important. After all, since the middle of the 19th century, generations of sociologists have carefully charted developments in Western capitalism. Classical Marxists, underdevelopment theorists, Weberians and non- Marxists may dis agree on a variety of topics but they do not disagree on the continual growth and extension of capitalism. Where debates *will* occur are over how these changes, at all levels of the system, are to be *interpreted.* Modernization is undoubtedly occurring and sociologists legitimately describe and try to explain its pattern. However, they are not especially qualified to evaluate whether or not changes arising from modernization constitute development (Harrison, 1988, pp. 57–62).

Arguably, the issues involved indicate that the sociology of tourism is closely linked to the sociology of modernization and development, as well as to sociological approaches to leisure, urbanization and migration. In fact, given the multi-faceted nature of tourism, there will always be overlap and the precise labels given to sociological perspectives favoured in studying tourism are not a problem. However, it *is* necessary for sociologists to clearly distinguish their own evaluations of the phenomena under study. Such a position is traditional within the discipline, but is no less valid for that. They can make a start by not confusing social *effects*, which are part of the legitimate sociological study of tourism, with social *problems* which, like other people, they evaluate from within the parameters defined by their own values and their cultural milieu.

References

Allcock, J. (1990) The Sociology of Tourism and the Sociology of Leisure: Notes on the Possibility of Theoretical Convergence. *Paper presented in the Tourism Thematic Group, International Sociological Association, XIIth World Congress*, Madrid, 9–13 July.

Becker, H. (1967) Whose side are we on? *Social Problems* 14, 239–247.

Bendix, R. (1967) Tradition and modernity reconsidered. *Comparative Studies in Society and History* 9(3), 292–346. DOI: 10.1017/S0010417500004540.

Bottomore, T.B., Harris, L., Kiernan, V.G. and Miliband, R. (eds) (1983) *A Dictionary of Marxist Thought.* Basil Blackwell, Oxford.

British Broadcasting Corporation (BBC) (1988) *Not in Front of the Tourists*. Written and presented by R. Chesshyre, BBC, London.

Cohen, E. (1972) Toward a sociology of international tourism. *Social Research* 39(1), 164–182.

Cohen, E. (1979) Rethinking the sociology of tourism. *Annals of Tourism Research* 6(1), 18–35. DOI: 10.1016/0160-7383(79)90092-6.

Cohen, E. (1984) The sociology of tourism: approaches, issues, and findings. *Annual Review of Sociology* 10(1), 373–392. DOI: 10.1146/annurev.so.10.080184.002105.

Crick, M. (1985) Tracing the anthropological self: quizzical reflections on field work, tourism and the Ludic. *Social Analysis* 17, 71–92.

Dann, G. and Cohen, E. (1991) Sociology and tourism. In: Graburn, N. and Jafari, J. (eds) *Tourism and Social Research* 18. Annals of Tourism Research, pp. 155–169. DOI: 10.1016/0160-7383(91)90045-D.

Errington, F. and Gewertz, D. (1989) Tourism and anthropology in a post-modern world. *Oceania* 60(1), 37–54. DOI: 10.1002/j.1834-4461.1989.tb00350.x.

Evans-Pritchard, D. (1989) How “They” see “Us”: Native American images of tourists. *Annals of Tourism Research* 16(1), 89–105.

Gouldner, A.W. (1975) The sociologist as partisan. In: Gouldner, A.W. (ed.) *For Sociology Renewal and Critique in Sociology Today*. Penguin, Harmondsworth, pp. 27–68.
Graburn, N.H.H. (1983) Tourism and prostitution. *Annals of Tourism Research* 10(3), 437–443. DOI: 10.1016/0160-7383(83)90068-3.
Graburn, N.H.H. (1984) The evolution of tourist arts. *Annals of Tourism Research* 11(3), 393–419. DOI: 10.1016/0160-7383(84)90029-X.
Greenwood, D. (1989) Culture by the pound: an anthropological perspective on tourism as cultural commoditization. In: Smith, V.I. (ed.) *Hosts and Guests: The Anthropology of Tourism*, 2nd edn. University of Pennsylvania Press, Philadelphia, pp. 171–185.
Harrison, D. (1988) *The Sociology of Modernization and Development*. Routledge, London.
Harrison, D. (1992a) Tourism to less developed countries: the social consequences. In: Harrison, D. (ed.) *Tourism and the Less Developed Countries*. Belhaven Press, London, pp. 148–162.
Harrison, D. (1992b) Tradition, modernity and tourism in Swaziland. In: Harrison, D. (ed.) *Tourism and the Less Developed Countries*. Belhaven Press, London, pp. 148–162.
Hewison, R. (1989) Heritage: an interpretation. In: Uzzell, R. (ed.) *Heritage Interpretation, Vol.I: The Natural and Built Environment*. Belhaven Press, London, pp. 15–23.
Hobsbawm, E. (1983) Introduction: inventing traditions. In: Hobsbawm, E. and Ranger, T. (eds) *The Invention of Tradition*. Cambridge University Press, Cambridge, pp. 1–14.
Jafari, J. (1987) The tourist system: sociocultural nodels for theoretical and practical application. *Problems of Tourism* 10(3), 3–17.
Lanfant, M.-F. (1960) Tourism in the process of internationalisation. *International Social Science Journal* 32(1), 14–43.
MacCannell, D. (1976) *The Tourist: A New Theory of the Leisure Class*. Macmillan, London.
MacCannell, D. (1992) *Empty Meeting Grounds: The Tourist Papers*. Routledge, London.
McDonald, M. (1987) Tourism: chasing culture and tradition in Brittany. In: Bouquet, M. and Winter, M. (eds) *Who From Their Labours Rest?* Avebury, Aldershot, pp. 120–134.
Picard, M. (1990) "Cultural tourism" in Bali: cultural performances as tourist attraction. *Indonesia* 49, 37–74. DOI: 10.2307/3351053.
Rudolph, L.I. and Rudolph, S.H. (1967) *The Modernity of Tradition: Political Development in India*. The University of Chicago Press, Chicago.
Sweet, J.D. (1989) Burlesquing "the other" in Pueblo performance. *Annals of Tourism Research* 16(1), 62–75. DOI: 10.1016/0160-7383(89)90030-3.
Urry, J. (1991) The sociology of tourism. In: Cooper, C. (ed.) *Progress in Tourism, Recreation and Hospitality Management*. 3. Belhaven, Lymington, pp. 48–57.
World Tourism Organization (WTO) (1991a) *Compendium of Tourism Statistics*. WTO, Madrid.
World Tourism Statistics (1991b) *Yearbook of Tourism Statistics*. 1, 27. WTO, Madrid.

3

Tradition, Modernity and Tourism in Swaziland[1]

Introduction

Although the social and economic contribution of international tourism to national development has clear theoretical ramifications and involves both modernization and underdevelopment perspectives, my aim in this chapter is not so much to demonstrate the accuracy of one or other of these arguments as to offer some observations which can make consideration of the social effects of tourism in Swaziland better informed. In particular, I focus on the ambiguous status of 'tradition', which simultaneously legitimates the existing political structure and provides a crucial theme in selling the tourist industry, thus contributing to the process of 'modernity'.

The Kingdom of Swaziland is a small, landlocked country with a population of some 700,000 people. Situated between the Republic of South Africa and Mozambique, it has been crucially affected by the social and political conflicts of the region. A member of the South African Customs Union (SACU) and the Rand Monetary Area (RMA), the country is economically dependent on the Republic which, not surprisingly, is Swaziland's major trading partner.

Throughout the 1980s, the main problem of the Swazi economy was keeping pace with population growth which, over the period 1976–1986, averaged 3.2% per annum (Department of Economic Planning and Statistics, 1987). In general, the economy has moved in fits and starts. Approximately two-thirds of its export income is derived from sugar, fertilizer and wood pulp, and imports are mainly accounted for by machinery and transport equipment, manufactured goods, mineral fuels and chemicals and chemical products. Swaziland is predominantly agricultural, with some 80% of its population in rural areas, and agriculture contributes almost 25% to the GDP. However, efforts to diversify the economy have had some success and, in 1986, the share of manufacturing in the GDP virtually matched that of the agricultural sector. In addition, since political independence in 1968, tourism has figured increasingly prominently in government plans and the number of visitors to the country has correspondingly increased. There is, indeed, considerable potential for tourism expansion, for Swaziland's 17,364 square kilometres contain a wide variety of scenically attractive geographical and climatic features.

In 1972, Swaziland was visited by 89,015 tourists, of whom 61% were on holiday and 20% were on business. By 1989 this number had risen, somewhat unevenly, to 257,997, with the proportions of holiday and business visitors at 63% and 22% respectively. However, the increase in numbers is not matched by increases in expenditure which, in real terms, has remained static. Most tourists come by road from the Republic of South Africa to take a short break, 2 or 3 days at most, and many stay for only one night, visiting the country to watch television programmes or sports activities which, because of international opposition to apartheid, have until recently been denied to the Republic.

The consequences of tourism

In some respects, the economic consequences of tourism in Swaziland can be stated quite clearly. By 1985, more than 2000 workers were directly employed in hotels and restaurants, making up almost 3% of the total wage earners in the Kingdom. By 1988, tourism was contributing about 3% to GDP and more than 4% of total exports. Most workers were Swazi, but the higher and better-paid positions in the hotels were generally occupied by foreigners.

In addition to those directly employed in hotels, numerous others also benefit from tourism, including taxi drivers, sculptors, carvers, wholesalers and retailers. However, the most widespread benefits probably accrue to rural workers, primarily women who produce handicrafts, and previous research has estimated that about a third of all households in the Kingdom have at least one individual involved in handicraft production (Russell, 1983, p. 16). Handicrafts tend to be sold by the producer, or by kinswomen, and there is also provision for government agencies to act as wholesalers. As in the Gambia, however (cf. Wagner, 1982, p. 61 and 69), expatriate involvement has been important. In Swaziland, several foreign retailers dominate the top end of the handicrafts market.

The tourist industry is concentrated in central Swaziland, especially in the Ezulwini Valley between Mbabane and Manzini, as indicated in Fig. 3.1. The bigger the hotel and the more costly its accommodation, the more likely it is to be in this corridor and the more likely it is to be foreign owned. In July, 1987, there were 1165 hotel rooms in Swaziland. Of these, non-Swazis owned 70% and over 90% of those in the Ezulwini Valley. Sun International dominated the upper sector of the industry, operating its own hotels and managing those in which the government, or the *Tibiyo Takangwane* (an investment arm of the monarchy), have substantial interests. This hotel chain owned or operated 616 hotel beds in Swaziland, some 53% of the total. More importantly, perhaps, it dominated the 'luxury' end of the market, owning or controlling 90% of all hotel rooms which (at 1987 rates) cost more than 40 Emalangeni a night.[2]

As Swazi tourism expanded, new formal institutions have been developed and others have been adapted. In the private sector, the Hotel and Tourist Association is the main pressure group, and within government ministries there are departments to promote Swazi tourism and handicrafts. Cooperation between the private sector and the Swaziland National Trust Commission has led to the establishment of several game and nature reserves, considered increasingly important as tourist attractions.

Other social consequences have also followed from tourist development. Swazis have increasingly interacted with whites who, by most standards, are relatively affluent. Not all such interaction may be considered beneficial and, as shown below, increases in begging, prostitution and other forms of 'immoral' activity have been laid at the tourist door.

'Modernity' and 'tradition' are often contrasted, both by modernization theorists, who focus on the beneficial effects of Westernization, and their opponents, who tend to regard tradition as helpless in the face of savage attacks by modernity. Nevertheless, such positions are belied by the reality of present-day Swaziland, where 'tradition' legitimates 'modern' government and is simultaneously threatened by that same government's involvement in and support for the tourist industry. In order to substantiate such assertions, however, it is necessary, first, to detail the major political and economic institutions of Swaziland and then to indicate their relationship to the tourist industry (Fig. 3.1).

The 'traditional' basis of 'modern' government

In theory, the Swazi constitution attempts to blend Swazi tradition with the features of a

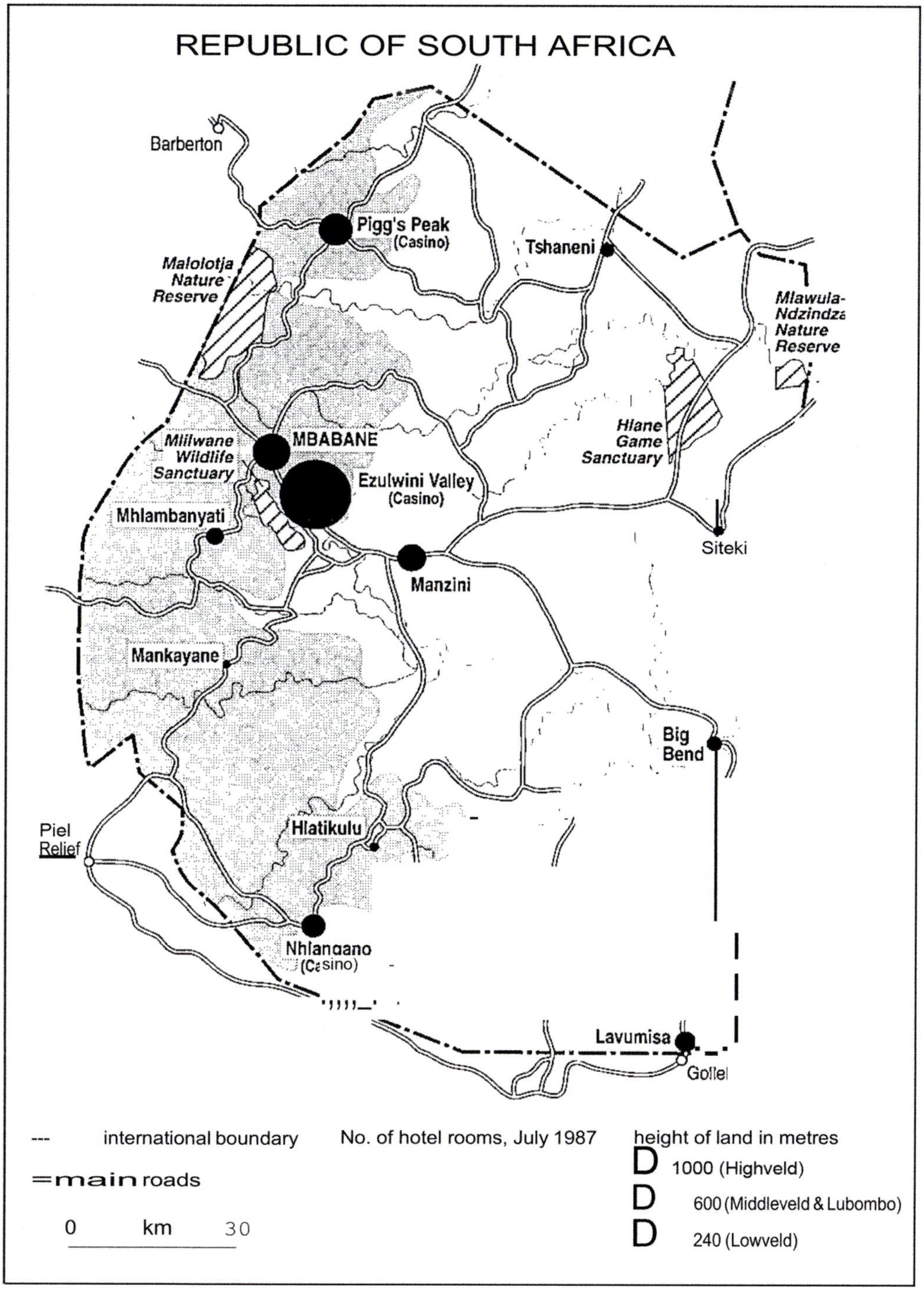

Fig. 3.1. Swaziland: ecological zones and tourist accommodation, 1987.

modern democratic society. The 'traditional' elements are described in the early ethnographies of Swaziland (Marwick, 1940; Kuper, 1947, 1961, 1963). The King (*Ngwenyama*), as the Head of State, shared power with the Queen Mother (*Ndlovukati*) and senior princes and princesses of the royal family. Chiefs were selected by the King from the royal family, and male kin of the Queen Mother were also given important positions in the administration. The King was also assisted by *Indvunas*, the foremost of whom was equivalent to a prime minister, and by *tinyanga*, specialists in ritual. Both categories of councillor were selected from the (male) members of clans which were not part of the royal family. The King's closest advisers formed the *liqoqo*, which was, in effect, an inner council of state. Conversely, all men in the nation, including councillors, chiefs and headmen, formed the *libandla*, the Swazi National Council.

This traditional political organization contains elements of feudalism, in that all land belongs to the Crown, with control vested in the chiefs. Clearly hierarchical, as described by anthropologists the distinctions were between aristocrat and commoner, young and old, and men and women. However, the system was undoubtedly complicated by the introduction of migrant labour at the end of the 19th century and work in the mines of the Republic allegedly led to a rudimentary consciousness of class (cf. Bonner, 1983, p. 211 ff.). By contrast, Kuper notes that, through migrant labour, Swazi men became increasingly aware of a pan-African identity (Kuper, 1947, p. 23). She also stresses the role of the 'uniform of colour' in effectively segregating Europeans from Swazis, a segregation which applied both in the Republic and in Swaziland.

In these circumstances, it is unnecessary to oppose ethnicity to class. Both forms of identification may have been present. Certainly, close economic and political involvement with the Republic of South Africa crucially affected Swaziland, penetrating deep into the rural hinterland and leaving its mark on the 'traditional' organization of Swazi homesteads (cf. Booth, 1986). At a constitutional level, too, the 'traditional' has not given way to the 'modern' in Swaziland. Rather, it has been adapted. The King remains Head of State, sharing power with the Queen Mother. Until its disbandment in 1987, the *Liqoqo* was the Council of State, in parallel to the modern Cabinet, and was chaired by a senior prince known as the Authorised Person who, in an interregnum, became Prince Regent to the Queen Mother's Queen Regent. Technically, the Swazi National Council also remains, even though the circumstances in which it might meet are unclear.

In fact, the Swazi monarchy is closely involved in the 'modern' sector of government. In 1973, 5 years after political independence had been gained from Britain, King Sobhuza II repealed the Westminster system and banned political parties. Since 1978, following a period of government by proclamation, a modified bicameral system has been in operation. Open (that is, not secret) elections are held every 5 years in 40 traditional constituencies (*Tinkhundla*), every one of which is controlled by an *indvuna*, who owes his position to the King. Every *Tinkhundla* elects two members to an electoral college which, in turn, selects 40 of its number to sit in the House of Assembly, where they are joined by ten of the King's appointees. The influence of the King is even more marked in the Senate, the second chamber, which is composed of 20 members, of whom half are elected by the House of Assembly and half appointed by the King. Finally, it is the King who selects the Prime Minister and members of the Cabinet. In short, even in the 'modern' sector of government, the 'traditional' monarchy is highly influential at all levels, from the nomination of *Tinkhundla* candidates, through the appointment of additional members of both chambers of parliament, up to the selection of the Prime Minister and members of the Cabinet.

The Swazi monarchy is no mere symbol. It holds real power, which is based on distinct but related sources. In the years preceding and following political independence, the monarchy purchased some of the land it had lost, through concessions made to Europeans, at the end of the 19th century. This land was added to the existing stock of Swazi Nation Land, held in trust by the King for the Nation, with control vested in the chiefs. As the position of *indvuna* is often held by a chief, it is these chiefs who control the *Tinkhundla* and, through the electoral college, elect members to the House of Assembly and the Senate. As already indicated, it is the King who appoints (and can dismiss) the Prime Minister and members of the Cabinet. This amounts to an impressive degree of political control.

At the economic level, and equally important, is the *Tibiyo Takangwane*, a commercial and investment company controlled by and operating to the direct benefit of the Crown (and not the Ministry of Finance) often in conjunction with foreign capital. Along with its offshoot, the *Tisuka Takangwane*, which receives mineral and other royalties on the King's behalf, *Tibiyo* represents a real, powerful and unaccountable royal stake in the 'modern' sector of the Swazi economy.

The process through which royal control of politics and the economy was established in Swaziland has been nicely described by Macmillan, who notes:

> The major peculiarity of decolonisation in Swaziland was that it was the only one of the African colonies and protectorates, with the arguable exception of Zanzibar, to come to independence as a monarchy in which real power rested with the King and a council of chiefs and elders in direct descent from those who had wielded power in the pre–colonial era (1986, p. 104).

To re-establish this 'traditional' power and authority, King Sobhuza II, who led Swaziland into independence, had to re-interpret Swazi tradition so that the colonial administration and his own people accepted him and his descendants as legitimate rulers of an independent state. According to Macmillan, over three decades the King and his allies, including such social anthropologists as Hilda Kuper, were able to establish the claims of the monarchy not only against the British, but also against an increasingly proletarianized work force, a growing Swazi intelligentsia, and pressures from elsewhere. The battle was not easy and, *inter alia*, involved the creation of national (as opposed to mission) schools and a revival of the age-graded regimental system, institutions through which King Sobhuza II and his allies could control the socialization of the young. However, by the late 1950s, class divisions and foreign influence seemed to threaten the monarchy:

In the eyes of the 'traditionalists', the unity of their nation was now threatened as never before by emerging class divisions, and by an influx of aliens who were in competition with the Swazi for jobs, women and space in the increasingly crowded Swazi Nation areas. At the same time, the inflow of educated black South Africans to key posts, especially in education, seemed to present more forcefully than before the risk of an alien intelligentsia not only monopolizing good jobs, but also 'subverting' or 'indoctrinating' – both words were used – Swazi youth (Macmillan, 1986).

According to Macmillan, the new threat from organized labour and African nationalism led King Sobhuza II into an alliance with white settlers and international business interests. This alliance took Swaziland into the period of political independence and beyond. During this period, too, political parties briefly appeared, including one led by the King, only to be banned by Parliament, on his direction, in 1979. They were replaced, in effect by the *Tinkhundla*, allegedly a traditional Swazi institution. However, as Macmillan remarks:

> It was clear that 'traditional', in this context, meant anything but old or inherited. It was as it had been so often in the past, a disguise for something thoroughly new and intended to keep power in the hands of a small section of the population (1986, pp. 121–122).

The process Macmillan describes as leading to 'the triumph of tradition' is essentially political, a quite deliberate attempt by the monarchy and its supporters over a long period to retain power in the face of competing claims from within and from outside Swaziland. There is no need to fully accept his argument (or the implicit definition of ideology as a tool for the ruling class) to recognize that, in Swaziland, 'traditional' institutions were re-formulated by King Sobhuza II and his supporters specifically to define Swazi nationhood and the political and economic place of the royal family within the newly independent society.

It should be emphasized that the Swazis have not directly engaged in the invention of tradition, noted in other parts of colonial Africa and elsewhere, where 'tradition' has been manufactured to confer what might be regarded as a spurious authenticity on either a colonizing power or a new ruling class (cf. Hobsbawn and Ranger, 1983). However, states may frequently attempt to control and manipulate custom for their own benefit, or in the interests of a wider 'unity', and this may lead to an increased focus on performance and display, perhaps exacerbated by tourism (cf. Acciaioli, 1985).

As Macmillan points out, it was necessary for King Sobhuza II and his allies to revive many practices which had fallen into disuse under the British:

> In the first 20 or so years of British administration there can, however, be no doubt that Swazi institutions decayed. The chiefs were stripped of most of their judicial functions; the age-grade or regimental system declined in the face of the competing demands of taxation and labour migration; and massive land alienation created further problems of jurisdiction for chiefs, most of whom were left in reserves while many of their followers remained on white farms....
>
> It should, however, be noted that the major Swazi ceremony, the *Ncwala*, a ritual of kingship and first-fruits ceremony, could not be fully performed between 1899 and 1920 owing to the absence of the King, and Sobhuza's initial *Ncwala* in 1921 was in itself something of a revival (Macmillan, 1986).

In effect, King Sobhuza II and his allies revived and interpreted Swazi tradition, for the colonial power and for the Swazi people. In so doing, they provided a package which legitimated the constitutional position of the monarchy and laid the functions for its political and economic dominance in the period following political independence. Importantly, as part of this process, there was an increasing focus on Swazi culture, a revival of interest in Swazi ceremonies and the promotion of Siswati as the national language. In effect, the monarchy became both the guardian and the epitome of Swazi culture. The King and the nation were one.

This section should not be regarded as a mere excursion into political economy. Tradition – revived, interpreted, even manipulated – is the basis of the Swazi constitution and its monarchy, and the monarchy is symbolically, politically and economically dominant. However, a focus on 'tradition' is also a key feature of Swazi tourism, which many Swazi consider responsible for the introduction or exacerbation of unwelcome elements of 'modernity'. For them, at least, 'tradition' and 'modernity' are in direct conflict. And as the monarchy is heavily involved in the tourist industry, its position may be considered decidedly ambiguous.

Tourism and the monarchy

As noted above, the Swazi monarchy occupies a key position in the 'modern' political order and in the Swazi economy. The government, dominated by the monarchy, is involved in tourism through its various ministries. In particular, the Division of Tourism, a part of the Ministry of Commerce, Industry, Mines and Tourism, supervises the activities of the Swaziland Tourism Development Company, which operates government shareholdings in the Protea Peak Casino Hotel (100%) and the Nhlangano Sun Hotel (49%), which also has a casino. These hotels, situated away from the popular Ezulwini Valley, represent major government investments and both are operated by Sun International. The involvement of the monarchy in tourism, however, is not restricted to its investment through government institutions. Its own investment agency, the *Tibiyo Takangwane*, has a 39% interest in the other Swazi operations of Sun International, the major transnational hotel chain in southern Africa which, as described above, dominates the Swazi tourist industry. And among its numerous other investments, *Tibiyo* owns the Swazi National Airline. Together, these give the monarchy, whether through the Government or its private investment agency, a colossal stake in the tourist industry. Its position may be legitimated by 'tradition' but, through its involvement in tourism, it also has a substantial interest in 'modern' trends and, as a consequence, in the packaging and selling of Swazi 'tradition'.

The growth of 'modernity'

If it is assumed that 'modernity' is what is currently considered as 'up-to-date', and development is regarded as change which is in some respects 'progressive', it should be evident that they are not synonymous (cf. Harrison, 1988). Certainly, in Swaziland it is commonly felt that foreign influences tend to corrupt the young – a view held over several decades and not restricted to Swaziland. For many older Swazis, alcoholic drink, drugs and discotheques amount to decadence and depravation, and are seen as being encouraged by tourism.

In Swaziland, discotheques are the preserve of the young and provide a setting where interaction between young Swazis and non-Swazis is most likely. In addition, alcohol is consumed, thus increasing the 'disreputable' nature of this form of social activity. Most of the larger hotels have discotheques and set out to attract young Swazis, especially in mid-week and periods of the year when the tourist trade is slack.

Equally contentious among some Swazis are the three casinos that now exist in Swaziland (all of which have some government or royal investment). The amount won on the jackpot on the fruit machines is regularly published in the newspapers, and it is felt that, while foreigners may 'play the machines', such behaviour is immoral if carried out by Swazis, who are likely to have less money to spend on gambling (and who, it is sometimes suggested, might be more inclined to become addicted to the activity).[3]

If Swazi men were considered most likely to gamble, it was young Swazi women who, in other respects, were held to be more inclined to adopt other 'foreign' patterns of behaviour. In particular, women who wear trousers, especially jeans, attract considerable hostility, which is regularly aired in newspaper articles:

> There is a man at the bus rank ... who takes it upon himself to enforce law and order where bus queues are concerned.... The person he picked on was an attractive teen-age girl wearing a pair of trousers. He hit her on her face and shoulders and would have had a field day if he had not been stopped by young men.... Other women rushed to form a line, chuckling to themselves, saying complacently, 'That's what you get from wearing trousers'....
>
> In another incident, a young woman wearing trousers was waiting for a bus and hitching at the same time on the road from Ngwenya. Four or five army vehicles drove by and stopped. One of the soldiers got off and asked her why she was wearing trousers. He then slapped her, climbed back into his truck and they all drove off.

Time and time again, the matter is brought up in parliament. Should there be a ban on women wearing trousers? The matter came up in 1979 and there was a motion against it in the House of Assembly (the lower house). The matter was again brought up last year by one of the MPs. At the time it was said that the matter would have to be taken to the *Tinkhundla*.

The reason the MP gave for wanting trousers to be banned was that it lowered Swaziland as a nation if the women wore trousers (and men permed their hair).

One of the reasons why there was some reluctance to ban trouser wearing was the fear that tourists, as in the case of Malawi, might be chased away from coming into the country if they were not free to wear what they wanted. Then the issue came up of banning trousers for local women. In essence, Swazi nationals. Unfortunately, this would lead to an apartheid-like situation ...

Some government officers will not allow a smart looking woman wearing even a business-like pants-suit into their ministries. Does what she is wearing limit her ability to do a job? (Gcinaphi Dlamini, in the *Times of Swaziland*, 21 July, 1987)

One tourist official I interviewed said that he did not wish Swaziland to become like Malawi and ban tourists if they came in jeans or similar clothing. When that happened, the tourist industry declined. However, he stressed that Swazi women *knew* that such clothing was wrong, especially if they were married, and he would not allow women in his office to wear such clothes, even if it were cold. What they did at home was a different matter. (Indeed, there have been times when such strictures have also been applied to white women in Swaziland; I know of at least two cases of white women being banned from entering government offices because they were wearing trousers.)

It would be a serious mistake to regard the issue of dress as trivial and numerous issues are raised in this debate. Obviously, one element is the authority of Swazi men over 'their' women, especially young women, who feel that they should be able to wear clothes of their own choosing. In defying (male) authority in such matters, women lay themselves open to several linked accusations. First, they are usurping male roles. Second, they are emulating tourists, and third, they are engaging in 'immoral' behaviour. In fact, their 'immorality' – for which they may suffer public admonition and humiliation – consists in defying Swazi tradition. The wearing of jeans by a young woman thus takes on considerable symbolic significance. The fact that such attire is often associated with having a good time at discotheques, at which drinking also occurs, is simply seen as further evidence of moral degeneracy.

For many Swazis, 'modernity' in its various manifestations is clearly something of a problem. The fact that some young women want to wear trousers, especially jeans, is clearly not due entirely to the development of tourism. Swazi citizens visit the Republic, and in Swaziland there is open access to magazines and journals with details of European and North American fashions. Furthermore, many white visitors to Swaziland, whose behaviour local people are said to be emulating, are not holiday tourists at all; instead, they are in the Kingdom to work, often for the United Nations and other international agencies, and their arrival can be seen as part and parcel of a much longer and more complex exposure to 'Westernization'. Such 'demonstration effects', which have been noted elsewhere (cf. de Kadt, 1979), and which have been described as instruments of 'slow torture' (Turner and Ash, 1975), do not arise from tourism alone. True, they have often been considered detrimental to members of the 'host' society, but such a perspective implies that 'traditional' patterns of behaviour are both static and morally preferable, a viewpoint which, to say the least, is somewhat contentious, if not downright patronizing.

In some respects, it would be surprising if discotheques were not popular in Swaziland and, although hotels undoubtedly provide opportunities for young Swazis to enjoy Western dancing, this applies as much to hotels owned and operated by Swazis as to those operated by expatriates. Swaziland is not isolated from the rest of the world and, with or without tourism, such influences will be brought to bear on its traditions, young people and established authorities. Government or Crown ownership of some of the hotels offering gambling, drinking and discotheques to the young of Swaziland merely highlights the dilemma. Indeed, it is probably the larger and more expensive hotels which try hardest to control access to alcohol. For class reasons, if for no other, they desire to protect their largely white residents from unlimited exposure to members of Swazi society. It is left to the cheaper, Swazi-owned hotels and bars to cater for the rest of the population.

What remains true is that such 'modern' activities are more likely to appeal to the young than the middle-aged or the elderly, and that the young are exposed most to those foreign influences which prevail in tourist establishments throughout Swaziland. Equally, young or single whites from the Republic who visit Swaziland may be more likely to reject the constraints of their own, racially divided society than their (older) compatriots who stay in the better hotels. The former's motives for visiting may not always be to everyone's taste, and in Swaziland there is a small but evident incidence of prostitution. It is not a new phenomenon, and is neither 'modern' nor a new form of social contact, but it has undeniably been associated with the arrival of Europeans in Swaziland.

Prostitution

It is often argued that tourism in the Third World encourages prostitution and other forms of immorality (cf. Wolfers, 1974, pp. 70–72; Turner and Ash, 1975, pp. 230–231). Indeed, it has been alleged that gambling, pornographic films and 'prostitution across the colour line became the cornerstones of Lesotho's and Swaziland's tourist attractions' (Crush and Wellings, 1987).

However, in the case of Swaziland, at least, such an assertion is highly questionable. This is not to say that, in Swaziland, prostitution (defined as payment specifically for sexual services) does not exist. It was already considered a problem in the late 1930s, when the young were allegedly being alienated from Swazi traditions and corrupted by increased European influence and urbanization (cf. Marwick, 1940, p. 301). By the mid-1950s, such social problems were considered sufficiently pressing for the colonial government to set up a Commission to investigate why they were occurring, and in 1956 it reported, arriving at very similar conclusions. (Commission to Investigate Juvenile Delinquency and Allied Problems, 1956). At that time, however, the men involved were considered to be Swazi migrant workers returning from mines of the Republic.

In 1970, prostitution was the subject of yet another report (Committee on Juvenile

Delinquency, Sexual Morality and the Maintenance of Illegitimate Children), which covered much the same ground as the 1956 report, again blaming increasing urbanization and the breakdown of social control, especially in the urban areas. However, it was then noted that prostitution was centred around a few specified hotels and, despite somewhat contradictory evidence, most of the women's clients were considered to be from outside Swaziland.

The situation was unchanged by the mid-1980s. However, the scale of prostitution in Swaziland is difficult to quantify. Published figures are rare and unhelpful, and the police are known to allocate this kind of offence a low priority. In addition, newspaper reports tend to sensationalize the issue. After all, it is 'good copy'. In April, 1984, the *Weekend Observer* carried out a '3 week investigation' and broadcast its findings under the banner headline, 'SEX FOR SALE'. It suggested that most of the women involved were dropouts from school, and quoted one as earning E200 a night from South Africans spending a public holiday in the Kingdom:

> For every shot, I charge the white man and other foreigners E30 and when it means spending a night with them in their hotel rooms, then the amount doubles,' she said (*Weekend Observer*, 28 April, 1984).

White and black tourists alike were quoted as being shocked at the immorality in Swaziland, and the opinions of the Vice-Chancellor of the University of Swaziland and the Anglican Bishop were sought. However, no estimate of the number of prostitutes operating in Swaziland was attempted and, by contrast, a spokesman for the police was more cautious:

> From the legal point of view there is 'nothing' like prostitution in Swaziland ... He said when more women are found in drinking places enjoying themselves, it is assumed they are prostitutes (*Weekend Observer*, 28 April, 1984).

Interestingly, in a companion article of the same issue, 'Why do local girls fall for Shangaans?', the paper quoted Swazi men asserting that the women wanted money and a good time, which they could obtain more easily from *Shangaans* (foreigners):

> But two senior workers in the hotel industry, both of them women, had different views: 'The Shangaans have that tender care. They know how to look after a woman and they know what a woman is, unlike our local guys, who only know how to booze.' (*Weekend Observer*, 28 April, 1984).

What seems to be at issue is not prostitution, where sexual services are provided specifically in return for financial gain, but an alleged preference of some Swazi women for foreigners. As in previous reports, the distinction is frequently ignored. It must also be recognized that such stereotypes are not unique to Swaziland. They exist in many societies where more wealthy, allegedly high status men from outside come into contact with local women. Much the same kind of accusation is levelled at women in other less developed countries and, indeed, was commonly directed at British women who became involved with American GIs during the Second World War.

The view that prostitution in Swaziland is increasing may be common but it is not universally held and, for reasons already provided, is difficult to prove. From the available evidence, the government appears to tread carefully. It is aware that tourists bring with them customs, habits and wants which, according to some criteria, are less than wholesome. Drinking alcohol, dancing at discotheques, the wearing of trousers by young women, using drugs, mixing with foreigners and prostitution are part of a pattern which, it should be noted, reflects almost entirely on the behaviour of women. Young men were rarely criticized tor adopting 'foreign' ways. However, officials were reluctant to formally restrict the behaviour of young Swazis, and not that of tourists, on the grounds that they would be seen to operate double standards, a consideration which, by contrast, apparently did not apply at informal levels of social control. 'The man in the street' was far more willing to persuade the woman in the street to conform to traditional norms.

To argue that immorality has actually increased with Westernization is quite different and requires a series of assessments and value judgements which would be out of place in this context. What has happened is that colonialism and Westernization have introduced and often imposed changes to which Swazi society has had to respond. In one way or another, 'tradition' has been threatened. In the final section, I suggest that one response has been to package this tradition for the benefit of tourists while quite consciously attempting to preserve it from other forms of attack.

Tradition and Tourism in Swaziland

Tradition is not a fixed entity. As suggested earlier, the monarchy deliberately adapted Swazi tradition for its own purposes, and in so doing was not unique. However it is defined, 'tradition' is eminently adaptable. This is perhaps especially the case when it is being portrayed by the tourist industry.

When Swaziland is advertised as a tourist venue, its traditions are emphasized, whether or not the publicity emanates from the private sector or government. We learn that in Swaziland

Ancient traditions mingle with up-to-the-minute amenities

> Swaziland is a fascinating blend of contrasts and contradictions, old traditions mingling with 20th-century technology ... and the conservative and tradition-loving people are proud to wear their national dress of which feathers, skins, cow-tails and brilliant colours are such a memorable feature (Swaziland Chamber of Commerce, 1987–pp. 94–95).

In the same publication, tourists are advised that the arts and crafts of Swaziland are one of the ways in which 'the culture and traditions of the Swazi people are kept alive' (Swaziland Chamber of Commerce, 1987).

In a sense, this is indeed the case. Material artefacts, whether produced for decoration or everyday use, for sale or for the producer, are tangible reflections and expressions of culture. However, it must be recognized that many of the allegedly 'traditional' items produced for sale to tourists have only recently been introduced into Swaziland. As the above publication recognizes, the grass woven by rural women into baskets and mats is a traditional material but the items themselves have little to do with Swaziland, and much to do with what it is hoped the tourists will buy. They are the 'curios' regarded by visitors, in Russell's words, as 'authentic evidence of their journey to foreign parts' (Russell, 1983, p. 25).

A similar argument might be made for carvers and sculptors. The skills exercised by Swazis currently engaged in the production of such items have been acquired relatively recently and cannot really be described as 'traditional'. The National Handicraft Training Centre, for instance, is a recent institution, quite specifically set up to develop skills appropriate for the tourist market. Early in 1987, it entertained two visiting experts on handicrafts and tourism. According to a newspaper report:

> They are from the EEC. They are Mr Ray Joyce, an expert on tourism and Mr Donald Wiels on handicraft. Mr Mkhonta, Principal Secretary for Commerce, Industry and Tourism, said the EEC wants Swaziland to spell out the needs of the two sectors, and therefore sponsored the tour of the two men. Mr Mkhonta said recently the Ministry advertised the posts of three experienced people within the two sectors. Two of them will be seconded to handicrafts while the third will go to tourism.
>
> We have advertised the post of a handicraft designer who will help us design our handicrafts for sale in the outside market. We think that if we can get somebody experienced, especially from Europe, we would be in a better chance to know how the Europeans want our products to look like (*Times of Swaziland*, 14 May, 1987).

The carved animals and busts of human figures so common throughout Swaziland are examples of how Swazis view tourist demand. They are part and parcel of what Africa is supposed to be. So, too, are the 'traditional' dances performed in the more expensive hotels, and the animals in the game parks, preferably viewed from comfortable surroundings, might also be regarded as performers, equally 'stage managed' for the benefit of tourists.

This may appear a somewhat cynical representation of the tourist industry, and such stage management is certainly not confined to Swaziland. It can also be argued that the production of any tourist artefact, indeed all tourist art, contains elements of performance, and that even the most obvious item of 'airport art' is the product of a highly complex tourist art system, through which 'the expectations of consumers are typified, mirrored, and represented to them by others' (Hobsbawn and Ranger, 1983). According to this perspective, the meaning of a product to all those involved in an apparently simple transaction is no easy matter to determine. The product itself is the result of a series of symbolic exchanges.

This is undoubtedly the case. However, cultural exchanges are collective as well as individual and both performer and audience may subscribe to social stereotypes, which then become part of the exchange. This point may be illustrated by reference to promotional material issued by the Tourist Division of the Swazi government. In July, 1987, it produced two posters advertising Swaziland, both of which focused on 'traditional' Swazi culture. One pictures a Swazi woman in traditional costume. She is young and bare-breasted, and her costume, a beaded apron and woollen tassels, is colourful and revealing. Such pictures must be seen in the context of other coverage of 'African' affairs, where 'the native' is similarly portrayed – for example, in postcards from South Africa and other colonies in the 1920s and 1930s, as well as many European news items on recent African political events. For many Swazis, such costumes are regarded as being quite different from the allegedly provocative forms of dress worn by some young Swazi women at discotheques.[4] For whites, at whom this promotional literature is directed, and who will have been socialized into specific racial stereotypes, the distinction might not be so clear. Preconceptions of another kind might also be involved in 'reading' the second picture, of three Swazi men, again in traditional costume. Carrying knobkerries, wearing skirts, with skins around their waists and beads across their chests, the men are indeed 'picturesque'. The commentary in the official tourist guide, in which several such pictures appear, provides additional information:

> Present-day Swazis delight in their customs and traditions. Swazis dress in colourful costume, Emahiya, the women wearing the traditions [sic] beehive hairstyle - Sicholo. Swazi warriors carry shields, knobsticks, spears and battleaxes as part of their dress. These are some of the outward signs of a deeply-rooted social system, unique in Africa (Swaziland Tourist Office, undated, p. 2).

In fact, as even a casual visitor to Swaziland discovers, such costumes are worn only on special occasions. The Swaziland Tourist Office, a government institution, is packaging and selling Swazi tradition. The message it is conveying to potential tourists, mostly whites from the Republic of South Africa, is that Swaziland is a country of exotic people, with unfamiliar customs and colourful costumes, which is locked in the past. Once again, we have returned to the 'picturesque occupants' of Swaziland, noted by John Buchan early this century (Buchan, 1903, p. 132).

In presenting the above images, the Swaziland Tourist Office is doing no more than similar organizations in other less developed countries. Indeed, its approach is recommended by many consultants to the industry:

> Reason enough for foreign tourists to come to Swaziland from afar to discover and have a look at 'Africa at its best'... Reason enough to consider this country as a whole to be a 'Nature and Heritage Reserve'. So, a future tourism management should do its best to create favourable conditions for visitors to witness, hear and experience the uniqueness of Swaziland. Any sidestep from traditional and national lifestyle would be fatal (G. Conradi, Tourism Manager and Consultant, 'Tourism rediscovers paradise', *Observer*, Swaziland, 6 April, 1985).

The logical conclusion of this policy is to consider the entire country a showpiece, a living, moving, breathing storehouse of tradition. In fact, if Macmillan is correct in his analysis of the development of 'traditionalism' in Swaziland, such a development would fit perfectly with the prevailing political ideology which, nevertheless, has prevented neither the Swazi government nor the royal family from increasingly participating in numerous sectors of a modern economy.

At its worst, focus on tradition as a selling point can lead to parody. Consider the following advertising feature in the in-flight magazine of the national airline, owned by *Tibiyo Takangwane*, the monarchy's investment arm, and designed to attract conference tourists to the Royal Swazi Sun Hotel, in which it has a substantial interest:

> The Royal Swazi Sun has a reputation for organising special evenings which are built around a particular 'theme'... Suggestions include: 'Out of Africa' evening - using the Swazi village in the grounds of the hotel, we can have lamb or ox on the spit, Swazi musicians and singers, huge fires and flares for lighting effects - a different type of evening for those of us who live in cities and towns.

As the conference organizer went on to explain,

> I remember when one big insurance company asked for this theme and delegates really entered into the spirit of the evening, dressing up in Swazi costumes or safari suits. One guy actually hired a gorilla suit in Johannesburg and brought it with him. This is the kind of crowd we really enjoy working for (Royal Swazi National Airways, June/July, 1987, p. 13).

Although tradition may be packaged and sold by the Swazi authorities to increase tourism, it is no mere marketing device. On 19 May, 1987, Pastor Phil Dacre of the Rhema Church, an evangelical sect, was arrested for sedition and detained under the law (designed to combat terrorism and political unrest) which allowed the authorities to commit him to prison for up to 60 days without trial. His offence was to write pamphlets criticizing, on religious and moral grounds, features of the traditional Incwala kingship and first fruits ritual. Dacre was subsequently joined in detention by another pastor, the church's administrator, and the General Manager of the company which printed the pamphlets. In July, the 60-day period was renewed, and Dacre's first appearance in court was set for 21 September, 1987, 4 months after his initial arrest.[5]

His arrest, under a law which can only be described as draconian, indicates a revival of opposition on the part of mission churches to some Swazi traditions, an opposition which, according to Macmillan, was 'discreetly dropped' in the late 1960s (1986, p. 121). It also supports Macmillan's argument for the importance of 'traditionalism' in Swazi politics. Not even pastors are allowed to rock the boat in paradise.

Conclusions

In Swaziland, the monarchy is dominant. As Head of State, the King controls the electoral process, appointing and dismissing leading government officials, and indirectly controlling all government investments. As monarch, he has a constitutional power base and direct control over his subjects through the chiefs, who hold and allocate Swazi Nation Land on his behalf. He also controls the royal finances and has a major stake in the country's economy, often in partnership with foreign capital. Government and the Crown have increasingly invested in tourism, especially the hotel sector, normally in partnership with the Sun International chain of hotels, which dominates Swazi tourism.

Overall, the political and economic dominance of the monarchy is legitimated by 'tradition', re-interpreted by King Sobhuza II and his allies in the years preceding independence and later enforced by further changes in the constitution in the post-independence period. And when the role of such tradition is questioned, the full power of the state is exercised and critics are charged with sedition. In short, it is 'tradition' or, at any rate, the appeal to tradition, which both established and continues to reinforce the 'modern' Swazi state.

However, the involvement of the state and *Tibiyo Takangwane* in tourism has also made the monarchy party to the presentation, packaging and selling of Swazi tradition to attract tourists to the country. In such circumstances, 'tradition' becomes an advertising gimmick, presenting potential tourists (mainly white South Africans) with exotic people in colourful costumes and spectacular ceremonies, thus reinforcing what many would regard as an unfavourable stereotype of Swazi society. To add to the contradiction, it is also possible to see tourism as but part of a continuing process of 'modernization', which for decades has allegedly undermined the authority of (male) traditional Swazi rulers over the young and (especially) over women, introducing forms of behaviour and dress into the country which are considered to threaten the traditional Swazi way of life.

In Swaziland (as elsewhere) 'tradition' and 'modernity' are not polar opposites. Furthermore, the relationship of tourism to tradition and modernity is in no sense straightforward. For the Swazi government, tourism may be a mixed blessing. Arguably, its power is based on a successful appeal to tradition and yet, in its support for and investment in tourism, it may be encouraging a continued process of modernization which ultimately will exoticise and undermine the foundation of its legitimacy.

When a country opens its doors to international tourism, its traditions (however marketable) are going to be changed, if not threatened.

And it is the young, both 'visitor' and 'host', who will be in the forefront of 'culture contact' and who will be the most susceptible to change. In Swaziland, this process did not start with international tourism; it commenced with the first arrival of the Europeans and has continued ever since. What tourism has done, as it has done elsewhere, is to emphasize the contrast between tourist and 'host' and reinforce the process. In Swaziland, so close to and dependent on the Republic of South Africa, it is unlikely that this pattern can be reversed.

Notes

[1] Fieldwork in Swaziland in July and August, 1987, was part of a visit funded by the British Council. I am grateful to Don Funnell and Ieuan Griffiths, colleagues at the University of Sussex, and to members of the Social Science Research Unit, University of Swaziland, for comments on a more comprehensive report on this research (cf. Harrison, 1990). Sue Rowland was good enough to provide the map and Malcolm Crick commented on an earlier draft.
[2] The value of the Emalangeni, the currency of Swaziland, is equal to that of the South African rand, to which it is directly linked.
[3] It was my *impression,* and that of Crush and Wellings (1987, p. 103), that male Swazis, perhaps employed by the casino, were considered to be most 'at risk', tending to spend their earnings gambling rather than on their wives and families.
[4] Nevertheless, an amused King Sobhuza II made no concessions to a white visitor's discomfort when she was shocked by the traditional costumes and 'arrogant yet enticing' deportment of young Swazi women dancers: 'They dance to excite the warriors,' said Sobhuza (Packer, 1953, pp. 169–170).
[5] Up to the time of writing, I have been unable to discover the outcome of these proceedings.

References

Acciaioli, G. (1985) Culture as art: from practice to spectacle in Indonesia. *Canberra Anthropology* 8(1–2), 148–172. DOI: 10.1080/03149098509508575.

Bonner, P. (1983) *Kings, Commoners and Concessionaires: The Evolution and Dissolution of the Nineteenth-Century Swazi State*. Cambridge University Press, Cambridge.

Booth, A. (1986) Homestead, state and migrant labour in colonial Swaziland. In: Daniel, J. and Stephen, M.F. (eds) *Historical Perspectives on the Political Economy of Swaziland: Selected Articles*. Social Science Research Unit, University of Swaziland, Kwaluseni, Swaziland, pp. 7–50.

Buchan, J. (1903) *The African Colony: Studies in the Reconstruction*. William Blackwood and Sons, Edinburgh and London.

Crush, J. and Wellings, P. (1987) Forbidden fruit and the export of vice: tourism in Lesotho and Swaziland. In: Britton, S. and Clarke, W.C. (eds) *Ambiguous Alternative: Tourism in Small Developing Countries*. The University of the South Pacific, Suva, pp. 91–112.

de Kadt, E. (ed.) (1979) *Tourism: Passport to Development*. Oxford University Press, New York and Oxford.

Department of Economic Planning and Statistics (1987) *Economic Review and Outlook*. Government of Swaziland, Mbabane.

Harrison, D. (1988) *The Sociology of Modernization and Development*. Unwin Hyman, London.

Harrison, D. (1990) Tourism in Swaziland. Research Paper No. 29. *Social Science Research Unit, University of Swaziland*.

Hobsbawn, E. and Ranger, T. (eds) (1983) *The Invention of Tradition*. Cambridge University Press, Cambridge.

Kuper, H. (1947) *The Uniform of Colour: A Study of White–Black Relationships in Swaziland*. Witwatersrand University Press, Johannesburg.

Kuper, H. (1961) *An African Aristocracy: Rank Among the Swazi*. Oxford University Press, London and New York.

Kuper, H. (1963) *The Swazi: An African Kingdom*. Rinehart and Winston, New York and Chicago.

Macmillan, H. (1986) Swaziland: decolonisation and the triumph of "tradition". In: Daniel, J. and Stephen, M.F. (eds) *Historical Perspectives on the Political Economy of Swaziland: Selected Articles*. Social Science Research Unit, University of Swaziland, Kwaluseni, Swaziland, pp. 104–125.

Marwick, B.A. (1940) *The Swazi: An Ethnographic Account of the Natives of the Swaziland Protectorate*. Cass, London.

Packer, J. (1953) *Apes and Ivory*. Eyre and Spottiswoode, London.

Russell, M. (1983) *The Production and Marketing of Women's Handicrafts in Swaziland, Report of a field survey in Swaziland, January-October, for the International Labour Organisation*. Social Science Research Unit, University of Swaziland, Kwaluseni, Swaziland.

Swaziland Chamber of Commerce (1987) *Swaziland: A Review of Commerce and Industry – 1987 edition*. B and T Enterprises, Manzini, Swaziland.

Turner, L. and Ash, J. (eds) (1975) *The Golden Hordes: International Tourism and the Pleasure Periphery*. Constable, London.

Wagner, U. (1982) *Catching the Tourist: Handicraft Traders in the Gambia, Department of Social Anthropology*. University of Stockholm.

Wolfers, M. (1974) *Black Man's Burden Revisited*. Allison and Busby, London.

4

Tourism and Prostitution: Sleeping with the Enemy? The Case of Swaziland

It is at least arguable that travel has often been associated with sexual freedom. Indeed, one writer has gone so far as to suggest that travel, a 'spermatic journey', may be 'stimulated by a male reproductive motive, a search for temporal extensions of self in children, only achievable through the agency of women' (Leed, 1991; O'Kane, 1994).[1] Be this as it may, the development of international tourism on a large scale, with the extension of the cash economy and widespread commoditization, has led critics of tourism to decry the apparent association of tourism with prostitution. For Turner and Ash (Turner and Ash, 1975), for instance, tourism in The Pacific is 'a corrupter of innocence', converting 'the moral nexus into a cash nexus – particularly with regard to sexual mores and traditions of hospitality', and they later assert that the West Indian, like the Tahitian or Fijian, 'soon comes to value his/her sexuality in money terms, hence prostitution and the "gigolo" soon appear in tourist-intruded communities, even where they had not existed before' (p. 197).

Similar claims have been made for other less developed countries (LDCs), for example the Gambia (Wolfers, 1974), and Haiti, where culture itself 'is in danger of being prostituted as tourism prospers' (Turner and Ash, 1975) and, more recently, tourism and prostitution have been closely linked in South and South-east Asia (Mackie, 1988; Meyer, 1988; Thanh-dam Truong Sex, 1990; Hall, 1992). Indeed, 'sex tourism' is a specific category of holiday-taking and numerous pressure groups now publicize and combat what is commonly regarded as a highly unacceptable aspect of international tourism (O'Grady, 1990, O'Grady, 1992; Srisang, 1991; Hall, 1992).

That tourism leads to or exacerbates prostitution has also become increasingly accepted in the academic community. Indeed, one eminent anthropologist approvingly quotes the assertion that, at least in South-east Asia, 'tourism is prostitution', with poor countries, and poor people within these countries, forced to sell themselves to the rich in order to survive (Graburn, 1983).

Nevertheless, the links between tourism and prostitution may not be as obvious as critics imply. Tourism is clearly predated by prostitution, which is not easy even to define (Perkins and Bennet, 1985). It may be regarded as a simple business transaction in which money or goods are exchanged for sexual services, or as a subset in the more general provision of sexual services within, for example, capitalist society, where 'prostitutes' are either regarded as a distinct vocational group or where the characteristics they share with other members of the same category, for example men, women and children, are most emphasized. To complicate matters even further, definitions of prostitution may vary across societies over time and place, and the term may be applied to a wide range of behaviour broadly defined as 'immoral'.

For present purposes, prostitution is defined as a business transaction where money or goods are given for sexual services, which are more or less clearly delimited in time and place. It follows from this definition that prostitutes are not the only providers of sexual services and, equally clearly, that prostitution and tourism may operate quite independently of one another. How far female prostitution, especially, is or was linked to tourism in Swaziland, in southern Africa, and why the nature of the link is important, is discussed in the remainder of this paper.

Swaziland and Tourism

With a population of about 700,000 people, Swaziland is bordered by the Republic of South Africa and Mozambique (Fig. 4.1). Although politically independent, the country's economy is tied to that of the Republic, and political stability is also largely conditional on events in neighbouring countries. Links with the Republic of South Africa are not new. During the 19th century, contact with Europeans led to the alienation of Swaziland and, from the 1880s, to the recruitment of Swazi men for work in the mines of South Africa (Booth, 1982; de Vletter, 1982; Fransman, 1982).

Since political independence in 1968, an allegedly 'modern' system of cabinet government has coexisted with a 'traditional' monarchy (Daniel, 1982; Macmillan, 1986; Funnell, 1991) and, with the scenic attractions of Swaziland, tradition has increasingly figured in the marketing of the country as a tourist destination (Harrison, 1992).

Swaziland's tourism industry really only developed after 1965, when southern Africa's first casino was opened near Mbabane (see Fig. 4.1). Even in 1969, there were only 42,000 tourist arrivals, increasing to nearly 80,000 in 1971 (Government of Swaziland, 1972). By 1975, there were 130,568 hotel visitors. The following years were less successful but numbers again increased in the early 1980s and stabilized by the end of the decade (Table 4.1). In 1991, more than 280,000 tourist arrivals were recorded, of whom 63% were estimated to be holiday makers, primarily from the Republic of South Africa (Central Statistical Office, 1991),[2] with most hotel visitors going to foreign-owned hotels in the Ezulwini valley.

Immorality and Prostitution in Swaziland

As in other countries, tourism is said to bring considerable economic benefits to Swaziland. However, it is also considered by some Swazi to have corrupted young people (Harrison, 1992), a view shared by some outside observers. Indeed, it has been alleged that gambling, pornographic films and 'prostitution across the colour line' have become the 'cornerstones' of the tourist industry in Swaziland and Lesotho (Crush and Wellings, 1983, 1987), an assertion repeated unquestioningly in more recent literature (Craik, 1991). In effect, the Swazi case is a well-known, specific variant of a more general argument in which tourism is first said to prompt the development of female prostitution and, as a consequence, is then heavily criticized for its detrimental effects on a nation's morality.

It is not the purpose of this paper to defend or attack prostitution, or to argue that tourism has no detrimental social or cultural effects. Instead, the intention is to examine the basis of the common claim that female prostitution in Swaziland is linked to tourism and to ask what general lessons, if any, can be drawn from the analysis. In fact, from the very outset, it is clear that such accusations of cultural pollution are not new. Even in the late 1930s, Swazi men working in the mines of South Africa were said to be acquiring 'the habit of homosexuality and ... other vicious habits which are rife in the urban areas' and women, too, 'flocked to urban areas inside and outside the territory, there to live as prostitutes or concubines' (Marwick, 1940). By the mid-1950s, such social problems were considered sufficiently pressing for the colonial government to set up a Commission to investigate what was happening in Swaziland, and it reported in 1956.

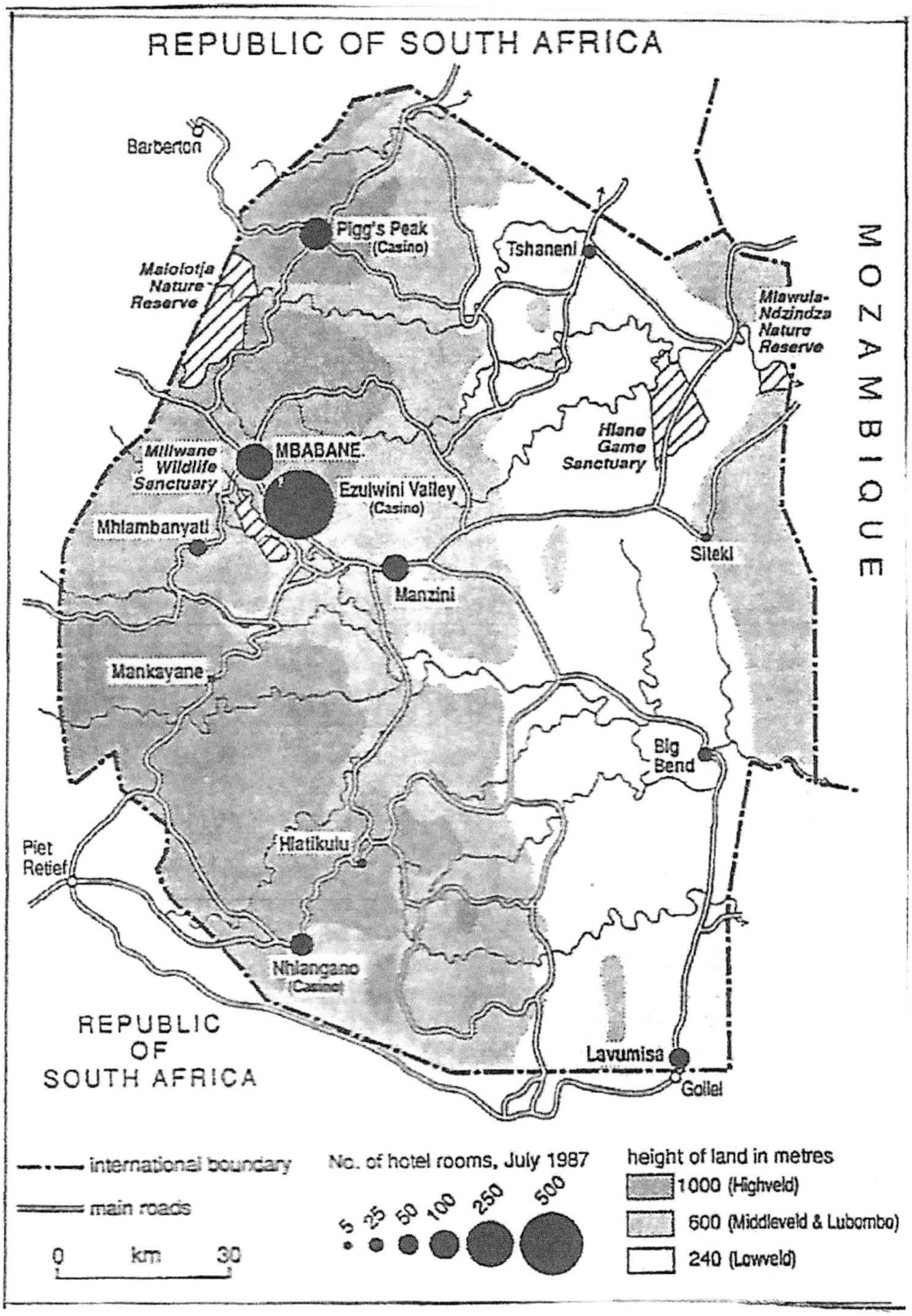

Fig. 4.1. Swaziland, ecological zones and tourist accommodation, 1987.

Sources: Funnell, D, *Under the Shadow of Apartheid: Agrarian Transformation in Swaziland,* Avebury, Aldershot, 36 and author's research 1987.

Table 4.1. Hotel tourists in Swaziland, 1969–1991, according to purpose of visit.

Year	Total	Holiday tourists	%	Business	%	Other	%
1969	42,000						
1970	Not available						
1971	79,840						
1972	Not available						
1973	89,153						
1974	96,123						
1975	130,568	90,408	69	24,871	19	15,289	12
1976	108,69	75,512	70	22,150	20	10,407	10
1977	103,159	67,715	65	26,387	26	957	9
1978	81,970	49,830	61	21,329	26	10,811	13
1979	9434	54,213	58	25,676	27	14,145	15
1980	98,531	48,623	49	30,273	31	19,635	20
1981	76,407	42,749	56	20,336	27	13,322	17
1982	107,657	63,691	59	27,630	26	16,336	15
1983	122,905	71,398	58	35,220	29	16,287	13
1984	154,095	89,516	58	44,158	29	20,421	13
1985	200,078	116,829	58	55,659	28	27,590	14
1986	234,034	132,216	57	47,175	20	54,643	23
1987	199,710	125,259	63	44,201	22	30,250	15
1988	201,438	126,343	63	44,584	22	30,511	15
1989	257,997	161,817	63	57,102	22	39,078	15
1990	287,796	180,507	63	63,697	22	43,592	15
1991	289,493	181,571	63	64,073	22	43,849	15

Sources: Government of Swaziland *Second National Development Plan: 1973–1977* Mbabane (1972) p. 110; *Third National Development Plan: 1978179–1982183* Mbabane (1977) p. 119; Central Statistical Office *Annual Statistical Bulletin* Mbabane (1979, 1983, 1986, 1989, 1991).

The 1956 Report (*Commission, 1956*)

First, the Commission presents a rosy picture of social life in precolonial Swaziland, emphasizing the social control once exercised by the elders, the role of age-groups and the close limits set on untoward behaviour by the informal system. It also noted the stability of marriage customs and was sympathetic to the institution of *lobola* ('bride price'), which gave male elders a vested interest in (and considerable control over) their children's marriages. The Commission was quite explicit that traditional Swazi family life was stable and regulated, as was the wider structure within which it operated.

Second, the Commission echoed the beliefs of Swazi authorities, described in the Report as 'thinking Swazi' (p. 4), by pointing to the disruptive (but allegedly inevitable) (Fig. 4.1) effects of Westernization, especially the development of the Witwatersrand gold fields, European settlement and the cash economy, on Swazi culture (pp. 4–5).

Third, and most relevantly for this paper, the Commission was concerned about, but quite vague on, the extent of prostitution, recognizing that although soliciting for gain was a punishable offence, there had been few, if any, convictions for this 'in recent times' (p. 11). Witnesses' accounts were confused, and prostitution, promiscuity and illegitimacy were frequently and indiscriminately lumped together. A Commissioner of Police, for example, asserted that prostitution had increased 'markedly' over the previous 15 years, but could provide no figures; a missionary felt it was 'widespread', blaming 'Africans' for having too lax an attitude towards illegitimate births; and a 'Swazi headman' agreed with the policeman but blamed missionaries and the civil authorities for introducing monogamy (p. 11). Other Swazi witnesses blamed the decline of old customs and the failure to punish offenders.

According to District Commissioners, however, prostitution, if defined as soliciting for gain, was not as widespread as generally believed. They, too, considered illegitimacy had increased, but differed on the social implications of such a change. As one Senior District Commissioner commented, implicitly drawing a distinction between legal and social illegitimacy:

> There are doubtless many children born out of wedlock among the Africans but these are either in the custody of the mother or the father, and treated as children born in wedlock (p. 12).

Fourth, and crucially, where precise examples are given of prostitution, the clients are Swazi men returning from the Republic of South Africa. At Siboteleni township, Stegi (Siteki), members of the Commission were shown:

> ... a brothel where men returning from the mines, who de-bus at the nearby Native Recruiting Corporation camp are lured, and where they find girls and liquor and lose much of their savings ... Frequently groups of girls may be seen waiting for the bus bringing the mineworkers home. But by the time these men reach their homes they are often poorer and, we hope, wiser, for the road taken leads past two of the very few brick buildings in Siboteleni – the local shebeen and the brothel (pp. 8–9).

The Commission was later to recommend, *inter alia*, that laws against vagrants should be revised to allow police to remove women from townships in which they did not normally reside – a practice already followed in other parts of colonial Africa (Cutrufelli, 1983; Robertson, 1984) and which lasted well into the post-independence period (Jacobs, 1989).

The Commission was primarily concerned with prostitution and immorality among the Swazi and prostitution involving white men and other 'aliens', although mentioned, was not discussed. By implication, it was considered relatively unusual. Nevertheless, the issue leads to an interesting example of attitudes then prevailing towards sexual interaction across the colour divide:

> Ordinance No. 46 of 1903 (The Immorality Ordinance of 1903) while more adequate in its purpose than the Vagrancy Law, would also appear to be in need of revision and amendment. It is interesting to note that while Section 19 of this Ordinance renders liable to imprisonment 'any white woman who voluntarily permits any native to have unlawful carnal connection with her', a white man is apparently free to indulge in similar practices without fear of prosecution. It seemed to the Commission that Section 19 is directed against something which probably does not exist, and leaves open a serious loophole of which advantage is believed to be taken by alien and other 'white' males (p. 19).

It appears to have been assumed that white women (of course?) would not voluntarily associate with 'native' men, whereas the existing opportunities for some (perverse?) 'white' men, sufficiently deviant to want to associate, with native women, should be denied by law. Such attitudes were to continue, as indicated below.

Finally, and despite its rather confused analysis of 'the problem', the Commission recommended that the colonial and Swazi authorities join to arouse the 'social conscience of the people' against prostitution and promiscuity (p. 13), and that laws relating to prostitution and vagrancy should be reviewed. In addition, the 'orderly development' of Swazi townships should be encouraged, with greater control exercised over people entering them.

In the mid-1950s, then, when tourism was virtually non-existent in Swaziland, prostitution (however poorly defined) was already considered a problem, allegedly caused by the failure of Swazi society to adapt to Westernization. And prostitutes' main clients, in so far as they could be identified, were Swazi men returning from the mines of the Republic of South Africa.

Such a situation was not unique to Swaziland. In sub-Saharan Africa generally, the links between prostitution and male migrant labour have been well documented. In mining areas, especially, where large numbers of men were employed and where women were denied wage labour, unstable sexual unions and prostitution were virtually inevitable and the roads of the migrant workers came to be known as 'whores tracks' (Cutrufelli, 1983, p. 30). The story was much the same for Zimbabwe (Jacobs, 1989), the Congo (MacGaffey, 1986), Zambia (Parpart, 1986), Kenya (Robertson, 1984) (p. 41) and Lesotho (Murray, 1980). However, the prior existence of prostitution in Swaziland, and its association with migrant labour, becomes important when considering later assertions of the links between prostitution and tourism.

The 1970 Report

It is not clear how far the 1956 Commission's recommendations were carried out, but prostitution and juvenile delinquency remained. Some 14 years later, after Swaziland had obtained its independence, another committee was given an identical brief and subsequently reported, in almost identical terms.

The 1970 Report covers the same ground as its predecessor but comes to rather different conclusions. Linking 'immorality' among Swazi youth to a breakdown of social order in urban areas (p. 5), it focuses firmly on the material roots of delinquency, especially unemployment, and notes the prevalence of begging by small boys in Swazi towns, especially from visitors from the Republic of South Africa (p. 8). Prostitution is discussed at length. As in 1956, there was doubt about the number of prostitutes in the Kingdom. The Assistant Commissioner of Police estimated that between 100 and 150 women regularly solicited for purposes of prostitution, some 50 in the Hhohho region (which includes Mbabane and Pigg's Peak), another 50 in Manzini and 'a handful in Stegi' (Appendix, p. 1), but it was recognized that such figures were approximate.

A survey carried out by a Miss Silber, quoted extensively in the Report, suggested four categories of women were involved in prostitution. In the first, a relatively small group, were professional prostitutes from the Republic of South Africa who visited Swaziland to avoid the Republic's immorality laws and keep liaisons with men across the colour line, either as the result of prior arrangement or from a contact made in the Kingdom.

Women in the second category, also from outside Swaziland, were those of so-called 'loose' behaviour who might be described as part-time prostitutes, soliciting for gain on an occasional basis. In a paper accompanying that of Silber, the Assistant Commissioner of Police estimated that the total number of women in these two categories was no more than 30 (Appendix, p. 1).

The two remaining categories of women were Swazi. The first were full-time prostitutes and the second, considered by far the largest category, were young, relatively uneducated women, often with illegitimate children to support, who imitated their more professional counterparts. This second group of Swazi women allegedly included those who, from their early teens, had been attracted to hotels as centres of music and dancing: 'These girls, because of the lack of recreational centres and their uninteresting home environment, look for some outlet to express themselves' (p. 16).

Although prostitution was known to be centred on specific hotels, there was no consensus on the Committee as to who the clients were. According to Silber, they were 'usually visitors from neighbouring territories who spend casual weekends in Swaziland as well as local residents working in the country either temporarily or permanently' (p. 80). In addition:

> Miss Silber said that both black and white men have approached her with indecent proposals while she has been conducting her survey on juvenile delinquency. She states that the white men making these proposals do not predominate but, in fact, more black men have approached her. The other members of the Committee agreed and said that civil servants were very much involved. After the 25th of each month, when civil servants received their pay, the hotel trade improves (p. 33).

By contrast, the Assistant Police Commissioner considered that most clients were white:

> The male Europeans who come to Swaziland from the Republic for the purposes of prostitution are, generally, either South African

> subjects, German nationals resident in South Africa on contract jobs, or in some cases, Czecks [*sic*] resident in South Africa ... Many of the Portuguese males resident in Swaziland are of the artisan class and some are unmarried or have no wives with them and they consequently seek female company (Appendix, p. 2).

Quite where this left the civil servants on pay-day is unclear. There was an evident tendency on the 1970 Committee to blame (black or white) outsiders, from neighbouring (Portuguese) Mozambique or the Republic of South Africa, who were working in Swaziland. However, although it was agreed that prostitutes' clients were from within and outside the country, the role of holidaymakers (as opposed to business tourists) was given little attention.

The atmosphere in which the Committee collected information can be gauged from evidence submitted to it in writing. Outright hostility to any inter-racial sexual relationships was obvious. Some respondents proposed a ban on all such relationships, and one recommended that Portuguese men should be allowed to enter Swaziland only if accompanied by their wives, adding that it should be a criminal offence for a Portuguese man to give a Swazi woman a ride in his car. Another, ignoring the focus on outsiders, suggested all young Swazi males should be sent to work in South African mines to be kept out of mischief, while a further correspondent proposed a ban on all mixed marriages. And one Committee member reported a request from the Republic's Special Branch for names and car numbers of men from the Republic of South Africa seen with Swazi women, 'and they have said that if they have these names they will be dealt with' (p. 35).

No indication is given as to whether or not he complied with the request, thus helping police from the Republic enforce the provisions of the Immorality Act then operating in that country. Of equal interest, perhaps, was the letter to the Committee from a local magistrate naming Mbabane and Pigg's Peak (then noted only for its asbestos mine and its timber production) as centres of prostitution.

The Committee recommended several measures to counter the prevailing 'immorality' and, if these failed, the police were encouraged to use their powers of arrest more rigorously and to report hotels which encouraged immorality. As part of the new drive against prostitution, it suggested that police should occasionally visit places like the Mantenga Falls, in the Ezulwini Valley, '(where we understand mixed nude bathing takes place) with a view to enforcing the laws against public decency' (p. 22). Hoteliers, too, were exhorted to ask those suspected of immoral behaviour or intentions to leave the premises rather than serve them, presumably emulating the somewhat draconian measures adopted by a hotelier who gave evidence to the Committee:

> Mr Sergeant told the Committee that quite often a man would come in to hire a room for himself and his wife and he would insist on seeing the wife before letting the room. He says he tells the man to fetch his wife from the car and to bring their passports so that he can check up on them properly before he lets the room and usually the man goes back to the car and never returns (p. 35).

When comparing the 1956 and 1970 reports, it is evident that the emphasis has changed. In 1956, when tourism was negligible, the clients of female prostitutes were Swazi miners returning from the Republic of South Africa. By 1970, prostitution was centred on a few down-market hotels in Mbabane, Manzini and in some smaller Swazi towns. Despite some contrary evidence, the women's clients were said to be foreigners working in Swaziland, but research also needs to be carried out on the role of the Gordon Highlanders, a contingent of which was sent from Kenya, in 1963, to put down industrial unrest, and which remaid in Swaziland until 1968. It would be surprising if this resident army had not contributed to prostitution during its stay in the country.

The crucial fact to note, however, is that in 1969, when most of the work on the 1970 Report was carried out, there were only 49,000 hotel visitors to Swaziland. Prostitution may indeed have been centred on the hotels but, in 1970 (5 years after the opening of the first casino in Swaziland) holidaymakers were conspicuous largely by their absence – as they had been in the 1950s.

Swazi tourism in the 1970s and 1980s

Between 1969 and 1975, the situation changed. Hotels of international standard were constructed in Swaziland and the tourism industry boomed. By 1975, tourist arrivals numbered more than 130,000, of whom more than 90,000 were holidaymakers, and it seemed that the prospects for Swazi tourism were good (see Table 4.1). However, against expectations, tourism then declined, despite the construction of a second casino hotel in 1978, and in 1981 there were less than 77,000 hotel visitors, of whom 42,749 were holidaymakers.

The reasons for the varied fortunes of the Swazi tourism industry have been detailed elsewhere (Harrison, 1994) but the decline is explained primarily by the world recession arising from the increased price of oil and economic sanctions against the Republic of South Africa. In addition, casino complexes in Bophuthatswana and the Transkei also provided alternative destinations for South African residents anxious to taste the 'forbidden fruit' of casino gambling, undoubtedly marketed as one of Swaziland's attractions (Crush and Wellings, 1987). However, it is vital to note that such facilities were provided *after* Swazi tourism had already entered a period of decline and that, after 1975, the decrease in the numbers of holidaymakers visiting Swaziland was not matched by a corresponding decline in visits for business or other purposes (see Table 4.1).

The tourism industry attempted to reverse the decline by promoting Swazi tradition and by laying on events (e.g. football matches, visits by popular entertainers and televising the Olympic Games) which, because of sanctions, were unavailable in the Republic of South Africa. Such policies were successful in increasing arrivals which, by 1991, reached almost 290,000. However, they were counteracted by a declining length of stay. Put differently, between 1974 and 1991 arrivals increased by more than 200% but bednights by less than 4%.

Prostitution and tourism in Swaziland: a blurred connection?

Where did prostitution figure in all of this? As in earlier periods, it was undoubtedly considered 'a problem', but, as before, the links with tourism were far from clear. Indeed, the complex situation in the 1960s and 1970s was nicely summarized by two informants who had lived in Swaziland for almost 30 years. They recall, even in the late 1950s, occasional visits by Boer farmers across the border for sex across the colour line, and are in no doubt that the casino, and the showing of films banned in the Republic of South Africa, increased the country's attractions. That there were prostitutes who catered for white visitors was undoubtedly the case. However, they also noted that many of the liaisons were stable relationships of white men, working in the Republic during the week and returning to Swaziland for the weekends, with black Swazi or South African partners. Indeed, one informant insisted that, throughout the 1960s and 1970s, white visitors were far less responsible for prostitution and casual sex than Swazi men, who found willing partners among younger Swazi women, many of whom had moved to the urban centres, subsequently becoming alienated from their families, a problem also highlighted in the 1970 Report, which recommended moral education, a national youth movement, community centres and girls' hostels as remedies for the situation (Committee, 1970) [pp. 15–21]). And as one informant (a white South African) remarked, 'dirty weekends' often involved only white visitors from the Republic, who conducted their extra-marital affairs across the border, in Swaziland.

In fact, it is noteworthy that although academic prominence is given in the 1980s to the links of Swazi tourism with prostitution, the little evidence presented related more to the 1970s than the 1980s, primarily to 1977, when soliciting was made a criminal offence, and to 1978 when tourism was at a particularly low ebb (Crush and Wellings, 1987). However, during the 1980s, newspaper evidence continued to link prostitution with hotels. In 1984, for example, 12 women (including two whites) were reportedly arrested for loitering for purposes of prostitution – four in Mbabane, six in Nhlangano, one at Hlathikhulu and one at an unnamed hotel – but no indication was given as to who their prospective clients might have been. Unsurprisingly, newspaper reports tend to sensationalize the issue. In 1984, one article debated whether 'shangaans' (foreigners) were preferred by some Swazi women because they were more wealthy, or because they were more considerate than Swazi men, suggesting that such a preference, for whatever reason, itself indicated immorality (Weekend Observer, 1984a, b.)

In the same issue, the results of one '3 week investigation' were broadcast under the banner headline 'SEX FOR SALE'. Suggesting that the women involved were dropouts from school, the article quoted one as earning E200[3] a night from South Africans on holiday in the Kingdom. She said:

> 'For every shot, I charge the white man and other foreigners E30 and when it means spending a night with them in their hotel rooms, then the amount doubles' (Weekend Observer, 1984a).

However, no estimate of the number of prostitutes operating in Swaziland was attempted and a police spokesperson was more cautious:

> From the legal point of view there is 'nothing' like prostitution in Swaziland ... He said when more women are found in drinking places enjoying themselves, it is assumed they are prostitutes (Weekend Observer, 1984a).

A similar position was later taken in another journal:

> Contrary to popular belief, prostitutes are not an issue at all in Swaziland. In fact, there could be less than thirty professional prostitutes, mainly concentrated between one seedy hotel and disco house in the valley, another discotheque in Manzini, and of course that other one at the back of Mbabane. The danger lies with the hundreds and even thousands of amateur enthusiasts who by popular demand will share it with any man in town just to be in the spirit of the party (Harrison, 1994).

In 1987, the writer visited Swaziland in July and August, and although prostitution was not the main object of research, its incidence was investigated (Harrison, 1989). The situation then seemed similar to that described in the 1970 Report. This was confirmed by hearsay, direct experience and newspaper reports.

In general, prostitution was restricted to a few nightspots, especially a motel in the Ezulwini Valley. It undoubtedly also occurred elsewhere, but on a very limited scale, and was officially discouraged by most hoteliers, who developed different strategies for dealing with prostitutes and their clients. The owners of one establishment, a notorious centre of prostitution in the 1970s, insisted that known prostitutes were accompanied by a male partner – and to the restaurant rather than the bar – a case of public spiritedness bringing its own reward. Some hoteliers simply directed male residents to the more disreputable nightspots, while a few tried to meet the demand themselves. A son of the owner of a small, inexpensive Swazi-owned hotel near the border with the Republic of South Africa, at which both blacks and whites stayed, stated that white men, on their way to or from Mbabane or Manzini, where they worked, for instance, as air-conditioning mechanics, sometimes asked him to obtain 'a nice, clean girl' for them, a request he was finding increasingly difficult to satisfy.

The scale of prostitution in Swaziland at the end of the 1980s remained difficult to quantify. Published figures are rare and unhelpful, and the police allocate the offence a low priority. However, personal observation, confirmed by informants, indicated that most holidaymakers were in couples or family groups, and thus in circumstances that made visiting prostitutes at least more difficult, and probably unlikely. By contrast, single males on business in Swaziland suffered no such constraints.

Conclusions

It was the intention of this study to investigate the links between tourism and prostitution in Swaziland, a country long reputed to be a centre for illicit sex and other 'immoral' activities. It was found that prostitution was considered a problem well before tourism was introduced into the country, that as tourism became important prostitution was focused more on hotels, but that even at the peak of the tourist industry in the mid-1970s there was considerable doubt as to who prostitutes' clients were – whether tourists (on business or holidaymakers) or local (black or white) residents. Furthermore, tourism in Swaziland declined drastically even *before* alternative entertainment centres were developed elsewhere in the region.

In such circumstances, it is crystal clear that tourism did not cause prostitution in

Swaziland, and that the closest association of the two occurred for a brief period in the 1970s when, because of repressive legislation in the Republic of South Africa, a wide range of sexual liaisons then banned in the Republic, undoubtedly including prostitution, were entered into in Swaziland. Furthermore, evidence has been provided that at least some residents of Swaziland considered all inter-racial sexual relations to be morally reprehensible, an attitude reflecting the prevailing ideology in the Republic.

The reputation of Swaziland as an orchard loaded with 'forbidden fruit' seems to have continued, at least in academic circles, into the 1980s and early 1990s, becoming even less justified. In Swaziland some female prostitution clearly continues but its extent is limited and, as in the past, clients include not only holiday-makers, who are probably in a minority, but also white, Asian and black businessmen. And it is still as difficult as in the past to distinguish full-time prostitutes from 'good time girls'. In Swaziland, the tendency is to regard them as synonymous and any young woman seeking freedom from 'traditional' controls, drinking or dancing at discotheques, or wearing jeans or trousers, is likely to be branded 'immoral' – a term which covers a multitude of sins (Harrison, 1992).

Several general lessons may be drawn from, this study of prostitution in Swaziland. First, the social effects of tourism, including those considered to be detrimental, need to be examined within a wider context. Tourism is but one aspect of a process of modernization, occurring in both developed and less developed countries and, as an industry, can hardly be held responsible for attracting some who earn a living by socially illegitimate means. As the Swazi data indicate, the services provided may simply be switched to tourism from other sectors. Putting it crudely, entrepreneurial activity is directed, first and foremost, to where profit can be made, and tourism (or *any* influx of new clients) may simply reinforce tendencies already existing in a host society (O'Kane, 1994).

Second, the actions of business or holiday tourists when away from home cannot necessarily be blamed on the tourist industry. Much depends on the social situation in the tourist-sending society. Even if the 'forbidden fruit' argument were totally correct – and the available evidence suggests otherwise – to berate Swazi tourism for providing an outlet for behaviour considered unacceptable in the Republic of South Africa is to enter a moral quagmire. From 1948, the policy of apartheid extended and enforced by the Nationalist government, along with strict controls on gambling and all forms of public entertainment, brought repression into all areas of the Republic's public and private life and, as indicated above, there were pressures on the colonial government in Swaziland to introduce a similar regime. 'Banned in South Africa' may indeed have been a label attached to such films as *Bonnie and Clyde* and the *Emmanuelle* series, as well as to casino gambling and virtually *any* social relations, other than master and servant, across the racial divides, but that does not make such entertainments and practices immoral.

Third, and following from the above, the Swazi case should lead to caution in prosecuting tourists in their home countries for actions committed while abroad. In December 1993, for example, an amendment to the French Penal Code made French citizens liable to prosecution in France if, while abroad, they were to commit sex acts illegal in France but not in the countries visited. Aimed at discouraging the activities of French paedophiles in South-east Asia, the motives behind the amendment were understandable, but the general principle is more contentious, as any consideration of the 'forbidden fruits' of Swaziland would indicate.

Finally, governments should be expected to control what goes on within their own borders if it offends against generally held national values. Successive governments in Swaziland could indeed have acted strongly against prostitutes and their clients, as can other legitimate governments. Arresting the clients (or tour operators) rather than the prostitute is not, in practice, any more difficult, even if the ramifications are likely to be more extensive. Similarly, strong action against Western tourists engaged in drug-running, or any other illegal activity, is not inherently problematic, even if it does result in intensive lobbying from their parliamentary representatives.

To argue that the links between prostitution and tourism in Swaziland are highly complex, and that care is needed in discussing 'immorality' across national boundaries, is not to advocate non-involvement, to justify a sanguine approach to social issues arising from tourism, or to be an apologist for the tourist industry. However, it does indicate that, in the frequently emotional debate over tourism's social effects, social scientists should focus on empirical evidence and aim to develop an understanding of the wider social and political context within which tourism occurs, rather than make sensational but unsupportable claims about the influence of international tourism on hapless communities in less developed countries or elsewhere.

Notes

[1] Among women, too, travel and sexual freedom may be more linked than commonly realized, cf. M. O'Kane, 1994.
[2] In fact, this proportion is an estimate by the CSO, based on the analysis of tourist arrivals in 1987.
[3] The Lilangeni (plural Emalangeni) is on a par with the South African rand.

Acknowledgements

This research was based on a visit to Swaziland in July and August 1987, which was funded by the British Council, whose support is gratefully acknowledged.

References

Booth, A. (1982) The development of the Swazi labour market, 1900–1968. *South African Labour Bulletin* 7(6–7), 34–57.

Central Statistical Office (1991) *Annual Statistical Bulletin*. Central Statistical Office, Mbabane, p. 20.

Commission (1956) *Commission Report of the Commission Appointed to Investigate Juvenile Deliquency and Allied Problems in Swaziland*. Mbabane.

Committee (1970) *Report of the Committee on Juvenile Deliquency, Sexual Morality and the Maintenance of Illegitimate Children in the Urban Areas of Swaziland*. Mbabane.

Craik, J. (1991) *Resorting to Tourism: Cultural Policies for Tourist Development in Australia*. Allen and Unwin, Sydney, p. 104.

Crush, J. and Wellings, P. (1983) The southern African pleasure periphery, 1966–83. *The Journal of Modern African Studies* 21(4), 673–698. DOI: 10.1017/S0022278X00024265.

Crush, J. and Wellings, P. (1987) Forbidden fruit and the export of vice: tourism in Lesotho and Swaziland. In: Britton, R. and Clarke, W.C. (eds) *Ambiguous Alternative: Tourism in Small Developing Countries*. University of the South Pacific, Suva, pp. 91–112.

Cutrufelli, M.R. (1983) *Women of Africa: Roots of Oppression*. Zed Books, London, pp. 22–23.

Daniel, J. (1982) The political economy of colonial and post-colonial Swaziland. *South African Labour Bulletin* 7(6–7), 90–113.

de Vletter, F. (1982) Labour migration in Swaziland: recent trends and implications. *South African Labour Bulletin* 7(6–7), 114–139.

Fransman, M. (1982) Labour, capital and the state in Swaziland, 1962–1977. *South African Labour Bulletin* 7(6–7), 58–89.

Funnell, D. (1991) *Under the Shadow of Apartheid: Agrarian Transformation in Swaziland*. Avebury, Aldershot, pp. 61–74.

Government of Swaziland (1972) *Second National Development Plan, 1973–1977*. Government of Swaziland, Mbabane, p. 110.

Graburn, N.H.H. (1983) ' Tourism and Prostitution' . *Annals of Tourism Research* 10(3), 437–443. DOI: 10.1016/0160-7383(83)90068-3.

Hall, C.M. (1992) Sex tourism in South-east Asia. In: Harrison, D. (ed.) *Tourism and the Less Developed Countries*. Belhaven Press, London, pp. 64–74.
Harrison, D. (1989) *Tourism in Swaziland Research Paper No 29*. Social Science Research Unit, University of Swaziland, Kwaluseni, pp. 45–55.
Harrison, D. (1992) Tradition, modernity and tourism in Swaziland. In: Harrison, D. (ed.) *Tourism and the Less Developed Countries*. Belhaven Press, London, pp. 148–162.
Harrison, D. (1994) Development of tourism in Swaziland. *Annals of Tourism Research* 21(4).
Jacobs, S. (1989) Zimbabwe: state, class and gendered models of land resettlement. In: Parpart, J. and Standt, K.A. (eds) *Women and the State in Africa*. Lynne Reiner, Boulder and London, pp. 161–184.
Leed, E.J. (1991) *The Mind of the Traveller: From Gilgamesh to Global Tourism*. Basic Books, New York.
MacGaffey, J. (1986) Women and class formation in a dependent economy. In: Robertson, C. and Berger, I. (eds) *Women and Class in Africa*. Africana Publishing, New York and London, pp. 161–177.
Mackie, V. (1988) Division of labour: multinational sex in Asia. In: McCormack, G. and Sugimoto, Y. (eds) *Modernization and Beyond: The Japanese Trajectory*. Cambridge University Press, Cambridge, pp. 218–232.
Macmillan, H. (1986) Swaziland: decolonisation and the triumph of “tradition”. In: Daniel, J. and Stephen, M.F. (eds) *Historical Perspectives on the Political Economy of Swaziland*. Social Science Research Unit, University of Swaziland, Kwaluseni, pp. 104–125.
Marwick, B.A. (1940) *The Swazi: An Ethnographic Account of the Natives of the Swaziland*. Protectorate Frank Cass, London, pp. 300–301.
Meyer, W. (1988) *Beyond the Mask*. Verlag Breitenbach, Saarbrücken.
Murray, C. (1980) ‘ The effects of migrant labour: a review of the evidence from Lesotho. *South African Labour Bulletin* 6(4), 37.
O’Grady, R. (1992) *The Child and the Tourist: The Story Behind the Escalation of Child Prostitution in Asia*. The Campaign to End Child Prostitution in Asian Tourism (ECPAT), Bangkok, in association with Pace Publishing, Auckland.
O’Grady, A. (ed.) (1990) *The Challenge of Tourism: Learning Resources for Study and Action*. Ecumenical Coalition on Third World Tourism, Bangkok.
O’Kane, M. (1994) All in a night’s work. *The Guardian* Section 2, 2–3.
Parpart, J.L. (1986) Women in Northern Rhodesian copper mining communities. In: Robertson, C. and Berger, I. (eds) *Women and Class in Africa*. Africana Publishing, New York and London, pp. 141–160.
Perkins, R. and Bennet, G. (1985) *Being a Prostitute: Prostitute Women and Prostitute Men*. 4. George Allen and Unwin, Sydney.
Robertson, C.C. (1984) Women in the urban economy. In: Hay, M.J. and Stichter, S. (eds) *African Women South of the Sahara*. Longman, London and New York, pp. 33–49.
Srisang, K. (1991) *Caught in Modern Slavery: Tourism and Child Prostitution in Asia*. Ecumenical Coalition in Third World Tourism, Bangkok.
Thanh-dam Truong Sex (1990) *Money and Morality: Prostitution and Tourism in South-east Asia*. Zed Press, London and New Jersey.
Turner, L. and Ash, J. (1975) *The Golden Hordes: International Tourism and the Pleasure Periphery*. Constable, London, p. 165.
Weekend Observer (1984a) Sex for sale. Weekend Observer, 28 April.
Weekend Observer (1984b) Why do our local girls fall for shangaans? Weekend Observer, 28 April.
Wolfers, M. (1974) *Black Man’s Burden Revisited*. Allison and Busby, London, p. 71.

5

Tourism to Less Developed Countries: The Social Consequences

Introduction

Numerous writers have tried to explain why people take holidays. For some, tourists are trying to escape, however unsuccessfully, the alienation of modern (or postmodern) society (MacCannell, 1976; Krippendorf, 1987; Urry, 1988, 1990), seeking an 'authenticity' which, if found, will inevitably be destroyed. Tourism may be a form of pilgrimage, a 'sacred journey' (Graburn, 1989), a move from structured *societas* to unstructured *communitas*, perhaps echoing rites of transition practised in some pre-industrial tribal– societies (Turner and Turner, 1978, pp. 34–39). At a more general (and reified) level, underdevelopment theorists argue that paid vacations provide recreation for workers to enable them to continue their tasks in the production system (Schillerr, 1976, p. 14; Britton, 1987b, p. 172).

Whereas, in these accounts, tourism performs functions for Western society as a whole, others have examined the social-psychological motivations of tourists (Crompton, 1979; Dann, 1981; Dunn and , 1991) and the roles of marketing agencies and the mass media (Britton, 1979; Moeran, 1983; Thurot and Thurot, 1983; Uzzell, 1984; van Raaij and Francken, 1984; Cohen, 1989a). However, for whatever reason, millions of tourist trips are made every year. In this chapter, the focus is on the consequences of such a vast, temporary migration for tourist-receiving societies, especially those regarded as less developed.

Consequences and problems

Domestic tourism is not unproblematic and, because of differences in class, status and ethnicity, its consequences may be similar to those of international tourism (Monk and Alexander, 1986). Domestic visitors can seem foreign, as did urban visitors to a *mestizo* village in Mexico (Nunez, 1963), and it was Indonesian and not Western tourists who desecrated an ancient burial site on the Indonesian island of Sulawesi (Crystal, 1989, p. 150). In such circumstances, perceived 'foreignness' may bear little relationship to political boundaries.

However, in international tourism, where the 'modern' and 'traditional' stand in stark contrast to one another, the effects of 'modernization', of which tourism is but one element, are most evident. While its economic consequences are not universally welcomed, the cultural and social effects of international tourism rouse the most passion, and the focus of this chapter is on these areas.

At the outset, a crucial distinction has to be drawn between social *consequences* and *social problems*. The former, arising from a well established pattern of modernization, include alterations to the physical environment to accommodate and entertain tourists, deep and possibly traumatic changes in the social structure, and alterations to the value system, either throughout society or in some part of it. Such changes can be described, demonstrated and, in principle, explained. By contrast, social problems are defined according to perceptions of 'development'. Like beauty, they are in the eye of the beholder and lists of social costs provide a spurious objectivity to a highly evaluative selection process. Such remarks are not intended to trivialize the issue, for tourism development strategies raise serious moral dilemmas. Indeed, how these are dealt with, and by whom, becomes part of the social consequences, and may lead to the perception of other problems. Clearly, consequences and problems are linked – but they are not identical.

The Physical Landscape

Perhaps the most obvious effect of tourism is on the physical landscape. Initially, existing properties may be adapted to accommodate visitors, but increasing numbers soon prompt the construction of new buildings, perhaps in tourist 'enclaves'. The distinction between physical and social impact soon becomes blurred. There are debates on style ('traditional' or 'modern'?) and appropriateness, and government may legislate against alleged excesses. Where old buildings become hotels, links with the past may be retained, albeit ironically, as where abandoned sugar mills in the Caribbean are given a new lease of life, adding weight to the idea that tourism is 'a new kind of sugar' (Finney and Watson, 1975; Harrison, 1975). Hotels may come to symbolize 'modernity' (Wood, 1984, p. 368) and, as centres of entertainment, they are attractive, especially to the young, even from rural areas, and their use as conference centres often performs a similar function for older and wealthier local patrons (Meyer, 1988). In such circumstances, 'unsuitable' local people may seek entry, not only to hotels and similar establishments, but also to 'public' beaches and other facilities, and questions of access, and its restriction, often arise. One answer, frequently implemented, is to charge high prices, both to increase revenue and to keep out the undesirable poor.

Increased Commoditization

Welcomed as necessary by advocates of modernization (cf. Parsons, 1964, pp. 349–350), the continuing spread of market relationships has been greeted more ambivalently by proponents of underdevelopment theory, who have nevertheless charted the continued incorporation of 'underdeveloped' regions into the capitalist world system. Undoubtedly, tourism contributes to this process, extending the formal, 'modern' sector by increasing wage employment or cash payments for items valued by tourists deep into the rural 'hinterland'.

Tourism clearly affects the production and nature (and more arguably the quality) of local art forms. It is easy to scoff at tourist taste and its effect on indigenous arts and crafts (Turner and Ash, 1975, pp. 141–148) and most people are aware of the derogatory connotation of 'airport art'. According to Graburn, indigenous art production has moved from 'functional traditional art', rooted in religious ritual, towards commercial production for the tourist market, ultimately in the form of souvenirs and novelties. While accepting that even 'functional traditional arts' may be the result of earlier diffusion (1984, pp. 400–401), he considers art forms developed for tourists to be divorced from local culture and less satisfying for the producer. They thus constitute a 'symbolically incomplete' experience (1984, p. 406) and result in the 'museumification' of traditional art forms and marginalization and alienation of the minorities who produce them (Graburn, 1984, p. 414).

A similar process is detailed by Nason (1984) in his study of the Truk district of the Caroline Islands of the Pacific, which were subjected to successive colonial and cultural influences from Spain, Germany, Japan and the United States. By the 1970s, handicraft production was increasing, especially among women, but there was a reduction in the production of *meetooch* goods, objects considered worthy of 'special aesthetic consideration' (1984, p. 427). Nevertheless, Nason credited tourism with promoting handicraft production and increasing local awareness and estimation of material and cultural products (Nason, 1984). Such issues have been more positively emphasized by others (Boissevain and Inglott, 1979; Cohen, 1988a; Richter, 1989).

Although tourism sometimes seems to have a cheapening effect, as in the near obscene transformation of the Berlin Wall into ear-rings, production for tourists is not necessarily less 'authentic' as a consequence. The African producers of popular art studied by Jules-Rosette, although constrained by the need to standardize and simplify their products, nevertheless searched 'for new designs, innovative combinations of styles, and bolder artistic messages' (Jules-Rosette, 1984, p. 236). What is deemed 'authentic', in fact, depends on the perspective of producer and buyer, both nodes in a highly complex network of continually negotiated meanings, rather than on allegedly 'objective' social scientists (cf. Cohen, 1988a), whose implicit assessments of good taste may be highly questionable, albeit evident from what they themselves have purchased on their travels, quite possibly at rates lower than market valuations (cf. May, 1975, p. 129).

That tourism need not lead to a diminution of standards can be seen from the Amish of Pennsylvania, islands of 'tradition' in an ocean of modernity. Traditionally, Amish women designed and produced quilts for local consumption. However, by the mid-1980s, 90% of all quilts produced were being sold to tourists. As a consequence, new designs and materials were obtained from outside and 'traded among Amish quilters' (Boynton, 1986, p. 460), apparently without any deterioration in quality.

The effects of tourism on ritual, the heart of 'tradition', have been much discussed. Viewing the tourist industry as a vast school for the modernization of a people's values (Greenwood, 1972, p. 90), Greenwood alleged that the Alarde, a major ritual in Fuenterrabia, northern Spain, had degenerated into a mere tourist show. By being commoditized, it was no longer regarded as significant by the town's inhabitants. Later, however, he retracted this position as an expression of 'anger and concern', conceding that the ritual held considerable political significance for the townspeople of Fuenterrabia (Greenwood, 1989, first published 1972, p. 181).

The example of Fuenterrabia, which has been widely quoted in the literature on tourism, is instructive. That the ritual became a tourist attraction, and was used as a source of income, undergoing changes in the process, is not at issue; that it lost meaning for the local people is highly debatable.

A similar argument has centred on the effects of tourism in Bali. Considering the view that tourism degrades culture, McKean and Noronha assert that it led to a strengthening of Balinese arts, crafts and tradition, and a similar position has been taken, somewhat more guardedly, by Crystal on Indonesia (McKean, 1989, first published 1979, p. 165). According to McKean, 'economic interaction between Balinese and tourists bind the two in a common field' (McKean, 1989, first published 1978, p. 98), whereas Noronha suggests that 'the Balinese want tourism; they also want to be involved in it' (1979, p. 201). For him, Balinese traditions are only marginally affected by tourism. Indeed, through profits made in tourism, they are strengthened:

> Tourism has thus affirmed these most important ties which link the past with the present and the future and which form the boundaries through which no outsider can penetrate, not even Balinese who are not members of the same *bandjar* [village section] (Noronha, 1979, p. 202).

The story may not be so simple. Dances traditionally performed in temples may become forms of entertainment, but the opposite also occurs (Picard, 1990, pp. 52 and 61), thus blurring the distinction between 'sacred' and 'profane', as when dancers continue to wear consecrated masks and obtain religious benefit when dancing specifically for tourist audiences. In short, while the regional authorities endeavour to 'disenchant' parts of the world, the dancers continue to move in a totally 'enchanted' world (Picard, 1990). The entire situation is made even more confusing because the Balinese authorities, faced with the continued threat of Indonesianization of their culture, use tourism to assert their national identity:

> In other words, their culture has become, for the Balinese, on the one hand what characterises

them as a specific society, and on the other hand what provides their tourist product with its distinguishing feature (Picard, 1990).

As part of a wider modernization process, international tourism contributes to the spread of market relationships and tourist-receiving societies must inevitably adapt. How such changes are interpreted by those they most affect cannot be postulated in advance, however disadvantageous they may appear to outside observers. Furthermore, perceptions change over time, as in the Catalan coastal town where, having replaced fishing as the main source of income, tourism came to be seen as the *traditional* industry, to be defended against the ravages of a later invasion of fishermen (Pi-Sunyer, 1989, first published 1978, pp. 196–197).

Changes in Social Structure

Along with other forms of modernization, tourism creates new opportunities in the formal and informal sectors of the economy, new criteria of social status and contributes to changes in such basic social institutions as the family, reinforcing the role of established political and economic institutions and leading to new social structures, often associated with increased state involvement. Finally, it facilitates the spread of Western norms and values, sometimes in direct contrast to 'tradition'.

Owners and entrepreneurs

As Butler suggests (Butler, 1980), tourist regions tend to develop in stages. Where tourist development has been most intense, the more extensive and expensive facilities have often been taken over by transnational companies, perhaps in partnership with the state, leaving the smaller and less profitable part of the market to local entrepreneurs. Although there have been exceptions to this pattern, as in centrally planned economies, where the state has dominated (Allcock and Przeclawski, 1990; Hall, 1991), or where previously non-utilized areas have been developed for tourism, the internationalization of tourism in LDCs, as elsewhere, is clearly the (statistical) norm. This is indicated (to cite but a small selection) in empirical studies of the Pacific (Britton, 1982; Oglethorpe, 1984; Britton, 1987a), The Bahamas (Debbage, 1990), India (Chandrakala, 1989), Kenya (Jommo, 1987; Bachmann, 1988), the Gambia (Wagner, 1982; Farver, 1984), the Caribbean (Bryden, 1973; Perez, 1975; Poon, 1987), southern Africa (Crush and Wellings, 1987; Harrison, 1990) and Thailand (Meyer, 1988). The pattern is confirmed by more general studies of the involvement of transnational companies (UNCTC, 1982; Dunning and McQueen, 1982).

Where indigenous entrepreneurs do emerge, they often come from those with existing links to the wider business community. Indeed, a few entrepreneurs can wield considerable influence, as among cushion producers in parts of Thailand (Parnwell, 1992). Some entrepreneurs may be 'marginal', for example ethnic Chinese in Malaysia (Din, 1982, p. 458; Bird, 1989) or returning migrants (or their descendants), whose experience elsewhere has increased their expertise and access to funds. Examples can be found in the gemstone and jewellery of Thailand (Parnwell and Khamanarong, 1990, pp. 14–17), in Negril, Jamaica (MacKay, 1987), and Brittany, France (Shurma-Smith, 1988). Retiree tourists, too, have been found to occupy important entrepreneurial roles, acting as 'change agents' in such diverse places as the Caribbean (Lowenthal, 1976) and the Scottish Inner Hebrides (Kohn, 1988). However, not all returning migrants are impressed by tourism and its effects. Temporarily leaving Vermont, USA, to re-visit the island of Antigua, Kincaid saw tourism as the successor to colonialism and bitterly defined every tourist as an 'ugly, empty thing, a stupid thing, a piece of rubbish passing here and there to gaze at this and that' (Kincaid, 1988, p. 17).

Structures inherited from the colonial period may prevent individuals or groups from developing entrepreneurial talents, or restrict them to specific market sectors. In Kenya, only local owners of up-country hotels catering for domestic tourists were successful, and Jommo noted '[the] collapse of almost every venture undertaken by indigenous large-scale investors in hotels' (1987, p. 89). Among tour operators, where there was substantial 'local' involvement, it was largely white or Asian (Jommo, 1987, pp.

97–98). In Nairobi, too, souvenir shops were owned and run by Asians, and restaurants and places of entertainment catering for international tourists were also owned primarily by non-Africans (Jommo, 1987). By contrast, in Fiji, although expatriates again held top administrative positions, Indians reportedly occupied low-status labouring jobs, while indigenous Fijians held more prestigious positions as waiters, barmen, receptionists and security guards (Samy, 1975).

In addition to competition among ethnic groups, the business community in tourist-receiving societies may be riven by divisions between local (small-scale) and metropolitan (large-scale) interests, with the latter perhaps held responsible for importing 'alien' foodstuffs and values. 'To hell with paradise', enunciated by a Caribbean politician (also involved in running a small hotel) then becomes a nationalistic rallying cry, aimed at discouraging large-scale tourism, exaggerated advertising and metropolitan 'hype' (Mitchell, 1972). Indeed, residents of Bequia, from which he comes, are reported to be divided on the benefits of tourism of any kind. Those with land to sub-divide (including former planters), housing to rent, hotels to run, and labour, goods or services to sell, legally or otherwise, favour the industry's expansion, even with foreign involvement. The landless and many of the poor are less impressed (cf. Price, 1988, pp. 206–244). In such circumstances, tourism may simply reinforce the pre-existing social structure.

Frequently, elites in LDCs become partners of overseas capital. Meyer, for instance, emphasizes a 'triple alliance' of Thai elites, including the royal family and the Chinese commercial class, the Thai government and transnational companies, which together dominate Thailand's tourism(Meyer, 1988, pp. 113–195). The UDT perspective is neatly summarized by Britton, who considers international tourism to be constituted by a three-tiered hierarchy, with metropolitan capital, based in developed societies, at the top, followed by local, comprador capitalists and, at the bottom, small-scale enterprises, whose success depends on that of the two higher categories, which cream off most of the profits (1982, p. 346). Power allegedly remains in the metropole – for Nash, another advocate of UDT, a sufficient definition of 'imperialism' (Nash, 1989, first published 1987, p. 38) – and LDCs themselves are left with 'a largely passive and dependent role' (Britton, 1982, p. 347).

As with UDT in general (cf. Harrison, 1988, Chapter 4), there are problems with this kind of analysis. First, it is not clear whether the objections are to capitalism as such, or to the dominance of international capital (cf. Phillips, 1977, p. 19). Both positions have their problems. Referring approvingly to Tonga's 'monarchic structure,' which 'sheltered the country from outside forces' and controlled the expansion of the tourist industry (1982, p. 349). Britton elsewhere suggests the country's tourism industry suffered because it 'was not exploited as a fully fledged colony' (1987a, p. 131). Second, hostility to tourism development often centres on the (alleged) increase in inequality across classes in LDCs. Indeed, successful entrepreneurship may even be regarded as inherently undesirable, as by Wagner, whose analysis ignores a woman entrepreneur whose dominance in handicraft production was achieved through good timing, use of her social network and advertising (1982, pp. 9, 13, 57–58).

If all participants in tourism obtain some benefit (Britton, 1982, p. 346), increased inequality is not necessarily problematic. However, tourism development can increase social inequality and lead to other consequences commonly regarded as unacceptable. Thai villagers, for example, have been evicted for outsiders to develop tourist resorts (Ekachai, 1990) and elsewhere in Thailand other development projects have increased rural poverty, sometimes increasing migration to urban centres and the less acceptable regions of the informal economy. In such circumstances, the extent of the disruption is conditioned by the response, at local, national and international levels, of the people, the state (Boissevain, 1977) or of other agencies.

Job opportunities in tourism

Elites in many LDCs actively promote tourism and the benefits accruing to LDCs through tourism (or, indeed, any development strategy) are rarely distributed equally. As with other labour-intensive industries, few jobs in tourism are well paid, as numerous critics have noted. Seen in the context of a colonial history, work in tourism may be a sensitive issue. When tourists are white and hotel employees are black, as in the Caribbean, 'service' may be interpreted as a continuation of servility (Perez, 1973–1974, p. 46). In such circumstances, calls for 'flexibility, segmentation and diagonal integration', a 'new tourism' (Poon, 1989, 1990), however understandable at the industry level, may be difficult to implement when lower-level participants in the industry are reluctant to 'serve'.

Whether or not jobs in other industries are (or could be) available at higher rates of pay is a moot point (Lin and Sung, 1984), but in most countries they probably pay more than agriculture. However, numerous studies draw attention to the 'inferior' nature of employment in tourism. and the tendency for administrative and other senior positions to be held by outsiders (e.g. Kent, 1975, pp. 182–190; Price, 1988, p. 209; Bird, 1989). Indeed, in the 'informal' sector, a mass of 'unofficial' and unwaged entrepreneurial activity is revealed. Unlicensed guides, pavement sellers, beach hawkers, and male and female prostitutes, adult and child, are all in this category which, as the following pages indicate, is highly comprehensive.

Women, especially, carry out many of the lower-status jobs in the tourist hierarchy (Enloe, 1989, pp. 33–34), not merely because they are cheaper than men but because, in many respects, 'travel for pleasure and adventure has been profoundly gendered' (Enloe, 1989, pp. 40–41). Nowhere is this more evident than in sex tourism and international tourism has undoubtedly contributed, directly or indirectly, to prostitution of all kinds.

But tourism has not caused prostitution. In Thailand, which 'enjoys' one of the sleaziest reputations in such matters (cf. Meyer, 1988), it was once confined mainly to Chinese women, increasing in the 1950s with rural–urban migration, developing further during the Vietnam war, and receiving added impetus as Thailand attracted foreign tourists. According to one 'guesstimate', up to one million women in Thailand are involved in prostitution, more than in any other occupation except for farming and housework. However, only about 7% are alleged to depend on overseas tourists for their livelihood (Meyer, 1988). If this is anywhere near the case, blaming international tourism for Thai prostitution is to indulge in scapegoating; the roots of prostitution may lie, instead, in 'traditional culture and Thai consciousness' (Meyer, 1988). A similar perspective is offered by Truong (1990), who links Buddhist ideas on sexuality to the development of the Thai state and the increased commoditization of sexual reproductive services. After the Vietnam War, when Thailand was used as a Rest and Recreation centre, these combined with international tourism, with its focus on hospitality and its advertisement of the 'otherness' of the East (Truong, 1990, p. 200), to produce a different kind of 'packaged' holiday.

By contrast, Graburn argues that in Korea, Taiwan, Indonesia, the Philippines, Vietnam and Japan, prostitution is closely linked with war and tourism (Graburn, 1983, p. 441), thus implying that, morally, all three are equally dubious. Indeed, following many proponents of UDT, he regards tourism itself as analogous to prostitution, where poor countries with few development alternatives must sell their beauty for foreign currency, being psychologically 'penetrated' by tourists from developed nations (Graburn, 1983, pp. 441–442).

Sex tourism is undoubtedly significant to some tourist destination regions, and in tourism advertising – as in advertising generally – images of women play an important role. Nevertheless, tourism may bring women some benefits. In Bequia, a Caribbean island administratively associated with St. Vincent, they gained increased independence from their menfolk as a result of part-time and seasonal work in tourism (Price, 1988, pp. 225–226), a feature also noted by Chant for women in Mexico (Chant, 1992, pp. 85–101). For such women, and for many men, tourism represents another set of opportunities, however limited, in a social world where options are few.

As well as prostitution, which is technically illegal in many tourist-receiving societies, including Thailand, other occupations in the informal sector operate on the borders of the law. Unlicensed guides, hustlers, beach boys, pavement sellers and many others live on the margin, eking out a precarious existence and often receiving the opprobrium of middle-class society for their pains. On the illegal side of the margin, too, it is often alleged that tourism increases crime rates. Given the relative wealth and obvious presence of tourists, it would be surprising if it did not do so, and crimes against property, for example bag snatching, theft and burglary, generally increase where there are heavy concentrations of tourists (Mathieson and Wall, 1982, pp. 150–152; Chesney-Lind and Lind, 1986). As 'outsiders', involved only in transitory contact with local people, tourists may be considered 'fair game' (as in some Sri Lankan shops, for example) and they are less likely to be able to rely on local support when in difficulties. In such respects, as in many others, LDCs are no different from developed countries, especially large cities, where crowds also attract a sizeable criminal element.

Tourism may specifically attract some criminal activities. Politically motivated crime is more likely to gain publicity when directed against tourist-populated areas, as by opposition groups, for example, in the Philippines during the 1980 conference of the American Society of Travel Agents (Richter, 1989, pp. 63–64) or by Basque separatists in Spain. At a different level, illegal currency dealings may be advantageous to all but the government concerned (Lehmann, 1980). They may even be a tourist attraction; according to one travel writer, in Burma 'changing money on the black market is temporary liberation from the economic madhouse' (Fielder, 1991, p. 27) and local dealers are drawn to foreign tourists, who possess valued currency but little knowledge of the 'going rate'. Drug dealers or pimps, too, recognize that tourists may engage in forms of behaviour regarded as socially unacceptable, perhaps criminal, in their own countries. Inevitably, such forms of behaviour may necessitate extra policing.

Finally, tourism may sometimes create categories of crime which previously did not exist, or were considered unimportant, again creating the need for more police. Before the Mexican village of Cajitlan was developed for (largely domestic) tourism in the 1960s, for instance, male villagers regularly expressed their machismo by racing horses and firing pistols in the streets. With tourism came three rural policemen and such activities were forbidden. In addition, 'unsuitable' clothing was discouraged, as was urinating in public, and stray dogs and livestock were cleared from the streets (Nunez, 1963). Elsewhere in some LDCs, for example Tunisia, police are posted on beaches, allegedly to protect tourists from over-zealous salesmen, but perhaps also to prevent local people becoming too agitated, in one way or another, at the sight of so much public near-nakedness – itself offensive to many members of the host population.

Changes in family structure

According to Peake (1989), before the development of tourism in Malindi, on the Kenyan coast, the Old Town was dominant. Economic and political power was held by the elders, who occupied important positions in mosques and headed extended family households. With tourism, the Old Town became marginalized and the authority of the elders declined as other family members, especially young men, gained independent sources of income. Some, the 'clerks', obtained office jobs in tourist establishments, using their income to form nuclear families and purchase consumer goods. Others, 'beach boys' disregarded the elders, turned their backs on their families, refused to attend the mosques and entered the informal sector, perhaps as male prostitutes. They thus rejected their Swahili identity in favour of the tourist 'scene', where they publicly displayed their dubiously acquired wealth. In response, the elders withdrew into ecstatic Islamic fundamentalism, romanticizing the Swahili past and worrying about the effects of tourism on the values of the young. However, the separation of beach boys was not necessarily permanent. Those who profited from tourism could later redeem themselves by taking an active role in the community, even to the extent of becoming elders themselves.

Although a similar pattern may have emerged with the introduction of another source of income, Peake's example shows how new job opportunities in tourism had ramifications for family organization, removed power from the traditional elders and influenced the political structure and status system of the community.

Similar changes have been reported elsewhere. After reviewing the evidence, Kousis (1989) notes that, in Drethia, rural Greece, the alienation of land to outsiders led to a loss of control by older household heads, which was exacerbated by increased government assistance to young people in the form of business loans and grants. Land was no longer given as dowry, being replaced by property, furnishings and fittings, and village endogamy declined, with some local men marrying foreign tourists (Kousis, 1989, pp. 327–328). Women played an important role in owning and running tourist businesses, but continued to be more restricted in their sexual activities than their male counterparts. However, as Kousis points out, extended family members continue to assist kin in tourist enterprises, for example, by looking after children, dishwashing and cleaning, or advertising the establishment to visitors (1989, p. 324).

The effects of tourism on family structure vary according to the extent and type of tourism and the 'traditional' base. In Thailand, tourism development has been implicated in the breakdown of family life and mental health problems, especially among women in the age group 'most over-represented in tourism-related activities' (Meyer, 1988). However, the causal connections are not demonstrated, and in Thailand (as elsewhere) agricultural decline and rural–urban migration occur independent of tourism. That many young female migrants enter tourist-related prostitution, often with the assistance of their families (Ekachai, 1990; Srisang, 1991), neither makes the practice morally acceptable nor shows that tourism is the cause of family breakdown.

New social institutions

As tourism expands, existing institutions are transformed and new ones created. In Bali, members of the old elites became influential in the hotel business and established dance academies. There, as elsewhere, a Tourism Development Board was created, along with numerous voluntary associations (Noronha, 1979, pp. 193, 197–199). Typically, the government set up the Board, either within a government department responsible for tourism or as a para-statal, to which organizations and groups with a direct interest in tourism are then affiliated. The breadth of representation, and whether or not it includes groups and individuals opposed to tourism, indicates the extent to which the government is prepared to tolerate internal dissent.

Hoteliers, too, set up their own business associations, which then act as pressure groups to further their perceived interest, sometimes differentially defined by indigenous and foreign owners. Other official or quasi-official organizations are also formed to promote or protect cultural and physical assets. The role of National Trusts, or their local equivalents, will be enhanced, to advertise, protect (and possibly even define) the nation's archaeological and natural heritage. Where relevant, organizations to promote national parks and wildlife reserves are formed or expanded, often with tourism as their major justification (and indigenous resentment as a major preoccupation). Arts Councils, perhaps aided or opposed by religious interests, may perform similar functions for the nation's culture.

If sufficiently organized, professional and worker organizations gear themselves to promote their members' interests – or protect them from 'exploitation'. They may be community-based, for example village councils, or occupationally based, as with guides, dancers and craft associations. Broadly speaking, their aims will be similar: to promote, oppose or modify tourism when it is in their perceived interest to do so.

Indeed, one result of tourism is the development of pressure groups specifically to oppose tourism, or aspects of it. Issues are then politicized, with local parties adopting an anti-tourist, nationalistic stance, as occurred in parts of the Caribbean during the 1970s, in the Philippines under Marcos and, more recently, in Goa with the Vigilant Goans' Army (JGF). At other times, overtly 'non-political' national organizations may be formed, with or without significant international support. In Thailand, for example, the Thai Development Support Committee, established in 1962, exists to raise awareness of the human costs of 'development', including tourism (Ekachai, 1990), whereas the Ecumenical Commission on Third World Tourism (ECTWT), based in Bangkok, is an international organization, regularly publishing accounts of the perceived excesses of tourism, in Thailand and elsewhere, and providing details of other pressure groups with similar aims (Holden *et al.*, 1985; O'Grady, 1990; Srisang, 1991).

In Nepal, the international context of such concern is evident in the work of the Annapurna Conservation Area Project (ACAP). Formed in 1986 by the King Mahendra Trust, it attempts to counter detrimental effects of trekking tourism, for example deforestation, pollution and degradation of the natural habitat, as well as increased crime, and to further the interests of local people. Publicized in Europe (Tuting, 1989), the programme promotes reforestation, improvements in the tourist infrastructure, and education in environmental awareness. With a national director, it involves numerous newly formed local organizations and receives assistance from several international bodies, including the German Volunteer Service (GVS), the German Alpine Association and an ecological institute in the USA (Tuting, 1989, pp. 15–18). Furthermore, it has been commended by at least one travel programme on British television for promoting environmentally sympathetic tourism – which may well increase the number of tourists visiting the country. The scheme, and the international support it receives, indicates how environmental and other problems raised by international tourism are increasingly met through international cooperation.

Despite the internationalization of tourism, the role of the state is crucial. Although not detailed here, the impact of tourism on the environment may often be considerable (Romeril, 1989). If it is to be protected, a thorough assessment of the possible consequences of tourism must be made (Boo, 1990, Vol. 2, Chapter 4). More generally, development plans, taxation, training, joint ventures, policies for agriculture and other sectors, education (of local people and tourists) and law enforcement all involve the state, as do controls on aspects of tourism development considered to be unsavoury.

Unfortunately, it is not that simple. As Wood notes for South-east Asia (1984), the state may plan and market tourism, arbitrate when conflicts arise and provide a venue for new forms of political action. Depending on the class and religious structure of any LDC, these roles may conflict with or contradict other functions of the state. Religion is a particularly delicate issue. In Muslim Indonesia, for example, Islam is absent in the marketing of tourism, which emphasizes the Hinduism of Bali (Wood, 1984, p. 365), and Malaysia reportedly operates a double standard, allowing tourists a moral laxity denied to the indigenous population (Din, 1989, p. 558). The position taken by other Islamic societies ranges from discouragement, through to isolation, accommodation and laissez-faire (Din, 1989).

In fact, some of the most violent debates in development theory have centred on the role of the state (Harrison, 1988, pp. 137–145) and numerous tourism studies (e.g. Jommo, 1987) suggest that in this sector, too, the state is frequently ineffective and inefficient. The absolute power and ability of a monarch to control tourism, as in Bhutan (Richter, 1989, pp. 176–179) is rarely found elsewhere. Even if there were a consensus on the most appropriate policies to be followed in tourism 'development', the state in most LDCs simply lacks the power, ability or will to formulate and enforce them. And yet, as de Kadt notes, (1979, pp. 27–32), state intervention in tourism is frequently essential. It is a crucial dilemma.

Interaction, values and dependency

The degree to which locals take on the values of tourists varies. Much depends on the extent of interaction between tourist and local, which is not always encouraged (Hall, 1984; Richter, 1989) and one has to ask, which tourist? Which local? Younger and less 'organized' backpackers, certainly, live among (some would say off) indigenous people and may bring economic benefits to rural areas (Meijer *et al.*, 1989). However, they may not be welcomed by the authorities and, in any case, their closest links are often with other backpackers (Riley, 1988). Other visitors, including such excursionists as cruise ship tourists and residents of more expensive hotels, especially in enclaves, meet locals only fleetingly, and then on well established tourist circuits. And almost by definition, tourists primarily interested in a country's history or its flora and fauna spend little time interacting with its residents. That said, however, in exceptional circumstances 'tourist–host' interaction may be transformed. In 1987, for example, independent Western tourists in Tibet witnessed the violent repression of demonstrations against Chinese rule. Although 'hippies' dissociated themselves from political involvement, other tourists were outraged and set up an informal network to provide medical assistance for the injured and actively collect and disseminate information on the events to outside Western news agencies. Further disturbances in 1988 and 1989 evoked a similar response (cf. Schwartz, 1991). In these circumstances, at least, tourists identified themselves with members of the 'host' society and became political activists.

Interaction is greater in sex tourism and home-stay tourism. Although the former is morally suspect, it actually covers a wide variety of relationships, not all of which are instrumental and affectively neutral (Cohen, 1982b; Meyer, 1988, pp. 344 and 355). Indeed, sex plays a role in many kinds of tourism and, for some hosts, interaction with tourists brings the possibility of increased status and the chance to leave home for more affluent and (allegedly) congenial surroundings (Cohen, 1971; Wagner and Yamba, 1986; Price, 1988, pp. 228–234). By contrast, home-stay tourism, although sometimes a supplement to hotel tourism, especially in Eastern Europe (Borocz, 1990, p. 22), has a more acceptable image of 'paying visitor' (Stringer, 1981; Frater, 1982; Cohen, 1989b, p. 136). However, among visitors from 'developed' countries, this kind of tourism is not highly favoured and it will probably remain a minority interest – as it must, if it is to survive as an 'alternative' to more organized forms of travel.

As Peake shows in his study of Malindi Old Town (1989), behavioural responses to tourism within a society may vary considerably. North American Indians make jokes at tourists' expense (Evans-Pritchard, 1989; Sweet, 1989), a reaction probably more common than imagined, whereas others may (quite literally) greet them with open arms or overt hostility. Tourist responses, too, are varied. Many holidaymakers make repeat visits – such business is crucial in most tourist-receiving societies – and some retire to the 'host' society, with considerable economic and social ramifications. However, although patterns of interaction may change over time (Machlis and Burch, 1983) and according to the purpose of the visit, the relationship between tourist and host is usually one of market symbiosis.

However, as an aspect of modernization, international tourism undoubtedly reinforces the transfer of Western values and patterns of behaviour to members of 'host' societies, a form of acculturation often subsumed under the term 'demonstration effects'. The issue was raised in early studies of tourism (UNESCO, 1976, pp. 93–94) and has been noted more recently. In Kenya, for example, tourism has reinforced the influence of the Westernized elite, with young people 'imitating not only Western clothing but also the behaviour and lifestyle of Europeans and American visitors, including their ethical and moral codes' (Bachmann, 1988) and Malaysia (Bird, 1989), where even young backpackers attract envy and unwanted attention through their possession of valued goods (Bird, 1989). However, as Bird also points out, new norms and values are frequently introduced by domestic tourists, who communicate easily and interact more closely with local people.

Demonstration effects are the cultural equivalent of the spread of market relationships and commoditization, an association leading some critics to refer to such acculturation as 'cultural imperialism' – a term more judgemental than descriptive. As LDCs have been subjected to economic domination, so the argument goes, their cultures have been undermined.

At the economic level, this argument has some merit: power differentials are a feature of the world system and some countries clearly depend on tourists from a few developed societies, often former metropolitan colonial powers. (Hoivik and Heiberg, 1980), as indicated by concentration ratios. However, at a political level, the argument is problematic. There is no direct evidence, for instance, that such a 'dependence' leads to political compliance, or that those who rely on tourists from the USA obtain a disproportionate share of grants or other forms of aid from that country (Francisco, 1983). Furthermore, as Erisman remarks for the Caribbean (Erisman, 1983), elite members are not equally influenced by Western culture (which, in any case, contains within itself the seeds of opposition) and nationalism, religion and other social movements, among elites and lower down the social hierarchy, may frequently counter the effects of Westernization.

Given the role of hotels and other tourist establishments as centres of entertainment, their attraction for the young is not surprising. Indeed, acculturation is undoubtedly occurring. However, there is more than a hint of hypocrisy, for example when members of developed societies bewail the desire for consumer durables among residents of LDCs, or when the middle classes of those societies criticize lower-class ambition. There is no inherent virtue, for example, in the extended family, which may as often be a source of repression and autocratic (and patriarchal) control as one of security and freedom. Similarly, the superiority of palm wine over Western beer, or of traditional dress over blue jeans, may be affirmed but, ultimately is a matter of taste.

Indeed, there is something quite patronizing in the view that the culture of many LDCs is weak and in dire need of protection from outside. In fact, the cultures of tourist-receiving societies, for example the Pacific islands, may possess some kind of 'deep structure' which allows them to adapt to new influences and yet retain – even reinforce – their vitality and coherence (cf. Ritchie, 1975). As Macnaught argues, for the same region, the 'fatal impact' thesis 'assigns Pacific islanders a helplessly passive role in the conflict of cultures, ignoring a mass of evidence that it was often a two-way process' (Macnaught, 1982, p. 363). He later remarks that changes so deplored by Western outsiders were enthusiastically welcomed by many islanders (Macnaught, 1982, p. 364), a finding certainly echoed elsewhere. Once again, the direction of social change is not an issue. Instead, the debate is centred on how desirable such changes are. We are once more in the realm of development.

Bibliography

Allcock, J. and Przeclawski, K. (eds) (1990) *Tourism in Centrally Planned Economies*. 17. Annals of Tourism Research.

Bachmann, P. (1988) *Tourism in Kenya: a Basic Need for Whom?* Peter Lang, Berne, Switzerland.

Bird, B. (1989) *Langkawi - From Mahsuri to Mahathir: Tourism for Whom?*. Institute of Social Analysis (INSAN), Kuala Lumpur.

Boissevain, J. (1977) Tourism and development in Malta. *Development and Change* 8, 523–538.

Boissevain, J. and Inglott, P.S. (1979) Tourism in Malta. In: de Kadt, E. (ed.) *Tourism: Passport to Development?* Oxford University Press, New York and Oxford, pp. 265–284.

Boo, E. (1990) *Ecotourism: The Potentials and the Pitfalls*. Vols. 1 and 2. World Wildlife Fund, Washington, DC.

Borocz, J. (1990) Hungary as a destination. *Annals of Tourism Research* 17(1), 19–35.

Boynton, L.L. (1986) The effects of tourism on Amish quilting design. *Annals of Tourism Research* 13(3), 451–465.

Britton, R. (1979) The image of the third world in tourism marketing. *Annals of Tourism Research* 6(3), 318–329.

Britton, S. (1982) The political economy of tourism in the third World. *Annals of Tourism Research* 9(3), 331–358.

Britton, S. (1987a) Tourism in Pacific island states: constraints and opportunities. In: Britton, S. and Clarke, W.C. (eds) *Ambiguous Alternative: Tourism in Small Developing Countries*. University of the South Pacific, Suva, pp. 113–139.

Britton, S. (1987b) Tourism in small developing countries: development issues and research needs. In: Britton, S. and Clarke, W.C. (eds) *Ambiguous Alternative: Tourism in Small Developing Countries*. University of the South Pacific, Suva, pp. 167–187.

Bryden, J. (1973) *Tourism and Development: A Case Study of the Commonwealth Caribbean*. Cambridge University Press, New York.

Butler, R.W. (1980) The concept of a tourist area cycle of evolution: implications for management of resources. *Canadian Geographer* 24(1), 5–12.

Chandrakala, S. (1989) *The Impact of Tourism on India's Environment*. Equations, Bangalore.

Chesney-Lind, M. and Lind, L.Y. (1986) Visitors as victims: crimes against tourists in Hawaii. *Annals of Tourism Research* 13(2), 167–191.

Cohen, E. (1971) Arab boys and tourist girls in a mixed Jewish-Arab community. *International Journal of Comparative Sociology* 12(4), 217–223.

Cohen, E. (1982b) Thai girls and Farang men: the edge of ambiguity. *Annals of Tourism Research* 9(3), 403–428.

Cohen, E. (1988a) Authenticity and commoditisation in tourism. *Annals of Tourism Research* 15(3), 371–386.
Cohen, E. (1989a) Primitive and remote: hill tribe trekking in Thailand. *Annals of Tourism Research* 16(1), 30–61.
Cohen, E. (1989b) Alternative tourism: a critique. In: Singh, T.V. (ed.) *Towards Appropriate Tourism: The Case of Developing Countries*. Peter Lang, Frankfurt and Berne, pp. 127–142.
Crompton, J. (1979) Motivations for pleasure vacation. *Annals of Tourism Research* 6(4), 408–424.
Crush, J. and Wellings, P. (1987) Forbidden fruit and the export of vice: tourism in Lesotho and Swaziland. In: Britton, S. and Clarke, W.C. (eds) *Ambiguous Alternative: Tourism in Small Developing Countries*. The University of the South Pacific, Suva, pp. 91–112.
Crystal, E. (1989) Tourism in Toraja (Sulawesi, Indonesia). In: Smith, V. (ed.) *Hosts and Guests: The Anthropology of Tourism*. University of Pennsylvania Press, Philadelphia, pp. 139–168.
Dann, G.M.S. (1981) Tourism motivation: an appraisal. *Annals of Tourism Research* 8(2), 187–219.
de Kadt, E. (1979) *Tourism: Passport to Development?* Oxford University Press, New York and Oxford.
Debbage, K.G. (1990) Oligopoly and the resort cycle in the Bahamas. *Annals of Tourism Research* 17(4), 513–527. DOI: 10.1016/0160-7383(90)90024-L.
Din, K.H. (1982) Tourism in Malaya: competing needs in a plural Society. *Annals of Tourism Research,* 9(3), 453–480.
Din, K.H. (1989) Islam and tourism: patterns, issues and options. *Annals of Tourism Research* 16(4), 542–543.
Dunn, R.E.L. and Iso-Ahola, S.E.. (1991) Sightseeing tourists' motivation and satisfaction. *Annals of Tourism Research* 18(2), 226–237.
Dunning, J.H. and McQueen, M. (1982) Multinational corporations in the international hotel industry. *Annals of Tourism Research* 9(1), 69–90.
Ekachai, S. (1990) *Behind the Smile: Voices of Thailand*. Thai Development Support Committee, Bangkok.
Enloe, C. (1989) *Bananas, Beaches and Bases: Making Sense of International Politics*. University of California Press, Berkeley, Los Angeles.
Erisman, H.M. (1983) Tourism and cultural dependency in the West Indies. *Annals of Tourism Research* 10(3), 337–361. DOI: 10.1016/0160-7383(83)90061-0.
Evans-Pritchard, D. (1989) How 'they' see 'us': Native American images of tourists. *Annals of Tourism Research* 16(1), 89–105.
Farver, J. (1984) Tourism and employment in the Gambia. *Annals of Tourism Research* 11(2), 249–265.
Fielder, M. (1991) Limited Burmese Days. Weekend Guardian, 9 November.
Finney, B.R. and Watson, K.A. (eds) (1975) *A New Kind of Sugar: Tourism in the Pacific*. East-West Center, Honolulu.
Francisco, R. (1983) The political impact of tourism dependence in Latin America. *Annals of Tourism Research* 10(3), 363–376.
Frater, J. (1982) *Farm tourism in England and Overseas, Research Memorandum 93*. Centre for Urban and Regional Studies, Birmingham.
Graburn, N. (1983) Tourism and prostitution. *Annals of Tourism Research* 10(3), 437–442.
Graburn, N. (1984) The evolution of tourist arts. *Annals of Tourism Research* 11(3), 393–419.
Graburn, N. (1989) Tourism: the sacred journey. In: Smith, V. (ed.) *Hosts and Guests: The Anthropology of Tourism*. University of Pennsylvania Press, Philadelphia, pp. 21–36.
Greenwood, D. (1972) Tourism as an agent of change: a Spanish-Basque case. *Ethnology,* XI(1), 80–91.
Greenwood, D. (1989) Culture by the pound. In: Smith, V. (ed.) *Hosts and Guests: The Anthropology of Tourism*. University of Pennsylvania Press, Philadelphia, pp. 171–185.
Hall, D.R. (1984) Foreign tourism under socialism: the Albanian 'Stalinist' model. *Annals of Tourism Research* 11(4), 539–555.
Hall, D.R(ed). (1991) *Tourism and Economic Development in Eastern Europe and the Soviet Union*. Belhaven, London.
Harrison, D. (1975) Social relations in a Trinidadian village. PhD. thesis. Department of Social Anthropology, University College London.
Harrison, D. (1988) *The Sociology of Modernisation and Development*. Routledge and Kegan Paul, London.
Harrison, D. (1990) *Tourism in Swaziland, Research Paper 29*. Social Science Research Unit, University of Swaziland, Kwaluseni, Swaziland.

Hoivik, T. and Heiberg, T. (1980) Centre-periphery tourism and self-reliance. *International Social Science Journal* 32(1), 69–98.

Holden, P., Horlemann, J. and Pfafflin, G.F. (eds) (1985) *Tourism, Prostitution*. Development: Documentation. Ecumenical Coalition on Third World Tourism (ECTWT), Bangkok.

Jommo, R.B. (1987) *Indigenous Enterprise in Kenya's Tourist Industry'*. Institut Universitaire d'Etudes du Developpement, Geneva.

Jules-Rosette, B. (1984) *The Messages of Tourist Art: An African Semiotic System in Comparative Perspective*. Plenum Press, New York and London.

Kent, N. (1975) A new kind of sugar,. In: Finney, B.R. and Watson, K.R. (eds) *A New Kind of Sugar. Tourism in the Pacific*. East-West Center, Honolulu, pp. 169–180.

Kincaid, J. (1988) *A Small Place*. Virago Press, London.

Kohn, T. (1988) Island Involvement and the Evolving Tourist. *Paper presented at the First Conference on Anthropology of Tourism*, Group for Anthropology in Policy and Practice (GAPP), Roehampton Institute, 22–23 April.

Kousis, M. (1989) Tourism and the family in a rural Cretan community. *Annals of Tourism Research* 16(3), 318–322.

Krippendorf, J. (1987) *The Holiday Makers: Understanding the Impact of Leisure and Travel*. Heinemann, Oxford.

Lehmann, A.C. (1980) Tourists, black markets and regional development in West Africa. *Annals of Tourism Research* 7(1), 102–120.

Lin, T. and Sung, Y. (1984) Tourism and economic diversification in Hong Kong. *Annals of Tourism Research* 11(2), 231–247.

Lowenthal, D. (1976) The return of the non-native: new life for depopulated areas. In: Kosinki, L.A. and Webb, J.W. (eds) *Population at Microscale*. New Zealand Geographical Society, New Zealand, pp. 143–148.

MacCannell, D. (1976) *The Tourist: A New Theory of the Leisure Class*. Macmillan, London.

Machlis, G. and Burch, W. (1983) Relations between strangers: cycles of structure and meaning in tourist systems. *Sociological Review* 31(4), 666–692.

MacKay, L. (1987) Tourism and changing attitudes to land in Negril, Jamaica. In: Besson, J. and Momsen, J. (eds) *Land and Development in the Caribbean*. Macmillan, London, pp. 132–152.

Macnaught, T.J. (1982) Mass tourism and the dilemmas of modernization in Pacific island communities. *Annals of Tourism Research* 9(3), 359–381. DOI: 10.1016/0160-7383(82)90019-6.

Mathieson, A. and Wall, G. (1982) *Tourism: Economic, Physical and Social Impacts*. Longman, Harlow.

May, R.J. (1975) Tourism and the artefact industry in Papua New Guinea. In: Finney, B.R. and Watson, K.A. (eds) *A New Kind of Sugar: Tourism in the Pacific*. East-West Center, Honolulu, pp. 125–132.

McKean, P.F. (1989) Towards a theoretical analysis of tourism: economic dualism and cultural involution in Bali. In: Smith, V. (ed.) *Hosts and Guests: The Anthropology of Tourism*, 2nd edn. University of Pennsylvania Press, Philadelphia, pp. 119–138.

Meijer, W.G. and Singh, T.V. (1989) Rucksacks and dollars: the economic impact of organized and non-organized tourism in Bolivia. In: Singh, T.V *et al*. (ed.) *Towards Appropriate Tourism: The Case of Developing Countries*. Peter Lang, Frankfurt and Berne, pp. 227–249.

Meyer, W. (1988) *Beyond the Mask*. Verlag Breitenbach Publishers, Saarbrücken, Port Lauderdale.

Mitchell, J.F. (1972) To hell with paradise: a new concept in Caribbean tourism. *Address at Caribbean Travel Association Press Conference*, Haiti, 21st September.

Moeran, B. (1983) The language of Japanese tourism. *Annals of Tourism Research* 10(1), 93–108.

Monk, J. and Alexander, C.S. (1986) Free Port fallout: gender, employment, and migration on Margarita Island. *Annals of Tourism Research* 13(3), 393–413.

Nash, D. (1989) Tourism as a form of imperialism. In: Smith, V. (ed.) *Hosts and Guests: The Anthropology of Tourism*, 2nd edn. University of Pennsylvania Press, Philadelphia, pp. 37–52.

Nason, J.D. (1984) Tourism handicrafts and ethnic identity in Micronesia. *Annals of Tourism Research* 11(3), 421–449.

Noronha, R. (1979) Paradise revisited: tourism in Bali. In: de Kadt, E. (ed.) *Tourism: Passport to Development*. Oxford University Press, Oxford and London, pp. 177–204.

Nunez, T. (1963) Tourism, Tradition and Acculturation: 'Weekendismo' in a Mexican Village. *Ethnology* 2(3), 347–352.

Oglethorpe, M.G. (1984) Tourism in Malta: a crisis of dependence. *Leisure Studies* 3(2), 141–161.

O'Grady, A. (1990) *The Challenge of Tourism*. Ecumenical Coalition on Third World Tourism (ECTWT), Bangkok.

Parnwell, M.J.G. (1992) Tourism and rural handicrafts in Thailand. In: Hitchcock, M., King, V.T. and Parnwell, M.J.G. (eds) *Tourism in Southeast Asia: Theory and Practice*. Routledge, London, pp. 234–257.

Parnwell, M. and Khamanarong, S. (1990) Rural industrialisation and development planning in Thailand. *Southeast Journal of Social Science* 18(2), 1–28.

Parsons, T. (1964) Evolutionary Universals in society. *American Sociological Review* 29(3), 339–357.

Peake, R. (1989) Swahili stratification and tourism in Malindi old town, Kenya. *Africa* 59(2), 209–220. DOI: 10.2307/1160489.

Perez, L.A. (1975) Tourism in the West Indies. *Journal of Communications* 25(2), 136–143.

Phillips, A. (1977) The concept of development. *Review of African Political Economy* 8, 7–20.

Pi-Sunyer, O. (1989) Changing perceptions of tourists and tourism in a Catalan resort town. In: Smith, V. (ed.) *Hosts and Guests: The Anthropology of Tourism*. University of Pennsylvania Press, Philadelphia, pp. 187–199.

Picard, M. (1990) Cultural tourism in Bali: cultural performances as tourist attractions. *Indonesia* 49, 37–74.

Poon, A. (1987) Information technology and innovation in Caribbean tourism: implications for the Caribbean tourist industry. Unpublished D. Phil. thesis. University of Sussex, Brighton.

Poon, A. (1989) Competitive strategies for a new tourism. In: Cooper, C.P. (ed.) *Progress in Tourism, Recreation and Hospitality Management*. Belhaven, London, pp. 91–102.

Poon, A. (1990) Flexible specialisation and small size: the case of Caribbean tourism. *World Development* 18(1), 109–123.

Price, N. (1988) *Behind the Planter's Back*. Macmillan, London and Basingstoke.

Richter, L. (1989) *The Politics of Tourism in Asia*. University of Hawai'i Press, Honolulu.

Riley, P.J. (1988) Road culture of international long-term budget travellers. *Annals of Tourism Research* 15(3), 313–338.

Ritchie, J.E. (1975) The honest Broker in the cultural market place. In: Finney, B.R. and Watson, K.A. (eds) *A New Kind of Sugar: Tourism in the Pacific*. East-West Center, Honolulu, pp. 49–58.

Romeril, M. (1989) Tourism - the environmental dimension. In: Cooper, P.C. (ed.) *Progress in Tourism, Recreation and Hospitality Management*. Vol. 1. Belhaven Press, London, pp. 103—1113.

Samy, J. (1975) Crumbs from the table: the workers' share in tourism. In: Finney, B.R. and Watson, K.A. (eds) *A New Kind of Sugar: Tourism in the Pacific*. East-West Center, Honolulu, pp. 111–121.

Schillerr, H. (1976) *Communication and Cultural Domination*. International Arts and Sciences Press, New York.

Schwartz, R.D. (1991) Travelers under fire: tourists in the Tibetan uprising. *Annals of Tourism Research* 18(4), 588–604.

Shurma-Smith, P. (1988) Kinship, Identity and Second Home Ownership on the Ile aux Moines, Brittany. *Paper presented at the First Conference on Anthropology and Tourism*, Group for Anthropology in Policy and Practice (GAPP), Roehampton Institute, 22–23 April.

Srisang, K. (1991) *Caught in Modern Slavery: Tourism and Child Prostitution in Asia*. The Ecumenical Coalition on Third World Tourism (ECTWT), Bangkok.

Stringer, P.F. (1981) Hosts and guests the bed-and-breakfast phenomenon. *Annals of Tourism Research* 8(3), 357–376. DOI: 10.1016/0160-7383(81)90004-9.

Sweet, J.D. (1989) Burlesquing 'the other' in Pueblo performance. *Annals of Tourism Research* 16(1), 62–75. DOI: 10.1016/0160-7383(89)90030-3.

Thurot, J.M. and Thurot, G. (1983) The ideology of class and tourism confronting the discourse of advertising. *Annals of Tourism Research* 10(1), 173–189. DOI: 10.1016/0160-7383(83)90121-4.

Truong, T.D. (1990) *Sex, Money and Morality: Prostitution and Tourism in South-East Asia*. Zed Press, London.

Turner, L. and Ash, J. (eds) (1975) *The Golden Hordes: International Tourism and the Pleasure Periphery*. Constable, London.

Tuting, L. (1989) Trekking tourism in Nepal. In: *Tourismus and Okologie*. Okozid 5, focus Verlag, Giessen.

UNCTC (1982) *Transnational Corporations in International Tourism*. United Nations, New York.

UNESCO. (1976) The effects of tourism on socio-cultural values. *Annals of tourism research* 4(2), 74–105. DOI: 10.1016/0160-7383(76)90100-6.

Turner, V. and Turner, E. (1978) *Image and Pilgrimage in Christian Culture*. Basil Blackwell, Oxford.

Urry, J. (1988) Cultural change and contemporary holiday-making. *Theory, Culture and Society* 5(1), 35–55. DOI: 10.1177/026327688005001003.

Urry, J. (1990) *The Tourist Gaze: Leisure and Travel in Contemporary Societies*. Sage, London.
Uzzell, D. (1984) An alternative structuralist approach to the psychology of tourism marketing. *Annals of Tourism Research* 11(1), 79–99. DOI: 10.1016/0160-7383(84)90097-5.
van Raaij, W.E. and Francken, D.A. (1984) Vacation decisions, activities and satisfactions. *Annals of Tourism Research* 11(1), 101–112.
Wagner, U. (1982) *Catching the Tourist: Women Handicraft Traders in The Gambia*. Department of Social Anthropology, University of Stockholm, Stockolm, Sweden.
Wagner, U. and Yamba, B. (1986) Going north and getting attached: the case of the Gambians. *Ethnos* 51(3-4), 199–222. DOI: 10.1080/00141844.1986.9981323.
Wood, R.E. (1984) Ethnic tourism, the state, and cultural change in Southeast Asia. *Annals of Tourism Research* 11(3), 353–374. DOI: 10.1016/0160-7383(84)90027-6.

6

Sustainability and Tourism: Reflections from a Muddy Pool

To be 'green' is now fashionable and politically correct but just who or what is green is not easy to determine. There are 'blue greens' (who favour a market approach), 'red greens' (who mix some form of environmentalism with some form of socialism), and 'really (or radical) greens', who claim approaches unadulterated by other political ideologies. One way or the other, it seems that greens are good for you. However, they are not necessarily good for one another. As Yearley points out (Yearley, 1991, pp. 4–6), numerous organizations, often with conflicting objectives and quite different memberships, claim to maintain or enhance the environment. In England, fox hunters and their opponents claim environmentalist credentials, as do hunting organizations all over the world. Not to be outdone, political groups adopt an environmentalist policy or, at least, its associated rhetoric, vying with one another to see who can be the most environmentally aware. Ironically, when such groups do achieve political power, as happened in (the former) West Germany for a brief period, consensus in decision making and its implementation proved impossible to achieve. However, despite political divergence and major differences of emphasis, the ideas of those concerned with the 'crisis of the environment' have become linked to the study of development, where a similar 'crisis' is said to be evident (Adams, 1990, pp. 7–8). Indeed, the two topics have become so closely linked over the last two decades that, as argued elsewhere (Harrison, 1992a, pp. 9–10), environmentalism has joined modernization theory (MT) and under-development theory (UDT) as a quasi-paradigm, or competing perspective on development (Redclift, 1984, 1987; Lélé, 1991). In fact, the three perspectives or schools have much in common: all were founded in the developed world, all are incontrovertibly based on some branch or branches of (Western) science and all attract adherents who, in time, become believers. In addition, all have a concept of a before and an after, a notion of progress inherent in the idea of development that is, by definition, a movement from one state or situation to another, which is more positively valued. Put simplistically, whereas MT advocates that less developed countries (LDCs) progress by Westernizing their economies and cultures, UDT urges them to delink from the West and then follow some kind of model of autonomous growth. By contrast, environmentalists focus on change from (real?) over-development to an ideal (and greener) future. Finally, all three perspectives incorporate some notion of a world system, the most complex of which is found among environmentalists who emphasize the interconnectedness and interdependence of human existence with the rest of the natural world (even though the postulated relationships will vary).

The Question of Sustainability

Current perceptions of sustainable development have been traced to a variety of sources, notably the growth of nature conservation, the rise in the science of ecology, and the developing awareness that environmental problems were of global significance (Adams, 1990, pp. 14–41). Interest has been focused and extended in a succession of international conferences and reports, notably the Stockholm Conference of 1972 and the World Conservation Strategy of 1980, with sustainability and development clearly linked in the Report of the World Commission on Environment and Development, the Brundtland Report (WCED (World Commission on Environment and Development), 1987).

This Report provided perhaps the best known of all definitions of sustainable development, regarding it as 'development that meets the needs of the present without compromising the ability of future generations to meet their own needs' (WCED (World Commission on Environment and Development), 1987, p. 43). This definition, based on the stated requirement to meet the basic needs of the world's population, has been much criticized, especially for its acceptance of economic growth and, like the Brandt Report, for its naïve assumption of mutuality (Adams, 1990, pp. 57–65). In addition, as will be argued below, its egalitarianism is open to question, as is the assumption that there is no contradiction between the needs of present and future generations.

Even in the most specific debates, there are problems with the notion of sustainability. As Beckerman, a known advocate of continued economic growth, has noted, few argue that the natural environment should be left completely alone; were such a policy to be followed, polluted water and insect infestation, for example, would remain and the basic needs of the poor would continue to be unmet. However, once it is accepted that environmental changes can be implemented to improve human welfare, that there are trade-offs, sustainability becomes unoperational. Quite simply, one cannot anticipate future preferences or technological possibilities, even when dealing with natural resources. What will future generations settle for? 'More trees and fewer insects? More machines and fewer fish?' (Beckerman, 1992, p. 492).

At least in one instance, the choice has been for fewer whales and better fed human beings. The decision of Norway, in May 1993, to recommence whaling, despite the continued ban by the International Whaling Commission, provoked much international condemnation. However, the perceived basic needs of Norwegian fishermen evidently took precedence for the Norwegian Prime Minister, none other than Mrs Gro Harlem Brundtland, who opposed moves to replace commercial whaling with whale watching – arguably a form of sustainable tourism.

Other critics argue that sustainable development is a contradiction in terms, carrying the misleading implication that some forms of economic growth do not damage the wider environment. It is thus 'essentially reformist, calling for a modification of development practice, and owes little to radical ideas, whether claiming a green or Marxist heritage' (Adams, 1990, p. 10).

It is noteworthy that, almost as a matter of course, social factors of development are often incorporated into discussions of sustainability. The Brundtland Commission, for example, states: 'Even the narrow notion of physical sustainability implies a concern for social equity between generations, a concern which must logically be extended to equity within each generation' (WCED (World Commission on Environment and Development), 1987). In the pages following this quotation it is made clear that, in this context, 'social equity' relates to access to such basic natural resources as water, forests and land, and that 'our ability to promote the common interest in sustainable development is often a product of the relative neglect of economic and social justice within and among nations' (WCED (World Commission on Environment and Development), 1987). Such references have not prevented critics from accusing the Brundtland Commission of ignoring the underlying political and economic causes of inequality in the distribution of scarce resources (Adams, 1990, p. 64).

Social factors are also given a high priority by Pearce, Barbier and Markandya, who list 'real income, health and nutritional status, educational achievement, access to resources, equity in income distribution and basic freedoms' as key characteristics of sustainable development (Bertram, 1993, p. 248). Indeed, Tisdell (1993, p. 216) suggests that little progress is likely to be made in formulating polides for sustainable development until the dimensions involved are specified: 'It may, for example, refer to economic sustainability, ecological sustainability, sustainability of the natural resource base and sustainability of cultural and community life'.

Sustainable Tourism

What does all this have to do with tourism in islands and other small states – or tourism anywhere? By combining development (inevitably a value-laden concept) with sustainability (which is allegedly non-operational and reformist) we thus arrive at the doubly vague concept of sustainable development, only then to focus on one aspect of this dubious process – that of sustainable tourism.

Clearly, anyone considering the idea of sustainable development to be reformist and inadequate will have little time for sustainable tourism. After all, the latter is a sub-branch of the former and they have common historical roots (Bramwell and Lane, 1993, p. 2). However, this derogatory view is not universally held and, with the rapid increase of international travel over the last thirty years, the concept of sustainable tourism has gained in popularity, if not in clarity.

The growth of tourism, which is sufficiently well known to need little elaboration in this context, has given cause for concern. By 1991, 455 million international tourist trips had been made (WTO (World Tourism Organization), 1993, p. 180), a figure likely to increase to more than 661 million by the year 2000 and 937 million by 2010 (WTO (World Tourism Organization), 1994, p. 1) and domestic tourism is about ten times this amount. In addition to its economic and social implications, which are not the direct concern of this paper, such a mass movement of population, often concentrated on tourist honeypots, has a considerable impact on the physical environment (Mathieson and Wall, 1982, pp. 107–132; Cater, 1991; Ryan, 1991, pp. 95–130; Eber, 1992; Jenner and Smith, 1992; Valentine, 1992). The normal and perhaps inevitable procedure is to treat the relationship of tourism and the physical environment on a kind of profit and loss basis. Such balance sheets generally indicate that the costs of mass tourism far outweigh its benefits, with highly detrimental effects on the atmosphere, water resources, soil and natural vegetation, wildlife habitats and on entire ecosystems, as well as on the built environment (Mathieson and Wall, 1982, p. 131; O'Grady, 1990).

Largely as a response to such balance sheets, emphasis has shifted to sustainable tourism which, among other objectives, is designed to bring about forms of tourism more compatible with the maintenance and even the enhancement of the physical environment. Impact studies (including the focus on various forms of carrying capacity), differential pricing, the education of tourists, and the development of all kinds of alternative tourism, incorporate attempts to develop sustainable tourism (Craik, 1991, pp. 107–132). Indeed, a major reaction to the perceived threat to the environment from tourism has been to promote ecotourism as a form of sustainable (and non-mass) tourism, allegedly combining (relatively high-cost) holiday-making and a concern for and education about the environment, elements of which actually constitute the tourist attraction. Indeed, ecotourism is increasingly seen as a profitable niche within the wider market, bringing economic benefits to small countries, for example Belize, Costa Rica and Ecuador (Boo, 1990), without the need to provide a colossal (and costly) infrastructure for huge numbers of tourists. At the same time, a growing body of alternative travel literature has exploited environmental concern by emphasizing ideologically sound forms of behaviour for 'good' tourists (e.g. Wood and House, 1992).

How far ecotourism (an extension of nature tourism and but one type of alternative tourism) and good tourists are genuine alternatives to mass tourism, rather than stages towards it, is open to question (Wheeller, 1993, pp. 78–79). As Adams and Carwardine (1991, p. 20) report from Bali: 'When we told our guide that we didn't want to go to all the tourist places he took us instead to the places they take tourists who don't want to go to tourist places. These places, of course, were full of tourists'.

In fact, mass tourism and alternative tourism may complement each other and there is evidence that, in some respects, 'ecotourism may well result in more profound and lasting changes than mass tourism' (Cater, 1993, p. 89). Indeed, members of the International Academy for the Study of Tourism considered the notion of alternative tourism to be 'of little scientific value', preferring to speak of 'alternative forms of tourism' (Smith and Eadington, 1992, p. xiv), a distinction which they nevertheless failed to explore (Harrison, 1993). However, despite this discrepancy, ecotourism and other forms of alternative tourism are increasingly claimed to be sustainable, allegedly enabling economic growth through tourism *to* occur without the attendant costs of mass tourism.

Certainly, the tourism and hospitality industry has jumped on the green bandwagon, if only to deflect the possibility of legislation. In May 1993, the International Hotels Environment Initiative (IHEI) was launched in London by the Prince of Wales, and simultaneously published a guide to best environmental practice for hotels (IHEI (International Hotels Environment Initiative), 1993). The event was accompanied by much advertisement of the industry's environmental awareness. Other major players in the travel industry make similar claims (British Airways, for instance, has an Environment Branch and some major tour operators have Environment Officers); it is clear that, at least in rhetoric, efforts are being made to 'green' the industry. It is equally clear, as in-house publications show, that such efforts bring the industry substantial economies. Nevertheless, it is worth pointing out that, whatever else it means, sustainable tourism does not necessarily denote a sustainable tourism industry. Tourism resorts and regions can and do change, sometimes following a path from development to decline (Butler, 1980). As the recession has shown, there is no guaranteed life for individual companies in tourism, and the industry as a whole has sometimes seemed to operate a kind of slash and burn policy, moving from one region to another.

There have also been other kinds of response. Mention has already been made of the increased literature on the threat to the environment from tourism, and the development of sustainable tourism as a topic for research is indicated by the 1993 launch in the UK of the *Journal of Sustainable Tourism*, which aims 'to develop both a theoretical base for the subject and reliable empirical evidence of its results and impacts' (Bramwell and Lane, 1993, p. 2). Important contributions to the debate have also been made by such pressure groups as Tourism Concern, a British-based organization seeking to minimize (what it sees as) the negative social, cultural and ecological effects of tourism (Eber, 1992).

However, despite these responses, the exact nature of sustainability is unclear, remaining undefined even in a well-regarded book on the subject (Redclift, 1987). Mathieson and Wall commented that there was considerable information on 'resort morphologies and infrastructural changes in urban environments but much less is known about the impacts of tourism on the natural environment' (1982, p. 130). To some extent, this still seems to be the case and if this is so for the relationship of tourism to the natural environment, it applies even more to its impact on the social and cultural environment, which is the focus of the bulk of this chapter.

The Vulnerability of Islands and Small States

It may be that islands and small states are especially vulnerable to environmental and other impacts of tourism. However, the terms 'island' and 'small state' are neither synonymous nor exact; within and across both categories there is a wide range of difference in culture, history, area, population, natural resource base, levels of economic development and political status.

Island status alone means little. Australia and Anguilla are islands, as are Greenland and Guernsey, but they have little else in common. Small islands, or groups of them, can vary considerably in the standards of living they support. Singapore, Bermuda and the Bahamas are wealthy when compared, for example, with Tonga, the Maldives or St Vincent and the Grenadines, and even by comparison with many mainland states (IBRD (International Bank for Reconstruction and Development), 1994, p. 228). Furthermore, islands are not necessarily more vulnerable to ecological degradation than mainland territories, as indicated by the concern many feel over the effects of tourism, or any other kind of industrial activity, on the fragile ecology of the continent of Antarctica (Smith, 1994).

If island status and geography alone provide insufficient criteria to assess the vulnerability of tourist receiving areas, other factors must be considered. Population density, another variable, is clearly not crucial. Indeed, extrapolating from World Bank figures (IBRD (International Bank for Reconstruction and Development), 1994, p. 228), some of the most developed areas of the world (for example, Japan, the Netherlands and Belgium) are among the most densely populated, whereas others (for example, Canada and Norway) have very low population densities.

In fact, little may be gained by focusing exclusively on islands and other small states that are formally politically independent. Guadeloupe and Martinique, departments of France, are not for this reason more at risk from tourism than, for example, the Seychelles, an independent state which experienced a similar background of colonialism. By the same token, independent Sri Lanka, which covers a much greater area and has a much larger population, may nevertheless be as ecologically, socially and culturally vulnerable to tourism as these smaller islands.

Ultimately, the alleged smallness of island or non-island states may often be as much a reflection of their structural position within the world economic and political system as an indication of their actual area, population and natural resource base. Although Wilkinson describes 'island microstates' as 'a relatively distinct subset of microstates which are worthy of study: smaller ... [with] many characteristics of Third World countries, and important (and often dominant) tourism sectors' (Wilkinson, 1989, p. 154), such structures of under-development are clearly widely distributed. Even within Third World states, centres and peripheries can be found. Bali, for example, which is heavily dependent on tourism, might be regarded as a periphery of Indonesia. More importantly, however, if the characteristics of dependency, especially a lack of economic and political power, are considered a defining feature in the analysis of islands and small states, it is evident that analysis cannot be restricted to these societies. Such features are also found in many regions of the developed world. As a consequence, the Shetland Islands of Scotland, the Principality of Wales, the Baleares and the Canary Islands, Belize, Sicily and Mykonos, along with any hitherto isolated, distinctive and unincorporated region may for some purposes be seen to share the social, cultural, economic and political characteristics of small states, whether or not they are islands. The degree of isolation, peripherality and vulnerability of the physical environment, the apparent distinctiveness of a culture, and the prevailing pressures to change, with the more objective features of geography, demography, population density and political status, then together determine which islands or regions may usefully and meaningfully be compared in assessing the impacts of tourism.

Thus, the notion of sustainable tourism is relevant not only to less developed countries. Like development more generally, it is capable of much wider application, and although some islands and geographically small states among the less developed countries may seem especially vulnerable to tourism, restricting attention to these alone would inevitably omit valuable sources of data and comparative material from other regions, which may not fall into these categories but may nevertheless be subject to the same ecological, economic, social and cultural processes.

Sustainable Tourism: The Social and Cultural Dimensions

It has been argued elsewhere that the social effects of tourism include an almost universal spread of market relationships and numerous changes across a wide range of economic, social and cultural institutions (Harrison, 1992b). However, the focus of this chapter is especially on sustainable tourism and its relationship to social and cultural phenomena. In fact, it is quite possible to define sustainable tourism without mentioning them – for example, as the development of tourism in such a way as not to prejudice, and preferably to increase, the ability of present and future members of host populations to meet their basic needs – a restricted definition which might not be universally approved.

Certainly, as in the debate over sustainable development, social and cultural dimensions are frequently slipped into arguments over sustainable tourism, often for reasons which seem to be related more to ideology than logic. By contrast, and for reasons given below, it is argued in this chapter that we should be cautious of tying sustainable tourism to forms of culture and social structure that are specifically favoured by some outsiders.

1. We cannot assume consensus in host societies.
2. Sustainability is a difficult concept to apply to cultures and social structures.
3. Considering sociocultural phenomena as part of a wider system does not solve the problem.

Consensus in host societies cannot be assumed

> Sustainable tourism is tourism and associated infrastructures that both now and in the future operate within natural capacities for the regeneration and future productivity of natural resources; recognize the contribution that people and communities, customs and lifestyles make to the tourism experience; accept that these people must have an equitable share in the economic benefits of tourism and are guided by the wishes of local people and communities in the host areas (Eber, 1992).

In this quotation, as in the report of the Brundtland Commission (WCED (World Commission on Environment and Development), 1987), equity is linked to sustainability: 'People and communities ... must have an equal share in the economic benefits of tourism'. This sounds both reasonable and highly moral, until one asks exactly what it means, and then problems arise. Is everyone in the community, man, woman and child, to receive an equal share? Does this include the lazy and the deliberately unemployed? Men and women? Are their shares to be the equal of one another and/or to those of everyone involved in the tourism industry in the region? Is this egalitarianism to be imposed even if it contravenes the norms of specific communities?

Such references to 'local people' and 'communities' imply that they are some kind of homogeneous mass and indicate a tendency, especially among middle-class liberals in the West, to glorify small communities, especially (but not only) in LDCs and, most especially, those considered 'untouched' by modernization. In fact, as every social anthropologist knows, small communities in most known societies frequently contain deep divisions, perhaps being so divided that they are 'back-to-back' rather than 'face-to-face' societies (Bailey, 1971, p. 4). Such divisions were so prominent in one Caribbean island that a social anthropologist wrote up his findings under the term 'Crab Antics', a reference to the idea that crabs placed in a basket continually try to pull down any of their number attempting to rise above the others (Wilson, 1973, p. 58). In such circumstances, where divisions of class, status and power (including gender) are great, it is hard to see how the wishes of local people and communities could ever be sufficiently unified to offer a practical guide to 'tourism and associated infrastructures' (an unwieldy term which nevertheless again implies a somewhat undifferentiated grouping).

Sustainability is a difficult concept to apply to cultures and social structures

> The sociocultural impact in terms of the demonstration effect and on local traditions and morality also has considerable implications for sustainability ... but it must be pointed out that traditional practices, evolved over generations, are frequently more sustainable than those with which they are supplanted, partially as a result of tourism (Cater, 1991).

The sustainable elements of so-called traditional practices (a highly general term in itself) are not easy to determine. Indeed, although there may be some evidence of the strength of traditional agriculture (Chambers, 1983, pp. 75–102), traditions tend to supplant one another, and the problem is then to decide which is the most authentic. Issues are even more complicated when it comes to social and cultural phenomena. By way of an example, few may doubt that the tartan kilt is traditional Scottish dress, whereas in fact tartan was a relatively late import into Scotland and the kilt was designed, first worn and then introduced to the Highlands of Scotland in the mid-18th century by an English Quaker industrialist. Particular tartan patterns did not become associated with specific clans until the early 19th century, a result of ingenious (not to say canny) marketing on the part of English tartan manufacturers (Trevor-Roper, 1984).

An example more relevant to tourism can be found in the experience of Cap Lloc on the Costa Brava, Spain. During the 1960s, the village ceased to rely on fishing and became an international tourist resort, a process apparently seen by villagers, and by Pi-Sunyer, as a mixed blessing.

> It is difficult to escape the conclusion that tourists, now the mainstay of the local economy, are disliked. Given the reality of mass tourism, residents no longer view an influx of temporary 'guests' as positive. At the risk of oversimplification, the dominant attitude is indifference and dislike. Outright hatred of tourists is rare. Tourists, like the weather, have become a fact of life, and the average villager feels that there is little that he can do to change the status quo (Pi-Sunyer, 1977).

Several years later, Pi-Sunyer's review of these conclusions makes interesting reading. Indeed, the above passage was omitted from the second edition. In 1987 a bountiful anchovy harvest had returned fishermen to the region in great numbers but, instead of rejoicing, villagers were upset. The waters of the harbour and the beaches were polluted and the tourist industry was threatened. 'Save Cap Lloc' leaflets were printed, in Catalan and Castilian, referring to 'our happy and prosperous tourist town' and bewailing the effects of the arrival of the fishing fleet: 'a homicide, attempted rapes, robberies, insults, quarrels, all of which led to undesirable publicity'. When a young Dutch tourist was knifed by a fisherman:

> Local and Barcelona papers carried angry letters describing the injured Dutch tourist as a friend and neighbour. Very similar judgements stressing the virtues of touristic harmony and warning of the dangers emanating from the anchovy fleet were voiced throughout the town in the summer of 1987. (Pi-Sunyer, 1977).

Although fishermen still came to the village, their significance to the villagers changed over time, as did the villagers' perceptions of tourists. In short, Cap Lloc had changed from a traditional fishing village to a traditional tourist town.

The basic issue is how far culture, a people's way of life passed on from generation to generation, can be regarded as a fixed quantity, to be maintained, sustained and defended against all comers. Two hundred years ago, England was predominantly rural and agricultural with a population that was relatively homogeneous. In all likelihood, the England of the 1990s would be unrecognizable to inhabitants of the former period. Does this mean that English culture and social structure are no more? Or have they retained crucial characteristics of Englishness? Similar questions can be asked about any modern society. Are such former fishing villages as Farol, on Spain's Costa Brava, so sympathetically described by Lewis (1985), or the much-maligned Benidorm, less Spanish as a consequence of tourism? Has Brighton or any other European seaside resort become less representative of its nation as a result of moving from fishing to tourism? And are livelihoods obtained from catering for tourists really less dignified or more onerous than, for example, the heavy and often dangerous manual labour involved in fishing or agriculture? Certainly, it would be difficult to argue that tourists are a less sustainable source of income than fish, or that workers in tourism are less able to meet their basic needs than those in more traditional occupations. To return to a previous example, are Norwegian whalers better employed carrying tourists to watch whales or killing them?

The picture becomes more complicated if we accept that the production of indigenous arts and crafts can receive a new impetus from tourism, as reported, for example, in Malta (Boissevain, 1977, pp. 532–534; Boissevain and Inglott, 1979, pp. 252–253). However, it is also frequently argued that items produced for sale rather than domestic use lose their cultural significance for producers, a reference to the highly charged debate over authenticity and commoditization (Harrison, 1992b, p. 21; Cohen, 1993). It is not that culture disappears; rather, with the advent of tourism, the meanings attributed to traditional activities by members of the host population are changed and, by implication, their culture has allegedly changed for the worse.

The difficulty of determining when one culture ends and another begins makes it hard to decide what the 'sustainability of cultural and community life' (Tisdell, 1993, p. 216) actually means. However, this is often advocated, along with the recommendation that traditional cultures of host societies should be protected against threats from tourism, as if they should somehow be preserved in aspic – a policy which, ironically, the heritage industry of the West has been accused of following as it attempts

> the restructuring and commodification of private memory itself, something that is very evident in the new folk and industrial museums which seek to produce 'the way we were', and where individual memories are erased by the collectively reconstructed image of a period or way of life. (Hewison, 1989, p. 20)

In fact, culture is highly flexible and, even where tradition may be regarded as an important guide to current behaviour, specific traditions change and are replaced by others, to which people become equally committed; the king is dead: long live the king.

It is not easy to decide which elements of traditional behaviour should be sustained. In Swaziland, women wearing jeans are subjected to negative sanctions, even though the style of the dresses and skirts they are expected to wear is equally Western, albeit less modern. The conflict is not as trivial as it appears: by wearing jeans young women defy Swazi tradition, allegedly usurping male roles in a society dominated, especially, by older men (Harrison, 1992c, pp. 155–156). The example (which is reminiscent of similar but earlier debates in the West) is but one of many 'demonstration effects', where tourism or other agents of modernization transfer aspects of Western culture to LDCs, but it raises important general issues. Who is to decide whether or not the present Swazi social structure should be maintained in the face of pressures to change? Should those in favour of sustainable tourism defend male elders against their young and against their women? Should traditional authorities always be maintained in power in the interests of sustainability?

Considering sociocultural phenomena as part of a wider system does not solve the problem

> Development (including tourism) is a wide-ranging activity that should never be divorced from a serious and concerned consideration of environment and society. In fact, the development system consists, at a minimum, of a mixed system of elements and linkages from the economy, environment and society (Farrell, 1992).

> The environment, considered here in its broadest sense as encompassing socio-economic and cultural phenomena as well as biophysical elements, represents not merely a constraint for tourism development but a resource and an opportunity... Tourism certainly can contribute to environmental degradation and be self-constructive; it also has the potential to bring about significant enhancement of the environment (Pigram, 1992).

To argue that social action occurs within a wider system is hardly unusual. Indeed, as indicated earlier, it is common to all major approaches to development, and it is also a key feature of ecology. However, this does not solve those problems of sustainability associated with social structure and culture.

By definition, systems are maintained around an equilibrium and if this process breaks down the overall system is threatened. However, even when analysis has been confined to social systems alone, it has proved virtually impossible to agree on the nature of the equilibrium, whether it is dynamic or in a steady state, and on the different roles of units constituting the system. Once the system is widened to include natural phenomena, in effect incorporating the universe and all therein, the situation seems to defy sensible analysis. What are the various subsystems? What are their needs? Whose needs are paramount? According to what criteria? How far should trade-offs be considered between the needs of (some part of) humanity and the functional imperatives (however these are to be defined) of other parts of the system, or of the system as a whole? How far can specific cultures or social structures be considered to have some kind of fit within the wider system?

The problems of sustainability might be broken down, to some extent, by focusing on specific sub-systems, in effect reducing the overall unit of analysis to predefined regions, and trying to ensure that development within them is both sustainable and self-sufficient. Theoretically, this might be possible, but it is not likely to occur in practice, especially in islands and small states. As Bertram (1993, p. 248) has noted:

> Many outside observers of these very small, very open economies have suggested that sustainable development is not a feasible option, in the sense that no conceivable degree of accelerated resource utilization could permit domestic product to reach levels consistent with actual standards of material welfare, let alone future aspirations.

His alternative is to focus on an indigenous people, wherever they happen to reside, and consider sustainability to have been achieved as long as they 'retain a set of entitlements sufficient to support material welfare standards over the foreseeable future, while preserving their collective identity and the natural environment of their home territory' (Bertram, 1993, p. 248). To accept such criteria, however, would be to abandon any notion of basic needs and to move away from the requirement, surely essential if sustainability of any kind is to be achieved, that individuals and collectivities are held accountable for damage to their immediate environment.

More important, perhaps, is the fact that to survive any system must be both sustainable and adaptable, incorporating elements of continuity and elements of change. Indeed, since its inception the science of ecology 'has been centrally concerned with the concept of adaptation' (Allen, 1985, p. 218) and many social scientists came to consider adaptation as a defining feature in their own concept of social system. At the risk of drawing groans from sociologists who first studied their subject between 1955 and 1970, this is particularly the case for Talcott Parsons, perhaps the best known theorist of social systems. It is perhaps no coincidence that as an undergraduate he studied biology.

The Parsonian system of action is well established in the literature. For Parsons, it was constituted by four major sub-systems: the biological organism, the personality, the social system and culture, all of which required four basic functions to be carried out if they were to survive, and they are summarized in the well-known acronym of AGIL. Put simply, all systems need an adaptive capacity (A), to meet goals (G), integration (I) to coordinate the behaviour of constituent parts of the system or sub-system, and latency (L), or pattern-maintenance, which basically refers to the need of a system to keep the constituent units able to act for the purposes of the system as a whole (Morse, 1961).

It is unnecessary, in this context, to formulate a sparkling new world system to facilitate analysis of the relationships of human society, including tourism, with the wider environment, and the work of Parsons is sufficiently well known to require no further commentary. However, this discussion does relate to the question of sustainable tourism. At first sight, tourism could be defined as a sub-system, complete with its AGIL functional imperatives, thus necessarily interacting with its social and physical environment. Unfortunately, there are no clear criteria by which to distinguish changes in culture that are adaptive, designed (or programmed) to enable the cultural sub-system, or the wider system, to meet new challenges in its environment, and cultural change which is inherently destructive, thus contributing to the demise of the system.

The argument can be applied directly to specific social and cultural phenomena: many changes introduced by tourism (and other forms of modernization) commonly regarded as destructive could, instead, be regarded as adaptive, thus contributing to the survival of the wider system. Seen in this way, changes in traditional political and familial authority patterns, individualism, commoditization and all the demonstration effects imaginable, can be redefined as functional for the sustained continuation of the wider whole. Indeed, although Parsons did not assume all pre-modern societies would become modern, he attached a positive value to modernization, believing it to enhance individual freedom 'from tradition, the narrow constraints of kinship and the dominance of hierarchical structures based on ascribed, particularistic criteria' (Holton and Turner, 1986, p. 22).

More outlandishly, perhaps, even such phenomena as prostitution, often exacerbated by tourism (and almost thought by some critics to have been caused by it), could also be considered functional and thus contributing to the sustenance of cultural forms. In short, all the social and cultural phenomena regarded by critics of tourism as negative impacts could then be gleefully recorded by tourism's advocates as highly functional for the continuation of host societies.

Conclusions

It has been argued that development and sustainability are problematic concepts and that debates over sustainable development are no less prone to confusion. Sustainable tourism, itself a form of sustainable development, is therefore likely to attract criticism, both from those who regard it as reformist and from those who consider the notion of sustainability to be non-operational, but it has been given wide coverage, both academically and within the tourism industry, and has been especially applied to islands and small states. However, while such regions may be considered vulnerable to the various impacts of tourism, they are not unique in this respect. Indeed, the processes involved in incorporating hitherto isolated and relatively self-contained regions into the modern world system also operate within developed societies, where the changes encountered by small communities intent on becoming tourist destinations may often be considered as traumatic as those occurring in less developed societies.

When the focus was shifted to the sociocultural dimensions of sustainability, three major areas of concern were identified. First, it was argued that consensus in host societies cannot be assumed, and that a vague advocacy of equity may reveal more of the ideology of outsiders than the host society's social and cultural practices. Indeed, social and cultural divisions in islands and small states, and in small communities more generally, may be reflected in major differences of opinion as to the desirability or otherwise of tourism development, or over the types of tourism preferred. In such circumstances, 'being guided by local people' is a nice example of populist rhetoric completely without substance.

Second, sustainability is an especially problematic concept when applied to sociocultural phenomena. There are, indeed, traditional societies where the activities of the present continue to be largely validated by the past, but specific traditions change and are supplanted by others. Generally, the culture of a people or a community is highly flexible and few criteria exist for assessing which elements of a culture or a social structure should be sustained. Indeed, it cannot be assumed that even the core institutions of a culture will be shared, as indicated in earlier debates over the nature of 'plural societies', especially in the Caribbean (Rubin, 1960; Smith, 1965, 1984).

Third, although there may be some value in depicting social and cultural phenomena as part of a wider natural system, this does not make them any more amenable to analysis. Systems theory in social science has long experienced difficulty in distinguishing between social and cultural attributes that are functional and dysfunctional for the continuance of a social system and such analytical difficulties are unlikely to be eased by considering social systems as but a part of the wider natural environment, even though there may be evident *a priori* reasons for postulating such links. As a consequence, there are no clear criteria for determining which of the sociocultural changes laid at the door of tourism represent an outright threat to a society's distinctiveness, which will help sustain the status quo, and which allow for adaptation in the face of external pressures to change. In fact, the difficulties currently experienced in assessing change in tourist-receiving societies are based at least as much on ideology and moral posturing as on sound social analysis, which has been mainly noticeable by its absence.

That there are difficulties in operationalizing the concept of sustainable tourism, especially in consideration of sociocultural phenomena, need not mean that the concept itself is valueless. However, if the term is to be used, the above problems need to be recognized. When applied to human societies, many questions of development and sustainability are inevitably and irretrievably evaluative and the same applies to the concept of sustainable tourism, which is but one aspect of the overall debate about the nature of development.

To argue that issues of sustainability inevitably involve moral judgements is not to belittle them, but simply to accept Max Weber's view that, in matters of ultimate meaning, social scientists are no more qualified than anyone else to make pronouncements (Weber, 1948, p. 143). This does not mean that they are bound to silence. After all, as an ideal, sustainable tourism has much to recommend it and social scientists, too, are moral beings.

In addition, and still following a Weberian line (1948, pp. 150–152), social science can help to clarify the issues involved, in this case of the sustainability of social and cultural phenomena, including the formulation of working definitions and objectives. As indicated earlier, it is possible to define sustainable tourism without explicit reference to social and cultural phenomena, that is, as the development of tourism in such a way as not to prejudice, and preferably to increase, the ability of present and future members of host populations to meet their basic needs – as long as basic needs are defined only in material terms. Moral evaluation is not totally absent from this definition, for judgements are required on the nature of basic needs, even at a material level, and the claims of future generations must be weighed against those of the present generation. However, if basic needs are defined more broadly to include self-reliance and cultural independence, for example, the definition of development has again shifted to more intangible, albeit equally important, criteria (Seers, 1977, p. 5).

At a practical level, there is no need to reject the notion of sustainable tourism. Sustainability is evidently in the interests of all who participate in the tourism industry. Members of host societies wish to continue to receive the material benefits of tourism, and tourists increasingly value destinations which are in harmony with the wider environment. If this is lacking, they are likely to take their custom elsewhere. Unsurprisingly in such circumstances, those on the supply side, often with investments and other interests in tourist destination regions (Sinclair and Sutcliffe, 1988; Debbage, 1990; Dieke, 1991; Sinclair and Vokes, 1993), rapidly increase their appreciation of the environment. For some, at least,

> it must be apparent that eventually private and public interests will coincide, sometimes more quickly than might be thought. Such seems to be the case in Torremolinos, where past neglect of environmental quality now has implications for tourism's viability and profitability (Pollard and Rodriguez, 1993, pp. 257–258).

The existing literature on sustainable tourism indicates a clear need to confront the problems involved in making and implementing decisions. In practice, policies are decided and implemented from particular points of view, furthering the perceived interests of some at the expense of others. What social scientific approaches to sustainable tourism may achieve is to clarify the contradictions, which will vary from case to case. In Swaziland, for example, international tourism, which is strongly supported by the Swazi monarchy, itself legitimated by tradition, probably helps to erode the authority traditionally exercised by the monarchy and the male elders over the young, especially women (Harrison, 1992c). Of course, whether or not we should seek to sustain such inegalitarian structures is a moot point. However, to retain control a number of options are open to the authorities. They can attempt to keep tourists and young Swazi apart (which would be virtually impossible), apply draconian punishments to the 'offenders' (which has sometimes happened), or abandon tourism as a development strategy (which is not likely to occur).

The point of the example is to indicate that the desire for equity and other forms of social justice may run counter to other criteria of sustainability. It is commonly argued that sustainable tourism must be managed (Craik, 1991, p. 131) and that only the state 'can provide the conditions for movement toward greater sustainability' (de Kadt, 1992). Yet to advocate a strong state may contradict the desire, even the perceived imperative, to listen to (follow? be guided by? consult with? obey?) the wishes of 'the local community', always assuming that a single view can be identified. Where this is possible, as Smith has noted for Boracay, (Smith, 1992, p. 153) we may discover that people value tourism for the spending power it brings, even though such consumerism may be much derided by advocates of sustainable tourism, and even if it is the result of top-down rather than bottom-up strategies of development.

Interestingly, one of the few examples given of successful alternative tourism, the case of Hana in Hawai'i, is of luxury tourism operated by the highly paternalistic and very wealthy

Rosewood Company of Texas, owned by Caroline Hunt and presided over by Robert Zimmer, of Rosewood Hotels. The case is far from proven but, as Farrell (1992) records:

> in Hana is found the unique set of features, all human, which is the essence of the community – a largely 'non-haole', potentially very friendly, little-developed Hawaiian community. The local people made it; it is theirs. Rosewood capitalized on that fact, while providing most of the economic base for its existence... Rosewood had done many admirable things which probably would to some degree meet the test in all elements of sustainability. However, even Rosewood, though masters in the tourism business, were relative neophytes at sustainable development.

Although full employment, public amenities and affordable housing were provided at Hana, little equity was in evidence. As Farrell also notes, by 1989, when the hotel and its estate were sold – to be developed as a golf resort for Japanese tourists – it was already unclear how sustainable this particular project had become (1992, p. 134).

A further twist in the argument over top-down and bottom-up development concerns the relative status of local knowledge and that of outside experts. As Tisdell indicates, we can assume neither that residents of islands and small societies exist in harmony with nature, nor that they are fully aware, or aware at all, of the outcome of development strategies that have been implemented elsewhere (1993, p. 214). As with the rest of us, their knowledge is constrained and defined by their environment, even though this, like the overall political, cultural and social environment, will vary from case to case. Furthermore, appeals to 'the people' tend to assume they will accept the position of the speaker, always provided (of course) that they have the relevant information. The question then arises as to who will inform (and from what angle) the residents of a small fishing village about to be placed on the international tourist map of the destiny that might await them, if 'the truth' is told, will it be believed? And what is the reaction of advocates of sustainable tourism if villagers, on hearing of Torremolinos, enthusiastically ask for a similar form of development? In short, are we to assume privileged communication on the part of the experts (if they have been able to agree among themselves) and the subsequent need for top-down development, with a strong and preferably benevolent state to sanction plans and their implementation? And where, in this position, is the case for local democracy?

This is not to denigrate the value of consultation and the widest possible publicity in the promotion of sustainable tourism. Any consultation is likely to be better than none and, for pragmatic reasons alone, '"the target groups" – the communities affected – must be receptive to change and see the policy as a constructive response to their priorities' (Pigram, 1992). The more comprehensive the consultation, the more involved are members of the local community at all levels, and the fairer (not necessarily the most equal) the distribution of benefits appears, the more successful the tourism industry is likely to be. In this way, there may indeed be a somewhat loose fit between local democracy and sustainable tourist development. By contrast, there are sufficient indicators to support the view that where substantial numbers of people do not want tourists, the feeling is soon reciprocated. In such circumstances, tourism of any kind becomes unsustainable.

References

Adams, W.M. (1990) *Green Development: Environment and Sustainability in the Third World*. Routledge, London.

Adams, D. and Carwardine, M. (1991) *Last Chance to See*. Pan Books, London.

Allen, R.F. (1985) Ecology. In: Kuper, A. and Kuper, J. (eds) *The Social Science Encyclopedia*. Routledge, London, pp. 218–220.

Bailey, F.G. (1971) Gifts and poison. In: Bailey, F.G. (ed.) *Gifts and Poison: The Politics of Reputation*. Blackwell, Oxford, pp. 1–25.

Beckerman, W. (1992) Economic growth and the environment: whose growth? whose environment? *World Development* 20(4), 481–496. DOI: 10.1016/0305-750X(92)90038-W.
Bertram, G. (1993) Sustainability, aid, and material welfare in small South Pacific island economies, 1900–1990. *World Development* 21(2), 247–258. DOI: 10.1016/0305-750X(93)90020-A.
Boissevain, J. (1977) Tourism and development in Malta. *Development and Change* 8(4), 523–538. DOI: 10.1111/j.1467-7660.1977.tb00754.x.
Boissevain, J. and Inglott, P.S. (1979) Tourism in Malta. In: de Kadt, E. (ed.) *Tourism: Passport to Development?* Oxford University Press, for the World Bank and UNESCO, Oxford, pp. 265–284.
Boo, E. (1990) *Ecotourism: The Potentials and the Pitfalls*. Vols. 1 and 2. World Wildlife Fund, Washington, DC.
Bramwell, B. and Lane, B. (1993) Sustainable tourism: an evolving global approach. *Journal of Sustainable Tourism* 1(1), 1–5. DOI: 10.1080/09669589309450696.
Butler, R. (1980) Alternative tourism: the thin end of the wedge. In: Smith, V.L. and Eadington, W.R. (eds) *Tourism Alternatives: Potentials and Problems in the Development of Tourism*. University of Pennsylvania Press, Philadelphia.
Cater, E. (1991) *Sustainable Tourism in the Third World: Problems and Prospects. Discussion Paper No. 3*. University of Reading, Department of Geography, Reading.
Cater, E. (1993) Ecotourism in the third world: problems for sustainable tourism development. *Tourism Management* 14(2), 85–90. DOI: 10.1016/0261-5177(93)90040-R.
Chambers, R. (1983) *Rural Development: Putting the Last First*. Longman, Harlow.
Cohen, E. (ed.) (1993) *Tourist arts*. 20. Annals of Tourism Research.
Craik, J. (1991) *Resorting to Tourism: Cultural Policies for Tourist Development in Australia*. Allen and Unwin, Sydney.
de Kadt, E. (1992) Making the alternative sustainable: lessons from development for tourism. In: Smith, V.L. and Eadington, W.R. (eds) *Tourism Alternatives: Potentials and Problems in the Development of Tourism*. University of Pennsylvania Press, Philadelphia, pp. 47–75.
Debbage, K.G. (1990) Oligopoly and the resort cycle in the Bahamas. *Annals of Tourism Research* 17(4), 513–527. DOI: 10.1016/0160-7383(90)90024-L.
Dieke, P.U.C. (1991) Policies for tourism development in Kenya. *Annals of Tourism Research* 18(2), 269–294. DOI: 10.1016/0160-7383(91)90009-Z.
Eber, S. (ed.) (1992) *Beyond the Green Horizon: A Discussion Paper on Principles for Sustainable Tourism*. Godaiming, Surrey, World Wide Fund for Nature.
Farrell, B. (1992) *Tourism as an element in sustainable development*. Hana, Maui.
Harrison, D.H. (1992a) International tourism and the less developed countries: the background. In: Harrison, D. (ed.) *Tourism and the Less Developed Countries*. Belhaven, London, pp. 1–18.
Harrison, D.H. (1992b) Tourism to less developed countries: the social consequences. In: Harrison, D. (ed.) *Tourism and the Less Developed Countries*. Belhaven, London, pp. 19–34.
Harrison, D.H. (1992c) Tradition, modernity and tourism in Swaziland. In: Harrison, D. (ed.) *Tourism and the Less Developed Countries*. Belhaven, London, pp. 148–162.
Harrison, D.H. (1993) Review of V.L. Smith and W.R. Eadington (eds) Tourism alternatives: potentials and problems in the development of tourism. *Journal of Development Studies* 30(1), 256–259.
Hewison, R. (1989) Heritage: an interpretation. In: Uzzell, D. (ed.) *Heritage Interpretation. Vol.1: The Natural and Built Environment*. Belhaven, London, pp. 15–23.
Holton, R.J. and Turner, B.S. (1986) *Talcott Parsons on Economy and Society*. Routledge and Kegan Paul, London.
IBRD (International Bank for Reconstruction and Development) (1994) *World Development Report 1994: Infrastructure for Development*. Oxford University Press, New York.
IHEI (International Hotels Environment Initiative) (1993) *Environmental Management for Hotels*. Butterworth-Heinemann, Oxford.
Jenner, P. and Smith, C. (1992) *The Tourism Industry and the Environment. (Special Report No. 2453)*. Economist Intelligence Unit, London.
Lélé, S.M. (1991) Sustainable development: a critical review. *World Development* 19(6), 607–621. DOI: 10.1016/0305-750X(91)90197-P.
Lewis, N. (1985) *Voices of the Old Sea*. Penguin, Harmondsworth.
Mathieson, A. and Wall, G. (1982) *Tourism: Economic, Physical and Social Impacts*. Longman, Harlow.
Morse, C. (1961) The functional imperatives. In: Black, M. (ed.) *The Social Theories of Talcott Parsons: A Critical Examination*. Prentice-Hall, Englewood Cliffs, pp. 100–152.

O'Grady, A. (ed.) (1990) *The Challenge of Tourism: Learning Resources for Study and Action*. Ecumenical Commission on World Tourism, Bangkok.
Pi-Sunyer, O. (1977) Through native eyes: tourists and tourism in a Catalan maritime community. In: Smith, V.L. (ed.) *Hosts and Guests: The Anthropology of Tourism*, 2nd edn. 1989. University of Pennsylvania Press (149–156), Philadelphia, pp. 187–202.
Pigram, J.J. (1992) Alternative tourism: tourism and sustainable resource management. In: Smith, V.L. and Eadington, W.R. (eds) *Tourism Alternatives: Potentials and Problems in the Development of Tourism*. University of Pennsylvania Press, Philadelphia, pp. 76–87.
Pollard, J. and Rodriguez, R.D. (1993) Tourism and Torremolinos: recession or reaction to environment? *Tourism Management* 14(4), 247–258.
Redclift, M. (1984) *Development and the Environmental Crisis: Red or Green Alternatives?* Methuen, London.
Redclift, M. (1987) *Sustainable Development: Exploring the Contradictions*. Routledge, London.
Rubin, V.(ed.) (1960) Social and cultural pluralism in the Caribbean. *Annals of the New York Academy of Sciences* 83(Art. 5), 761–916.
Ryan, C. (1991) *Recreational Tourism: A Social Science Perspective*. Routledge, London.
Seers, D. (1977) The new meaning of development. *International Development Review* 19(3), 2–7.
Sinclair, M.T. and Sutcliffe, C. (1988) The economic effects on destination areas of foreign involvement in the tourism industry: a Spanish application. In: Goodall, B. and Ashworth, G. (eds) *Marketing in the Tourism Industry: The Promotion of Destination Regions*. Croom Helm, London, pp. 111–132.
Sinclair, M.T. and Vokes, R.W.A. (1993) The economics of tourism in Asia and the Pacific. In: Hitchcock, M., King, V.T. and Parnwell, M.J.G. (eds) *Tourism in South-East Asia. Routledge, London*, pp. 200–213.
Smith, M.G. (1965) *The Plural Society in the British West Indies*. University of California Press, Berkeley.
Smith, M.G. (1984) *Culture, Race and Class in the Commonwealth Caribbean*. Department of Extra-Mural Studies, University of the West Indies, Mona, Jamaica.
Smith, V.L. (1992) Boracay, Philippines: a case-study in 'alternative' tourism. In: Smith, V.L. and Eadington, W.R. (eds) *Tourism Alternatives: Potentials and Problems in the Development of Tourism*. University of Pennsylvania Press, Philadelphia, pp. 135–157.
Smith, V.L. and Eadington, W.R. (eds) (1992) *Tourism Alternatives: Potentials and Problems in the Development of Tourism*. University of Pennsylvania Press, Philadelphia.
Smith, V.L. (ed.) (1994) *Special issue: Antarctic tourism*. 21. Annals of Tourism Research.
Tisdell, C. (1993) Project appraisal, the environment and sustainability for small islands. *World Development* 21(2), 213–219. DOI: 10.1016/0305-750X(93)90016-3.
Trevor-Roper, H. (1984) The invention of tradition: the highland tradition of Scotland. In: Hobsbawm, E. and Ranger, T. (eds) *The Invention of Tradition*. Cambridge University Press, Cambridge, pp. 15–41.
Valentine, P.S. (1992) Review: nature-based tourism. In: Weiler, B. and Hall, C.M. (eds) *Special Interest Tourism*. Belhaven, London, pp. 105–127.
WCED (World Commission on Environment and Development) (1987) *Our Common Future*. (The Brundtland Report), Oxford University Press, Oxford.
Weber, M. (1948) Science as a vocation. In: Gerth, H. and Wright Mills, C. (eds) *From Max Weber*. Routledge and Kegan Paul, London, pp. 129–156.
Wheeller, B. (1993) Contemporary tourism issues. *Tourism Management* 14(1), 78–79.
Wilkinson, P.F. (1989) Strategies for tourism in island microstates. *Annals of Tourism Research* 16(2), 153–177. DOI: 10.1016/0160-7383(89)90065-0.
Wilson, P. (1973) *Crab Antics: The Social Anthropology of English-Speaking Negro Societies of the Caribbean*. Yale University Press, New Haven.
Wood, K. and House, S. (1992) *The Good Tourist: A Worldwide Guide for the Green Traveller*. Mandarin, London.
WTO (World Tourism Organization) (1993) *Compendium of Tourism Statistics*, 13th edn. WTO, Madrid, pp. 1987–1991.
WTO (World Tourism Organization) (1994) WTO news 1(January–February).
Yearley, S. (1991) *The Green Case: A Sociology of Environmental Issues, Arguments and Politics*. HarperCollins, London.

7

Learning from the Old South by the New South? the Case of tourism

Some years ago, the 'Third World' seemed a useful form of shorthand which distinguished a rather heterogeneous collection of colonies, ex-colonies and nation states from the First World of the USA and its allies and the Second World of the USSR and its satellites. Economic criteria apparently supported the distinction, for Third World societies seemed characterized by poverty, low GNPs, and a focus on the export of primary products. Politically, too, their experience as ex-colonies made them a distinct group, and their social structures, typically highly inegalitarian, also distinguished them from societies in the First and Second Worlds (Harrison, 1988).

Times have changed. With the Second World no longer an actor on the geopolitical stage, it is now simply illogical to posit the existence of a Third World. Furthermore, there are other signs that the distinction is breaking down: Eastern Europe in its current phase of decolonization is duplicating some of the experiences of African and Asian societies of the 1950s and 1960s. At the political level, for example, privileged classes (previously sustained by outside support) voluntarily relinquish power – or seem to relinquish power – during (relatively) bloodless revolutions in a generally supportive international political environment (Tiryakian, 1991).

A slightly different version of this argument, focusing on economic criteria, is put forward by Killick and Stevens, for example, who suggest that the experience of the (old) South of structural adjustment programmes during the 1980s can be directly applied to Eastern Europe in the early 1990s:

> In varying degrees, these countries were confronting a similar task to that of the countries of Eastern Europe: to reform their economic institutions and policies so as to reduce (often very large) distortions in their economies, increase the efficiency of resource use, cope with often large debt burdens and reverse deteriorations in production, investment and living standards (Killick and Stevens, 1992).

Second, social change follows economic reform. Economic rationalization is accompanied by social polarization and, consequently, unemployment and poverty become increasingly evident at the bottom of the social scale, just as new groups of entrepreneurs (or groups once hidden) emerge at the top. There is also, most tragically in the former Yugoslavia but also in the former Soviet Union, a renewed emphasis on ethnicity and nationalism as the restraints of the old order are swept away. Is it too fanciful, for example, to compare Bosnia with Biafra?

In fact, the reform process in Eastern Europe is debated in the language of modernization, an idiom too readily written off in the 1970s and

1980s. Economic growth through capitalism is the means and the end: the way to a 'better life' and the symbol of the achievement of that life. The talk (again) is of innovation, of entrepreneurs, of the diffusion of technology and 'modern' (or Western) values. Once again, the 'new' is superimposed on the 'old': in Third World societies of the 1960s, Western rationality (allegedly) replaced the dead hand of 'tradition', whereas in the Eastern Europe of the 1990s the market is the emancipatory replacement of a defunct, debased and degraded Communist Party bureaucracy which stifled enterprise and democracy throughout the USSR and its satellites. And in both sets of circumstances, market and administrative rationalization precede the 'liberal-enlightened rationalisation of conduct and world views' (Outhwaite, 1992). Put differently, old ways of thinking and behaving continue within newly introduced, 'modern' structures. We have the prismatic society, mark II (Riggs, 1964), in which institutions manifest the same form in very different kinds of societies but, underneath, perform very different functions.

Third, like those of the Third World 30 years ago, the resurrected states of Eastern Europe compete with one another for aid from the West. Encouraged by the West to reject their former masters and seek economic and political freedom, they now wait, mainly in vain, for the pay-off. Ironically, by bringing the Cold War to such a sudden conclusion, they have correspondingly reduced their own bargaining power and have simply added themselves to an already long list of countries waiting more-or-less patiently for aid of one sort or another from the West – which, of course, never includes the right of entry into, or completely free trade with, Western Europe or North America.

What is being argued here is that the removal of Russian 'colonialism', the internal polarization of Eastern European societies, the drive for economic growth and modernization, and the resurgence of ethnicity and nationalism, are all symptomatic of an increased convergence of the Second and Third Worlds. In a world system that is unmistakeably capitalist, societies are more or less 'developed' and Albania, Bulgaria and Romania, for example, must take their place (however reluctantly) alongside Egypt, Venezuela and India in the ranks of the less developed countries (LDCs). Schematically speaking, the Old South has been joined by the New South, and both are looking Westward.

Quite where they are to be placed in the ranks is still unclear. Although statistical indicators are highly unreliable, it has been suggested that 'both economic and social indicators place the eastern European countries, as a group, in an intermediate position between most LDCs and the OECD industrial countries' (Stevens, 1991). This is certainly arguable; the World Bank, for example, categorizes Romania (placed between Jamaica and Namibia), Poland and Bulgaria (between Turkey and Costa Rica) as Lower Middle Income Countries, and Estonia, Hungary, Belarus, the Russian Federation and Latvia as Upper Middle Income Countries (World Bank, 1993). Indeed, the very existence of this debate, unthinkable 5 years ago, suggests that Eastern European countries can no longer be regarded as highly distinct from other LDCs – and that they are a long way from the achievements of the NICS.

To argue that Eastern European countries can now be categorized as LDCs is not to suggest that they are in all respects identical to one another or to countries of the Old South, which was always characterized by considerable economic, political, social and cultural diversity. As Killick and Stevens also remark, in most LDCs of the Old South, the market economy was 'well entrenched' (Killick and Stevens, 1992), whereas in Eastern Europe a command economy was the norm. Other differences may also be found. In the manufacture of machinery and transport equipment, for example, Eastern Europe is in advance of most LDCs, as in providing formal education for its children (World Bank) and, possibly, in the provision of vocational training. Less tangibly, in religion, language and cultural orientation generally, much of Eastern Europe may identify more with Western Europe than with Asia. Clearly, there are differences in all these respects even within specific Eastern European countries, and events in the former Yugoslavia chillingly indicate the perils of 'nation building' when it involves the destruction of the state, along with many of its citizens – a lesson already learned by newly independent LDCs during the 1960s and 1970s, only to be repeated more recently in Rwanda.

Tourism and Development

With economic restructuring in Eastern Europe, parts of the region have experienced rapid industrial decline. Since 1989, for example, Bulgaria's industrial output has declined by an estimated 50% (Robinson, 1993). As a consequence, the service sector takes on renewed significance and, as in parts of the Old South, tourism is increasingly regarded as a 'fast track' to economic growth, employment and foreign exchange. However, compared with most newly independent states of the 1960s, when mass international tourism was in its infancy, tourism in much of Eastern Europe is already well established, albeit focused more on the domestic market than on international arrivals (Buckley and Witt, 1990).

How far the existence of a substantial tourism sector at the outset of economic restructuring is an advantage is open to question. Domestic tourists, and visitors from other centrally planned economies, were less demanding (and less spoiled for choice) than their Western counterparts. As a result, the accommodation and other tourist facilities, along with the quality of service offered, inherited by the new market-oriented economies were considerably below international standards (Hall, 1991b). Of Bulgaria's 254 seaside hotels, for instance, 80% have only two stars (Stanchev, 1993) and a high proportion of these require substantial refurbishment. In such circumstances, it is arguable whether or not hotels with two stars are better than none. Countries having to develop their tourism industries from a zero basis can at least benefit from the advantages of late development.

Although, for reasons already given, the pattern of tourism development in the 'new' LDCs of Eastern Europe will not be identical to that followed by the LDCs of the 'old' South, the question nevertheless arises as to whether or not the former can learn from the experience of the latter. However, this is not a simple question to frame for, like 'development' in general, tourism is always going to be assessed from specific evaluative positions (Harrison, 1992b). Who might be doing – or ought to be doing – the learning? Governments, with an interest (apart from staying in power) in 'developing' economies and/or satisfying their constituencies with economic growth and/or generally improved standards of living? Or 'the people' – a broad and singularly unhelpful category – whose interests may vary widely across class, ethnic, gender and other divisions? Or do we aim to instil some learning into another amorphous and equally disparate group – the outside 'experts', whose own disagreements indicate an equally wide range of political, ideological and theoretical positions?

Clearly, to ask if the New South can learn from the Old South is a misleadingly simple question to pose, especially as there is also wide disagreement over the benefits from and costs of international tourism (Harrison, 1992a). As a consequence, views on the lessons to be learned will be largely conditioned by the stance taken on tourism development overall. However, for the purposes of this paper, 'tourism development' can be said to occur when tourism makes a significant contribution to a state's economy without any major, obvious and highly-publicized sociocultural ill-effects. Any figures will be arbitrary but once international tourism contributes 5% of exports or 2% of GDP it might be regarded as an important factor in economic growth. The definition of 'major obvious sociocultural effects' is even more problematic and, for some critics, tourism anywhere is more trouble than it is worth. However, it is generally recognized that tourism in such countries as Thailand and Kenya, for example, has generated more hostility than in Singapore or Barbados. Whether or not this is simply because the opponents of tourism in the former countries have been more vociferous or more adept at gaining publicity is, of course, another matter, but the fact remains that tourism in some countries, and some kinds of tourism generally, have received a particularly bad press. It is also the case, as indicated by even a cursory glance at conventional 'development' indicators, such as those produced by the World Bank, that the poorest countries in the world system are also those who, for whatever reason, have the least appeal to international tourism. In short, although tourism is generally considered an 'engine of economic growth', the economic benefits of the industry are sometimes considered to be outweighed by the social and cultural costs.

If these assumptions are accepted, the degree to which there are lessons from the Old South can be reduced to more manageable issues: where, in the Old South, have policies for tourism apparently managed to avoid social and cultural 'problems'? Where such problems have occurred, what might have been done to avoid them? How might the experience of the Old South, in both its negative and positive aspects, be of some value to those involved in tourism in Eastern Europe?

Tourists do not bring 'instant development'

This may seem obvious, but it was not evident to many policy makers in LDCs in the 1960s and it is not necessarily clear to everyone in Eastern Europe now. The clichés are well known: tourism is 'an industry without chimneys' whose main assets are 'plentiful natural resources' and the 'legendary hospitality of the people', along with their 'varied cuisine' and the 'unforgettable' bargains to be found in the local marketplace. Of course, the reality is somewhat different and while tourists may indeed discover 'instant sunshine', their hosts do not discover instant development. What went wrong?

First, it was soon realised that if tourists (as opposed to independent, lone travellers) are to be attracted to a region, they have to be (quite literally) transported to and from their destination, in some number, in relative comfort and safety, and then to be accommodated and fed in ways to which they have become accustomed. What this means is, in effect, that a certain level of economic development is required even before tourists in any number are attracted to a country. Tourism thus becomes a symptom *of* development rather than a *cause* of development. It is no coincidence, then, that most tourist travel is between 'developed' countries and that few 'low-income economies', as defined by the World Bank, have a significant tourism industry. Exceptions to such a rule include Tanzania, Kenya, Sri Lanka and (possibly) China and India. It is also no coincidence, then, that the most popular 'long-haul' destination for tourists from the UK is Florida.

Second, if a decision to encourage tourism is taken (an issue to which I shall return below), this may necessarily entail upgrading airport and road facilities. Such items are expensive, demand foreign exchange and in themselves represent a commitment to further invest in the industry. It is really at a very early point in the proceedings that one has to decide whether or not investment (from local resources, from outside, or from both) in tourism is a case of throwing good money after bad. And all this expenditure may meet with increasingly vocal local resistance.

The need to know the market

Sociologists and anthropologists have spent much time attempting to categorize tourists and, indeed, in March 1993, the United Nation's Statistical Commission endorsed a new (and highly complex) definition of tourism. The key point, though, is that there is no such product as a typical tourist (even though there are stereotypes, which are rather different). A potential receiving region needs to determine in advance the kind of tourist it (or groups within it) want to attract. This is no easy matter and the advice of 'development consultants' is not necessarily (if ever) accurate or disinterested, as was shown most spectacularly by the Zinder Report on tourism in the Caribbean (Levitt and Gulati, 1970). This reference is not merely of historical interest: with the proposed development of Eastern Europe, economic consultants (deprived by the recession of business in the West) are queuing at the doors of Eastern European embassies. In the search for scarce funds, they are unlikely to recommend courses of action which leave them without a crucial role. Although tourism planning is indeed an 'emerging specialisation' (Inskeep, 1988), planners need to be given a clear brief with which to work and (I suggest) their conclusions need to be treated with a healthy scepticism. Even examples of 'best practice' in planning can only mirror the current state of knowledge and, in matters of *social* impact, especially, this is virtually non-existent.

The impact and contribution of tourism changes over time

There is an old story of the Bedouin who allowed a camel, suffering from the cold of the desert night, to put its head in his tent. As the story goes, the Bedouin is persuaded to allow more and more of the camel in the tent, until the entire animal is inside and the Bedouin has been ejected. There are those who argue that this is precisely what happens when tourism develops and, although this is not my position, the impact of tourism clearly changes over time. The work of Butler (1980) is perhaps best known in this respect. His model suggests that destination regions typically 'evolve' through several stages, from exploration, involvement, development, consolidation and stagnation to either decline or rejuvenation, with every stage bringing different economic and (implicitly) social consequences.

Butler's model, or ideal type, has been applied to the growth of tourism in both 'developed' and 'underdeveloped' societies (Meyer-Arendt, 1985) and, although many more case studies are needed, there are some obvious lessons for Eastern Europe. It is clear, for example, that once mass tourism is introduced, at the development and consolidation stages the role of international capital increases, advertising is at its most intense and, with tourist numbers swelling to what might seem an invasion, local opposition is likely to be at its greatest. How far some of these 'problems', if such they are, can be foreseen and countered in Eastern Europe is largely dependent on economic and political circumstances in the countries concerned. The experience of the Czech Republic, for example, is unlikely to be the same as that of Romania or Albania. However, at the outset, there may simply be an argument for caution: it is not imperative that tourism development plans aim at the highest possible numbers. Where this *has* happened, in coastal Spain or the Balearics, for example, there has followed a painful period of reappraisal: belated legislation to control the style and sites of hotel construction, the preservation of areas devoted to agriculture and wildlife, and the 'greening' of the concrete urban tourist jungle. To some extent, such policies are having some kind of success but how far they need to go can be demonstrated simply by mentioning Benidorm – an urban centre of tourism which, however unjustly, has become for some the epitome of all that is commonly regarded as tacky, mass and vulgar in international tourism.

The need for state involvement

Arguably, the key lesson to be learned from the Old South is the need for continued state involvement in tourism. This may not be an easy lesson to learn, especially in the current ideological climate. In the West, the role of the state (including para-statal organizations) in tourism development has been largely restricted to marketing and visitor servicing and, in general, it has played little part in planning tourism development or researching its effects (Pearce, 1992), a position which has been regarded unfavourably by many in the tourism industry. By contrast, until comparatively recently, many countries of the Old South followed a different pattern, in tourism as in other sectors, with the state far more involved in actually operating hotels and other tourist establishments. Ironically, this kind of 'developmental state' owed much to the positive perception of many politicians in the South, and their advisers, of the command economies of the Soviet Union and its allies. By the early 1980s, such policies were under attack by the 'New Right' (Harrison, 1994) and were already discredited when the radical reorientation of Eastern Europe commenced at the end of the 1980s. At the same time, again ironically, opinion was shifting to yet another re-evaluation of the role of the state in the Newly Industrializing Countries and it was increasingly being suggested that totally rejecting the role of the state in development was a 'reaction too far' (Killick, 1992).

Arguing for a continued role for the state in Eastern Europe is no easy matter when the talk is of privatization, of innovation, of entrepreneurship and capitalist values and investment, and where post-communist regimes seem to lack the ability (or the will) to dismantle command economies (unless, in so doing, the sectional advantage of the former masters is served). However, it is commonly recognized, even within the industry, that too rapid and opportunistic a development of tourism runs the risk of destroying its intrinsic assets. If 'sustainable tourism' is to be developed, so that resources are protected, so that the scale of development is consistent with the resource base, and so that attention is paid to the distribution of benefits, there must be clear consideration and formulation of policy through national plans and an efficient and successful implementation of that policy. Of course, such planning could be carried out by the private sector; however, while individual capitalist companies rationally appraise and plan their own operations, they have neither the brief nor the inclination to consider the wider or long-term interests of the industry as a whole or of the receiving society. Indeed, it is arguable that, in any capitalist society, it would be against the interests of any individual firm to do so. If this is the case, the state must step in. The issue of state involvement thus moves away from the question of whether or not it should be involved, to the kind of role it might usefully play. Reducing the role of the state in the ownership and operation of tourism facilities (an approach supported by experience in the Old South and currently being exercised in Eastern Europe) is not to deny it a key role in an enabling framework in which the aims, types and limitations of tourism development are clearly specified.

There are numerous examples of this kind of state involvement in developed countries and the South. Mention has already been made of Spain and the Balearics, where (belated) legislation has been introduced to protect areas of scenic beauty, preserve indigenous flora and fauna, conserve agricultural areas, and to reduce noise and traffic pollution in towns. In the Old South, there are marked variations across and within regions: in East Asia, the state plays a strong supportive role in development generally and relatively few problems have been reported for tourism. However, in parts of South-east Asia, notably Thailand and the Philippines (where the state is relatively weak), tourism development has been much criticized. By contrast, Myanmar's tourism is under rigid state control, in South Asia, tourism seems to cause no notable problems in India, it is strictly limited by the monarchy in Bhutan, and is confined to island enclaves in the Maldives. However, tourism in Nepal is alleged to have been ecologically damaging but the government-inspired Annapurna Conservation Area Project – formed to protect the region's ecology and the interests of local people from previously unrestricted tourism development – seems to have been a well publicized success. In the region defined by the World Tourism Organization (WTO) as 'West Asia', the state's influence varies from the almost-absolute (Israel) to the tenuous (Turkey) and laissez-faire (Cyprus).

Similar variations are found across Africa, not a major tourist-receiving area. However, where tourism is important, it has allegedly created problems. To the east, Kenya's tourism policies have been much criticized, as have those of The Gambia to the west, and of Egypt, Morocco and Tunisia to the north. In these regions, relatively 'weak' states and marked cultural differences between 'guest' and 'host' have apparently resulted in a less than stable tourism industry, a situation avoided in the Seychelles, where tourism has been restricted to a high-spending elite (Wilson, 1994).

Although some governments have deliberately targeted high-spending tourists, others seek a different tourist 'mix' and, clearly, policies vary over time. However, in general, the experience of the Old South suggests that there is a role for government in setting the limits of tourism development and in enforcing legislation controlling tourist behaviour, especially where there are wide differences between 'host' and 'guest' (a situation probably less common in Europe). Experience also suggests that, to be effective, policy implementation requires an equally efficient range of supporting institutions.

The need for appropriate supportive institutions

As de Kadt recognizes, formulating policy is one task and implementing it quite another. Public policy is implemented through institutions, run either by the state or by private bodies. And yet 'the most formidable task on the road to sustainable development, and tourism development, is that of building the institutions needed for policy implementation' (de Kadt, 1992). Of course, what one considers an 'appropriate' institution is a fine point and, in the end, conditional on a wider perspective of what constitutes 'development'. However, experience in the Old South suggests that tourism institutions are at their most effective when the general level of efficiency of all public institutions is high. This may be particularly so for institutions run by the state: sex tourism in Thailand, for example, has been much criticized, but since prostitution is technically illegal, it has continued and flourished only with the knowledge of and tacit acceptance by the Thai government and with the active support of the agencies of law and order. Whatever the morality of sex tourism, in Thailand it exists primarily because the state authorities have been unwilling to deal with it, for reasons which may have as much to do with Thai culture and history as with the desire for international tourism (Meyer, 1988).

Other agencies of the state make similarly important contributions to tourism development. It can be learned from the Old South, for instance, especially from small island societies, that 'leakages' from tourism increase if the industry is not closely linked to other sectors, for instance, agriculture – another indication that tourism development is most successful when integrated with development elsewhere. Furthermore, while employment may be generated by tourism, it tends to be skilled and semiskilled. Only a deliberate policy to provide a good general educational background and skills specific to tourism (through, for example, hotel and training colleges, management programmes, and so on) will help ensure that locals take positions in the higher management of tourism accommodation and other facilities. Providing such opportunities does not simply reduce leakages and increase multiplier effects; it also enhances the self-image of members of the host population, who might otherwise resent becoming a 'nation of bell hops and chamber maids' – an argument often used against tourism development in numerous countries – especially where there has been a history of colonialism, including Kenya, Malaysia, Tanzania and the Caribbean.

Many non-governmental institutions are also of direct relevance to tourism. Hotel and guest house associations, national trusts, arts and crafts associations and cultural associations all need to play a role if tourism development is to be successful. They can contribute to a better organized tourism industry but, in addition, they give ordinary people a greater stake and more influence in the success of the local tourism industry.

In a sense, one of the major lessons that can be learned from the Old South and the West is that tourism development does not occur in a vacuum. Although some island societies are, almost inevitably, heavily reliant on tourism for development, the economic benefits (and, in all probability, the minimum of social costs) are found in societies where tourism has been integrated into other economic sectors or, indeed, where it has been an offshoot of development in other spheres.

'Social problems' can arise from tourism

In the Old South, tourism development has wrought evident changes in the physical landscape and has brought considerable economic benefits, notably jobs, income and foreign exchange. Such development has involved migration to tourist 'honeypots', a closer integration into the world system, the introduction or further spread of a cash economy, the commoditization of ritual and (sometimes) the confrontation of 'tradition' with Western values (Harrison, 1992a). Indeed, it is the social 'problems' arising from tourism development that have attracted most attention and persuaded many critics (from within and outside tourist-receiving societies) that the social costs of tourism outweigh the economic benefit. They point, for instance, to the abandonment of established values by the young, and their replacement by 'immoral' Western patterns of behaviour, and to the loss of authority by 'traditional' groups. Are there lessons here for Eastern Europe?

The answer to this question is 'Yes, perhaps'. It has to be remembered that the contrast between the culture of 'host' and 'visitor' is possibly greater in the Old South than in much of Eastern Europe. As a consequence, there was perhaps a greater risk of conflict when tourism was first introduced. In addition, although Eastern Europe is now adopting the principles of a free market economy, commoditization has long been established. A market economy, albeit one distorted by a period of central control, is not new to Eastern Europe. Nevertheless, it has to be recognized that tourism, as part of an overall process of modernization, exposes the young to new values and encourages resentment in traditional groups and among those who feel deprived of a slice of the tourist cake.

Just as the economic impact of tourism may vary over time, so, too, will its social impact. As a consequence, attitudes to tourism as an industry and to individual tourists may change drastically. From research carried out in the Caribbean and Canada, Doxey concluded that attitudes to tourists change from euphoria, at the initial stage of development, to apathy, as tourism becomes more commercial, then to annoyance when a community's identity appears to be threatened by a vast influx of tourists, finally ending with antagonism and outright hostility, when visitors are held responsible for all the problems being visited on the community (Murphy, 1985). As Murphy recognizes, Doxey's model 'suggests a unidirectional sequence' (Murphy, 1985), whereas much may depend on other factors, for example a region's level of development and the characteristics of visitors (Butler, 1975). In addition, the role of the wider political environment cannot be discounted. Doxey's first research in Barbados, for instance, followed periods of unrest in Jamaica and Trinidad and throughout the Caribbean there was heightened sensitivity to matters of race and politics. In such circumstances, (white) tourists among (black) hosts might have seemed particularly vulnerable. In fact, even in the early 1970s, most Barbadian residents interviewed were favourably disposed to tourists (Doxey and Associates, 1971). Furthermore, it might also be argued that, depending on steps taken to mollify local opinion, there may be an accommodation of tourists and hosts, with the latter exhibiting a more professional (or resigned) attitude to the former. As a consequence, there may be stages beyond that of antagonism even in a unidirectional model like that of Doxey. Indeed, this may well be the case for Barbados, which continues to be one of the foremost tourist-receiving parts of the Caribbean (Caribbean Tourism Organisation, 1990).

Nevertheless, social reactions to tourism in Eastern Europe may approximate to patterns in the Old South. In Varna, Bulgaria, in the early 1990s, there was reportedly resentment that, when residents of the town suffered cuts in electricity supply, the hotels were unaffected, and informants volunteered the information that tourists were no longer welcomed with the same openness as when the industry first commenced. In so far as such feelings are widespread (and this should be expected), there is a need, and it is clearly politic, to ensure that the benefits of tourism are shared across the resident population.

In addition, it has long been noted that, under communism, many Eastern Europeans went to considerable lengths to obtain scarce foreign currency. With a greater focus on the market and attempts to attract more tourists, especially from the West, an increase in consumerism, especially among the young, is distinctly possible. How far this – and the accompanying hostility to the old communist leadership – can be seen as analogous to a rejection of 'tradition' in the Old South is open to question, and the extent to which local 'traditions' in Eastern Europe are likely to be 'threatened' by an increase in tourism cannot be prophesied. Much will depend on the type and numbers of tourists involved and on the sociocultural background of the receiving region. However, where such conflict has been perceived in the Old South, for example where tourists have offended local sensibilities in matters of (un)dress, drunkenness, drug-taking and other forms of unacceptable behaviour, it has been open to the state to take action. Topless or nude sunbathing would soon be eliminated if offenders were rewarded with 2 or 3 days in jail. Of course, there would be an economic cost to pay with the instant decline of tourism, but that is a problem for the authorities – and further indicates that tourism policies need to be planned sensibly and implemented efficiently.

The desirability of community participation and accountability

Social conflict involving tourists arises from a variety of sources, only one cause of which may be a general or specific resentment at their presence. As Ryan makes clear (1993), their association with crime may be as coincidental victims, as temporary residents of a tourist region, as deliberate targets of criminal or terrorist activity, and as generators of demand for illegal services and/or goods. No set of remedial measures is likely to change this situation entirely but there is perhaps a case for involving as many members of the host society in tourism as possible. This suggestion is made not from some implicit ideological position on the nature of democracy and a preferred relationship of rulers and ruled, but rather because there is more chance of sustaining a viable tourism industry and viable communities if the communities it most affects are in sympathy with and benefit from tourism development. Indeed, perhaps the largest single cause of hostility to tourism is the perception that its benefits are restricted only to an elite. In such circumstances, unrest is a distinct possibility and there is ample evidence that tourism and socio-political unrest do not mix (Richter, 1992).

In the Old South, it is undoubtedly the case that some tourism development has been actively pursued in the face of considerable opposition from local groups. Many tourist-receiving societies (not only in LDCs) find it necessary to employ 'tourist police'. On the beaches of Tunisia, for example, tourists are protected not only from traders, the official explanation for their presence, but also from angry or over-interested Tunisians roused by the near-nakedness of many visitors. Indeed, objections to tourist 'immorality' are common throughout LDCs. In some, there are definite 'no go' areas which tourists enter only at their peril: this is so for Jamaica, where many beachfront hotels are heavily protected, and perhaps in Egypt, where the targeting of tourists by fundamentalist terrorists has reportedly led the state to provide escorts for tourists visiting prominent sites. The fact that similar measures are also required in Florida indicates that such problems are not restricted to LDCs, and at least one example in popular fiction suggests that tourists are not the only ones who need to take care in large North American cities (Wolfe, 1988).

The situation should not be exaggerated: despite well publicized objections to some aspects of tourism in the Old South, and the development of numerous pressure groups to bring about change, existing research suggests that local people most involved in tourism wish to see further promotion of the industry. Objections tend to come either from those who feel they (or others) are being denied some of the fruits of the industry or from those who consider the integrity of 'the nation' or 'tradition' is threatened by tourism development. Such complaints might be minimized if local communities are consulted early in the planning process and clearly have their wishes taken into account. Unfortunately, there are few examples from the Old South of where this has successfully occurred and virtually all of these come under the heading of 'alternative tourism'. In Indonesia, Bini Swada (Community Self-Reliance Development Agency), an NGO committed to encouraging local participation in development projects, has apparently been offering '... specific interest tours for openminded, well-informed and adventurous travellers both domestic and foreign' (Ismawan, 1992). An example of enlightened 'top-down' tourism development, this project is similar to the rural tourism projects pioneered in Lower Casamance, Senegal, where tourists live with Senegalese villagers for a few days in fairly basic accommodation units.

In practice, it is usually difficult to define which members of the community should be most involved in the 'participation' process. Considerable efforts seem to be under way in Bali to involve villagers in decisions over tourism development (Wall, 1993), but most calls for 'local involvement' tend to assume the existence of an acceptable and indigenous local democracy. Experience of community development programmes elsewhere, especially the ujamaa experiment in Tanzania (Hyden, 1980), suggests that the main beneficiaries tend to be relatively wealthy and male. A similar process is likely to occur in tourism development which focuses on closer involvement with community leaders. By contrast, less 'planned' schemes may (quite unintentionally) be more radical and egalitarian in their effects, as well as being more disruptive of the social structure (Harrison, 1992a). One wonders what would have happened in Malindi Old Town, Kenya, studied by Peake (Peake, 1989), for example, if local people had first been consulted about tourism: would the elders, whose authority was so reduced when the young gained jobs in hotels, have expressed themselves willing to entertain the prospect of tourism? And could or should concessions have been made to meet their objections (had they had the foresight to make them) in advance?

In this respect, Eastern Europe may hold lessons for the Old South. Home-stay tourism, where tourists rent rooms in private homes, has long been common in Bulgaria, Hungary and Poland (Harrison, 1993). In the past, this has been a way of dealing with highly seasonal demand, especially in the summer months, bringing into play a kind of 'reserve army' of rooms which, when the season is over, revert to family use. Although used as an adjunct of package tourism, and directed at tourists who are not likely to spend vast amounts of money, this kind of tourism leads to a wider and more egalitarian spread of income, and has fewer 'leakages' than other forms of tourism. Ironically, its survival in the face of recent changes in Eastern Europe is now under threat.

Summary

My argument is that although the Old South and Eastern Europe are historically distinct regions, in recent years there have been signs of convergence in their social, economic and political characteristics. Both regions are constituted by a wide range of less developed countries (LDCs), all of which are now situated in a world capitalist system. Within this system, international tourism, oriented primarily to a capitalist market and organized primarily on capitalist lines, generally developed later in Eastern Europe than in the South, and it is suggested that LDCs in the former region may learn from the experience of the latter, thus gleaning some of the advantages (and avoiding as many of the constraints as possible) from the 'late development effect' (Dore, 1972).

The first lesson is that tourism is best seen as a symptom rather than a cause of economic development. Very poor societies do not attract tourists and investment in infrastructure and international-standard accommodation and other facilities are prerequisites for a successful tourism industry.

Second, different types of tourism have different kinds of requirements and it is essential that tourism is planned rationally, in accordance with a region's natural and human resources. In this connection, the advice of economic consultants should be both sought and treated with some scepticism.

Third, mass tourism, which brings social costs as well as economic advantages, is not the only possible policy. Indeed, some societies in the Old South have deliberately aimed to attract only high-spenders, a policy which is clearly elitist, and other societies (e.g., in the Caribbean) cater to a 'mix' of visitors. A highly visible elite tourism enclave, especially where the overall standard of living is low, can create unfavourable comparisons, which are sometimes avoided (as in the Maldives) by creating tourist resorts away from the centres of population. The essential point, though, is that (depending on the receiving society) all tourism will have some social effects.

Fourth, it is essential that tourism development occurs within an overall framework provided and monitored by the state. This is not to suggest that the old communist regimes are resurrected, but that policy-making is the responsibility of politicians and should not be left, by default, to the market. Planning should occur with due regard to knowledge of the market and the receiving society, and its subsequent development controlled within limits set by the state.

Fifth, an efficient and relevant tourism industry does not occur in a vacuum. Instead, it requires a high level of supporting institutions, both within and outside the state. Unless it can be enforced, legislation is useless, and building regulations are irrelevant unless they are obeyed. Indeed, to be successful, tourism development should be integrated with other economic sectors, thus avoiding 'leakages' from the economy and bringing benefits to a wider range of the population.

Sixth, tourism inevitably has social consequences, some of which may be regarded as 'problems', even 'moral problems'. The distinctions are important: social consequences may include changes in social institutions, for example the family, and the creation of entirely new institutions, such as Tourist Boards, hotels and so on. Migration, too, may occur to tourist centres. This is likely to create practical problems for government, local authorities and residents, but these are not necessarily moral issues. However, depending on the perception of those involved, some changes may be considered to have a bearing on moral issues. To take an apparently trivial issue: when young people adopt tourists' fashions, this is a social consequence of tourism; if their elders object on moral grounds, it has been defined by them as a moral problem and, if they then create a social disturbance, it becomes a political problem for the forces of law and order. Whether or not tourism leads to such disturbances, of course, will depend on the type of tourism and the societies involved. However, one of the lessons from the South is that tourism has the *potential* of contributing substantially to social unrest, especially where there are considerable differences in culture and wealth between 'hosts' and 'guests'.

Finally, experiences in the South indicate a practical need to involve as wide a range as possible of members of tourist-receiving societies in tourism development. Just as local hostility to tourism can prevent its development, forms of tourism considered beneficial and appropriate to local people can prosper. Arguably, with a few poorly researched exceptions, in most of the Old South this lesson has been learned the hard way – by default. Is it too optimistic to hope that, in Eastern Europe, it will be recognized that tourism has social as well as economic impacts, and that these should be considered, as far as possible, in advance, and not in a rather painful retrospect?

References

Buckley, P.J. and Witt, S.F. (1990) Tourism in the centrally-planned economies of Europe. *Annals of Tourism Research* 17(1)7, 7–18. DOI: 10.1016/0160-7383(90)90111-4.

Butler, R.W. (1975) Tourism as a factor in national and regional development. Occasional Paper 4. Dept of Geography, Trent University, Ontario.

Butler, R.W. (1980) The concept of a tourist area cycle of evolution: implications for management of resources. *The Canadian Geographer/Le Géographe canadien* 24(1), 5–12. DOI: 10.1111/j.1541-0064.1980.tb00970.x.

Caribbean Tourism Organisation (1990) *Caribbean Tourism Statistical Report, 1989*. CTO, Christchurch, Barbados.

de Kadt, E.and (1992) Making the alternative sustainable: lessons from development for tourism. In: Smith, V.L. and Eadington, W.R. (eds) *Tourism Alternatives: Potentials and Problems in the Development of Tourism*. University of Pennsylvania Press, Philadelphia.

Dore, R. (1972) The late development effect. In: *IDS Communication 103*. Institute of Development Studies, Brighton.

Doxey, G.V., Associates (1971) *The Tourist Industry in Barbados: A Socio-Economic Assessment*. Dusco Graphics, Kitchener, Ontario, pp. 64–67.

Hall, D.R. (1991b) Eastern Europe and the Soviet Union: overcoming tourism constraints. *Hall Tourism and Economic Development*, 62–63.

Harrison, D. (1988) *The Sociology of Modernization and Development*. Unwin Hyman, London, p. 161.

Harrison, D. (1994) *The Sociology of Modernization*, pp. 167–175.

Harrison, D. (1992a) Tourism to less developed countries, 20–34.

Harrison, D. (1992b) Tourism to less developed countries, 26–27.

Harrison, D. (1993) Bulgarian tourism: a state of uncertainty. *Annals of Tourism Research* 20(3), 519–534.

Hyden, G. (1980) *Beyond Ujamaa in Tanzania*. Heinemann, London.

Inskeep, E. (1988) Tourism planning: an emerging specialization. *Journal of the American Planning Association* 54(3), 360–372. DOI: 10.1080/01944368808976497.

Ismawan, B. (1992) Participatory sustainable tourism for grassroots income generation, Indonesia. In: Eber, S. (ed.) *Beyond the Green Horizon*. Tourism Concern/World Wide Fund for Nature, London, pp. 46–48.

Killick, T. (1992) *A Reaction Too Far: Economic Theory and the Role of the State in Developing Countries*. Overseas Development Institute, London.

Killick, T. and Stevens, C. (1992) Lessons from the Third World. In: Stevens, C. and Kennan, J. (eds) *Reform in Eastern Europe and the Developing Country Dimension*. Overseas Development Institute, London, p. 23.

Levitt, K. and Gulati, I. (1970) Income effect of tourist spending: mystification multiplied: a critical comment on the Zinder report. *Social and Economic Studies* 19(3), 326–334.

Meyer-Arendt, K.J. (1985) The grand Isle, Louisiana resort cycle. *Annals of Tourism Research* 12(3), 449–465. DOI: 10.1016/0160-7383(85)90009-X.

Meyer, W. (1988) *Beyond the Mask*. Verlag Breitenbach, Saarbrücken.

Murphy, P.E. (1985) *Tourism: A Community Approach*. Methuen, London, pp. 124–125.

Outhwaite, W. (1992) Critical theory and modernization. In: Sogomonov, A. and Danilova, E. (eds) *The Social Critique of the Theory of Modernization*. British Sociological Club in Moscow, Moscow, p. 66.
Peake, R. (1989) Tourism and Swahili identity in Malindi old town, Kenyan coast. *Africa* 2, 209–220.
Pearce, D. (1992) *Tourist Organisations*. Longman, Harlow.
Richter, L.K. (1992) Political instability and tourism in the Third World. In: Harrison, D.. (ed.) *Tourism and the Less Developed Countries*, pp. 35–46.
Riggs, F.W. (1964) *Administration in Developing Countries: The Theory of Prismatic Society*. Houghton Mifflin, Boston, MA.
Robinson, A. (1993) Re-emerging—if slowly. In: *Financial Times Survey: Bulgaria*. Financial Times, London, p. 29.
Ryan, C. (1993) Crime, violence, terrorism and tourism: an accidental or intrinsic relationship? *Tourism Management* 14(3), 173–182.
Stanchev, P. (1993) *Tourism: niches of high return*. 6. Bulgarian Economic Review, Sofia.
Stevens, C. (1991) Eastern Europe and the developing countries. *ODI Briefing Paper* June, 2.
Tiryakian, E.A. (1991) Modernisation: Exhumetur in pace (rethinking Macrosociology in the 1990s). *International Sociology* 6(2), 165–180. DOI: 10.1177/026858091006002004.
Wall, G. (1993) International collaboration in the search for sustainable tourism in Bali, Indonesia. *Journal of Sustainable Tourism* 1(1), 38–47. DOI: 10.1080/09669589309450699.
Wilson, D. (1994) Unique by a thousand miles: Seychelles tourism revisited. *Annals of Tourism Research* 21(1), 20–45.
Wolfe, T. (1988) *The Bonfire of the Vanities*. Picador, London.
World Bank (1993) *World Development Report. Investing in Health*. Washington, DC, pp. 238–239.

8

The World Comes to Fiji: Who Communicates What, and to Whom?

Introduction

Fiji is comprised of more than 300 islands, of which 100 are inhabited. Situated in the South Pacific, almost 3000 km east of Australia and 1800 km north of New Zealand, it is often described as the 'gateway' to the Pacific. Along with its topographical variety and demographic diversity, Fiji is, indeed, a cultural 'potpourri', the result of patterns of migration going back hundreds, even thousands, of years. Initially, the movement was from Indonesia to New Guinea and the Solomons, later to Vanuatu and New Caledonia, with the first arrivals to Fiji, some 3000 years ago, probably coming from Vanuatu. Expansion continued further east, taking in Tonga, Niue, and Samoa, until the Marquesas were reached, about 2000 years ago, thus providing a further basis for even more migration to Hawai'i and southernmost Polynesia (Crocombe, 1989; Nunn, 1994).

The Fijian islands were central in this process of migration and were influenced by Melanesian and Polynesian cultures. Much later, European exploration and settlement added a new stratum to the diversity: missionaries and beachcombers, visiting traders and resident traders intermittently combined and splintered but nevertheless formed a distinct segment of Fijian society, relating in numerous and often contradictory ways to the Fijians to whom they preached, with whom they mated, and from whom they attempted to take land (France, 1969). From 1879, the situation was further complicated by the introduction in Fiji (as previously in Mauritius, Guyana, and Trinidad and Tobago) of East Indian indentured labourers, who were to comprise the third important cultural section within the region, initiating a series of problems in Fijian politics that, as yet, have not been resolved.

The world has been coming to Fiji for a long time. As a consequence, the population of Fiji is exceptionally diverse, with some 51% of its population of 770,000 considered indigenous Fijian, 44% Indian and 5% in the apparently vague, anomalous but nevertheless significant category of 'other'.

Fiji: A Plural Society

These migrations, especially those of the 19th and early 20th century, resulted from improvements in communications and the global influence of British colonialism. They led to the formation of a society that, at independence from Britain in 1970, was characterized by

Fisk (1970) as 'the three Fijis' – that is, Fijians, Europeans and Indians. In Furnivall's famous phrase, social and cultural sections in such 'plural' societies may come together in the market place but, in other respects, remain separate. 'They mix, but do not combine.'

This is not the place to defend the notion of 'plural society' in detail, and (in the context of the Caribbean in the 1960s and 1970s) it has often come under consistent and angry attack (Rubin, 1960; Despres, 1967; Morris, 1967; Cross, 1968). There are indeed difficulties in applying the concept to Caribbean societies, which are (and for a long time have been) characterized by a high degree of consensus, including a use of a common language, across ethnic lines. Arguably, though, such problems are markedly less in their application to Fiji.

Despite the use of English in commerce and education, indigenous Fijians and Indo-Fijians are separated by language and, even after more than a century of living together, few know more than a few words of each other's language. Language differences are compounded by religious divisions. Whereas most Indo-Fijians are Hindu or Moslem, the overwhelming majority of indigenous Fijians are Christian. At least in part, such factors contribute to the fact that, relative to the Caribbean, there is little mating and even less marriage across the ethnic divide. Miscegenation is rare, especially between indigenous Fijian and Indo-Fijian. Indeed, at the social level, there is an almost universal rule that members of the different ethnic groups socialize primarily with one another – a principle carried over to whites (both local and expatriate) and the amorphous category of 'other'. Such an unwritten rule can be seen in operation in most parts of Viti Levu, the largest and most populated island in the Fiji group. Whites have their own institutions and patterns of interaction (the Fiji Arts Club, wine tasting sessions, the Rucksack Club), their own drinking clubs (e.g. the Fiji Club and the Yacht Club), and indigenous Fijians have theirs (for upper-class Fijians, for example, the Union Club, with informal 'grog' sessions a standard institution throughout Fiji).

Politically, too, the major ethnic groups are largely represented by different parties, a situation now threatened (to an extent as yet unknown) by the partial acceptance of the present (largely Fijian) government of the major recommendations of the Fiji Constitution Review Commission (Reeves *et al.*, 1996). Indeed, the major divergence from the recommendations of the Commission has been the retention of a higher number of 'communal seats' than it initially recommended.

Ever since the introduction of the Indians into Fijian society, they have consistently and deliberately been denied full civil and political rights. In particular, they have been largely denied access to land, of which more than 80% is inalienably in the possession of Fijians, largely administered through the Native Land Trust Board. At the same time, while indigenous Fijians currently hold political power, a situation until recently ensured by successive constitutions, Indians dominate the formal economic sector and the demand for major products (including sugar and tourism) is clearly and irrevocably situated outside Fiji. One need not be a neo-Marxist to argue that, at least in one sense, the economy is governed from outside. Indeed, any focus on tourism rapidly reveals that the hotel sector and other parts of the tourism industry are dominated by owners and managers from overseas, and that the market for Fiji's key foreign currency earners, notably tourism and sugar, is largely determined by conditions overseas.

Exceptions to such a pattern of social and cultural pluralism are rare. The Fiji Club, for example, does have Indo-Fijian members, and some indigenous Fijians, and there is a sprinkling of whites at the Union Club. Other social institutions indicate a degree of inter-ethnic co-operation – the trades union movement (Norton, 1990), various aspects of the women's movement, and the University of the South Pacific, a regional institution, does have staff from all ethnic groups (albeit unequally at different levels of seniority). More informally, 'grog' sessions, where the ubiquitous yaqona is drunk, are not restricted to Fijian participation. In addition, personal experience, observation and anecdotal evidence suggest that links between resident whites (both Fijian-born and expatriate, especially males) and people in the 'other' category (especially females) are common. Among young people, too, of all ethnic groups, the clubs of Suva, especially', provide a meeting point where liaisons of at least a temporary nature can span the ethnic divide.

Research would undoubtedly suggest other examples of interaction across ethnic divisions, possibly more so in the rural than the urban areas. The fact remains, however, that any examination of the social, cultural, economic and legal structures of the Fiji islands provides strong evidence of a plural society, where there is an unequal and institutionalized access to rights in the public sphere (leading to structural pluralism), and where social and cultural practices are firmly segregated, thus entailing social and cultural pluralism between groups primarily identified (by their own members and others) through ethnic markers (cf. Smith, 1974). This is not to suggest that alternative forms of identification (e.g. class, gender, age or religion) are impossible, or that the situation is immutable. It *does* indicate, though, that the 'host' society of Fiji is characterized by marked and consistent divisions based on ethnicity, and that close communications between and across participating ethnic groups in Fiji are, at best, an exception rather than the rule.

Table 8.1. The growth in Fiji tourism: international arrivals for selected years

Year	No. arrivals
1961	14,722
1965	40,128
1970	110,042
1975	162,177
1981	187,939
1982	203,636
1983	191,616
1986	257,824
1987	189,866
1988	208,155
1989	250,565
1992	278,534
1993	287,462
1994	318,874
1995	318,495

The Visitors

It is into this plural society, founded through successive waves of migration, with its divisions, tensions, ethnic exclusiveness and inequalities, that a new, modern, form of migrant arrives: the tourist. Tourism is not new to Fiji. Indeed, the Fiji Visitor's Bureau was developed in the 1920s to show passengers changing ships in Suva something of Fiji and to encourage them to consider settling in the territory (Scott, 1970). However, like international tourism generally, Fijian tourism really developed after the Second World War, aided by wartime construction of Nadi airport and advances in aircraft technology and carrying capacity (Table 8.1).

Between 1961 and 1970, visitor arrivals increased by more than 600%, but from 1971 to 1980 the increase was less than 25%. In the early 1980s, progress was erratic but the industry was dealt a body blow in 1987 with two successive coups and, inevitably, visitor arrivals plummeted. By the mid-1990s, the situation had again improved but, by international standards, Fiji and other island states in the South Pacific were struggling to retain their market share. Whereas between 1986 and 1996 tourism arrivals in East Asia and the Pacific grew by 171% (WTO, 1990, 1997), for example, tourist arrivals in Fiji increased by only 31% (WTO, 1990; Bureau of Statistics, 1997, 22 October). Prospects for the industry were not helped by the dramatic collapse of some East Asian economies at the end of 1997 and although that year saw a rise in the numbers of tourists coming to Fiji, growth is unlikely to be sustained, even with the 20% devaluation of the Fijian dollar in January 1998.

Nevertheless, tourism in Fiji is the country's major export earner, contributing about 21% of GDP, an estimated 40,000 jobs (direct, indirect and induced) in a total workforce of 260,000 and a formal sector of approximately 100,000.

For the record, and subject to all the usual riders about the unreliability of tourism statistics, if current figures are any indication, some 80% of these arrivals are likely to be in Fiji for holiday purposes (Bureau of Statistics, 1997, 22 October). More to the point, though, it is perhaps stating the obvious to note that visitors to Fiji, a 'traditional' or 'less developed' country, overwhelmingly come from societies that are, in general, more 'developed', more 'modern', and more 'Western' (Table 8.2). Indeed, Australia, New

Table 8.2. Visitor arrivals to Fiji, 1997: by country of residence and per cent of all arrivals (Bureau of Statistics, 1997).

Area	Number	% of arrivals
Australia	80,351	22.4
New Zealand	68,116	19.0
USA	44,376	12.3
Canada	13,359	3.7
UK	35,019	9.7
Continental Europe	32,806	9.1
Japan	44,783	12.5
Taiwan	567	0.1
Malaysia	317	0.1
South Korea	12,181	3.4
Rest of Asia	5461	1.6
Pacific Islands	20,381	5.7
Others	1724	0.4
Total	359,441	100.0

Source: Bureau of Statistics (1998, 26 January).

Zealand, North America, Europe, Japan and South Korea account for more than 90% of all international arrivals to Fiji. As modernization perspectives have consistently emphasized (most recently, Rojek and Urry, 1997), such tourists bring not only their passports and foreign exchange, but also their cultures.

Without probing the complex issue of tourist motivations too far, visitors to Fiji are persuaded to come, at least according to the Fiji Visitors' Bureau, because of what they see in brochures and are told by travel agents (43%), what they read in travel books (25%) and what they hear from friends or relatives who have already visited Fiji (14%) (Fiji Visitors' Bureau [FVB], 1995).

But what are they told? What are the messages that travel agents and their advertising material convey? What is the 'language of tourism' (cf. Dann, 1996) that is directed at visitors to the 'plural society' of Fiji where, according to a recent survey, some 25% of the population lives under the poverty line? (Government of Fiji and United Nations Development Programme, 1997). For reasons detailed elsewhere (Harrison, 1997a), it cannot be assumed that potential tourists believe everything they are told by travel agents, but it is nevertheless worth examining the content of tourism promotional material, where there is an *intention to* communicate some kind of message about, or an image of, the destination area.

Example 1: 'Fiji Islands: the way the world should be'

This is a recent promotional leaflet issued by the Fiji Visitors' Bureau, and several fairly obvious points can be made about it:

The majority of the pictures are of 'nature': flowers, waterfalls, lagoons, an atoll and coral. Most pictures of people are of Fijians, mainly in 'traditional' clothing, engaged in 'traditional' activities (yaqona ceremony, showing mats, etc.). There are also pictures of tourists, either with Fijians (on the *bill bili*, a raft) or together, as a couple. There are no Indo-Fijians in the pictures.

The text is also instructive. Fiji is described as: 'the world's friendliest paradise,' 'a beautiful dream and yet a reality,' 'a nature lover's paradise,' 'a land of age-old rituals,' 'With such a richness of cultures living in harmony, the hospitality of our people is legendary.' (There is also a mention of the Hibiscus Festival and *Diwali*, the Hindu Festival of Light.)

Exercises in content analysis are standard enough these days (Selwyn, 1993; Dann, 1996), but it is still necessary to make the point that such images, conveyed through tourism promotional literature, are a far cry from the social reality of the destination area. To suggest that Fiji is a role model for the rest of the world, 'with such a richness of cultures living in harmony', is surely enough to make the claimant, at least in local eyes, instantly certifiable.

Example 2: Fiji islands official travel guide

Issued by the Fiji Visitors' Bureau in October 1997, this Guide is distributed to travel agents throughout the world. Not surprisingly, perhaps, it largely reinforces the message of the previous example. Eighty-seven pages of pictures and information about the Fiji group are packed with pictures and details of the Fiji islands, its attraction and its accommodation. Nevertheless:

1. Among the dozens of pictures and hundreds of faces, only *one* is Indian.
2. Where the built environment is presented, it is nearly always the hotels.
3. Where Fijians are depicted, they are in 'traditional' dress, and their homes are the 'traditional' thatched bures that, in fact, are now highly unusual in Viti Levu.

The Hosts

It might be argued that even if visitors to Fiji are provided with a highly selective image of Fiji before their arrival, they will soon develop a more rounded view of 'reality' once they reach their destination. This is unlikely to be the case.

First, most hotel accommodation in the Fiji group is away from centres of population. Indeed, Suva, the capital, contains only 16% of all hotel rooms (which are primarily for business visitors, as Suva is not a highly attractive tourist destination). Most rooms are in the SW of Viti Levu on the Coral Coast (19%) and in Nadi and the NW Mamanuca and Yasawa islands (42%) (Deloitte & Touche, Ministry of Transport and Tourism, and Tourism Council of the South Pacific, 1997).

Second, opportunities to meet Fijians outside of the hotel context are few and far between. Physical distance from centres of population is reinforced by hotel design and by the relative exclusivity of indigenous Fijian villages. Some larger and more 'up-market' hotels (e.g. in the Denerau complex near Nadi, in the west of Viti Levu) already form something of a tourist ghetto, and current plans to extend such developments will simply reinforce this tendency. By contrast, hotels situated in rural areas, along the Coral Coast or in the Mamanucas and the Yasawas, may be closer to Fijian villages but (very reasonably, perhaps) access to these villages is strictly controlled through Fijian custom, which requires a formal welcoming ceremony (*sevusevu*) involving the presentation of yaqona to the chief. In general, those tourists who visit villages will do so on an excursion organized through one of the hotels or tour operators, stay a relatively short period in the village (sometimes much less than half an hour) and make minimal social contact with its residents.

Third, the hotels themselves are centres of a lifestyle that, in many respects, is still foreign to most residents of Fiji, whether indigenous Fijian or Indo-Fijian. Exact figures are not available, but an examination of the 74 members of the Fiji Hotel Association (as listed in August, 1997) indicates that no less than 80% of their membership (the bulk of all hotels in the Fiji group) is either foreign owned or foreign managed. In effect, they are running establishments where the food, the beverages, the accommodation, even sometimes the plumbing, and certainly the entire ethos, are quite foreign to the average resident of Fiji. As a consequence, Fijians employed by the hotels do not simply have to be trained in the necessary job-related skills; they also have to be *socialized* into a completely new environment.

Fourth, whether the hotels are relatively large and up-market (e.g. the Sheraton) or small but expensive island resorts (e.g. Turtle Island in the Yasawas), the frontline staff coming into immediate contact with the guests are indigenous Fijian. Although they contribute to operations behind the scenes, to all intents and purposes Indo-Fijians are generally conspicuous by their absence. By contrast, indigenous Fijians are accorded a prominent profile. They are presented in every respect to the guests as the providers of hospitality and, indeed, there appears much in Fijian culture that accords with this aim. At its best, the tourism product in Fiji can appear quite remarkable. In particular, guests at some of the small, island resorts, which set out to promote an intimate, family atmosphere, develop great attachments to the Fijian staff. Time and again, for instance, departing guests refer to one such resort, in the Yasawas, as 'magical', and 'wonderful', a 'paradise', and one departing guest, a New Zealander, said of the island, 'this is Fiji'. In fact, it is owned by a US citizen, has no indigenous village, and provides an international cuisine, owing little to Fijian cooking, for its guests. Its 'Fijianness' comes from its physical situation, its appearance as a stereotypical coral atoll, and from its indigenous Fijian staff.

Fifth, the marginalization of Indo-Fijians extends beyond the hotels to the absence of their culture from the scheduled 'attractions' on offer to visiting tourists. *Diwali* (the Hindu Festival of Light) is accorded a passing mention in the Fiji

Visitor's Bureau brochure, and there is a reference to Indo-Fijian firewalking ceremonies in Fiji's official travel guide (FVB, 1997) but they are not promoted as tourist attractions. Whether or not this neglect is to the liking of the Indo-Fijian population is debatable; the fact remains that *their* culture, their 'exoticism', is rarely communicated to visitors. After all, everyone knows that Fiji is all about Fijians!

Finally, although many links between guests and their indigenous Fijian 'hosts' are formed, and are often characterized by great mutual affection, they are not symmetrical. Fijians enjoy meeting visitors and learning about their societies but, as indicated below, the communication tends to be one-way. At the same time, as in similar resorts elsewhere, the physical situation and organization of many island resorts in Fiji tend to ensure that the closest social interaction occurs between and among guests, rather than between the 'guests' and their official 'hosts'.

Inward and Outward Acculturation

Clearly, tourism is not the only agent of acculturation in Fiji or the rest of the South Pacific. Processes of globalization in Fiji can be traced (at least) to the beginning of European settlement (Harrison, 1997b). 'Traditional' Fijian religion (Methodism) and dress are, in fact, introductions from outside, as was Indian migration and all that it entailed. Elements of globalization are also seen in urban centres in the region (to which people are increasingly migrating), where some facilities, at least, are similar to those on offer in Australia or New Zealand. Suva, for example, has numerous clubs (at least half a dozen within 3 minutes walking distance of one another) where the offerings are at least as modern, as deafening and (arguably) as tuneless as anything to be found elsewhere. The city has several cinemas, including a Village 6 complex, where the latest films from Hollywood and Bollywood can be seen, often before they go on general release elsewhere. On the smaller screen, over the last few years Fiji has moved from having no TV station to three, with Sky TV offering a daily diet of world sport and North American sitcoms, to accompany those from New Zealand and Australia found on the 'local' channel. In addition, Fijian citizens, especially Indo-Fijians, have a long history of emigration from the area and residents receive regular visits from friends and relatives overseas, a category of tourist largely ignored by promoters of tourism, despite the fact that they stay, on average, for 18 days, as opposed to the 8.3 days of other holidaymakers (Bureau of Statistics, 1997, 22 October). The University, too, is an example of what might be called a Western 'intrusion', where a global culture is promoted as a necessary accompaniment to formal coursework in a range of subjects matching that available in many universities in developed societies, and other higher education institutions offer their own varieties of modernity. In fact, much 'human resource development' inevitably entails the preparation of (some of) Fiji's youth for their place in the formal, wage-earning sector of an economy where most economically active men and women are still in subsistence agriculture.

Despite the difficulty of separating the role of tourism from other 'change agents', however, it can at least exacerbate the acculturation process. The acculturation of members of the 'host' society (inward acculturation), for example, may occur through the introduction of capitalistic concepts of ownership of land and other 'natural' resources, and through modifying elements of 'traditional' Fijian culture. By contrast, there are possibilities of what might be referred to as 'outward acculturation', whereby tourists learn something of Fiji and Fijians and (perhaps with less likelihood) incorporate these experiences into their own lifestyles.

Inward Acculturation

Land disputes

The issue of access and ownership of land in Fiji has long been the subject of much debate (cf. Norton, 1990). In 1941, with the formation of the Native Lands Trust Board (NLTB), conflict between indigenous Fijians and Indo-Fijians was removed from the local scene and centralized. The NLTB administers native-held land, which is more than 80% of the area of Fiji, on behalf of local patrilineages, known as *mataqali*, which it is empowered to lease (but not to sell) on behalf of the owners (Norton, 1990). Despite this centralization of administration, however, land disputes involving local owners, lessees and the NLTB are common, and there is considerable evidence many Fijians, at least, are often reluctant to accept the curtailment of access to 'their land', even though leases have been signed to this effect.

Current regulations about land ownership, and the system of leasing native land, affect tourism in several ways. First, they affect investment decisions about the siting of hotels. Fiji is unique as a tropical island destination in having an extensive hotel stock that is close to the international point of arrival but that is not by a beach. This has been exacerbated by investment decisions based more on land ownership and control than on conventional criteria, which suggest that no hotel or resort should be more than 1.5 hours from the international point of entry. Pacific Harbour is well outside this distance and is also in the wetter zone of Viti Levu. It saw early development largely because it was freehold land (Deloitte & Touche, Ministry of Transport and Tourism, and Tourism Council of the South Pacific, 1997).

Second, the structure of land ownership influences the employment practices of hotels constructed on land leased from the NLTB. Fijians from the landowning units have priority for jobs provided by the hotel – a stipulation that undoubtedly reinforces the existing tendency to employ indigenous Fijians in frontline tourism positions.

Third, there are numerous examples of disputes between landowners and local landowning units over access to the leased land, and others where hotel owners are innocent but suffering bystanders in conflicts between the NLTB and the landowners, or between *mataqali* with competing claims to the land (cf. Harrison, 1997b). Indeed, arguments sometimes centre on who rules the waves! One dispute, for instance, has been simmering, and occasionally boiling over, for months over who should take surfers to a reef off the Coral Coast. Because two villages claim the area is part of their traditional fishing grounds, they *also* claim to be the only ones entitled to carry surfers to the area. While they are (not unnaturally) supported by an upmarket hotel providing them with tourists (and a per capita payment), the claim is disputed by another, cheaper surfers' hotel, which employs boatmen from a *different* village. As a result, there have sometimes been pitched battles on the high seas, involving boatmen from the warring villages and surfers, who do not take kindly to being ordered out of the water! (Baba and Rubin, 1996, 30 November)

Such problems are not new, and they predate the onset of tourism. Problems over access to land also emerge with alternative forms of development (e.g. commercial logging). However, when hotel operations involve the use of a resource that is allocated a cash value, when previously it had none, the process of commoditization is enhanced. And insofar as local landowners (or officials in the Ministry of Fijian Affairs) prefer interpretations based on precolonial custom, it is legitimate to refer to 'incomplete globalization' or 'incomplete commoditization'.

Demonstration effects

The phenomenon of 'demonstration effects', especially among the young, is a matter of common concern and is frequently raised by critics of international tourism (Harrison, 1992). Once again, it is difficult to show that tourism per se causes changes in values and behaviour, given pressures of 'modernity' from other sources. It is even more difficult to argue objectively that such changes are for the worse. However, there is an apparent tendency for young Fijians, especially, to adopt more casual Western dress, develop a taste for alcohol and cigarettes, chew gum and generally indicate a liking for 'Western' cultural icons.

It is not only the young who are affected by the various processes of globalization. Throughout Fiji, for example, there was considerable grief at the death of Princess Diana, and much discussion of the numerous conspiracy theories (which is unlikely to prevent an invitation to the Queen to once again become head of state in Fiji!).

How far all this amounts to 'cultural pollution', and is due to tourism, can be debated. Certainly, hotel staff in most resorts are encouraged to talk to the guests, and (in the interests of hospitality) to learn their interests and backgrounds. In such circumstances, it would be surprising if acculturation did not occur. At the same time, there is evidence that dress codes in Fijian villages generally are, in practice, being relaxed to accommodate tourists. Formally, at least, in Fijian culture women are not supposed to bare their legs, and some guidebooks to Fiji specifically encourage men and women in shorts to cover up with a sulu when visiting villages. In fact, this injunction is frequently ignored, and without comment from the Fijian hosts. It is also the case that such ceremonies, which can often take hours when tourists are not involved, are greatly speeded up to accommodate successive groups of tourist arrivals.

If this dilution of some aspects of indigenous Fijian culture is, in fact, occurring, it is likely to affect the authority of the chiefs. Although this is still considerable, some signs indicate that more universalistic criteria are being applied to them. The new constitution, for example, was accepted despite the reservations of many members of the Great Council of Chiefs, and informants indicate that the status of a chief will be considerably enhanced if he (or sometimes she) has the benefit of a modern education.

Clearly, these issues need to be investigated further. However, insofar as the culture of indigenous Fijians is under pressure to change, from tourism and other 'change agents', there is little evidence that *Indo-Fijian* culture is facing a similar challenge. This leads to an interesting paradox, which also merits further research: by virtue of their role as tourist 'attractions', through their possession of land which may be leased to hotel owners, and through their employment in the hotel sector, indigenous Fijians are being socialized into a global culture faster than Indo-Fijians, whose relative marginalization from the tourism industry, both as 'attractions' and employees, serves to distance their culture from the onslaught of globalization.

Outward Acculturation

Acculturation, however, can be a two-way process, and tourism is frequently put forward as a means of increasing international understanding. This leads to the question of how much tourists are able to learn of Fijian culture. At present, the answer seems to be that their exposure to it (and thus their learning experience) is relatively little. First, as indicated earlier, village visits are generally brief, 'one-off' excursions. Second, when they do occur, there is little attempt, if any, to explain Fijian culture, the social and economic organization, the lineage (*mataqali*) system or the system of agriculture. Third, as a general rule, villagers are not trained to interpret their own culture to visitors.

This situation need not remain, and there are some indications that government is beginning to understand how 'ecotourism' could bring both economic and other benefits to many villages in Fiji. Indeed, it can be conservatively estimated that with 74 members of the Fiji Hotel Association, numerous inbound tour operators, and several other agencies, well over 100 villages will be receiving visits from tourists on a regular basis. After all, a visit to a Fijian village is an attraction commonly offered by hotels. In addition, there are currently rather vague plans by one inbound tour operator to develop 'home-stay' tourism in more than 80 villages, which, if implemented, will have a considerable knock-on effect throughout Fiji. At present, knowledge of, and policy towards, ecotourism in Fiji is somewhat confused (Harrison and Brandt, 1997), but the government has recently accepted guidelines for the development of village-based tourism in Fiji (Harrison, 1997c, October) and, if these are followed, the next few years could see the development of a more consistent, aware and equal exchange between tourists from the 'developed' world and their indigenous Fijian hosts.

Conclusions

It has been suggested that Fiji is a society characterized by structural, social and cultural pluralism, which inevitably conditions communications across as well as within cultural sections. Over the last 35 years, international tourism has become increasingly important in the Fijian economy, and one result of this has been the promulgation in tourism promotion literature of a highly romanticized image of Fiji and Fijians, who are invariably seen as *indigenous*

Fijians, with the corresponding marginalization of other ethnic groups, especially Indo-Fijians. Despite such images, however, most tourists stay in hotels owned and/or run by expatriates, who preside over an environment that is predominantly non-Fijian, albeit one where the frontline staff, who deal directly with guests, are indigenous Fijians.

In the process, there is the possibility that indigenous Fijians, through their ownership of land and their privileged position in obtaining hotel employment, might be more socialized into global culture than Indo-Fijians, who have no privileged access to hotel employment and whose culture is not portrayed as an 'attraction' for tourists.

How resistant indigenous Fijian culture will be to the pressures suggested here remains to be seen. A sensible, well-planned and controlled development of 'ecotourism' would facilitate a greater, and more equal, cultural interchange across the host and guest divide, enable indigenous Fijians to develop skills in interpreting their own culture, and give their guests a more worthwhile experience of life in Fiji. In this context, as in any other, 'ecotourism' must be considered an ideal and, like other ideals, not fully attainable. However, it is undoubtedly the case that more could be done to present indigenous Fijian culture, in all its richness, to tourists who currently receive, at best, a glimpse of it. If Indo-Fijians were also involved in the tourism industry to a greater extent than they are, at present, the 'tourist product' of Fiji would be immeasurably enhanced. However, this would entail a greater recognition of their role in Fijian society than either those who promote Fiji tourism, or Fijians themselves, may care to admit. The divisions within plural societies are indeed difficult to remove. For the foreseeable future, it is likely to be generally the case that indigenous Fijians and Indo-Fijians will continue to mix but not to combine.

References

Baba, P. and Rubin, V. (1996, 30 November) Battle rages over cloud-break. *Fiji Times* Weekend Section, 2.

Bureau of Statistics (1997, 22 October) *Statistical News*. No. 42.

Bureau of Statistics (1998, 26 January) *Statistical News*. No. 2.

Crocombe, R. (1989) *The South Pacific: An introduction*, 2nd edn. University of the South Pacific, Suva.

Cross, M. (1968) Cultural pluralism and sociological theory: a critique and re-evaluation. *Social and Economic Studies* 17(4).

Dann, G. (1996) *The Language of Tourism*. CAB International, Wallingford.

Deloitte & Touche, Ministry of Transport and Tourism, and Tourism Council of the South Pacific (1997) *Fiji: Tourism Development Plan*. Author, Suva.

Despres, L. (1967) *Cultural Pluralism and Nationalist Politics in British Guiana*. Rand McNally, New York.

Fisk, E.K. (1970) *The Political Economy of Independent Fiji*. Australian National University Press, Canberra.

France, P. (1969) *The Charter of the Land: Custom and Colonisation in Fiji*. Oxford University Press, Melbourne.

Government of Fiji and United Nations Development Programme (1997) *Fiji Poverty Report: A Summary*. Author, Suva.

Harrison, D. (1992) The social consequences of tourism. In: Harrison, D. (ed.) *Tourism and the Less Developed Countries*. Unwin Hyman, Chichester, pp. 19–34.

Harrison, D. (1997a) Barbados or Luton? Which way to paradise? *Tourism Management* 18(6), 393–398. DOI: 10.1016/S0261-5177(97)00031-9.

Harrison, D. (1997b) Globalisation and tourism: some themes from Fiji. In: Oppermann, M. (ed.) *Pacific Rim Tourism*. CAB International, Wallingford.

Harrison, D. (1997c, October) *Ecotourism and Village-based Tourism: A Policy and Strategy for Fiji*. Ministry of Tourism, Transport and Civil Aviation.

Harrison, D. and Brandt, J. (1997) Ecotourism in Fiji: Making sense of the muddle. *Paper presented at the XIIth Inter-Congress of the Pacific Science Association*, University of the South Pacific, Suva, 13–19 July, 1997.

Morris, H.S. (1967) Some aspects of the concept plural society. *Man* 2, 169–184. DOI: 10.2307/2799485.

Norton, R. (1990) *Race and Politics in Fiji*, 2nd edn. University of Queensland Press, St. Lucia, Queensland.
Nunn, P. (1994) *Environmental Change and Early Settlement in Pacific Islands*, Environment series, East-West Center Working Papers, No. 39.
Reeves, P.., Vakatora, T.R. and Lai, B.V. (1996) *Towards a United Future: Report of the Fiji Constitution Commission*, Parliament of Fiji Paper No. 34 of 1996. Government Printer, Suva.
Rojek, C. and Urry, J. (1997) Transformations of travel and theory. In: Rojek, C. and Urry, J. (eds) *Touring Cultures: Transformations of Travel and Theory*. Routledge, London, pp. 1–19.
Rubin, V. (1960) Social and cultural pluralism in the Caribbean. *Annals of the New York Academy of Sciences* 83.
Scott, R.J. (1970) The development of tourism in Fiji since 1923. *Transactions and Proceedings of the Fiji Society* 12, 40–50.
Selwyn, T. (1993) Peter pan in south-east Asia: views from the brochures. In: Hitchcock, M., King, V.T. and Pamwell, M.J.G. (eds) *Tourism in South-East Asia*. Routledge, London, pp. 117–137.
Smith, M.G. (1974) *Corporations and Society*. Duckworth, London.
WTO (1990) *Compendium of Tourism Statistics 1985–1989*. Author, Madrid.
WTO (1997) *Tourism highlights, 1996*. Author, Madrid.

9

Islands, Image and Tourism

Many islands are hugely popular tourist destinations, and islands regularly feature in tourism promotion. Such a high profile is not new, for tropical islands, especially, have figured prominently in Western culture for centuries, and have long been regarded with ambivalence. They have been variously regarded as Edens of bounteous plenty, or mystical and ominous unknowns, peopled by natives of childlike innocence and simplicity, or by savages of barely disguised ferocity and barbarism. Colonial exploration reinforced such images, and from the 16th century a genre of island literature emerged, in fiction and non-fiction. Works by such writers as Defoe, Ballantyne and Stevenson again reinforced these perceptions and crucially influenced the way islands and their inhabitants were perceived by children and adults. In the 20th century, as mass international tourism developed, the feature film, the guidebook and the tourist brochure helped establish many islands as popular holiday destinations, again evoking a series of established images that seem much at variance with the multiple realities experienced by island residents.

Even to the casual observer, it should be clear that islands, especially tropical islands, are hugely popular tourism destinations. They seem to be the ideal holiday venue: warm and sensuous, 'away from it all' providing stressed-out visitors the much-needed opportunity to relax, escape, recharge their batteries, and generally appreciate a way of life that, sadly, has been lost in the too-busy commercial environment of the globalizing, post-industrial Western world.

The evidence is there in the travel statistics. By 1991, small island states and dependencies were receiving more than 21 million international tourists. '4.6% of all international stayovers, [and] receipts from international tourism in 1991 accounted for at least 15% of gross national product (GNP) in no less than 15 of the 31 small island states' (Weaver, 1998, p. 49).

Dependence on tourism tends to be most marked in islands and small states that possess few natural resources, and the very features that prevent them from diversifying their economies – their climate and topography – often attract tourists. In 1995, the 25 countries where international tourism made the greatest percentage contribution to Gross National Product were all islands, as indicated in Table 9.1, and contributions ranged from 83% in the Maldives to 21% in Cyprus (World Tourism Organization, 1998, p. 71).

The Uses and Images of Islands

Mass tourism has come to islands and micro-states relatively recently. Colonial island possessions were usually developed to supply such

Table 9.1. International tourism receipts as a percentage of the GNP: The Top 25, 1995

	% of GNP	Rank		% of GNP	Rank
Maldives	83.3	1	Jamaica	33.8	14
Anguilla	80.3	2	St. Kitts & Nevis	31.1	15
Antigua, Barbuda	72.3	3	Gibraltar	29.8	16
British Virgin Isles	69.5	4	Bermuda	29.6	17
US Virgin Isles	60.8	5	Montserrat	28.0	18
Cook Islands	56.3	6	Netherlands Antilles	27.9	19
St. Lucia	51.0	7	Martinique	27.4	20
Macau	42.0	8	Vanuatu	27.2	21
Guam	41.7	9	St. Vincent/Grenadines	23.2	22
Bahamas	41.6	10	Grenada	21.7	23
Aruba	41.0	11	Seychelles	20.9	24
Barbados	40.4	12	Cyprus	20.8	25
Guadeloupe	38.2	13			

Source: *World Tourism Organization*, 1998, p. 71.

export crops as tobacco, sugar and cocoa to the metropoles, frequently involving the enforced movement and enslavement of African populations and, later, indentured labour from China and South Asia. Other islands, for example Elba, St. Helena and, more recently, Robben Island (Deacon, 1995), have been places of imprisonment and exile, as has the Fijian atoll of Nukulau, which served as a quarantine and processing station for Indian indentured labourers brought to Fiji for work in the sugar plantations (Naidu, 1980, pp. 10–11). More recently, it has been used to isolate rebels who overthrew the elected government of Fiji in May 2000.

Islands and their native populations have been colonized, settled, and used to produce crops for export. They have been scenes of bondage and enslavement, of great suffering and cruelty, of poverty and pain, and they have been used to isolate the sick, the dangerous, the rebel and the dissident. They have also been used as experimental targets and venues for testing weapons of mass destruction (Dibblin, 1988; Teaiwa, 1994; Hussein, 1996; Adams, 2001), and – as in Puerto Rico – to gauge the effectiveness of medicines and contraceptives before their introduction in the West (Hanson, 1962, pp. 56–57). Unsurprisingly, perhaps, many islanders aim to escape the vagaries of their island troubles, the limited opportunities and the pettiness of communities riven by gossip, by emigrating to a favoured metropole.

Despite these harsher realities of island life, islands remain popular destinations for holidaymakers. Relatively little research has been carried out on *why* this is so, but such features as physical isolation and cultural and climatic differences are frequently cited. These characteristics, along with a sense of adventure in making the journey and the slower pace of island existence, may indeed appeal to the modern traveller (Butler, 1993; Baum, 1997, p. 5; Lockhart, 1997). However, it is unlikely that so prosaic a list of attractions can explain the mystique and significance that islands have had for so long in Western culture. Even today, they conjure up conflicting images and stereotypes, ranging from that of romantic and idyllic havens, offering escape and redemption from a busy world, to the dangerous and untamed unknown, where climate and hostile natives combine to threaten the hapless visitor.

In fact, such images are not new. They predate Western colonialism, reaching as far back as the literature of ancient Greece, Rome and even Southern India. Greek writers, for instance, idealized 'idyllic and lush conditions of the Isles of the Hesperides ... at a time when some of the Greek islands were still much better forested than the mainland' (Grove, 1995, p. 21), and also imagined other distant, mythical landscapes, apparently situated in India (Grove, 1995, p. 21). As described by Grove, 'the search for Eden' was

> a phenomenon whose roots lay in a complex mixture of European, Arabic and Indian philosophical traditions. With the Renaissance and the growth of European maritime expansion, this search was sharpened by the pursuit of gold and profits, the settlement of distant island colonies, the garnering of botanical knowledge and, increasingly, the lure of India and China and their riches (1995, pp. 23–24).

Sir Thomas More's *Utopia* was an island. Its fictional inhabitants are introduced by More (in 1516) through the report of an equally fictional sailor (More, 1975, first published 1516) and accepted as real by many who hungered for details of the newly discovered regions. Throughout the Tudor period, voyages of exploration led to further colonization and settlement, and public fascination with islands grew apace, even though the peoples of the New World came to be regarded with considerably more ambivalence than More's sensible and socialistic Utopians. Indeed, by the time Shakespeare wrote *The Tempest*, in about 1610, there was much scepticism about the potential of native islanders to follow a Utopian way of life. While both Marx (1964, pp. 34–72) and Grove (1995, pp. 32–38) analyse the play in detail, their overriding interpretation is similar: Prospero's island is both a metaphor for the Americas, conceptions of which may alternate from the lushness of a new Eden to that of 'a howling desert' (Marx, 1964, p. 43), and 'the meeting place between indigenous inhabitant and European colonist' (Grove, 1995, p. 34). Again, emphasis is on the tension between the image of islands as locations of Utopia and, by contrast, places of great insecurity and danger. As Grove notes, 'the tropical island was, at a basic level, not a known environment' (1995, p. 37). And in general, whether stranded on magical islands or in the Forest of Arden, Shakespeare's leading characters returned with alacrity to the city and the comforts of civilization as soon as they were given the chance to go home.

Grove suggests that as colonialism progressed, expeditions increasingly included surgeons. These independent scientists, whose medical interests focused on native flora and fauna, were responsible for the development of botanical gardens in metropoles and the colonies and, through their links with the academies and scientific societies of Europe, helped diffuse information on and theories about the new colonies, in so doing, they provided the basis for what Grove describes as early forms of environmentalism. Indeed, by the end of the 18th century, islands had taken on an even wider significance:

> In many respects the isolated oceanic island, like the frail ships on the great scientific circumnavigations of the seventeenth and eighteenth centuries, directly stimulated the emergence of a detached self-consciousness and a critical view of European origins and behaviour, of the kind dramatically prefigured by Daniel Defoe in *Robinson Crusoe*. Thus the island easily became, in practical environmental as well as mental terms, an easily conceived allegory of a whole world. Contemporary observations of the ecological demise of islands were easily converted into premonitions of environmental destruction on a more global scale (Grove, 1995, pp. 8–9).

Mention of *Robinson Crusoe* (Defoe, 1994, first published 1719, first published 1719), which was vastly influential throughout the 18th and nineteenth centuries (Shinagel, 1994), serves to shift attention to the rich vein of fiction dealing with those who, by accident or design, found themselves colonists of tropical islands. Interestingly, although Crusoe's island was situated in the Caribbean, the book was based on an account of a sailor shipwrecked in the South Pacific (Day, 1987), and islands of both regions figured prominently in 19th and 20th century fiction (as well as in 20th century travel writing). As Loxley has indicated, *Robinson Crusoe* (itself a re-writing of *The Tempest*) was the forerunner of a genre of island literature, the collective works which came to serve, for children and adults, as 'the dream-texts of English colonialism' (1990, p. 11).

Special consideration should perhaps be accorded books about (and read by) children (Loxley, 1990, pp. 73–126). *The Coral Island* (Ballantyne, 1995, first published 1858, first published 1858) is a prime example. Again, the tensions of island life are indicated at the beginning of the book, where seamen report

> thousands of beautiful fertile islands that had been formed by a small creature called the

> coral insect, where summer reigned nearly all the year-round; where the trees were laden with a constant harvest of luxuriant fruit; where the climate was almost perpetually delightful; yet where, strange to say, men were wild, bloodthirsty savages, excepting in those favoured isles to which the gospel of Our Saviour had been conveyed (Ballantyne, 1995, first published 1858).

One child much impressed with Ballantyne's story was young Robert Louis Stevenson (Rennie, 1998), whose own fascination with islands was evident by the age of twelve, continued with his novel *Treasure Island* (Stevenson, 1994, first published 1883, first published 1883) and ended with his death in Samoa in 1894. *Treasure Island*, a swashbuckling tale of piracy, intrigue and treachery, was first serialized in a magazine for young people and remains a children's classic to this day. According to one commentator:

> The island itself is described with such particularity that it is a significant contribution to the 'desert island' myth which has haunted English literature since the publication of *Robinson Crusoe*... It has the fascination of all such islands; remote, mysterious, virginal, isolated from human contact (Hammond, 1984, p. 107).

The influence of works such as *The Coral Island* and *Treasure Island*, and other works within the island genre, is difficult to assess but it is likely that they prompted many young readers to travel overseas. 'Asterisk', author of letters from the South Pacific published as *Isles of Illusion* (1986, first published 1923), was clearly influenced by Stevenson, as were the friends to whom the letters were addressed (Lynch, 1986, first published 1923, p. xiii) and Gavin Young, who introduces a more recent publication of the letters:

> For my grandmother's attic in Flexbury Lodge was full of dusty, exciting books of adventure and travel, and it was my reading of these books by Captain Marryat, C.A. Henty, Stevenson, Melville and Conrad – coupled with the proximity of the Atlantic at the bottom of the road – that inspired in me a dream of one day Running Away to Sea (1986, p. ix).

Another indication of the extent of Ballantyne's influence is Golding's *Lord of the Flies* (1960, first published 1954), a latter-day version of *The Coral Island*. However, there are crucial differences in emphasis. Ballantyne's central characters, Ralph, Jack and Peterkin, remain close friends in a kind of boy-scout companionship from the beginning, and evil is introduced into their lives by other men, either island 'savages' or pirates, with redemption quite literally coming from missionaries. By contrast, in Golding's story, which focuses on a group of boys whose plane has crashed on a deserted island, (another) Jack and Ralph are featured, but now on opposite sides. The veneer of order and civilization quickly disintegrates into murder and mayhem, characterized by savagery and evil that emanate from *within* the boys themselves. Redemption, although again from external agencies, comes this time in the Leviathan-like guise of a British naval officer, whose uncomprehending comments to this 'pack of British boys' reveal the irony of the situation and the link with Ballantyne: 'Jolly good show. Like the Coral Island' (Golding, 1960, first published 1954, p. 192).

By the early years of the 20th century, travel writing was developing into a separate literary genre, albeit one that was still closely associated with colonialism. Travel writers varied considerably in their perception of island life. Some, like Froude (1888) firmly believed island peoples were incapable of governing themselves; others focused on voodoo, *obeah* and the *magical* nature of island existence, as in Haiti (e.g. Seabrook, 1929), and yet others were utterly disillusioned at the disjunction of the *images* of islands (often obtained from writers such as Stevenson, Marryat and Melville) with their harsh reality ('Asterisk', 1986, first published 1923, first published 1923, pp. 78, 137). Nevertheless, the popular view persisted that, at least until the onset of colonialism, tropical islanders were like children, poor but happy, living languidly, exotically, erotically, sometimes violently, in circumstances which, in effect, endowed them with the tropical equivalents of milk and honey. Such views of island society were strongly gendered, for most travel writers were male, and published accounts of male islanders are less likely to be eroticized. However, as Jolly points out, Beatrice Grimshaw readily sexualized *Polynesian* women in Tahiti, reserving very different judgements for Melanesians (Grimshaw, 1907, pp. 22–29; Jolly, 1997) – a perspective likely to be shared, more often than not, by male commentators (Young, 1986). Bali was similarly exoticized and eroticized by Western writers, and during the 1920s and 1930s it hosted its share of anthropologists, artists, and tourists (Hitchcock and Norris, 1995, p. 12), including many homosexuals, who reportedly believed 'they had found their earthly paradise in Bali' (Hitchcock and Norris, 1995, p. 29).

In many respects, descriptions of island life in the South Pacific and the Caribbean were similar, and were often repeated in educational texts for children. Young readers were informed, for example, that

> Jamaica has a great future before it, and only needs energetic, capable men with capital to regain its old position. Nine-tenths of the people are coloured, either black or yellow. They are very contented, and very lazy, though if forced to work they will do so cheerfully enough. The island is so fertile that no man need ever be hungry, a mud-covered wattle hut suffices for shelter, and little clothing is necessary. The native is a good husband and father. He is not very truthful, but he is very good-humoured, and makes a pleasant and agreeable servant (Parrott, 1908, p. 181).

Later literary figures were also fascinated by islands, especially in the Pacific (Day, 1987, pp. 1–12), and exercised an influence, which was reciprocated, on such artists as Gaugun (Day, 1987, pp. 188–194). However, at the start of the 20th century, travellers of a somewhat different orientation were also publicizing island life, for studies of island communities were to feature strongly in the development of social anthropology. Interestingly, Malinowski's path-breaking fieldwork among the Trobriand islanders was carried out only because he was, in effect, exiled there for the duration of the First World War (Malinowski, 1922; Firth, 1957, p. 3). Radcliffe-Brown, another stalwart in the history of social anthropology, had already been to the Andaman Islands (Radcliffe-Brown, 1922), and Mead went to Samoa (Mead, 1943, first published 1928, first published 1928), to describe an intrinsic casualness of social life (1943, pp. 159–160) that was later strongly challenged (Freeman, 1984). In addition, Firth studied the Tikopia (Firth, 1936) and Sahlins was later to focus on island life in Polynesia (Sahlins, 1958, 1962). This is a roll call (which could be substantially expanded) of some of the most influential social anthropologists of the 20th century and it can be no coincidence that so many focused on island societies. However, what motivated them *personally*, as opposed to academically, to concentrate on islands is but one of many questions that would repay further research. How far were they, too, influenced by travellers' tales, and the myths of the island paradise?

Although sociological and anthropological approaches are now firmly embedded within social science, links with travellers' tales remain. The hugely popular books of Paul Theroux and Bill Bryson, for instance, reveal considerable sociological insights and, by contrast, some anthropological studies have had an extensive popular appeal (Turnbull, 1961; Castenada, 1968). At the same time, non-scientific accounts of the vagaries of anthropolgical fieldwork have sometimes become best-sellers (Bowen, 1964; Barley, 1983).

Islands and Tourism

It is argued here, then, that even before the advent of mass international tourism, Western society had long been exposed to the myths and legends of island societies. And when tourism did get under way, in the period after the Second World War, very different 'texts' emerged to reinforce the images that had already been established. A key development was the production of the feature film, the 'movie', that brought the rest of the world in glorious technicolour to the cinema screens of Europe and North America. While exotic destinations generally were prominent – the Thailand of *The King and I*, for example (Lang, 1956) – islands continued to have a special appeal. This was so for comedy, as in *The Road to Bali* (Walker, 1952) and, especially, in the story of wartime romance, most notably in the musical *South Pacific* (Logan, 1958), which was based on Michener's Pulitzer prize-winning *Tales of the South Pacific* (Michener, 1973, first published 1946, first published 1946). More recent variants on the same theme include *Mediterraneo* (Salvatores, 1991) and *Captain Morelli's Mandolin* (Madden, 2001). Other films, repeating themes from the past, include *Blue Lagoon* (Kleiser, 1980) and *Cast Away* (Zemeckis, 2000) and are much used in advertising the locations at which they were filmed. Equally interesting, if less popular, are films that use island locations as setting for stories of horror and suspense, again bringing out the dark side of island life. A topic associated with islands is that of the beach, which figures prominently in recent social scientific literature (Shields, 1991; Corbin, 1994, pp. 250–281) and novels (Garland, 1996). Indeed, the beach embrace of Deborah Kerr and Burt Lancaster in *From Here to Eternity* achieved notoriety in the early 1950s (Zinnemann, 1953) and Garland's novel led to a major feature film for Leonardo de Caprio (Boyle, 1998).

Travel literature has been further supplemented by the increasing popularity of the tourist guidebook and, especially, holiday brochures. As Cohen (1982) makes clear, the latter have perpetuated myths about island 'paradises' and utopias, especially in the Polynesian islands, painting pictures for potential visitors which, at least for many islanders, stand in stark contrast to their perception of their own existence. Others have come to similar conclusions (Selwyn, 1993; Dann, 1996, p. 1993; Harrison, 1998).

The extent to which such texts influence choices of holiday destinations is difficult to establish, but the continued appeal of islands, especially tropical islands, in Western culture is undeniable. They are promoted in radio programmes and on television. They provide the erotic, sunny, sultry, sensuous backdrop in advertisements for a vast range of products, from chocolates to alcoholic drinks, from shampoos to fashion goods, and trips to them are given as valued prizes in quiz programmes and sundry other competitions. That the tourist *image* of tropical island life is different from that of most islanders almost goes without saying, but such images have a remarkable tenacity, and often continue to be held by repeat tourists who have visited islands on many previous occasions.

While the tourism and advertising industries busily re-create island images for tourists and other Western consumers, the very boundaries which are the defining feature of islands are becoming increasingly irrelevant. Even by the end of the 19th century, colonial expansion and the ensuing globalization had ensured that most islands were already incorporated into the world social and economic system. More recent developments, for instance modern electronic communications, 'out sourcing' and migration, have been little constrained by strips of ocean, however wide. The irony is that while island status, for practical purposes, is no longer meaningful, the *myths* of islands and their peoples remain as vibrant and potent as ever.

References

Adams, D. (2001) Kennedy held in Puerto Rico bomb exercise protest. *The Times* 30 April,12.

'Asterisk' (1986, first published 1923) *Isles of Illusion*. Century, London.

Ballantyne, R.M. (1995, first published 1858) *The Coral Island*. Penguin, London.

Barley, N. (1983) *The Innocent Anthropologist*. Penguin, London.

Baum, T. (1997) The fascination of islands: a tourist perspective. In: Lockhart, D.G. and Drakakis-Smith, D. (eds) *Island Tourism: Trends and Prospects*. Pinter Press, London and New York, pp. 21–35.

Bowen, E.S. (1964) *Return to Laughter: An Anthropological Novel*. Doubleday, New York.

Boyle, D. (1998) *The Beach*.

Butler, R. (1993) Tourism development in small islands: past influences and future directions. In: Lockhart, D.G., Drakakis-Smith, D. and Schembri, J. (eds) *The Development Process in Small Island States*. Routledge, London, pp. 71–91.

Castenada, C. (1968) *The Teaching of Don Juan: A Yaqui Way of Knowledge*. Ballantine Books, New York.

Cohen, E. (1982) The Pacific islands – from utopian myth to consumer product: the disenchantment of Paradise. *Centre des Hautes Etudes Touristiques, Serie B., 27*, Aix-en-Provence.

Corbin, A. (1994) *The Lure of the Sea*. Polity Press, Cambridge.

Dann, G. (1996) Images of destination peoples in brochures. In: Butler, R. and Hinch, T. (eds) *Tourism and Indigenous Peoples*. Thomson, London, pp. 349–375.

Day, A.G. (1987) *Mad About Islands*. Mutual Publishing, Honolulu.

Deacon, H. (ed.) (1995) *A History of Robben Island*. David Philip, Cape Town.
Defoe, D. (1994, first published 1719) *Robinson Crusoe*. Penguin, Harmondsworth.
Dibblin, J. (1988) *Day of Two Suns: US Nuclear Testing and the Pacific Islanders*. Virago, London.
Firth, R. (1936) *We the Tikopia: A Sociological Study of Kinship in Primitive Polynesia*. George Allen and Unwin, London.
Firth, R. (1957) Introduction: Malinowski as scientist and as man. In: Firth, R. (ed.) *Man and Culture: An Evaluation of the Work of Bronislaw Malinowski*. Routledge and Kegan Paul, London, pp. 1–14.
Freeman, D. (1984) *Margaret Mead and Samoa: The Making and Unmaking of an Anthropological Myth*. Penguin, Harmondsworth.
Froude, J. (1888) *The English in the West Indies*. Longmans, Green and Co, London.
Garland, A. (1996) *The Beach*. Viking, London.
Golding, W. (1960, first published 1954) *Lord of the Flies*. Penguin, Harmondsworth.
Grimshaw, B. (1907) *In the Strange South Seas*. Hutchinson, London.
Grove, R.H. (1995) *Green Imperialism: Colonial Expansion, Tropical Island Edens and the Origins of Environmentalism, 1600-1860*. Cambridge University Press, Cambridge.
Hammond, J.R. (1984) *A Robert Louis Stevenson Companion: A Guide to the Novels, Essays and Short Stories*. Macmillan, London.
Hanson, E.P. (1962) *Puerto Rico: Ally for Progress*. Van Nostrand, New Jersey.
Harrison, D. (1998) The world comes to Fiji: who communicates with whom? *Tourism, Culture and Communication* 1(2), 129–138.
Hitchcock, M. and Norris, L. (1995) *Bali: The Imaginary Museum: The Photographs of Walter Spies and Beryl de Zoete*. Oxford University Press, Oxford.
Hussein, B. (1996) The end of French fallout. *Pacific Islands Monthly* 66(10), 17–19.
Jolly, M. (1997) From Point Venue to Bali Ha'i: eroticism and exoticism in representations of the Pacific. In: Manderson, L. and Jolly, M. (eds) *Sites of Desire, Economies of Pleasure*. University of Chicago Press, Chicago, pp. 99–122.
Kleiser, R. (1980) *The Blue Lagoon*.
Lang, W. (1956) *The King and I*.
Lockhart, D.G. (1997) Islands and tourism: an overview. In: Lockhart, D.G. and Drakakis-Smith, D. (eds) *Island Tourism: Trends and Prospects*. Pinter Press, London and New York, pp. 3–20.
Logan, J. (1958) *South Pacific*(dir.).
Loxley, D. (1990) *Problematic Shores: The Literature of Islands*. St. Martin's Press, New York.
Lynch, B. (1986, first published 1923) *Foreword to 'Asterisk,' Isles of Illusion*. Century, London, pp. vi–0.
Madden, J. (2001) *Captain Corelli's Mandolin*.
Malinowski, B. (1922) *Argonauts of the Western Pacific*. Routledge, London.
Marx, L. (1964) *The Machine in the Garden: Technology and the Pastoral Ideal in America*. Oxford University Press, London and Oxford.
Mead, M. (1943, first published 1928) *Growing Up in Modern Samoa*. Penguin, Harmondsworth.
Michener, J. (1973, first published 1946) *Tales of the South Pacific*. Fawcett Crest, New York.
More, T. (1975, first published 1516) *Utopia*, Translated and edited by Robert M. Adams. W.W. Norton and Company, New York.
Naidu, V. (1980) *The Violence of Indenture in Fiji*, Fiji Monograph Series No. 3. World University Service in Association with the School of Social and Economic Development, University of the South Pacific, Suva, Fiji.
Parrott, J.E. (1908) *Britain Overseas: The Empire in Picture and Story. The Young Folk's Book*. Vol. 1. Nelson, London.
Radcliffe-Brown, A.R. (1922) *The Andaman Islanders*. Cambridge University Press, Cambridge.
Rennie, N. (1998) Introduction. In: Stevenson, R.L. (ed.) *In the South Seas*, edited with an introduction and notes by Neil Rennie. Penguin, London, pp. viii–0.
Sahlins, M. (1958) *Social Stratification in Polynesia*. University of Washington Press, Seattle.
Sahlins, M. (1962) *Moala Culture and Nature on a Fijian Island*. University of Michigan Press, Ann Arbor.
Salvatores, G. (1991) *Mediterraneo*.
Seabrook, W.B. (1929) *The Magic Island*. Harrap, London.
Selwyn, T. (1993) Peter Pan in south-east Asia. In: Hitchcock, M., King, V. and Parnwell, M.J.G. (eds) *Tourism in South-East Asia*. Routledge, London, pp. 116–137.
Shields, R. (1991) *Places on the Margin: Alternative Geographies of Modernity*. Routledge, London.
Shinagel, M. (1994) *Preface to D. Defoe, Robinson Crusoe*, 2nd edn. Norton and Company, New York.

Stevenson, R.L. (1994, first published 1883) *Treasure Island*. Penguin, London.
Teaiwa, T. (1994) Bikinis and other s/pacific n/oceans. *The Contemporary Pacific* 6(1), 87–109.
Turnbull, C. (1961) *The Forest People*. Chatto and Windus, London.
Walker, H. (1952) *The Road to Bali*.
Weaver, D. (1998) *Ecotourism in the Less Developed World*. CAB International, Wallingford.
World Tourism Organization (1998) *Tourism Economic Report*. WTO, Madrid.
Young, G. (1986) Introduction in 'Asterisk'. In: *Isles of Illusion*. Century, London, pp. v–x.
Zemeckis, R. (2000) *Cast Away*.
Zinnemann, F. (1953) *From Here to Eternity*.

10

Tourism in Pacific Islands

For almost two millennia, tropical islands have been portrayed in Western literature as latter-day examples of the Garden of Eden (Loxley, 1990; Grove, 1995; Harrison, 2001a). By the 19th century, the stereotype was well established, and Melville's mid-19th century account of his stay in the Marquesas, for example, describes 'softly swelling plains ... delicious groves ... and ... purling brooks, and the entire country but little elevated above the surrounding ocean' (Melville, 1996, first published 1846, p. 21).

Like all stereotypes, that of the tropical Pacific island paradise contains some truth but much inaccuracy. Melville's suggestion, especially in the first edition of *Typee*, that colonialism had brought more harm than good to the region (Melville, 1996, first published 1846, p. 153) was but one contribution to an apparently endless (and usually simplistic) debate over the 'fatal impact' of colonial expansion on the more or less 'noble' savage of the Pacific islands (Moorhead, 1966, pp. 3–85; Lal, 1994, pp. 439–455; Harrison, 2001a). The tension between inherent nobility and savagery continues in today's literature: tourism brochures focus on the warm hospitality of friendly islanders and encourage tourists to 'discover paradise', while press reports and development agencies paint a harsher picture of poverty, domestic and political unrest, law and order problems and racial discrimination.

In fact, Pacific islands vary considerably in almost every imaginable dimension. Coral atolls are common in Polynesia, where a total population of some half a million is widely dispersed, but the more populous Melanesian islands tend to be volcanic and more densely populated, and two-thirds of their six million people are in Papua New Guinea. Even within Pacific island national societies, there are considerable differences. The relative ethnic homogeneity of Tonga, for example, contrasts with the fascinating but troubled ethnic mix that so dominates Fiji's politics. It is thus not surprising that generalizing about Pacific islands is fraught with difficulty (Berno and Douglas, 1998, pp. 65–68).

Nevertheless, some crude generalizations are possible. First, external influences have been crucial, especially the introduction of Christianity and colonialism. By the end of the 19th century, 'only one Pacific island group, Tonga, was able to retain a semblance of its traditional sovereignty – but as a British Protectorate' (Lal, 1994, p. 438). All the great colonial powers have been involved in the region (along with some that are not so great), and few islands can now be regarded as truly independent.

Second, the impacts of colonialism on Pacific island life varied across time and territory. Some indigenous inhabitants paid a dreadful price for colonial expansion, especially in Australia, New Zealand and Hawai'i. Elsewhere, populations and cultures were more resilient or the invasive forces less rapacious. In Melanesian Papua New Guinea, Vanuatu and Solomon Islands, in Fiji (which effectively straddles the Melanesia/Polynesia divide), and in Polynesian Samoa and the Cook Islands, for example, large proportions of land are still communally owned. By contrast, in French Polynesia and New Caledonia the land of indigenous people was generally commoditized during the process of colonialism. This is not to say that communal ownership was unproblematic. Governors generally regarded as among the most enlightened have been criticized for wittingly or unwittingly misrepresenting traditional institutions and it is arguable that communal patterns of land tenure have often acted as a brake on 'rational' exploitation of natural resources (Lal, 1994, pp. 440–443).

Third, once colonial expansion was consolidated in the region, first through 'beachcombers' and then missionaries, traders and settlers (Ralston, 1978), 'development' (in so far as it occurred at all) focused primarily on such export crops as copra and sugar. In Fiji, the perceived need for labour on sugar plantations was so great that, following his experience in Trinidad, Governor Sir Arthur Gordon introduced indentured labourers from India, who were imported in great numbers from 1879 until 1920. In stark contrast to the policies he followed in Trinidad (Wood, 1968, pp. 272–280), Gordon ensured that most land remained in indigenous communal hands, thus *also* laying down a legacy of bitter conflict for the post-independence period (Akram-Lodhi, 2000; Ratuva, 2000).

Fourth, colonial links and geographical position ensured that Pacific islands were caught up in wars they themselves did not instigate. The Second World War brought intense confrontation, bloody battles, and hundreds of thousands of service personnel to the region, along with ports, airports, systems of communication and entertainment, and Western material goods, the like of which most islanders had never seen (Laracy and White, 1988; Lal, 1994, pp. 445–446; Laracy, 1994; Douglas, 1996c, pp. 27–28; Grant, 1996, pp. 78–79). One result was that existing 'cargo cults' in Melanesia were reinforced and others emerged (Worsley, 1968; Kaplan, 1995).

After 1945, the United States, Britain and France all tested nuclear devices in the region, notably in the Marshall Islands, especially Bikini atoll, the (then) Gilbert and Ellice Islands Colony (now the separate countries of Kiribati and Tuvalu), and Moruroa, in French Polynesia (Vidal, 1995; Johnson, 1999; Narsey, 1999; Niedenthal, 2001, pp. 2–13, 92–93). Damage of a different kind occurred through the exploitation of mineral resources, and phosphate mining on Ocean Island (Banaba, now part of Kiribati) made the island uninhabitable and forced its population to move to Fiji. Phosphates are also mined in Nauru, Pelau and French Polynesia, nickel in New Caledonia, and gold in Papua New Guinea and Fiji (Lal, 1994, p. 444).

Finally, despite (some would say because of) the interests of the great powers, Pacific island states have few development options. This is irrespective of their current political affiliation, which ranges from independence (as in Fiji, most of Polynesia and all of Melanesia except New Caledonia) to full statehood (Hawai'i). In between there are numerous forms of association with 'big' powers, including the USA (American Samoa, the Federated States of Micronesia, the Marshall Islands, Palau, Guam, the Northern Marianas), France ('French' Polynesia, the Wallis and Futuna Islands and New Caledonia) and New Zealand (Cook Islands, Tokelau and Niue). Unfavourably placed in the world trading system, most remain MIRAB economies, living on the proceeds of migration, remittances, overseas aid and employment in government bureaucracies (Bertram and Watters, 1985). Migration and remittances, in particular, have been crucial (Berno and Douglas, 1998, p. 67), as is overseas aid, especially from Australia and New Zealand. However, none have prevented periodic crises and virtual bankruptcy in some small states, particularly where the desire to create employment in the public sector has sometimes far outstripped prudent financial management.

Some states have employed novel methods to obtain foreign exchange, often in combination with a mixture of sheer opportunism and good fortune. Tuvalu earns money by licensing the sale of websites with the suffix *tv* (Waqa, 1999), the Federated States of Micronesia markets the domain name of *fm*, and Niue, Tuvalu, Nauru and Tokelau have all obtained income from licensing telephone lines for international phone sex services. Offshore banking facilities, too, have been developed in Nauru, the Cook Islands, Niue, Vanuatu and the Marshall Islands, allegedly making their governments unwitting accomplices in money-laundering (Keith-Reid, 1999, 2000a, p. 38; Henley, 2001), while Kiribati has dabbled in the sale of passports, and seeks to provide Japan ground facilities for its space programme (New industry for Kiribati, 1998, p. 11). Most bizarrely of all, perhaps, in September 2001 Nauru was 'persuaded' by Australia to establish a detention centre for several hundred asylum seekers and refugees seeking entry to Australia, in return for fuel, generators and other economic benefits (Barkham, 2001).

Tourism in Pacific Islands

Faced with limited development opportunities, most island states in the Pacific have welcomed tourism. Hotels of a kind were first made available to accommodate early traders and settlers, and by the 1930s several steam ships were operating in the South Pacific (Douglas, 1996a, pp. 39–67). In Fiji, up-market hotels catered for passengers changing ships, and the Suva Tourist Board (1923) grew out of the White Settlement League, which was founded to persuade temporary visitors of the economic opportunities awaiting them in Fiji (Scott, 1970, p. 1). With the Second World War came servicemen and women. Roads and airports were constructed, and their presence encouraged the sale of souvenirs, as well as mass recreation and sexual services (de Burlo, 1989, pp. 307–310; Bailey and Farber, 1992, pp. 35–61; Douglas, 1996c, pp. 27–28). However, it was the post-war improvements in aircraft technology, enabling aircraft to travel farther, faster, more economically and with more passengers, that really prompted the development of Pacific island tourism. By the end of the 20th century, tourism was the major industry for many Pacific islands, and a priority for many more.

The 'attractions' of Pacific islands are well known. The climate, especially in Polynesia, is ideal for those seeking sun, sea and sand and, as indicated earlier, the lasting (if inaccurate) composite image of coral atolls, waving palms, white sand beaches, balmy breezes and (of course) gentle, exotic, erotic and welcoming people, is well established in Western thought (Harrison, 2001a). It is consistently reinforced in tourism brochures, which also emphasize the slow, unhurried pace of island life. For those dissatisfied with mere sun and sand, the night life of Papeete (Tahiti) or Honolulu (Hawai'i) provides urban excitement; and Melanesia, especially, offers the ethnic and cultural diversity of 'cultural' tourism, overlaid with the frisson of meeting and photographing modern 'savages' descended from cannibals (Douglas, 1996a). Alternatively, there are ample opportunities to experience the natural attractions of marine life, or the tragic beauty of the wreckage (and carnage) of the Second World War, all neatly packaged for 'niche markets'.

The type of tourism found in Pacific islands is contingent on numerous factors, including the attractions on offer, distance, ease and cost of access from tourist-generating markets, their economic buoyancy, and general perceptions of health, safety and 'development' in the destination area. On a continuum, with mass tourism (quite rare in the Pacific) at one extreme and no tourism at the other, the former is represented by Hawai'i, with 1.2 million residents and more than 6 million visitors a year (WTO, 2001a). Tuvalu (population 10,000) is close to the latter end and receives about 1000 tourists annually, mainly on business or to visit friends and relatives. It attracts only about 200 non-Tuvaluan holidaymakers, because it is difficult to reach, has few recreational facilities, no quality accommodation or restaurants, and a highly irregular air service (TCSP, 1992, 1997a, pp. 18–47; WTO, 2001a).

The constraints to tourism development in Pacific islands generally have been discussed at length by others (Britton, 1987a; Fletcher and Snee, 1989; Milne, 1992; Pearce, 1995; Fagence, 1997; TCSP, 1997b; Ryan, 2001). They include limited natural resources, small size, poor accessibility and inadequate air links, especially where the domestic market is small or non-existent, an absence of skilled personnel and local capital, and hotel operations that are relatively small in scale. In many islands, too, the communal ownership of land is considered to hinder the rational development of tourism. All of this is exacerbated by out-migration, a process whereby younger and relatively well-educated people further their careers elsewhere. As a consequence, there is a reliance on international capital, *especially* where there is marked tourism development. In the Pacific, the more developed the tourism industry, and the more it caters for high-spending tourists or for tourists in great numbers, the more likely it is to be owned and operated by overseas interests.

A further constraint is political upheaval and domestic unrest. Over the last two decades the image of 'paradise' in the Pacific has undoubtedly been tarnished. In 1987 and 2000, for instance, coups in Fiji removed elected governments allegedly dominated by Indians, and in January 1998 riots occurred in Vanuatu (Decloitre, 1998). Papua New Guinea has seen a decade of corruption and inter-ethnic conflict, with secessionist rebels from Bougainville opposing the government, until the signing of a peace accord in April 1998 began the resolution process. Ethnic conflict, still unresolved, also surfaced in Solomon Islands in the late 1990s, culminating in a coup in July, 2000 and the threat of total breakdown of the national state, prompting 'friendly' Australian intervention in 2003 (Robie, 1989, pp. 198–253; Field, 1998; Anon, 1999; Keith-Reid, 2000b; Keith-Reid, 2000c). The first years of the new millennium suggest that such problems are not yet over. More generally, too, the terrorist attacks in the USA in September 2001 were immediately followed by a radical decline in air travel, and destination areas reliant on visitors from the USA, including parts of the Pacific, were especially affected. Not for the first time, Pacific island tourism was crucially influenced by external factors far beyond local control. The invasion of Iraq in 2003, and generalized post-9/11 'terrorism jitters', have done nothing to improve the situation.

The Pattern of Pacific Tourism

The general pattern of international tourist arrivals in the Pacific has been discussed at length elsewhere (Harrison, 2003) and (subject to the usual caveats about the problems of collecting tourism statistics) is broadly indicated in Table 10.1. Mass tourism is primarily associated with islands in the North Pacific most closely linked to the USA. Until 2001, Hawai'i, the most popular, was consistently attracting nearly seven million tourists a year. A third of its visitors are from East Asia (especially Japan), and nearly all the remainder are from North America (WTO, 2001a), which explains the reduction in arrivals in 2001. By contrast, visitors to Guam and the Northern Marianas are primarily from East Asia and the Pacific (89% and 95% respectively) (WTO, 2002, p. 79 and WTO, 2002).

In the South and Central Pacific, mass tourism has not yet developed, but there are concentrated facilities in Fiji and French Polynesia, which together account for some 50% of the region's arrivals. Most visitors to French Polynesia are from the Americas or Europe (especially France) and, until the bankruptcy of the leading tour operator (following the 11 September 2001 terrorist attacks in the USA) an increasing number flew to Tahiti to transfer to cruise vessels. By contrast, most tourists to Fiji (and many other South Pacific islands) are from New Zealand, Australia or the USA. In neither case, however, is the urban-based tourism of Hawai'i much in evidence, and tourist accommodation tends to be away from local centres of population.

Fiji's tourism development would undoubtedly have been maintained in 2000 had the May coup not occurred. However, the political uncertainty it generated led to a catastrophic decline in arrivals of nearly 30%. Ironically, the industry recouped some losses in 2001, despite a downturn in US visitors, and that year's figures paint a misleadingly successful picture. Fiji's loss was others' gain. In 2000, the Cook Islands received a record 73,000 arrivals, putting pressure on its limited stock of accommodation, and there were increased arrivals in French Polynesia, Samoa, Vanuatu and (despite problems with the national airline) Tonga. Unrest in Solomon Islands – yet to be established as a tourist destination – led arrivals to decline to less than 4000, and conflict and corruption in Papua New Guinea also restrict growth in that country's tourism industry.

Like Tuvalu, the small island states of Kiribati and Niue attract only a few thousand tourists a year, who nevertheless make an important

Table 10.1. Tourist arrivals in the Pacific, selected years, selected islands: 1990–2001[1].

Year	1990	1995	1996	1997	1998	1999	2000	2001	%change 1990–2000	%change 2000–2001
Cook Islands	33,882	48,500	48,819	49,964	48,628	55,599	72,994	74,575	115.4	2.2
Fiji	278,996	318,495	339,560	359,441	371,342	409,955	294,070	348,014	5.4	18.3
French Polynesia	132,361	172,129	163,774	180,440	188,933	210,800	240,450	227,658	81.7	–5.3
Kiribati	3040	3318	3940	4925	5450	4695	4578	4831	50.1	0.2
New Caledonia	85,213	86,256	91,121	105,137	103,835	103,352	109,587	100,515	28.6	–8.3
Niue	649	2161	1522	2041	1736	1778	2010	2069	209.7	2.9
Papua New Guinea	40,742	32,578	61,215	66,143	67,465	67,357	58,429	54,280	43.4	–7.1
Solomon Islands	9195	11,795	11,217	15,894	13,229	17,395	5753	3418	–37.4	–40.6
Tonga	20,919	29,520	26,642	26,162	27,132	30,883	37,694	32,386	80.2	–14.1
Tuvalu	671	920	1039	1029	1077	770	1504	976	124.1	–35.1
Vanuatu	35,042	43,721	46,123	49,605	52,100	50,746	57,360	53,203	63.7	–7.2
Samoa	47,642	67,954	73,155	67,960	77,926	85,124	87,688	88,263	84.1	0.7
Marshall Islands	4856	5500	6116	6254	5727	4622	5246		8.0	-
TOTAL	693,208	822,847	874,273	931,849	964,680	1,043,076	977,363	990,188	41.0	1.3
American Samoa	(?)26,000	49,060	41,050	29,997	41,201	39,749				
Federated States of Micronesia				28,000	27,000	28,000	33,000			
Guam	780,000	1,362,000	1,363,000	1,362,000	1,137,000	1,162,000	1,288,000		65.0	
Hawai‘i	6,971,000	6,629,000	6,829,000	6,876,000	6,738,000	6,741,037	6,948,595	6,303,791	–0.3	–9.3

Continued

Table 10.1. Continued

Year	1990	1995	1996	1997	1998	1999	2000	2001	%change 1990–2000	%change 2000–2001
N.Mariana Isles	426,000	669,000	728,000	685,000	481,000	496,000	529,000		16.4	
E.A–Pacific (m)	54.6	81.3	89.0	88.3	87.4	96.8	109.1	115.1	100.0	5.5
World Arriv. (m)	457.2	552.3	597.4	618.2	636.0	652.2	696.7	692.7	52.7	−0.6
%EAP of world	11.9	14.4	14.9	14.3	13.7	15.0	16.0	16.6		

Sources: Tourism Council of the South Pacific (u/d, 1994 and 1999); http://www.amsamoa.com/statistics.htm (accessed 26 August 2002); National tourism organisations; , World Tourism Organization (2001a, b, 2002) and http://www.world-tourism.org/market_research/facts&figures/latest_data.htm; (accessed 25 August 2002); South Pacific Tourism Organization (2001, 2002, personal communication).

contribution to those small economies. Clearly, the record of different island groups in the Pacific over the last decade is variable. However, the overall average increase in arrivals (up to 2001) in the Central and South Pacific of 41% needs to be situated in a wider context. It was less than the growth in all world arrivals over the same period (which was more than 52%) and *much* less than the increase in arrivals in the East Asia–Pacific region as a whole (almost 105%). Over the last 5 years, the numbers of tourists visiting the Federated States of Micronesia, Guam, the Northern Marianas and Hawai'i have remained virtually static. Seen in this light, and assuming that increased tourist arrivals *are* a measure of success, the progress of the tourism industry in the Pacific gives some cause for concern.

Tourist numbers alone are not especially helpful. Attention must also focus on the *nature* of visits to different destinations, the overall *purpose* and, if on holiday, the *type* of holiday being taken. Visitors to French Polynesia and Fiji are most likely to be holidaymakers, whereas the largest category of tourists to Samoa and American Samoa is made up of those visiting friends and relatives (Samoa Visitors Bureau, 2001; WTO, 2001a, pp. 70, 144). By contrast, business tourists predominate among arrivals to Papua New Guinea (WTO, 2002). As a consequence, throughout the Pacific a wide variety of accommodation is on offer, even in such established destinations as Fiji and French Polynesia, to cater for different types of tourists. And niche markets abound. Many islands, especially those with stereotypical white sand beaches and swaying palms, compete for wedding parties and honeymooners, perpetuating the image of Bali Hai and the 'romantic' Pacific, and there is also a tendency to focus on such niche markets as diving.

Pacific Island Tourism, 'Development' and Sustainability

The debate as to whether or not tourism *is* a tool for 'development' in less developed countries has generated much passion and has been articulated at length, both for less developed countries generally (Harrison, 1992, Harrison, 2001b, pp. 28–39) and in the Pacific (Harrison, 2003). Against tourism, it is usually argued that transnational companies cream off the economic benefits and leave local companies, employees and populations with relatively little; and that the industry produces negative impacts on the family, religion and local tradition, as well as polluting the physical environment. Others point to the increased income and training provided by transnational companies, suggesting that tourism is but one of many 'modern' influences in Pacific societies, where chiefs and elders have a vested interest in controlling 'their' women and young people. They also note the coincidence of interest of the tourism industry and environmentalists in ensuring that environmental damage is minimized. Furthermore, as Macnaught (1982, p. 363) and Douglas (1996b) point out, it is perhaps both dangerous and patronizing to treat Pacific islanders as hapless and helpless victims of modernization. They are no less able than others to arrive at their own moral judgements and, for good or ill, have long been willing participants in the continuous process of globalization (Harrison, 1997).

Indeed, it is sometimes argued that indigenous economic and social institutions actually *prevent* tourism development from occurring. This is especially so for 'traditional' forms of land tenure, which, most notably in Melanesia, are allegedly antithetical to tourism. How far this *is* the case is difficult to determine. Principles of land tenure varied considerably in the precolonial period, and changes introduced by colonizing powers were similarly varied. Although often made in the *name* of tradition, they frequently consolidated power in the hands of groups whose previous hold on land was tenuous and dependent on particularities of time and place (Crocombe, 1987, 1995). Precolonial structures were undoubtedly relevant to colonial and postcolonial development. The fact that land in precolonial Hawai'i was under the absolute control of a relatively small number of chiefs, for example (Meller and Horvitz, 1987), facilitated the later process of commoditization and helped ensure that land ownership in modern Hawai'i remained concentrated in relatively few hands (Meller and Horvitz, 1987). By contrast, whereas much land in Polynesia seems to have been controlled by chiefs, the leadership powers in

Melanesia apparently did not extend to control of land (Crocombe, 1987).

Throughout the Pacific, disputes over land, access to land and its resources are common, and involve not only tourism, but also other forms of development (Sofield, 1996; Thomson, 1996; Harrison, 1997; Johnston, 1998; Waqa, 1998). In addition, conflicts may sometimes be extended to the claimed ownership of marine resources. In Viti Levu, Fiji, for instance, for years there has been a (sometimes violent) dispute over which villagers have the right to ferry surfers to Cloudbreak Reef, one of the world's most famous surfing spots (Johnston, 1998) – an argument, in effect, over who rules the waves.

These kinds of problem are not going to go away, for while there are attempts to move from a narrow focus on sun, sea and sand (TCSP, 1998, p. 21), 'nature-based tourism' is likely to remain central to the marketing of Pacific islands for the foreseeable future. In fact, the criticisms levelled at mass tourism in such destinations as Hawai'i (Minerbi, 1992) should be regarded as warnings of what might happen in other Pacific islands, rather than indictments of what has already occurred. In any case, the mantle of such critics has been taken up and their ideas given much wider currency by those now focusing on 'globalization' (Waters, 1995) and the need for 'sustainable' development.

In effect, environmentalism, modernization theory and underdevelopment theory have converged. Much of what is now debated under the overall title of 'globalization' theory is, in fact, a re-working of much older but still relevant concerns. It was therefore inevitable that questions would be raised about how far tourism – an increasingly popular *tool* for development – should or could be 'sustainable' or (less often) how it fitted into more comprehensive programmes of sustainable development, especially in less developed countries (Wheeller, 1993; Harrison, 1996; Wahab and Pigram, 1997; Mowforth and Munt, 1998, pp. 84–124) and, even more specifically, in islands and microstates (Briguglio *et al.*, 1996a, b).

The tourism industry responded (Brown, 1994; Wight, 1994; International Hotels Environment Initiative (IHEI), 1996). European resorts started to re-evaluate their tourism 'product'. Books on 'sustainable tourism' multiplied (Inskeep, 1991; WTO 1994; Middleton and Hawkins, 1998), as did codes of practice (Mason and Mowforth, 1995; Neale, 1998). Pressure groups were formed (O'Grady, 1990; Botterill, 1991) and less damaging varieties of alternative tourism were proposed, including ecotourism (Butler, 1992; Cater and Lowman, 1994; Fennell, 1999; Honey, 1999). Tourism was going green.

At least, tourism *seemed* to be going green. In fact, as critics have pointed out, some responses from within the tourism industry are little more than cynical attempts to capitalize on the latest political correctness, and others openly doubt that there are realistic alternatives to mass tourism (Wheeller, 1993). There was also a distinct reluctance in some quarters to go beyond populist rhetoric and ask how criteria of 'sustainability' might actually be implemented in practice (Harrison, 1996).

In the Pacific region, a similar pattern emerges. Critiques of tourism development, especially mass tourism, provided the backdrop to the debate (Farrell, 1977, Farrell, 1992; Britton, 1982, 1983, 1987a, b; Kent, 1983; Milne, 1990; Hall, 1992; Minerbi, 1992; Tran, 1996; King, 1997, pp. 37–50) and, more than a decade ago, the Tourism Council of the South Pacific (TCSP) recommended legislation to protect the region's environment and assist in tourism development (TCSP, 1988, TCSP, 1990). Indeed, the TCSP's recognition of the value of environmental conservation was noted in a series of studies appearing in 1995 (Bourdais, 1995a, b; Kendell, 1995; McVey, 1995a, b, c; Schuller, 1995a, b, c), as well as by others more specifically focused on conservation issues (Watling and Chape, 1992, pp. 82–87; Weaver, 1992; Weaver and King, 1996). Still at the regional level, concern over the effects of mass tourism in Asia and the Pacific prompted the United Nations to issue guidelines on the development of coastal tourism (Economic and Social Commission for Asia and the Pacific (ESCAP), 1995).

Others have concentrated more on the extent to which *social* sustainability can be maintained during tourism development. Examples are Ranck's (1987) study of village-based tourism in Papua New Guinea, Farrell's work on Hana, on the island of Maui, Hawai'i (Farrell, 1992; Wyllie, 1998), and Sofield's work on

Vanuatu and Solomon Islands. Discussing the Pentecost land dive in Vanuatu, Sofield proposes several principles for sustainable ethnic tourism and emphasizes the importance of local control and government support (Sofield, 1991). This should be compared to de Burlo's later account of the same area (de Burlo, 1996). By contrast, Sofield's work is ample evidence that local involvement in 'ecotourism' in Solomon Islands is far from straightforward (Rudkin and Hall, 1996; Sofield, 1996).

As elsewhere, commitment to sustainable tourism development in Pacific islands is not always translated into practice. National tourism organisations (NTOs) pay lip service to the concept but their success is often evaluated (by themselves and others) primarily through their ability to increase tourist numbers. At the local level, resorts claiming 'green' credentials still promote 'reef walking', encourage guests to feed fish, and advertise shark feeding! On the credit side, throughout the Pacific islands greener and more sustainable forms of tourism are being encouraged (Harrison, 2003). Since 1997, a voluntary Cook Islands Accreditation Scheme has rewarded sound environmental management practices and, as Louise Twining-Ward's chapter in this special issue shows (Twining-Ward and Tuailemafua, 2004), sustainable tourism development is now being taken seriously in Samoa. In fact, several NTOs give annual ecotourism awards, as does the South Pacific Tourism Organization (SPTO) (the re-named TCSP). Changes in the way the SPTO is funded by the European Union – hitherto not known in the region for its knowledge of or commitment to sustainable tourism development – may facilitate the process (Keith-Reid, 1999), and SPTO's closer working relationship with the South Pacific Regional Environment Programme (SPREP) –which has developed initiatives for ecotourism in the region (Martel, 2000) – is especially encouraging. Finally, NZODA (New Zealand Official Development Assistance) has for some time actively promoted ecotourism in Pacific islands (Tourism Resource Consultants (TRC), 1999).

In some respects, the lack of tourist development in many Pacific islands is an advantage. It is still necessary to emphasize the need for sustainable (as opposed to unsustainable) mass tourism (SMT) (Weaver, 2001), but island states where mass tourism has not yet been introduced can still avoid its negative impacts. In the meantime (lacking the *facilities* for large-scale tourism), they can concentrate on implementing smaller scale *ecotourism*, guidelines for which have already been provided from within the region (Hay, 1992; Bushnell, 1994; Liu, 1994; Slot, 1996; Harrison *et al.*, 2003).

It should no longer be necessary to argue a case for promoting sustainable tourism development in Pacific islands. Properly planned and managed, tourism can help economic diversification and reduce dependency on primary export crops. It can assist in the conservation of natural resources and bring benefits to local communities, including valuable skills in entrepreneurship (often through training initially provided in-house by transnational hotel chains). However, if the ideal of sustainable tourism development is to be approached, it must be through increasing *awareness* of the issues involved, and through *partnership*, of government and private sector, of local populations and the tourism industry, of transnational operators and local business. It is to be hoped that this special issue will make a contribution to this process, which is really only just starting in many Pacific island states.

Issues Arising

It has been argued elsewhere that when tourism in less developed countries is under consideration, three key issues are prominent: the relationship of tourism to migration, the processes by which capitalism has expanded in the region, and the role of the state (Harrison, 2001b). To some extent, all these issues figure in Pacific tourism, but to a greater or lesser extent, depending on the level of tourism development in the destination area.

The paper by Ngaire Douglas on the early stages of tourism in Solomon Islands sets a rather bleak scene (Douglas, 2009). Tourism was instigated by expatriates of European origin, who catered primarily for visitors of European origin, and at least at some hotels in these early years denied entry to any Pacific Islanders. Instead, the 'locals' were 'attractions', enacting the parts of savage natives (men) or dusky, near naked beauties (women), in ethnographically-framed performances that had been widely advertised through travelogues in Australia, where still photographs and cinematography whetted the appetite for Melanesian destinations. Shipping agents capitalized on the demand, and advertised the prime shooting and fishing adventures awaiting the explorer-tourist in the Solomons. However, although cruising developed during the first decades of the 20th century, tourism remained a minor feature of the domestic Solomons economy, and continued so throughout the 20th century. Local participation in the industry was minimal, and Solomon Islanders generally lacked (and still largely lack) the financial and cultural capital to participate in tourism in any other than a passive sense. Attempts to involve them more directly have frequently failed (Sofield, 1996) and yet they continue, often through the good offices of non-governmental organizations (Russell and Stabile, 2003). Even at the time of writing, rather than having 'taken off', tourism is better described as having crash landed! Years of domestic unrest have taken their toll, and in 2001 less than 4000 tourists visited Solomon Islands. In such circumstances, small-scale 'ecotourism' is virtually the *only* option, but one that enables a more thoughtful approach to tourism planning and development. However, this will require a more active, organized and efficient state apparatus, which has yet to emerge in Solomon Islands. At present, the only way Solomon Island tourism can go is up!

Having foreign ownership and a (relatively) settled political situation is no guarantee of commercial success. Fiji may be the most 'developed' tourism destination in the South Pacific, but it still has a 'human resources' problem. As described by Dwyer and Kemp (2009), *Fiji's Hidden Paradise* (FHP), an ecolodge on the island of Vanua Levu, Fiji, was worthy soap opera material. Their account is of two philanthropic Australian businessmen who decide to 'do good' by financing a small hotel development, in which shares are to be held by hospitable, friendly villagers. However, there was a 'lack of sound, strategic management' on the part of the Australians, whose failure 'to conduct audits of its internal and external environments contributed to the implementation of strategies which ... resulted in failure to achieve its mission'. FHP was badly organized. Its mission statement was vague and confused, its owners lived in Australia, its local management was inexperienced and part-time, as were other staff members, who went to the resort only when there were guests. There were no recreational facilities, shops, radio or television. No business or marketing plans had been prepared, no marketing research had been carried out, and the marketing that did occur was haphazard. Partly as a result of these deficiencies, FHP was susceptible to better-organized competition from other companies, suppliers had little confidence in its ability to deliver a satisfactory tourism 'product' and it was unable to counter a general image of Fiji tourism as jaded and stale. Also, as a small and new venture in Vanua Levu, which is peripheral to Fiji tourism, it did not attract state support. Not surprisingly, the venture failed, to the disappointment of all the stakeholders. Philanthropic motives on the part of the Australians, and warmth and friendliness on the part of the villagers, were simply not enough. It cannot be argued that capitalism *failed*. Reportedly, the owners were not primarily interested in *making* a profit. Arguably, FHP failed because no one was capitalistic enough! Putting the issue crudely, it went bust because no one knew what they were doing!

For different reasons, neither Solomon Islands nor FHP can be considered satisfactory examples of sustainable tourism development – or even of sustainable *tourism*. However, the scenario in Samoa is more promising. Over the last decade, the number of international arrivals to Samoa has risen, albeit unspectacularly, as indicated earlier in Table 10.1. However, as Louise Twining-Ward and Tapulolou Tuailemafua indicate (2004), more than one-third of these arrivals are Samoans residing overseas, who go to Samoa to visit friends and relatives (a further indication of the importance of migration in small Pacific islands). At first sight, these visitors seem to have no negative impact on their society of origin, but Samoans were, and are still, anxious to avoid what many see as international tourism's negative effects.

They wanted to monitor tourism development and ensure that it was sustainable, and the paper focuses on a three-stage process, involving first the identification of sustainable tourism development (STD) *objectives*, then appropriate STD *indicators* and, finally, the formulation of an implementation *framework* to act on the results. More widely, however, the paper is *also* about the cooperation of numerous stakeholders in Samoan tourism, including the state, NGOs, the tourism industry and other sections of Samoan society, and how concepts of sustainable tourism development and stakeholder involvement are brought into sharper focus by being operationalized in a particular political context. Sustainable tourism development is thus seen to be linked to wider social and developmental concerns, and the *state* – most notably the Division of Planning and Development and the Samoa Visitors' Bureau – plays a key role.

Of course, questions remain. As the authors recognize, considerable subjectivity is involved in identifying indicators and deciding what constitutes an acceptable level of change (or continuity), and status and power differences on the Stakeholder Committee must surely have emerged or been subsumed. Indeed, the role of Louise Twining-Ward, whose PhD fieldwork was centred on the development of STD indicators, must have been relatively significant, as was *fa'aSamoa*, the underlying value system of Samoa, which would at least have conditioned the discussions and decision-making process. However, these are issues for other papers, and the Samoan experiment seems promising. If it continues to be successful, it could provide a sound and cooperative base for future tourism development in Samoa, and provide ideas for similar activities elsewhere.

A close reading of the paper by Staiff and Harrison (2004) on 'healthy tourism' in Fiji reveals marked similarities with the previous paper. Both were linked to other research projects, and both involved the quest for social indicators. However, Staiff and Bushell were (clearly) constrained by the fact that their research was part of a wider *Healthy Islands* initiative by the World Health Organization (WHO). From the outset, health and tourism were linked as aspects of sustainable development, and the research was 'framed and contextualised' not only by the WHO project, but also by *Agenda 21 for the Travel and Tourism Industry*. Furthermore, their research was disrupted by the coup and constitutional crisis of May 2000, which was decidedly *unhealthy* for Fiji tourism. Aiming to 'develop a conceptual framework for healthy tourism' they, too, encouraged stakeholder participation, and brought together 'discursive frameworks' from health and tourism, which seemed to them to coincide in a Fijian emphasis on village-based ecotourism. From two workshops held in Suva, involving tourism and non-tourism professions and interests, it emerged that participants from all sectors increasingly realised that tourism and health were far more interlinked than at first appeared, and needed to be integrated in a wider community context. However, it was felt that communities *themselves* should negotiate and formulate measurable indicators of health and tourism, rather than have them imposed by outside agencies. As Staiff and Bushell argue, *Agenda 21*, the Healthy Islands initiative and village-based ecotourism discourses have 'a similar conceptual language', and (ideally?) guidelines 'need to be "holistic", community-centred, locally "owned", empowering, consultative, culturally, environmentally, historically and gender sensitive and "scientifically" endorsable'. The authors conclude (adding a suitably open caveat) that any framework for healthy tourism should be 'enmeshed in a Fijian way of thinking and seeing (however it is locally defined and/or negotiated)'.

Reference to the 'operating environment' of a tourism venture – and suggestions that indicators of various kinds should be 'negotiated' and developed by communities themselves, rather than imposed from outside – imply that sustainable tourism development in Pacific islands should be positively related to local peoples' sense of belonging, to their psychological, collective and geographical sense of place. Such themes are developed, in their different ways, by Greg Ringer (2004) and Tina Jamieson (2004). The former focuses on Kosrae, in the Federated States of Micronesia, an island of devout Christians where 80% of the population is under 15. Among older people, relatively few have left the island, whereas younger members of the population are more widely travelled and more exposed to tourism. This is evident from the 'cognitive maps' provided by Ringer's respondents. Tourists envisaged the 'place' of Kosrae according to the services they received. Not surprisingly, it was perceived within a framework comprised of tourist-oriented buildings, sites and sights, often derived from guidebooks. Rarely venturing inland, they knew little of the island's communities. By contrast, Ringer suggests that islanders' perceptions were based on their historic experiences of place. As one would expect, they were much better informed than tourists about who lived where, and what they did. But there were differences among them. Responses of older islanders indicated clear gender differences, apparently related to their occupational patterns, whereas responses of younger people showed no gender variation (perhaps because of their greater involvement with tourists). They were, however, more knowledgeable than their seniors about what was on the island, having travelled around it more frequently. Unlike their elders, they were unaware of sites associated with the Second World War (which ended long before they were born) and they knew *nothing* about the Lelu Ruins, the historic centre of Kosrae.

Clearly, tourists bring to the destination area a partly-formed sense of 'place', which may (or may not) be reinforced by later experiences. However, through providing them with services and generally interacting with them, younger islanders were found to have developed a sense of 'place' for Kosrae that in some regards differed markedly from that held by the older generation. It could be argued that they started to see Kosrae through tourist eyes. How far this is indeed the case, in Kosrae or elsewhere, should be a matter for further research but, as Ringer notes, it is a tendency that tourism planners and managers would do well to understand and counter.

Tina Jamieson's discussion of the niche tourist market of Cook Island weddings and honeymoons encompasses some of these themes and raises other issues (2009). The 'romance' of the Pacific island is a long-established stereotype, and is reiterated and reinforced by travel agents and tour operators promoting the Cook Islands as a favoured venue for weddings. It is a 'conspiracy' in which Cook Island 'hosts' and the brides and grooms willingly collude, as Cook Islanders busily construct the (social and physical) context for the weddings, and couples project on to the islands all their own feelings of romance. Away from the daily grind and arduous social relations, relieved of the wedding list and the onerous preparations, couples project on to landscape, seascape, beachscape and local inhabitants the warmth they themselves feel (or wish to feel?). The natural and social worlds revolve around them. Rugby players become gentle, the hitherto unreal becomes real, and time and space are transformed in a new and romantic (sur)reality. In purchasing the wedding, it is argued, tourists assert not only their right to be on the island, but their *ownership* of it and all that is therein. This is not bad faith, for *place* – any place – does not exist 'out there', as some kind of objective reality. As elsewhere, in the Cook Islands it is *constructed and negotiated, through social interaction.*

Of course, this is not the complete picture. While wedding tourists and honeymooners are happily being helped along their romantic road, Cook Islanders are constructing very different scenarios, often involving emigration from the Cook Islands to live elsewhere. Equally, it might be argued that lovers visiting the Cook Islands are no more dewy-eyed than lovers marrying or honeymooning elsewhere. They might not be dewy-eyed at all, but going along with a romantic game, knowing full well that other realities will intrude soon enough. If this were not the case for at least some of them, we would be close to branding them all 'cultural dopes', an implication that sits uneasily on the kind of phenomenological approach underpinning the paper. Or is it simply that when 'true love' and the myth of the Pacific island paradise combine, one can really discover Bali Hai, that special island? In your dreams? In your dreams!

The paper by Levinson and Milne (2004) brings us down to earth, while yet remaining with tourism in the Cook Islands. It reminds us that persuading visitors to come to your destination is usually hard work, even if they are already predisposed to come to a tropical island. After all, it does not have to be *your* island, and there are plenty of competitors in the Pacific and further afield – in the Caribbean and the Indian Ocean, for example. The authors compare the marketing of the Cook Islands in a selection of tourism brochures (issued in New Zealand) and Internet websites. Can Internet technology help the small operator and hospitality provider counter the advantages of big hotels and international chains? The answer is positive, to some extent, for it is relatively inexpensive to develop websites, and contact with clients and potential clients is thenceforth cheap, quick and flexible. In this respect, small hotels and guest houses can 'catch up' with their more powerful competitors. However, the playing field is not yet level. Compared to small operators, large hotels are still more likely to have their own websites, which are better and more comprehensive and, because there is a shortage of IT skills in the Cook Islands, most such sites are developed and maintained overseas. Indeed, as was seen earlier in the case of Fiji's Hidden Paradise, mere possession of a website is no guarantee of success. In any case, one can perhaps make too much of the advantages of modern technology; Levinson and Milne discovered that, in most cases, the Internet merely reproduced the images and texts of more 'traditional' forms of advertising – in effect, no more than 'virtual brochures'.

Whether marketed through standard brochures or the Internet, Cook Islands tourism may, if anything, be too successful. In 2000, much aided by the May coup in Fiji, the tiny country was struggling to accommodate all its visitors (Keith-Reid, 2001), with subsequent complaints of inadequate sewerage and other facilities. In some circles, at least, this excess demand was attributed to successful advertising on the Internet (Brown, 2001). Such problems, though, would be welcome in Levuka, the first colonial capital of Fiji, a town which has struggled for more than a century to find a role and an economic basis for its development. Its lack of success is not through want of trying. Over the years, many have considered tourism to be an answer but, despite the intermittent efforts of the tourism industry, by 2000 only a few thousand visited the town every year. Would it be different if Levuka was formally listed as a UNESCO World Heritage site? Perhaps, but the answer is not as simple as that. Issues of place, and competing conceptions of history and place, emerge in this account, which shows that while UNESCO, the tourism industry and some pressure groups in Fiji would willingly see Levuka become the first World Heritage site in Fiji – and one of very few in South Pacific islands – others (including Levuka's most recently elected Town Council) are less convinced. How far should a town founded by people of *European* origin, and still associated in the minds of many indigenous Fijians with them and their descendants, be considered worthy of this honour? What is UNESCO's interest in promoting Levuka's cause? And how far could any government claiming to represent indigenous Fijians offer financial and other support to Levuka's claim?

It will take time for these questions to be answered. Meanwhile, it is again evident – as all the papers in this special issue indicate – that neither 'heritage' nor tourism is politically neutral. What is marketed, who does the marketing, and how 'attractions' are presented are profoundly political and contentious issues. Tourism can indeed bring many benefits to Pacific islands, but sustainable tourism development needs to be carefully planned, efficiently organized and implemented, and consistently monitored. If this does not occur, the benefits may be short-lived and the price may be high.

Note

[1] Figures for most destinations vary according to the source of the statistics. Those for American Samoa, in particular, show marked divergence.

References

Akram-Lodhi, A. (2000) Stop press: confronting the present: the coup of May 2000. In: Akram-Lodhi, A.H. (ed.) *Confronting Fiji futures*. Asia Pacific Press, The Australian National University, Canberra, pp. i–0.

Anon (1999) Getting rid of guns and heading for lasting peace: how Bougainville's preparing for this. *Islands Business* 26(6), 18.

Bailey, B. and Farber, B. (1992) *The First Strange Place: Race and Sex in World War II Hawaii*. Johns Hopkins University Press, Baltimore & London.

Barkham, P. (2001) Paradise lost awaits asylum seekers. *The Guardian* 11 September(3).

Berno, T. and Douglas, N. (1998) Tourism in the South Pacific: a Polynesia/Melanesia discussion. *Asia Pacific Journal of Tourism Research* 2(2), 65–73. DOI: 10.1080/10941669808721998.

Bertram, I.G. and Watters, R.F. (1985) The MIRAB economy in South Pacific microstates. *Pacific Viewpoint* 26(3), 497–519. DOI: 10.1111/apv.263002.

Botterill, T.D. (1991) A new social movement: tourism concern, the first two years. *Leisure Studies* 10(3), 203–217. DOI: 10.1080/02614369100390191.

Bourdais, D. (1995a) *Tahiti: Tourism Sector Review*. Tourism Council of the South Pacific, Suva.

Bourdais, D. (1995b) *New Caledonia: Tourism Sector Review*. Tourism Council of the South Pacific, Suva.

Briguglio, L., Archer, B., Jafari, J. and Wall, G. (eds) (1996a) *Sustainable Tourism in Islands and Small States: Issues and Policies*. Pinter Press, London.

Briguglio, L., Butler, R., Harrison, D. and Leal Filho, W. (eds) (1996b) *Sustainable Tourism in Islands and Small States: Case Studies*. Pinter Press, London.

Britton, S.G. (1982) The political economy of tourism in the Third World. *Annals of Tourism Research* 9(3), 331–358. DOI: 10.1016/0160-7383(82)90018-4.

Britton, S. (1983) *Tourism and underdevelopment in Fiji, Monograph 31, Development Studies Centre, The Australian*. National University, Canberra.

Britton, S. (1987a) Tourism in Pacific island states: constraints and opportunities. In: Britton, S. and Clarke, W.C. (eds) *Ambiguous Alternative: Tourism in Small Developing Countries*. The University of the South Pacific, Suva, pp. 113–139.

Britton, S. (1987b) Tourism in small developing countries: development issues and research needs. In: Britton, S. and Clarke, W.C. (eds) *Ambiguous Alternative: Tourism in Small Developing Countries*. The University of the South Pacific, Suva, pp. 167–187.

Brown, M. (1994) Environmental auditing and the hotel industry: an accountant's perspective. In: Seaton, A.V., Jenkins, C.L., Wood, R.C., Dieke, P.U.C. and Bennett, M.M., *et al.* (eds) *Tourism: the State of the Art*. Wiley, Chichester, pp. 675–681.

Brown, J. (2001) Tourism booming, so why are Islanders leaving? *Pacific* June, 22–23.

Bushnell, S.M. (1994) *The Ecotourism Planning Guide for Ecotourism Operators in the Pacific Islands*. College of Business Administration, University of Hawai'i, Honolulu.

Butler, R. (1992) Alternative tourism: the thin end of the wedge. In: Smith, V.S. and Eadington, W.R. (eds) *Tourism Alternatives: Potentials and Problems in the Development of Tourism*. University of Pennsylvania Press, Philadelphia, pp. 31–46.

Cater, E. and Lowman, G. (eds) (1994) *Ecotourism: a Sustainable Option?* Wiley, Chichester.

Crocombe, R. (1987) Overview: the pattern of change in the Pacific. In: Crocombe, R. (ed.) *Land Tenure in the Pacific*, 3rd edn. The University of the South Pacific, Suva, pp. 1–24.

Crocombe, R. (1995) Overview. In: Crocombe, R. (ed.) *Customary Land Tenure and Sustainable Development: Complementarity or Conflict?* Institute of Pacific Studies, The University of the South Pacific, & South Pacific Commission, Suva, Fiji & Noumea, New Caledonia, pp. 5–21.

de Burlo, C. (1989) Islanders, soldiers, and tourists: the war and the shaping of tourism in Melanesia. In: White, G.M. and Lindstrom, L. (eds) *The Pacific Theater: Island Representations of World War II*. University of Hawai'i Press, Honolulu, pp. 299–325.

de Burlo, C. (1996) Cultural resistance and ethnic tourism on South Pentecost, Vanuatu. In: Butler, R. and Hinch, T. (eds) *Tourism and Indigenous People*. International Thomson Business Press, London, pp. 255–276.

Decloitre, P. (1998) Riot rocks Port Vila. *Pacific Islands Monthly* 69(3), 8–9.

Douglas, N. (1996a) *They Came for Savages: 100 Years of Tourism in Melanesia*. Southern Cross University, Lismore, New South Wales.

Douglas, N. (1996b) Social and cultural impact of tourism in the Pacific. In: Hall, C.M. and Page, S.J. (eds) *Tourism in the Pacific: Issues and Case*. International Thomson Business Press, London, pp. 49–64.
Douglas, N. (1996c) Tourism in the Pacific: historical factors. In: Hall, C.M. and Page, S.J. (eds) *Tourism in the Pacific: Issues and Cases*. International Thomson Business Press, London, pp. 19–35.
Douglas, N. (2009) Towards a history of tourism in Solomon Islands. In: Harrison, D. (ed.) *The Journal of Pacific Studies*. 26, pp. 1–2.
Dwyer, L. and Kemp, S. (2009) Closure of an ecolodge: poor strategic management. In: Harrison, D. (ed.) *The Journal of Pacific Studies*. 26, pp. 51–74.
Economic and Social Commission for Asia and the Pacific (ESCAP) (1995) *Guidelines on Environmentally Sound Development of Coastal Tourism*. United Nations, New York.
Fagence, M. (1997) An uncertain future for tourism in microstates; the case of Nauru. *Tourism Management* 18(6), 385–392. DOI: 10.1016/S0261-5177(97)00036-8.
Farrell, B.H. (1992) Tourism as an element in sustainable development. In: Smith, V.S. and Eadington, W.R. (eds) *Tourism Alternatives: Potentials and Problems in the Development of Tourism*. University of Pennsylvania Press, Philadelphia, pp. 115–134.
Farrell, B.H. (ed.) (1977) *The Social and Economic Impact of Tourism on Pacific Communities*. Center for Pacific Studies, University of California, Santa Cruz.
Fennell, D.A. (1999) *Ecotourism: an Introduction*. Routledge, London.
Field, M. (1998) The Lincoln Agreement. *Pacific Islands Monthly* 69(3), 16–18.
Fletcher, J. and Snee, H. (1989) Tourism in the South Pacific islands. *Progress in Tourism and Hospitality Management* I, 114–124.
Grant, G. (1996) *Waikiki yesteryear*. Mutual Publishing, Honolulu.
Grove, R.H. (1995) *Green Imperialism: Colonial Expansion, Tropical Island Edens and the Origins of Environmentalism*. Cambridge University Press, Cambridge.
Hall, C.M. (1992) Environmental impact of tourism in the Pacific. In: Hall, C.M. and Page, S.J. (eds) *Tourism in the Pacific: Issues and Cases*. International Thomson Business Press, London, pp. 65–80.
Harrison, D. (1992) Tourism to less developed countries: the social consequences. In: Harrison, D. (ed.) *Tourism and the Less Developed Countries*. Wiley, Chichester, pp. 19–34.
Harrison, D. (1996) Sustainability and tourism: reflections from a muddy pool. In: Briguglio, L., Archer, B., Jafari, J. and Wall, G. (eds) *Sustainable Tourism in Islands and Small States: Issues and Policies*. Pinter Press, London, pp. 69–89.
Harrison, D. (1997) Globalisation and tourism: some themes from Fiji. In: Oppermann, M. (ed.) *Pacific Rim Tourism*. CAB International, Wallingford, pp. 166–183.
Harrison, D. (2001a) Islands, image and tourism. *Tourism Recreation Research* 26(3), 9–14. DOI: 10.1080/02508281.2001.11081194.
Harrison, D. (2001b) Tourism and less developed countries: key issues. In: Harrison, D. (ed.) *Tourism and the Less Developed World: Issues and Case Studies*. CAB International, Wallingford, pp. 23–46.
Harrison, D. (2003) Themes in Pacific island tourism. In: Harrison, D. (ed.) *Pacific Island Tourism*. Cognizant, New York.
Harrison, D., Malani, M. and Sawailau, S. (2003) Ecotourism and village-based tourism: a policy and strategy for Fiji. In: Harrison, D. (ed.) *Pacific Island Tourism*. Cognizant, New York.
Hay, J. (1992) Tourism business in the Pacific: providing a sustainable experience. *Proceedings of ecotourism conference*, Auckland, New Zealand, 12–14 October.
Henley, J. (2001) Money-laundering states get deadline. *The Guardian* 15.
Honey, M. (1999) *Ecotourism and Sustainable Development*. Island Press, Washington, DC.
Inskeep, E. (1991) *Tourism Planning: an Integrated and Sustainable Development Approach*. Van Nostrand Reinhold, New York.
International Hotels Environment Initiative (IHEI) (1996) *Environmental Management for Hotels: the Industry Guide to Best Practice*, 2nd edn. Butterworth-Heinemann, Oxford.
Jamieson, T. (2004) Making a scene: tropical island weddings: destination creation and the experience of place. In: Harrison, D. (ed.) *The Journal of Pacific Studies*. 26, pp. 151–171.
Johnson, G. (1999) Marshall islands attempts to explode nuclear test myths. *Pacific Islands Monthly* 69(7), 42–43.
Johnston, R. (1998) The explosive land issue. *Islands Business* 24(4), 36–38.
Kaplan, M. (1995) *Neither Cargo nor Cult: Ritual Politics and the Colonial Imagination in Fiji*. Duke University Press, Durham & London.
Keith-Reid, R. (1999) Why Europe is changing its approach. *Islands Business* 25(7), 22.

Keith-Reid, R. (2000a) Banking on millions through finance centre. *Islands Business* 26(6), 24–25.
Keith-Reid, R. (2000b) How Speight pushed too far: the rise and fall of a hostage taker. *Islands Business* 26(8), 29–31.
Keith-Reid, R. (2000c) How Morauta's getting economy back on its feet. *Islands Business* 26(8), 36–38.
Keith-Reid, R. (2001) Cooks booming. *Pacific* June, 17–18.
Kendell, P. (1995) *Tonga: Tourism Sector Review*. Tourism Council of the South Pacific, Suva.
Kent, J. (1983) *Hawai'i: Islands Under the Influence*. Monthly Review Press, New York.
King, B. (1997) *Creating Island Resorts*. Routledge, London.
Lal, B.V. (1994) The passage out. In: Howe, K.R., Kiste, R.C. and Lal, B.V. (eds) *Tides of History: the Pacific Islands in the Twentieth Century*. Allen & Unwin, St Leonards, New South Wales, pp. 435–461.
Laracy, H. (1994) World War Two. In: Howe, K.R., Kiste, R.C. and Lal, B.V. (eds) *Tides of History: the Pacific Islands in the Twentieth Century*. Allen & Unwin, St Leonards, New South Wales, pp. 149–169.
Laracy, H. and White, G. (1988) Taem blong faet: World War II in Melanesia. *'o'o* Number 4(Special Issue).
Levinson, J. and Milne, S. (2004) From brochures to the Internet: tourism and marketing and development in the Cook Islands. In: Harrison, D. (ed.) *The Journal of Pacific Studies*. 26, pp. 175–198.
Liu, J.C. (1994) *Pacific Islands Ecotourism: a Public Policy and Planning Guide*. College of Business Administration, University of Hawai'i Press, Honolulu.
Loxley, D. (1990) *Problematic Shores: the Literature of Islands*. St. Martin's Press, New York.
Macnaught, T.J. (1982) Mass tourism and the dilemmas of modernization in Pacific island communities. *Annals of Tourism Research* 9(3), 359–381. DOI: 10.1016/0160-7383(82)90019-6.
Martel, F. (2000) Ecotourism in conservation areas of the Pacific islands. *Summary for the WTO/UNEP Conference on Sustainable Tourism in Asia–Pacific Islands*, Sanya, Hainan, China, 6–8 December.
Mason, P. and Mowforth, M. (1995) Codes of conduct in tourism. *Occasional papers in Geography, 1*, Department of Geographical Sciences, University of Plymouth, Plymouth.
McVey, M. (1995a) *Papua New Guinea: Tourism Sector Review*. Tourism Council of the South Pacific, Suva.
McVey, M. (1995b) *Niue: Tourism Sector Review*. Tourism Council of the South Pacific, Suva.
McVey, M. (1995c) *Vanuatu: Tourism Sector Review*. Tourism Council of the South Pacific, Suva.
Meller, N. and Horvitz, R.H. (1987) Hawai'i: themes in land monopoly. In: Crocombe, R. (ed.) *Land Tenure in the Pacific*, 3rd edn. The University of the South Pacific, Suva, pp. 25–44.
Melville, H. (1996, first published 1846) *Typee*. Oxford University Press, Oxford.
Middleton, V.T.C. and Hawkins, R. (1998) *Sustainable Tourism: a Marketing Perspective*. Butterworth-Heinemann, Oxford.
Milne, S. (1990) The impact of tourism development in small Pacific island states: an overview. *New Zealand Journal of Geography* 89(1), 16–21. DOI: 10.1111/j.0028-8292.1990.tb00283.x.
Milne, S. (1992) Tourism and development in South Pacific microstates. *Annals of Tourism Research* 19(2), 191–212. DOI: 10.1016/0160-7383(92)90077-3.
Minerbi, L. (1992) Impact of tourism development in Pacific islands. *A paper for Greenpeace Pacific Campaign*, San Francisco, California.
Moorhead, A. (1966) *The Fatal Impact*. Hamilton, London.
Mowforth, M. and Munt, I. (1998) *Tourism and Sustainability: New Tourism in the Third World*. Routledge, London.
Narsey, W. (1999) British nuclear tests in the Pacific. *Pacific Islands Monthly* 69(7), 34–37.
Neale, G. (1998) *The Green Travel Guide*. Earthscan, London.
New industry for Kiribati (1998). *Fiji Times Weekend* 11(19 September).
Niedenthal, J. (2001) *For the Good of Mankind: a History of the People of Bikini and Their Islands*. Micronitor Publishing, Majuro, Marshall Islands.
O'Grady, A. (ed.) (1990) *The Challenge of Tourism*. Ecumenical Coalition on Third World Tourism, Bangkok.
Pearce, D.G. (1995) Tourism in the South Pacific: patterns, trends and recent research. *Progress in Tourism and Hospitality Research* 1, 3–16.
Ralston, C. (1978) *Grass Huts and Warehouses: Pacific Beach Communities of the Nineteenth Century*. University of Hawai'i Press, Honolulu.
Ranck, S.R. (1987) An attempt at autonomous development: the case of the Tufi guest houses, Papua New Guinea. In: Britton, S. and Clarke, W.C. (eds) *Ambiguous Alternatives: Tourism in Small Developing Countries*. The University of the South Pacific, Suva, pp. 154–166.
Ratuva, S. (2000) The Fiji coup of May 2000 and the indigenous question. *Indigenous Affairs* 3(July–August–September), 52–57.

Ringer, G. (2004) Geographies of tourism and place in Micronesia: The 'sleeping lady' awakes. In: Harrison, D. (ed.) *The Journal of Pacific Studies*. 26, pp. 131–150.
Robie, D. (1989) *Blood on Their Banner: Nationalist Struggles in the South Pacific*. Zed Books, London.
Rudkin, B. and Hall, C.M. (1996) Unable to see the forest for the trees: ecotourism development in Solomon Islands. In: Butler, R. and Hinch, T. (eds) *Tourism and Indigenous Peoples*. International Thomson Business Press, London, pp. 202–226.
Russell, D. and Stabile, J. (2003) Ecotourism in practice: trekking the highlands of Makira Island, Solomon Islands. In: Harrison, D. (ed.) *Pacific Island Tourism*. Cognizant, New York, pp. 38–57.
Ryan, C. (2001) Tourism in the South Pacific—A case of marginalities. *Tourism Recreation Research* 26(3), 43–49. DOI: 10.1080/02508281.2001.11081198.
Samoa Visitors Bureau (2001) *Annual Report (unpublished)*. Samoa Visitors Bureau, Apia.
Schuller, W. (1995a) *Kiribati: Tourism Sector Review*. Tourism Council of the South Pacific, Suva.
Schuller, W. (1995b) *Tuvalu: Tourism Sector Review*. Tourism Council of the South Pacific, Suva.
Schuller, W. (1995c) *Solomon Islands: Tourism Sector Review*. Tourism Council of the South Pacific, Suva.
Scott, R.J. (1970) The development of tourism in Fiji since 1923. *Transactions and Proceedings of the Fiji Society* 12, 40–50.
Slot, J.R.V. (1996) *An introduction to the Business of Ecotourism: Participants' Booklet*. College of Business Administration, University of Hawai'i, Honolulu.
Sofield, T.H.B. (1991) Sustainable ethnic tourism in the South Pacific: some principles. *The Journal of Tourism Studies* 2(1), 56–72.
Sofield, T.H.B. (1996) Anuha island resort, Solomon Islands: a case study of failure. In: Butler, R. and Hinch, T. (eds) *Tourism and Indigenous People*. International Thomson Business Press, London, pp. 176–202.
Staiff, R. and Bushell, R. (2004) Healthy tourism in a Fijian context. In: Harrison, D. (ed.) *The Journal of Pacific Studies*. 26, pp. 103–130.
TCSP (1988) *Nature Legislation and Nature Conservation as a Part of Tourism Development in the Island Pacific*. TCSP, Suva.
TCSP (1990) *Vanuatu: Evaluation and Development Needs*. Tourism Sector Report, TCSP, Suva.
TCSP (1992) *Tuvalu Tourism Development Plan, 1993–2002*. TCSP, for the Government of Tuvalu, Suva.
TCSP (1997a) *The Islands of Tuvalu*. TCSP, Suva.
TCSP (1997b) *Tourism in the South Pacific*. TCSP, Suva.
TCSP (1998) *South Pacific Regional Tourism Marketing Plan, 1998–2002*. TCSP, Suva.
Thomson, A. (1996) Tourism: prospects for the year 2000 and beyond. In: Grynberg, R. (ed.) *Economic Prospects for the Pacific Islands in the 21st Century*. School of Social and Economic Development, The University of the South Pacific, Suva, pp. 101–111.
Tourism Resource Consultants (TRC) (1999) *Report on the Pacific Ecotourism Workshop, Taveuni, Fiji Islands, 28–31 July*. TRC, Wellington.
Tran, M. (1996) Smiles vanish in the fight for rights. *The Observer* 25 August.
Twining-Ward, L. and Tuailemafua, T. (2004) Monitoring sustainable tourism development in Samoa. In: Harrison, D. (ed.) *The Journal of Pacific Studies*. 26, pp. 77–103.
Vidal, J. (1995) Rats of the rubbish society fight back. *The Guardian 9 August* 23.
Wahab, S. and Pigram, J. (eds) (1997) *Tourism, development and growth: the challenge of sustainability*. Routledge, London.
Waqa, V. (1998) The land problem. *Islands Business* 24(4), 32–33.
Waqa, V. (1999) Waiting (and waiting) for the TV millions. *Islands Business* 25(2), 20.
Waters, M. (1995) *Globalization*. Routledge, London.
Watling, D. and Chape, S. (1992) *Environment: Fiji: The national state of the environment report*. IUCN-The World Conservation Union and the Government of Fiji, Suva.
Weaver, S. (1992) Nature tourism as a means of protecting indigenous forest resources in Fiji. *Journal of Pacific Studies* 16, 63–73.
Weaver, D. (2001) Mass tourism and alternative tourism in the Caribbean. In: Harrison, D. (ed.) *Tourism and the Less Developed World: Issues and Case Studies*. Wiley, Chichester, pp. 161–174.
Weaver, S. and King, B. (1996) Environmental management and the Fiji tourism industry. *Journal of Pacific Studies* 19, 127–144.
Wheeller, B. (1993) Sustaining the ego. *Journal of Sustainable Tourism* 1(2), 121–129. DOI: 10.1080/09669589309450710.

Wight, P. (1994) The greening of the hospitality industry: economic and environmental good sense. In: Seaton, A.V., Jenkins, C.L., Wood, R.C., Dieke, P.U.C., Bennett, M., *et al*. (eds) *Tourism: the State of the Art*. Wiley, Chichester, pp. 665–674.
Wood, D. (1968) *Trinidad in Transition: the Years After Slavery*. Oxford University Press, Oxford.
Worsley, P. (1968) *The Trumpet Shall Sound: a Study of 'Cargo' Cults in Melanesia*, 2nd edn. MacGibbon & Kee, London.
WTO (1994) *National and Regional Tourism Planning: Methodologies and Case Studies*. Routledge, London.
WTO (2001a) *Compendium of Tourism Statistics: 2001 Edition*. World Tourism Organization, Madrid.
WTO (2001b) *Tourism Highlights, 2001*. World Tourism Organization, Madrid.
WTO (2002) *Compendium of Tourism Statistics: 2002 Edition*. World Tourism Organization, Madrid.
Wyllie, R.W. (1998) Hana revisited: development and controversy in a Hawaiian tourism community. *Tourism Management* 19(2), 171–178. DOI: 10.1016/S0261-5177(97)00109-X.

11

Contested Narratives in the Domain of World Heritage

The focus of this chapter is on the various relationships of world 'heritage' and tourism, and the purpose of this introduction is to raise issues that are germane to the topic, and in the process briefly relate them to the papers that follow.

The scene can be set with an anecdote. Many years ago, one of my neighbours was a Scotsman. He was stereotypically dour, in appearance and manner. A man of few words at the best of times, who struggled to respond to a greeting, he was never one to approach for a cup of sugar. However, once in a while he was transformed. On St Andrew's day, which is of great significance to Scotsmen, he would appear resplendent, immaculate, in kilt, sporran, and all the bright and shiny accessories that accompanied what, to this somewhat ignorant Sassenach, appeared to be a quite royal outfit. On one such day, I congratulated him on his appearance and casually (and undiplomatically) suggested, in passing, that the kilt and the tartan were actually English inventions the Scots had been persuaded to adopt, to the great profit of English tailors and the English cloth trade.

As anyone familiar with the work of Trevor-Roper (1983) would realise, the comment was not original. Nevertheless, it seems to have been unappreciated. I say it seems so, because although we remained neighbours for several more years, it was the last conversation we held. From that time on my greetings in the street were utterly ignored.

Clearly, heritage is no joking matter. Wearing national costume is a mark or statement of at least two kinds of identity: one that is national or collective, and the other that is individual. And for Scots living outside Scotland, assertions about identity may be more important than for those in Scotland. For those far from 'home', the exile, the refugee or the expatriate, it may be imperative to preserve links with another 'place'. More collectively, in immigrant communities, the things of 'home' take on an additional poignancy, a bittersweet remembrance of things past, which are present, and yet are not. Faced with foreignness, with alienation, the appeal to a collective and individual 'heritage' takes on an added momentum.

Conflicting Inter-subjectivities

Landscapes, too, are often incorporated into individual and collective 'heritage'. The work of poets may reinforce feelings of association, as when Wordsworth writes of the daffodils of the English Lake District or R.S. Thomas describes the bleak Welsh hills, and removal from the 'place' of home may result in alienation and restlessness. Dutch travel writer Cees Nooteboon (1993) recounts the story of an old sailor, recently retired, who was persuaded by his wife to take a holiday in Switzerland. They arrived at night, and in the morning he climbed out of bed and drew the curtains to look outside. Seeing the mountains, he disappointedly turned to his wife, and said 'I want to go home'. 'Why?' she asked, 'we have only just arrived.' 'I can't see the view', he said.

There is nothing 'natural' in our appreciation of landscape. We learn to appreciate it through our backgrounds and socialization, but the socialization of the expert may differ from that of the layman, and thus interpretations of what is natural will vary. Another example can illustrate this point. Near Brighton, on the south coast of England, there is a *camera obscura*, a darkened room, situated on a tower, with a lens on the roof through which is projected an image of the surrounding countryside. Through this, with the aid of a commentary provided by an interpreter, can be viewed the South Downs – an Area of Outstanding Natural Beauty. During one presentation, it was explained that South Downs rangers must ensure that the Downs remain pristine, and as part of this task must root out any gorse bushes (spiny yellow shrubs) they see growing, as they are not appropriate. Gorse bushes, it seems, are 'wrong' bits of nature, more like 'weeds', and have no place in an area of outstanding natural beauty.

What precisely is 'natural' about the South Downs can be disputed. A few hundred years ago it was heavily wooded, until the combined needs of the English navy for wood and farmers requiring grazing for their sheep conspired to leave them in their current 'natural' state. However, conflicts over the definition and exploitation of other 'natural' sites and resources are commonplace (Brandon and Wells, 1992), and examples involving tourists and residents can be cited from Newfoundland (Overton, 1980), Australia (Brown, 1999), Canada (Ritchie, 1998), the English Lake District (Clark *et al.*, 1994), South Pacific islands (Baines, 1987) and Kenya (Sindiga, 1999). What is frequently found is that the residents of such areas want to build houses, develop their businesses, harvest the forest or the wildlife, and generally put their surroundings to work. By contrast, town dwellers or tourism promoters, who may exert considerable influence with the politicians, prefer the wilderness untouched, to be visited on vacations, at weekends, for fishing or photography, perhaps, or to enable outsiders to commune with nature.

Although positions over the use of natural and built environments may seem entrenched, they do change. When tourists first arrived at a Catalan seaside resort in Spain, for example, many local residents regarded them as interlopers, who interfered with the fishing industry. Later, when tourism became established, it was the fishermen who were considered the outsiders (Pi-Sunyer, 1989, pp. 196–197).

There are, then, conflicts over what is natural and what is not, and how 'natural' sites are to be used. Something similar occurs over the built environment, particularly when buildings imbued with special meaning and significance are competed over by different groups, be they ethnic, religious or national. Nevertheless, a territory does not have to be considered sacred for violent conflicts to occur. Indeed, it is perhaps worth considering how many of these conflicts over territory, religious or not, are actually between groups of people who are in most ways very similar. Jews and Moslems, Protestants and Catholics, Turkish Cypriots and Greek Cypriots: is it because they *are* so similar that the conflicts are so bitter, so violent, so destructive?

Faking It: Heritage, Performance, Commoditization and Authenticity

Much of what we consider 'heritage' is a form of performance, and there are frequent debates about what is 'authentic' and what is 'fake' (MacCannell, 1976, pp. 91–107; Harrison, 1992, pp. 20–2; Olsen, 2002; Shepherd, 2002). In this context, however, it might be argued that we – and the term is used advisedly – 'perform' heritage for the benefit of people who are not 'us'. Some time ago, for example, in a study of tourism in Brittany, it was reported that there was a renewed emphasis on rural Brittany as a centre of tradition, where the Breton 'heritage', its language, customs, and cuisine, were very much alive. Eco-villages were started, villagers dressed in 'traditional' costume and old buildings were restored. Middle-class tourists from Paris, seeking a world they thought lost, spent holidays with traditional Breton families, learning their language as they carried on their daily duties, and returned to modern Paris refreshed in the knowledge that Breton tradition was alive and well. For their part, at the end of the tourist season these rustic Bretons breathed a huge sigh of relief, put away the uncomfortable old furniture, and returned to speaking French again. The performance was over (McDonald, 1987, pp. 131–132).

In MacCannell's terms, the Bretons were putting on a performance of 'staged authenticity' (1976, p. 98). Was the experience 'fake'? One assumes the hapless visitors were unaware of the deception and, if so, it might be described as a 'genuine fake' (Brown, 1996). Does it matter if it was? Does it matter if buildings considered to be part of heritage are also fake? The Pavilion in Brighton, England, for example, once considered a monstrosity and very nearly pulled down by an irate council earlier last century, is now a crucial feature of Brighton's 'heritage', and some years ago was duly restored. Most visitors to Brighton go there, but at least some of what they see is definitely 'fake'. The turrets, for example, are made of fibreglass – not an 18th-century product – because they are lighter and more resilient to the weather. Does it matter?

Clearly, the debate over what is or is not 'authentic' applies as much to historical objects as to buildings and rituals. It is highly unlikely that most visitors to the British Museum – including Greek visitors – could distinguish the genuine Elgin/Parthenon marbles from plaster casts. Nevertheless, there is continuing dispute about where they should be housed. The debate is not simply about whose *culture* is represented – Greek, British or the world's – but about who is entitled to possess the genuine article. In the West, certainly, much store is set on something being 'the real thing'. A fake Rolex may keep time as accurately as one which came directly from the manufacturer, but (if its provenance is known) is valued far less, and not simply because of the difference in monetary price. It is the *sense* of the real, the authentic. It is what it symbolizes. Similarly, no one is going to be that impressed by a fine photograph of the Mona Lisa, and stories about discovered or rediscovered masterpieces are common.

Ultimately, perhaps, what matters are the meanings that people project on to these inanimate objects, these 'things men have made' – to quote D.H. Lawrence. Such meanings are the result of a complex and continuous process of socialization, symbolic interaction and negotiation, where we learn the value of x, y or z. Even under conditions of capitalism, these meanings continue to be defined and redefined.

That social scientists may underestimate the ability of members of cultures to adapt and change can be illustrated by reference to Fuentarrabia, in Spain. As reported by Greenwood in the 1970s, the town's annual festival of the *Alarde* celebrated a historic victory over the French, but when it was shortened for the benefit of tourists, and then performed twice rather than once, he considered it had 'lost meaning'. This was symptomatic of a commercialized yet devalued culture, bought and sold in the tourism industry as 'culture by the pound'. In fact, re-visiting Spain several years later, he found that despite tourism, the festival was thriving. However, he *also* discovered its meaning to the participants had changed, and was now perceived by them to be a celebration of Basque nationalism. It was an honest reassessment, an admission that what he had previously written 'was an expression of both anger and concern' (Greenwood, 1989, p. 181), but one that is rarely mentioned by many tourism commentators, who find his initial analysis more politically correct and convenient.

Tourism may indeed introduce or exacerbate major social change, but it is easy to forget the hardships of work in agriculture in less developed countries, which is often depicted as rustic and idyllic, when life might more accurately be described as nasty, brutish and short. A similar process occurred in 19th century England, when the invisibility of the rural poor 'helped perpetuate and uphold the myth of Arcadian beauty which could be seized upon and utilised in selective imagery for a great variety of purposes' (Short, 1992). And Short continues:

> Thus to be truly English by the beginning of the twentieth century was to be rural. But 'rural' did not really mean rural *people*, and especially not *poor* rural people, who might not even fit stereotypes of poverty by living in picturesque cottages. The countryside was made by working people, but the rural idyll of pastoral [life] from the eighteenth to the twentieth century, itself an

urban product, has largely banished them from the scene (Short, 1992).

While tourism development is likely to promote or exacerbate the process of commoditization, this is not necessarily a new phenomenon. The museums and art galleries so beloved of cultural tourists include much that was traded, even in small-scale societies, and it was the market (albeit sometimes a restricted one) that prompted their production and encouraged the use of skills that might otherwise have fallen into disuse. Furthermore, while indigenous arts and crafts have sometimes been undermined by tourism (Graburn, 1984), the literature is replete with examples of where tourism has prompted a creative reaction in local people and has been instrumental in reviving arts and crafts that would otherwise have disappeared (Popelka and Littrell, 1991; Cohen, 1993; Daniel, 1996; Silverman, 2000).

To the Victor the Spoils

According to Hewison, though, it is the promoters of tourism who sanitise the past. In re-presenting it as entertainment, the 'heritage industry' mocks the dark, grim reality of what actually happened. Mines are cleansed of their dirt and their danger, nuclear test sites become 'heritage centres', and museums compete to 'exploit the opportunities for the sale of souvenirs, refreshments and so forth' (Hewison, 1989, p. 19). He continues:

> History is gradually being bent into something called Heritage, whose commodity values run from tea towels to the country house. My criticism is not simply that it is largely focused on an idealised past whose social values are those of an earlier age of privilege and exploitation that it serves to preserve and bring forward into the present. My objection is that Heritage is gradually effacing History, by substituting an image of the past for its reality. Our actual knowledge and understanding of history is weakening at all levels, from the universities to the primary schools (Hewison, 1989, p. 21).

This is a powerful argument, but it does rather assume that someone, somewhere, has privileged access to *real* knowledge, to a proper understanding, of what history and heritage are *really* about. By contrast, though, whatever elements of the past are presented as heritage – arts and crafts, events, rituals or buildings – they have already passed through a complex filtering process whereby someone, or some group, has *selected* them. Nothing – but nothing – is automatic heritage material. As Weber put it, 'all knowledge of cultural reality is always knowledge from particular points of view' (Weber, 1949, p. 81).

In the United Kingdom in 1983 it may have seemed sensible enough for the Natural Heritage Conference to define 'heritage' 'as that which a past generation has preserved and handed on to the present and which a significant group of the population wishes to hand on to the future' (Hewison, 1989, p. 16). The definition begs several questions. Why was something considered worth 'handing on'? Who or what is the '*significant group*'? Who does the selecting? As the case studies discussed in this volume show, and as Lowenthal has discussed at length (Lowenthal, 1985, 1997), when such questions are posed it is apparent that what is presented from the past, and how it is portrayed and interpreted, is a crucial ingredient in the continuous formation and re-formation of perceptions of the present (Uzzell, 1989a, b). It is a process in which numerous groups and agencies jockey for influence and power, and where dealers in 'authenticities' range from casual excursionists to 'objective' ethnographers and museum curators (Cohen, 1988).

If this view is accepted, it becomes much easier to explain the key feature of the process whereby sites are selected for special treatment as 'heritage' sites of some kind or another. It can be crudely summarized as 'to the victor the spoils'. The achievements of vanquished peoples are rarely accorded the luxury of heritage status, and Robben Island, for instance, would never have been nominated for World Heritage Listing by a white government in South Africa.

However, the cultural contributions of the defeated and the subordinate are not necessarily lost. They can be absorbed and/or re-interpreted. The 'Turkish' belly dance, for instance, was an import from Turkey's Arab possessions. Tikka Masala, a dish served in Indian restaurants in the UK and said by Robin Cook, a prominent British politician, to be an essential component of a 'traditional' Saturday night out in Britain, is a 'South Asian' dish invented specifically for the English palate! At other times, a dominated group may take on at least the semblance of the colonizer's culture. We thus find that Methodism is accorded a key place in 'traditional' indigenous Fijian culture, to the extent that some Fijian nationalists want Fiji (with 45% of its population of Asian origin, and primarily Hindu) to be formally declared 'a Christian country'. By contrast, the cannibalism once practised by indigenous Fijians is little remembered, and when it *is* recalled, it is with a degree of embarrassment.

Let Us Forget

In fact, where heritage is involved, collective amnesia is common. What is remembered, as tradition or heritage, is selected from a vast range of built, natural and cultural environments, to celebrate the past and bolster the present. Shameful episodes are rarely given prominence. At least initially, the memorial in Jerusalem to the massacred Jews, *Yad Vashem*, was only possible in Israel – and deals but little with non-Jews who shared the same fate. And as a memorial, like so many other memorials, it serves a dual purpose, enabling visitors to remember the fallen and simultaneously symbolizing Israeli nationalism (Bowman, 1992, p. 129). By contrast, the death camps of Auschwitz-Birkenau in Poland evoke much ambivalence among residents who live in their shadow, and are virtually ignored in some tourist promotional material (Lennon and Foley, 2000, pp. 63–4). Noticeably, while there are many examples of thanatourism, or dark tourism, this complex is rare among World Heritage Sites in that it is a memorial to infamy and shame rather than a celebration of past glories.

Class, status, power and nationalism: all are involved in the presentation and re-presentation of 'heritage', and as groups and classes rise and fall, so, too, do the claims for attention of different sites. In Cambodia, the Khmer Rouge did their best to destroy what they saw as the intellectual element of the society and its achievements and the current regime has to decide how far this resurgent nation should go in publicizing what happened in the 'killing fields'. The sites of torture, death and burial are clearly part of the nation's heritage, and in Phnom Penh the former high school now known as *Tuol Sleng*, where the Pol Pot regime tortured its victims, and the nearby Killing Fields, are graphic and gruesome reminders of what happened. Both are much visited by tourists, but there are still many in Cambodia who would prefer to forget – or have others forget – this period of torture, massacre and ignominy.

By contrast, Angkor Wat, the significance of which is debated by Tim Winter in this volume (Winter, 2005), is a different matter, and there seems to be less of a problem in celebrating this magnificent complex (against Thai assertions) as the achievement of the Khmer nation.

Similar issues arise when one considers how periods of colonialism are portrayed. In this volume, examples are given of the dilemmas faced by governments and tourism promotion agencies in depicting the colonial 'experience' in, for example, Singapore and Fiji. If colonialism was as damaging to the development of the colonized as is usually claimed, why should its buildings and anything else associated with it be celebrated at all, far less glorified as a form of achievement? Are the quaint old buildings in the centre of very modern cities – as in the case of Kuala Lumpur or Singapore – simply there as tourist attractions? If this is so, it might be argued that designating them as 'heritage' is little other than a ploy to attract tourists.

Ambiguities also surround the role of sites related to African slavery in the UK, USA, Caribbean and West Africa. Initially, their existence seems to confirm what Dann and Seaton (2001, p. 19) describe as a 'domination critique', which 'suggests that the determination of slavery heritage was mainly by powerful business institutions, their commercial agenda and the mainly white audiences they targeted'. If so, this would be a further exploitation of black slaves for white profit. However, as the authors recognize, the available facts are inconsistent with such a hypothesis. Not all sites are financed by private capital or geared to profit, and some are geared to black as well as white consumption. There was also a perception that some black people opposed slavery heritage sites on the grounds that they accentuated and perpetuated social disadvantage. Those who promoted them were in a no-win situation: 'If slavery heritage is not memorialised, it can be read as suppression; if it is commemorated, such heritage may be construed as unethical or compromised truth' (Dann & Seaton, p. 20).

Narratives of heritage, and the domain that heritage covers, are contested because there is nothing intrinsically sacrosanct about any building, any part of nature, or any cultural practice. As social relations ebb and flow, as one class or pressure group takes ascendancy over another, new perceptions, new views on the past and what was of *value* in the past, also take over. Previous accounts are challenged. Old statues are removed and new ones installed; Marx and Engels are replaced by new icons.

Putting it more phenomenologically, what is considered 'heritage' is continually subject to interpretation and re-interpretation, claim and counter claim, and negotiation. Whether we are dealing with formal categorizations of heritage – on the World Heritage List, or in any national hierarchies – the outcomes will depend on the balance of status and power at any one time and on who among the numerous stakeholders (if, indeed, all can be delineated) has the loudest voice. Who *are* the stakeholders? And how much are those who have the least power *really* involved in decisions about who or what constitutes heritage, or how it should be managed? Studies in this volume of heritage sites in the Netherlands, the USA and Mexico, in Kyrgyzstan, Cambodia, Vietnam and India, and Fiji strongly suggest that their participation is often minimal. The meanings such sites hold for them may be quite different from those propounded by national tourism marketing agencies and UNESCO. Similarly, taking a more historical view of heritage, where can the voices of the previously dispossessed be heard?

Who Decides?

It has been suggested that heritage is about individual and collective identity, that conflict occurs because we are socialized into cultures which predispose us to favour one set of indicators that include us in a specific group or category – hence excluding others from members of our group and ourselves from membership of theirs. Culture includes and excludes. It has been suggested that heritage may be about performance, 'fake' or 'genuine', and that this is linked to notions and judgements about the role of the *market*. Most of all, what is defined as heritage is linked to *power*: the power to impose a view of the world, especially of the past, on others. Perceptions of the past are closely linked to present hierarchies, and the voices of those at the top are often the most likely to prevail. By contrast, the voices of the most lowly stakeholders – whose houses were built of wood or thatch rather than stone – are muted or silent, and those whose lives were blighted by massacre, torture and discrimination are also less remembered. The unknowns of the past lie in unmarked tombs, where no flame burns.

It would be comforting if museum curators and archaeologists could define once and for all what is of world importance, of universal value. Unfortunately, despite occasional claims to the contrary, this is not possible. Like 'lay' members of society, they too are social products of their time, of their cultures, and they too are in the business of articulating *stories*, from their own perspective.

This also applies to guides, whose performance invariably has political ramifications, which may sometimes be quite explicit. Moshe Dayan, an Israeli General, reportedly said he would prefer to contend with an Arab bomber pilot than a Palestinian guide (Bowman, 1992, p. 131). Even if the comment were apocryphal, it contains the germ of accuracy, as the tourist guide – along with the guidebook – is often the visitors' main source of information about a destination area. Many states have recognized this, and Israel is not alone in ensuring that, as far as possible, international visitors 'see' the sights from the perspective of a guide who has been trained and licensed by the state. By contrast, as Gemma McGrath indicates later in this volume when referring to Peru (McGrath, 2005), there is much to be said for exposing tourists to local guides, and their distinct narratives.

It would be tempting, perhaps, to view the World Heritage Committee as a ruling body, the cultural equivalent of FIFA in association football, that could stipulate what is and what is not 'world heritage'. However, despite its title, it cannot determine which sites should be included on the World Heritage List. The Committee itself does not make nominations. Instead, these come from nation-states, and an application for World Heritage Listing is far from the end of a complex political process. Even when tentative lists are submitted, and preliminary conditions met by the submitting state party, it is difficult for the Committee to decide which sites are of universal value.

In any case, applications for World Heritage Status are neither made nor received in a global vacuum and ... Europe and Judaeo-Christian monuments and sites continue to dominate the List. Such an international imbalance has been recognized by the World Heritage Committee, and there is now a *political* imperative to go out to other parts of the world and find more sites! Understandable though the desire to widen representation on the List might be, however, such a quest may sit uneasily with the establishment of criteria that clearly establish sites to be of universal significance.

Thirdly, many UNESCO employees are part of an international elite, which others are anxious to join. Academics, archaeologists, surveyors and restorers, for example, have their own interests in working for UNESCO, and inevitably (and quite properly) they develop links with others of like mind in specific nation-states. One consequence of this international network is that UNESCO officials may deliberately or unwittingly seek to influence which built, natural or cultural sites are selected for possible inclusion on the World Heritage List. Indeed, if they can persuade countries or regions with little representation on the World Heritage List to submit other sites for consideration, they might do so at considerable benefit to their careers.

Finally, in UNESCO-organized activities, 'supervision' by experts can sometimes come to mean domination by experts. UNESCO support is valued, prestigious and important, and in many respects UNESCO sets the agenda. Where, even if mistakenly, it is felt inclusion on the World Heritage List might bring more tourists, and would thus increase economic prosperity and status, it might be considered politic to do what the experts suggest.

While it would again be comforting if these dilemmas could be resolved, this is unlikely to occur. The idealistic quest for universals in heritage will always be conducted within the inter-subjective and highly political process in which World Heritage Listing takes place. As Tomke Lask and Stefan Herold suggest (Lask and Herold, 2005), there would be clear benefits in setting up what they term 'tourism observation' stations in tourist destination areas, where all stakeholders could come together in a continuous exchange of information and concerns, and where intercultural interaction and exchange become genuine possibilities. However, the difficulties in such a project should not be underestimated. Debates and conflicts over 'heritage' take place in an ever-shifting scenario, where the 'achievements' of one class, one ethnic group, one nation-state, one era, are always negotiated and re-assessed by the next. History will be doctored, presented and re-presented to suit the demands and the imperatives of the present. The barbarians are always at the gate, but today's barbarians are tomorrow's establishment. And much as we should like it to be otherwise, there is no objective referee 'out there', in Paris, or in any UNESCO office, who can decide for us.

References

Baines, G.B.K. (1987) Manipulators of islands and men: Sand-cay tourism in the South Pacific. In: Britton, S. and Clarke, W.C. (eds) *Ambiguous Alternative: Tourism in Small Developing Countries*. University of the South Pacific, Suva, pp. 16–24.

Bowman, G. (1992) The politics of tour guiding: Israeli and Palestinian guides in Israel and the occupied territories. In: Harrison, D. (ed.) *Tourism and the Less Developed Countries*. Belhaven, London, pp. 120–134.

Brandon, K.E. and Wells, M. (1992) Planning for people and parks: design dilemmas. *World Development* 20(4), 557–570. DOI: 10.1016/0305-750X(92)90044-V.

Brown, D. (1996) Genuine fakes. In: Selwyn, T. (ed.) *The Tourist Image: Myths and Myth Making in Tourism*. Wiley, Chichester, pp. 33–47.

Brown, T.J. (1999) Antecedents of culturally significant tourist behavior. *Annals of Tourism Research* 26(3), 676–700. DOI: 10.1016/S0160-7383(99)00004-3.

Clark, G., Darrall, J., Grove-White, R., Macnaghten, P. and Urry, J. (1994) *Leisure Landscapes: Leisure, Culture and the English Countryside: Challenges and Conflicts*. Council for the Protection of Rural England, London.
Cohen, E. (1988) Authenticity and commoditization in tourism. *Annals of Tourism Research* 15(3), 371–386. DOI: 10.1016/0160-7383(88)90028-X.
Cohen, E. (ed.). (1993) Tourist arts. *Annals of tourism research* 20.
Daniel, Y.P. (1996) Tourism dance performances authenticity and creativity. *Annals of Tourism Research* 23(4), 780–797. DOI: 10.1016/0160-7383(96)00020-5.
Dann, G. and Seaton, A. (2001) Slavery, contested heritage and thanatourism. In: Dann, G. and Seaton, A. (eds) *Slavery, Contested Heritage and Thanatourism*. Haworth, New York and London, pp. 1–29.
Graburn, N.H.H. (1984) The evolution of tourist arts. *Annals of Tourism Research* 11(3), 393–419. DOI: 10.1016/0160-7383(84)90029-X.
Greenwood, D. (1989) Culture by the pound: an anthropological perspective on tourism as cultural commoditization. In: Smith, V. (ed.) *Hosts and Guests: The Anthropology of Tourism*, 2nd edn. University of Pennsylvania Press, Philadelphia, pp. 171–185.
Harrison, D. (1992) Tourism to less developed countries: the social consequences. In: Harrison, D. (ed.) *Tourism and the Less Developed Countries*. Belhaven, London, pp. 19–34.
Hewison, R. (1989) Heritage: an interpretation. In: Uzzell, D. (ed.) *Heritage Interpretation: Volume 1. The Natural and Built Environment*. Belhaven, London and New York, pp. 15–23.
Lask, T. and Herold, S. (2005) An observation station for culture and tourism in Vietnam: a forum for world heritage and public participation. In: Harrison, D. and Hitchcock, M. (eds) *The Politics of World Heritage: Negotiating Tourism and Conservation*. Channel View Press, Clevedon, pp. 119–131.
Lennon, J. and Foley, M. (2000) *Dark Tourism: The Attractions of Death and Disaster*. Continuum, London and New York.
Lowenthal, D. (1985) *The Past is a Foreign Country*. Cambridge University Press, Cambridge.
Lowenthal, D. (1997) *The Heritage Crusade and the Spoils of History*. Cambridge University Press, Cambridge.
MacCannell, D. (1976) *The Tourist: A New Theory of the Leisure Class*. Macmillan, London and Basingstoke.
McDonald, M. (1987) Tourism: chasing culture and tradition in Brittany. In: Bouquet, M. and Winter, M. (eds) *Who From Their Labours Rest*. Avebury, Aldershot, pp. 120–134.
McGrath, G. (2005) Including the outsiders: the contribution of guides to integrated heritage tourism management in Cusco. In: Harrison, D. and Hitchcock, M. (eds) *The Politics of World Heritage: Negotiating Tourism and Conservation*. Channel View Press, Clevedon, pp. 137–146.
Nooteboon, C. (1993) *The art of travel*. Radio 4, British Broadcasting Corporation.
Olsen, K. (2002) Authenticity as a concept in tourism research: the social organization of the experience of authenticity. *Tourist Studies* 2(2), 159–182.
Overton, J. (1980) Tourism development, conservation, and conflict: game laws for Caribou protection in Newfoundland. *The Canadian Geographer/Le Géographe canadien* 24(1), 40–49. DOI: 10.1111/j.1541-0064.1980.tb00974.x.
Pi-Sunyer, O. (1989) Changing perceptions of tourism and tourists in a Catalan resort town. In: Smith, V. (ed.) *Hosts and Guests: The Anthropology of Tourism*, 2nd edn. University of Pennsylvania Press, Philadelphia, pp. 188–199.
Popelka, C.A. and Littrell, M.A. (1991) Influence of tourism on handcraft evolution. *Annals of Tourism Research* 18(3), 392–413. DOI: 10.1016/0160-7383(91)90048-G.
Ritchie, J.R.B. (1998) Managing the human presence in ecologically sensitive tourism destinations: insights from the Banff-Bow valley study. *Journal of Sustainable Tourism* 6(4), 293–313. DOI: 10.1080/09669589808667318.
Shepherd, R. (2002) Commodification, culture and tourism. *Tourist Studies* 2(2), 183–201. DOI: 10.1177/146879702761936653.
Short, B. (ed.) (1992) *The English Rural Community: Image and Analysis*. Cambridge University Press, Cambridge.
Silverman, E.K. (2000) Tourism in the Sepik river of Papua New Guinea: favoring the local over the global. *Pacific Tourism Review* 4(2–3), 105–120.
Sindiga, I. (1999) Alternative tourism and sustainable development in Kenya. *Journal of Sustainable Tourism* 7(2), 108–127. DOI: 10.1080/09669589908667330.
Trevor-Roper, H. (1983) The invention of tradition: the highland tradition of Scotland. In: Hobsbawm, E. and Ranger, T. (eds) *The Invention of Tradition*. Cambridge University Press, Cambridge, pp. 15–41.

Uzzell, D. (1989a) *Heritage Interpretation, Volume 1: The Natural and Built Environment*. Belhaven, London and New York.

Uzzell, D. (1989b) *Heritage Interpretation, Volume 2: The Visitor Experience*. Belhaven, London and New York.

Weber, M. (1949) *The Methodology of the Social Sciences (translated and edited by E.A. Shils and H.A. Finch)*. Free Press, New York.

Winter, T. (2005) Landscape, memory and heritage: new year celebrations at Angkor, Cambodia. In: Harrison, D. and Hitchcock, M. (eds) *The Politics of World Heritage: Negotiating Tourism and Conservation*. Channel View Press, Clevedon, pp. 50–65.

12

Lao Tourism and Poverty Alleviation: Community-based Tourism and the Private Sector [Co-authored with *Stephen Schipani*]

Tourism as 'Development'

The study of tourism's role as a tool for development is not new. From the 1960s, at least, academic approaches to tourism phenomena have included what might be described as a 'development' stream (de Kadt, 1979a; Peters, 1969; Smith, 1978a). This is seen in the earliest of contributions. Focusing on less developed countries and regions (LDCs), de Kadt recognized tourism could create jobs, develop backward linkages with agriculture and other economic sectors, create opportunities for young people and women, encourage local entrepreneurial activity, and improve the quality of life of the poor through funding basic utilities, training and education. He also noted a tendency for expatriates to obtain the best-paid jobs, and for tourism to exacerbate economic and other inequalities, and emphasized that the extent to which tourism would benefit residents in tourist destinations depended considerably on local conditions and was certainly not inevitable (de Kadt, 1979b, pp. 11–12).

De Kadt did not focus specifically on poverty, but clearly saw it as a feature of development, noting that in the 'development thinking' of the time, there was

> a realization that growth alone may not suffice to overcome poverty within a reasonable time, and that the distribution of the material benefits of development among the poorest countries and the poorest population groups within individual countries requires special attention (de Kadt, 1979c, p. xii).

Several contributors to de Kadt's volume considered the impact of tourism on the poor. Wilson, for example, noted that tourism had brought few benefits to the urban and rural poor of the Seychelles (Wilson, 1979, p. 229) and Andronicou (1979) that – as with capitalism generally – the economic benefits from tourism were not distributed equally to all Cypriots (Wilson, 1979, p. 258). By contrast, Groupe Huit suggested that the poor of Tunisia *did* benefit, especially through employment at the construction stage of tourism development (Groupe Huit, 1979, p. 291).

The overall impression of tourism's economic benefits is similarly mixed in Smith's edited volume. Summarizing the views of her contributors, she suggests the extent to which tourism brought economic benefits to different sectors of the population depended on 'the type of tourism, the expectations of the tourists, and the host's ability to provide appropriate facilities and destination activities', and, she continued, 'the effects of tourism can be assessed along a continuum from a highly positive relationship that benefits all, to a highly disruptive, negative interaction fraught with conflict' (Smith, 1978b).

During the 1980s and 1990s, tourism studies broadened across a wide range of academic interests and disciplines. There was a general move to a more empirical 'knowledge-based' orientation (Jafari, 1989) and, among planners, to a wider, more holistic planning process (Hall *et al.*, 1997, pp. 18–22). Simplistic models of modernization and underdevelopment were rejected for a more comprehensive globalization framework (Harrison, 2001; Scheyvens, 2002; Telfer, 2002; Mowforth and Munt, 2003); environmentalist perspectives joined economic, social and cultural analysis; top-down development was (largely) rejected (theoretically) in favour of bottom-up development and local participation, and locally-owned, small-scale tourism enterprises were preferred (at least by academics) to transnational companies and mass tourism (Rodenburg, 1980). As a consequence, sustainable tourism development, especially ecotourism, became the preferred future of tourism in LDCs (Boo, 1990; Cater and Lowman, 1994; Weaver, 1998; Fennell, 1999; Honey, 1999).

During this period, too, some did investigate tourism's impacts on the poor, often as part of wider anthropological tourism studies (e.g. Freitag, 1994; van den Berghe, 1994, pp. 50–52; Chambers, 2000, pp. 32–39) and other researchers spent many years investigating such phenomena, as indicated, for example, by Momsen's continued interest in tourism's linkages to the agricultural sector in the Caribbean (Momsen, 1972, 1986; Torres and Momsen, 2004).

The 1980s and 1990s saw the increasing popularization of variants of sustainable tourism development, including ecotourism (with 2001 being the 'year of ecotourism') and other forms of alternative tourism, but just as the notion of sustainable development has itself been criticized (Adams, 1990), so too has its application to tourism. Ecotourism and other forms of 'alternative' or 'sustainable' tourism were said to be 'the thin end of the wedge' (Butler, 1992), to lack the potential to replace or improve on mass tourism, and to have negative impacts of their own (Cohen, 1989; Butler, 1999; Duffy, 2002), perhaps even leading to mass tourism (Wheeller, 1993). In addition, it has been argued that 'sustainability' is difficult to measure or apply in practice (Harrison, 1996) and that it has often been 'hijacked to give moral rectitude and 'green' credentials to tourist activities' (Mowforth and Munt, 2003, p. 80). Of special relevance to those with an interest in 'pro-Poor tourism', it has been suggested that claims that sustainable tourism development, especially ecotourism, specifically benefits local people (Fennell, 2001, p. 407) have yet to be substantiated. 'Relatively little quantitative analysis of ecotourism's success in achieving conservation and development objectives has been reported' (Lindberg *et al.*, 1996, p. 544), and inputs from the informal sector are especially difficult to measure (Timothy and Wall, 1997, pp. 323–325). Referring explicitly to community-based tourism (CBT), one critic has commented that 'the literature is full of claims but short on data and quantitative analysis' (Kiss, 2004, p. 234). Indeed, studies demonstrating how economic benefits are *distributed* (from any type of tourism) are rare and, where assessments *have* been made, conclusions are contradictory. Some have argued that large-scale tourism can bring economic benefits, that leakages are low, and that it is an important tool in poverty alleviation for LDCs (Page, 1999, pp. 62–63; Butcher, 2003, pp. 89–90), while others claim local benefits are few and leakages high (Walpole and Goodwin, 2000, pp. 571–572). In fact, whether the focus is on mass (mainstream) tourism or small-scale ecotourism or community-based tourism, economic data are rarely available and extremely hard to collect, either for such apparently precise measures as the tourist income multiplier (Archer, 1996) or of the overall economic and other benefits achieved by the many aid agencies working at community level.

In recent years, academic studies of tourism and development have been enriched by a much more specific focus on the benefits that tourism might bring to the poor. Such approaches, often instigated by practitioners outside the academic mainstream, have brought a distinctive, empirical and more economic orientation to the study of tourism and development. The term 'pro-poor tourism' emerged in 1999 in a report by de Loitte Touche to the UK's Department for International Development (DFID), which then funded a series of research projects into ways which tourism could enhance the livelihoods of the poor (Department for International Development (DFID), 1999). The Pro-poor Tourism Partnership[1] defined pro-poor tourism as 'tourism that generates net benefits for the poor' (Ashley *et al.*, 2001, p. 2) and implemented a series of case studies designed to test experimental methods of poverty alleviation by stimulating local involvement, partnerships and procurement (Ashley, 2000). More particularly, drawing on experiences in Southern Africa, Uganda, the Gambia, Nepal, Ecuador, the Caribbean and the Czech Republic, case studies explored the benefits (and limitations) of partnerships involving the private sector and communities and community-based tourism enterprises (including non-government organizations). They show that companies can derive business advantages from closer links with communities which, for their part, receive enhanced economic benefits through direct employment, or (with training) by providing goods and other services the company needs in the course of its normal trading relationships (Ashley *et al.*, 2001; Pro-Poor Tourism Partnership, 2006).

The above pro-poor tourism publications were supplemented and influenced by initiatives from numerous other organizations. Many national aid agencies, for instance, have devoted resources to 'community-based tourism'. The Canadian Agency for International Development (CAID), for instance, funded the Bali Sustainable Development Project (Nash, 1996, p. 139); the New Zealand Agency for International Development (NZAID), formerly New Zealand Official Development Assistance (NZODA), financed ecotourism projects in Fiji (Ministry of Foreign Affairs and Trade, 1995), and AusAid, the Australian Development Agency, also has an interest in tourism development in the South Pacific.

Researchers from the Partnership also contributed to discussions in the World Tourism Organization, which since 2001 has promoted its own 'Sustainable Tourism: Eliminating Poverty (ST-EP)' initiative (World Tourism Organization (WTO), 2002; Denman and Denman, 2004). Other UN organizations, too, have become involved. Both the United Nations Commission on Sustainable Development (UNCSD) and the United Nations Environmental Programme (UNEP) have highlighted the role of tourism in poverty alleviation (Gerosa, 2003, p. 18), as has the United Nations Economic and Social Commission on Asia and the Pacific (ESCAP).

Defining 'poverty' extremely broadly, and moving from the very specific projects of the Pro-Poor Tourism Partnership, ESCAP, in effect, returned to tourism as a development 'tool'. Following a seminar on tourism and poverty alleviation, held in 2003, it issued a massive manual, aimed primarily at government officials, with practical information to enable the planning and management of sustainable tourism, and details of 'techniques for achieving pro poor tourism' (Jamieson, 2003). It later published a series of country studies, and examples of good practice in pro-poor tourism, also presented at the 2003 seminar (UN Economic and Social Commission for Asia and the Pacific (ESCAP), 2005).

The main point to note here is that while interest in tourism as a tool for 'development' started in the 1970s, the concentration on its role in alleviating poverty emerged only in the late 1990s and, since then, 'pro-poor tourism' has quickly become a recognized, worthwhile and more self-consciously *moral* approach to tourism, especially when practised by community-based organizations under the broader rubric of 'sustainable tourism'.

Despite, or perhaps because of, their current popularity, 'pro-poor tourism' approaches have not been without their critics. Noting that pro-poor tourism initiatives have emerged from very different kinds of organizations, with very different ideological orientations, Scheyvens suggests that at least in some cases they have been prompted by a 'neo-liberal' (rather than a 'critical' or 'alternative') agenda, which focuses on trade liberalization, market-led growth and private sector development (Scheyvens, 2004). Furthermore, she asserts (on her own admission, with very little evidence) that in such instances they have *not* benefited the poor. By contrast, in Bhutan and

Samoa, local factors, including indigenous ownership and control of the tourism sector, have led to economic benefits being distributed more widely (Scheyvens, 2004).

A similar position is taken by Mowforth and Munt. They suggest, first, that pro-poor tourism initiatives (of the kind advocated by the UK Pro-Poor Tourism Partnership) inevitably take place in developing countries that are economically disempowered in the world economic system. Quoting Ashley *et al.* (2001), they remark that most 'critical decisions that affect this sector are made outside the country or by a few powerful local interests' (Mowforth and Munt, 2003, p. 271). Second, in so far as pro-poor initiatives *are* successful – and they note the uncertainties in many reports and case studies – pro-poor tourism does not necessarily eliminate or alleviate absolute poverty; rather, it is 'principally a measure for making some sections of poorer communities 'better-off' and reducing the vulnerability of poorer groups to shocks (such as hunger)' (Mowforth and Munt, 2003, p. 272).

Third, if it *does* have any effect, pro-poor tourism is likely to result in 'geographically uneven and unequal ... development' and 'may well display sharp regional and social disparities' (Mowforth and Munt, 2003, p. 272). Indeed, any support for pro-poor initiatives is simply repackaging familiar themes in development 'paradigms' and exacerbates existing power inequalities:

> In addition, it will prove necessary in the long run to consider the cumulative effect of supporting (through multilateral and bilateral aid programmes focused on economic growth) the expansion of capitalist relations and the manner in which this may undercut 'sustainable livelihoods' and exacerbate, rather than alleviate, poverty (Mowforth and Munt, 2003, p. 273).

This is the nub of their complaint: pro-poor tourist initiatives take place within a capitalist context and, in effect, help support the system.

When faced with this kind of criticism – itself a crude repackaging of a perspective frequently stated in the 1970s and 1980s – a limited number of responses are possible. First, one can simply agree or disagree with what is, in essence, an ideological position; second, attempts can be made (with the support of much unassailable historical evidence) to show that capitalist 'development' can indeed bring benefits and, third, it can be argued that, in the absence of any current socialist role models (a situation not universally deplored), the only choices today centre on the *kinds* of capitalist development that ought to be promoted. After all, as was noted in a similar debate more than two decades ago, no one has ever asserted that capitalism leads directly to equality, or that it has ever been 'nice' (Smith, 1983, p. 74).

Clearly, there is still a view that tourism, even ecotourism, does *not* bring about development and is, instead, a form of imperialism or colonialism (Nash, 1977, 1996; Johnston, 2006, pp. 66, 28–32). However, notwithstanding this debate, many LDCs, supported by international organizations and aid agencies, continue to promote tourism as a tool for development and poverty alleviation. In the remainder of this paper, the focus will be on one country, the Lao People's Democratic Republic (PDR), where such policies are prominent, to examine how they are promoted, and to see what light, if any, can be shed on the general debate over tourism's role as a development tool. In this context, it will be argued that the distinction between 'pro-poor, community-based tourism' and tourism enterprises operating in the private sector, both in Lao PDR and elsewhere, may need to be re-assessed, as donor-supported, community-based tourism projects may be sustainable only in partnership with the private sector, while the pro-poor credentials of tourism enterprises in the private sector are frequently underrated.

Lao PDR: The Background

The landlocked south-east Asian country of Lao PDR (commonly known as Laos) has a population of about 6.2 million, of which (depending on the criteria of categorizing them) nearly two-thirds are ethnic minorities, that is, not in the Lao-Tai linguistic group (Chazée, 2002, p. 14).

It emerged in the 1970s from a long period of war and civil unrest, prompted largely by the involvement of outside powers in its affairs. In 1893, it became a French Protectorate, was briefly occupied by Japan during the Second

World War, and later gained limited autonomy from the French in 1949. In 1953 it became fully independent, only then to enter a protracted civil war, which largely reflected big power interests in the region. In the north east, the communist Pathet Lao was supported by the Viet Minh, and to the south the anticommunist government forces were supported primarily by the USA.

Laos was secretly but massively drawn into the Vietnam/American war. During this 'secret war', US forces were supported by Thai and Hmong recruits, and US pilots carried out bombing missions from bases in Thailand and Viet Nam, assisted by non-uniformed military personnel and 'advisors' based in western Laos. The Ho Chi Minh trail in Laos and North Viet Nam were subjected to sustained bombing. In the East, the Pathet Lao were trained by and fought with the North Vietnamese army, which had tens of thousands of troops stationed in Laos, and proved more than a match for the US forces, despite the latter's superior weaponry (Warner, 1997; Hamilton-Merritt, 1999).

A ceasefire was agreed in 1973 and was followed by 2 years of coalition government. However, in 1975 the Pathet Lao gained control, removed the monarchy, and formed the Lao People's Democratic Republic. From then until the mid-1980s, Pathet Lao government was based largely on communist principles. Since 1986, however, especially after the collapse of the Soviet Union, there has been a sustained attempt to move from a command economy to a more capitalist, market-oriented system. One-party government continues, with only a slow emergence of a more open democratic system, but despite occasional domestic unrest up to 2000, the stability of the region has been reflected in Lao PDR and, with considerable international assistance, the poverty that is widespread throughout the country is being addressed.

Since the mid-1990s, the government has set itself the target of eradicating poverty, reducing its dependence on overseas development assistance, and moving out of the category of 'less developed country' by 2020 (Lao PDR, 2004, pp. 1–4). Substantial poverty reduction has undoubtedly occurred, partly because of increased social stability, but also because government policy and substantial overseas aid have had some success. From 1991 to 2000, for instance, real Gross Domestic Product grew at an annual average of 6.3%, and those living in poverty declined overall from 46% of the population in 1992 to 33% in 2002–2003. Most social indicators confirm the trend (Asian Development Bank (ADB), 1999a; World Bank, 2005, pp. 4–6).

Much remains to be done. There are stark differences across and within Provinces (especially between north and south), between urban and rural areas, and across ethnic minorities (World Bank, 2005, pp. 4–6). And although in 2003 the United Nations Development Programme upgraded Lao PDR to a medium developed country, poverty is still prevalent and the country relies heavily on overseas aid. In 2003, overseas development assistance amounted to more than 14% of GDP, and Lao PDR remained lower on the Human Development index (133) than neighbours Thailand (73), China (85), Viet Nam (108), Cambodia (130) and even Burma (Myanmar) (129) (United Nations Development Programme (UNDP), 2005, p. 282).

The Development of Tourism in Lao PDR

International tourists have been welcomed in Lao PDR only since 1989 (Hall, 2000), and the country's first national tourism plan was published in 1990 (Lao PDR, 1990). At that time, the emphasis was on small, tightly-controlled groups of package tourists. However, in 1995 tourism became a priority for economic development (Schipani, 2002, p. 18) and the second *National Tourism Development Plan*, published in 1998 (Lao PDR, UNDP and WTO, 1998), emphasized the value of four major types of tourism: conventional sightseers, special interest tourists – for example, eco- and adventure tourists – cross border tourists and domestic tourists (thus including, in effect, most categories of international tourist). All were valued as a source of foreign exchange, and it was thought that together they would bring benefits that would spread across the population and help conserve the natural and built environment (Lao PDR, UNDP and WTO, 1998). The policy succeeded, and the 1999 *National Tourism Marketing Plan* reported that tourism was the country's most important export (Lao PDR, UNDP and WTO, 1999).

By the opening years of the 20th century, there was further recognition of tourism's potential to alleviate poverty but, closely mirroring the priorities espoused by the Asian Development Bank, as indicated later in this paper, the focus seemed to shift to smaller-scale tourism. As the Government's *National Growth and Poverty Eradication Strategy* (NGPES) was to note:

> Tourism is now a major contributor to national income (7%–9% of GDP) and employment. Tourism is a labour intensive industry and contributes to poverty reduction. The Lao PDR's tourism strategy favours pro-poor, community-based tourism development, the enhancement of specific tourism-related infrastructure improvements, and sub-regional tourism co-operation (Lao PDR, 1990, p. 104).

A year later there was a return to a wider, broad-brush approach in the *National Tourism Strategy for Lao PDR*. It was to emphasize the role of properly planned tourism in reducing poverty and promoting national development, and the appeal to tourists of ethnic minority groups and traditional cultures. It also noted the wide range of archaeological and religious sites, and recommended promotion of the country's arts, crafts and its numerous natural attractions to visitors (Adams, 1990, pp. 18, 43).

Finally, yet another year later, a *separate* strategy was put forward for ecotourism, where it was defined as a 'tourism activity based in rural areas that is geared towards: conservation of natural and cultural resources, local socio-economic development, and visitor understanding of, and appreciation for, the places they are visiting' (National Tourism Authority of Lao PDR, 2005, p. 8).

There was evidently some ambivalence as to the kind of tourist most favoured in Lao development plans, and it is questionable how far increased tourist arrivals can be attributed to these plans, rather than political and social stability and the increasing popularity of south-east Asia as a regional destination. However, while early statistics of tourist arrivals to Lao PDR were unreliable and varied considerably according to their source, recent national data are more consistent and the trends are evident. Although punctuated by the impact of terrorist attacks in the USA in 2001, the war in Iraq and, more regionally, by terrorist attack in Indonesia and the SARS epidemic in Asia (2003), since the early 1990s tourist numbers have increased markedly, as indicated in Table 12.1. In 2005, there were more than 1 million international arrivals in Lao PDR, an increase exceeding 50% on the 2000 figure, and the total revenue generated by tourism in 2005 is estimated to have been more than US $146 who enter the country by road, and in fact they may stay one or two nights. By contrast, those classified as 'overnight visitors' are mainly from outside the region.

Most visitors (82%) are nationals from within the region, overwhelmingly from other

Table 12.1. International tourism in Lao PDR.

Year	Day visitors[a]	Overnight visitors[a]	Total visitors
1990	n/a	14,000	n/a
1991	n/a	38,000	n/a
1992	n/a	88,000	92,000
1993	67,000	36,000	103,000
1994	110,000	36,000	146,000
1995	286,000	60,000	346,000
1996	310,000	93,000	403,000
1997	270,000	193,000	463,000
1998	300,000	200,000	500,000
1999	355,000	259,000	614,000
2000	546,000	191,000	737,000
2001	501,000	173,000	674,000
2002	521,000	215,000	736,000
2003	440,000	196,000	636,000
2004	658,332	236,484	894,806
2005	807,550	287,765	1,095,315

Source: World Tourism Organization (WTO), 1996, p. 96; World Tourism Organization (WTO), 1999, p. 104; World Tourism Organization (WTO), 2001, p. 103; World Tourism Organization (WTO), 2005, p. 103; Lao National Tourism Administration, 2006.

[a]Those categorized by the World Tourism Organization as 'day visitors' are regional tourists million, undoubtedly making tourism the country's main source of foreign exchange (Lao PDR, 2006, p. 18).

ASEAN countries. This includes those categorized by the World Tourism Organization as 'day visitors' coming by road (who may, in fact, stay one or two nights), and 'overnight visitors', mainly from outside the region, as indicated in Table 12.1. However, significant numbers of relatively high-spending tourists were from Europe (1%) and the Americas (5.5%) (Lao National Tourism Administration, 2006, p. 6). Recognizing the value of these small but growing markets, government has prioritized Japan and Australia (with Thailand) in Asia, France, the UK and Germany in Europe, and the USA and Canada in the Americas (Lao National Tourism Administration, 2006, p. 14).

Much of Lao PDR has only very basic tourism infrastructure, and tourism development is concentrated in the Municipality and Province of Vientiane, at the UNESCO World Heritage Sites Luang Prabang Town and Vat Phou in Champassak, and in Savannakhet City (Lao PDR's second largest), with Savannakhet province serving as a major entry point for visitors from Viet Nam and Thailand. As indicated in Table 12.2, in 2005 Vientiane Municipality (21%), Vientiane Province (11%), Luang Prabang (15%), Champassak (12%) and Savannakhet (6%) together accounted for 65% of all accommodation establishments in Lao PDR, while the Vientiane-Luang Prabang corridor alone, which includes Vang Vieng, accounted for about half of all establishments (Lao National Tourism Administration, 2006, pp. 18–19).

The Organization of Tourism in Lao PDR

Numerous stakeholders have an interest in tourism in Lao PDR. However, at the time of writing, the main institutional actors are the Asian Development Bank, UNESCO, SNV (the Netherlands development organization), and the Lao government, primarily through the Lao National Tourism Administration (formerly the National Tourism Authority of Lao PDR). Their respective roles, discussed in detail in a forthcoming publication (Harrison, 2001), are briefly outlined here to situate the ensuing discussion on the place of ecotourism, community-based tourism and the private sector in Lao tourism. However, as indicated below, their initiatives in tourism have been especially influenced by the Nam Ha Ecotourism Project, one of the earliest ecotourism projects in the country, and certainly the largest, best organized and most important, considered by government and donors alike as the role model for ecotourism throughout the Lower Mekong Basin.

The Nam Ha Ecotourism Project

The Nam Ha Ecotourism Project (NHEP) was started in 1999 in the town of Luang Namtha, in Luang Namtha Province. Funds were provided by the New Zealand Official Development Assistance Programme (NZODA), now the New Zealand Agency for International Development (NZAID), the Japanese government, through the International Finance Corporation's Trust Funds Programme, and additional technical assistance by UNESCO and other agencies, as indicated in Fig. 12.1. The aim was to develop a series of treks to ethnic minority villages, with trained local guides, to provide income to the villagers and facilitate conservation efforts within a National Protected Area (NPA).

At the time of writing, the Project is into its second phase (2005–2008). It now operates under the supervision of the Luang Namtha Provincial Tourism Office, and several treks to Akha, Tai Daeng, Khamu and Lahu villages are on offer, ranging from less strenuous 1-day forest walks to more challenging 2- or 3-day treks, which include overnight stays in the villages. Over recent months, it has worked increasingly with Green Discovery, the tour operator in Lao PDR most involved with community-based ecotourism products (Fig. 12.1).

As well as running the treks, the NHEP cooperates with the public sector in developing links with the private sector, strengthening natural resource and protected area management in the Nam Ha NPA, and helping formulate a tourism master plan for Luang Namtha Province.

Financial assistance from NZAID was crucial to the continuation of the NHEP. Indeed, with an overall commitment of up to US$1

Table 12.2. Concentration of tourism in Lao PDR, 2005: selected indices.

Province	Visitors	No. of hotels and guest houses	Per cent of total	No. of rooms	Occupancy rate (%)	Average no. of rooms
Attapeu	13,740	12	1.1	190	50	16
Bokeo	89,027	24	2.2	309	67	13
Bolikhamxay	63,579	26	2.4	435	65	17
Champassak	99,044	126	11.6	1616	29	13
Houaphanh	3175	39	3.6	338	29	9
Khammouane	13,633	18	1.6	394	43	22
Luang Namtha	49,258	50	4.6	536	57	11
Luang Prabang	133,569	163	15.0	1722	70	11
Oudomxay	54,721	63	5.8	703	52	11
Phongsaly	9452	36	3.3	273	29	8
Saravanh	8000	22	2.0	230	42	10
Savannakhet	133,569	67	6.2	1257	60	19
Sayabouli	15,914	47	4.3	431	37	9
Sekong	6526	17	1.6	172	37	10
Vientiane Municipality	653,212	224	20.6	4891	64	22
Vientiane Province	92,657	115	10.6	1807	52	16
Xieng Khouang	24,174	32	2.9	441	40	14
Saysomboun	n/a	7	0.6	83	n/a	12
Total	–	1088	100.0	15,828	–	15

Source: Lao National Tourism Administration, 2006, pp. 18–19.

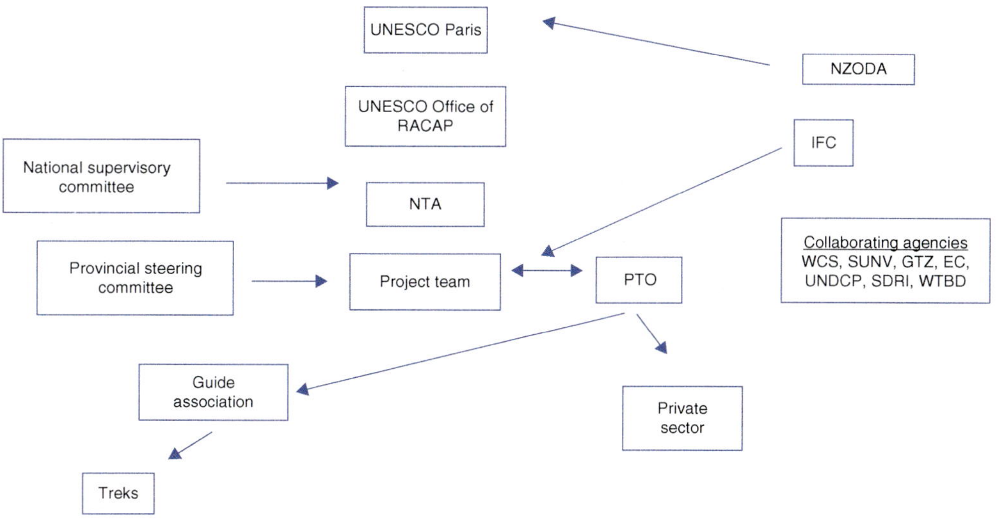

Fig. 12.1. The organizational structure of the Nam Ha Ecotourism Project.

million a year until 2010, New Zealand has recently emerged as one of Lao PDR's main pro-poor tourism donors. As part of its Lao country strategy, it is supporting not only the NHEP, but also a similar community-based ecotourism programme in Xieng Khouang Province, to begin in mid-2006. In addition, both Nam Ha Phase II and the Xieng Khouang Heritage Tourism Programme utilize technical assistance and monitoring sourced through the Office of the UNESCO Regional Advisor for Culture in Asia and the Pacific (RACAP).

The Asian Development Bank

The Asian Development Bank (ADB) is a major presence in the Greater Mekong region and in 2002 funded a feasibility study of priority tourism infrastructure projects in Lao PDR, Cambodia and Viet Nam (Asian Development Bank (ADB), 1999b). Committed to developing tourism throughout the region as a means of poverty reduction (Asian Development Bank (ADB), 2002a, b, p. 24), it has pledged up to US$30 million in low interest loans and technical assistance to the Mekong Tourism Development Project (MTDP), the overall aims of which are as follows:

> The Project's goal is to reduce poverty through increasing economic growth and employment opportunities, increasing foreign exchange earnings, and promoting the conservation of the natural and cultural heritage. The specific objective of the Project is to promote sustainable tourism development in the lower Mekong River Basin through infrastructure improvements, increased community and private sector participation, and sub-regional cooperation. The Project is targeted to improve high priority tourism-related infrastructure in the lower Mekong basin, promote pro-poor, community-based, sustainable tourism in rural areas, and strengthen sub-regional cooperation. It will mitigate environmental degradation from unplanned and unsustainable development, develop human resources, and promote cooperation between private and public sectors within the sub-region through the establishment of national tourism boards. (Asian Development Bank (ADB), 2002c).

About a third of the funds allocated by ADB to the MTDP have been allocated to Lao PDR (Asian Development Bank (ADB), 2002a) and, as in Cambodia and Viet Nam, will be directed to four distinct spheres of activity, over a 5-year period. The first emphasis is on loans to improve tourism-related infrastructure, and primarily involves building or improving roads and airports.

Currently, three projects are under way, in the Provinces of Luang Namtha, Khammouane and Luang Prabang.

The second focus is on the development of pro-poor, community-based tourism, and for this purpose ADB has a team of four international and four national consultants working closely with local project implementation units (PIUs), which are comprised of staff from provincial tourism offices in Luang Namtha, Luang Prabang, Champassak and Khammouane. These PIUs cooperate with private and public sector agencies, oversee guide training and awareness programmes, and ensure, where appropriate, that women and ethnic minorities are empowered to participate in tourism activities. ADB is also planning to extend its support for 'pro-poor demonstration projects' in other remote, poor provinces (Asian Development Bank (ADB), 2005).

The third emphasis is on strengthening regional cooperation, improving cross-border tourism facilities and harmonizing standards, and developing human resources in the Lower Mekong region, while the final sphere of activity is providing institutional support to implement the three major project components.

SNV (The Netherlands Development Organisation)

Over the last few years, the ADB has worked increasingly closely with SNV, an independent NGO which has traditionally received most of its income from the Dutch government. It has been in Lao PDR since 2000, initially operating as SUNV (through a cooperative programme with United Nations Volunteers), and is especially committed to providing technical advisers for the development of community-based ecotourism in rural areas. It supported the Nam Ha Ecotourism Project by providing a Handicraft Production and Marketing Adviser, and acknowledged Nam Ha's status as a role model in the *National Ecotourism Strategy and Action Plan*, which it prepared for the Lao National Tourism Administration (LNTA) (National Tourism Authority of Lao PDR, 2005, p. 6).

SNV focuses primarily on small-scale, community-based tourism in parts of Lao PDR not currently on the main tourist trail – notably Luang Prabang, Houaphanh, Savannakhet and Khammouane – and also cooperates with central government in developing national tourism policies (SNV, 2006). In Luang Prabang, for instance, its advisers work with the provincial authority in trying to extend the benefits of tourism, currently concentrated in the World Heritage city of Luang Prabang, to outlying villages in the Province.

In Houaphanh they helped the PTO formulate a tourism development plan, and are assisting in improving information and services at the network of caves once used by the Pathet Lao as command centres during the Indo-China wars. In Savannakhet, SNV advisers are developing treks to three protected areas in conjunction with villagers and local guides, and similar activities are occurring in Khammouane, as part of the Mekong Tourism Development Project.

All such activities are designed to increase earning opportunities for the rural poor, diversify their sources of income, build local management capacity and expertise in tourism (e.g. in guiding and heritage conservation), and empower local communities. They are complemented by technical assistance provided by SNV to the Lao National Tourism Administration in Vientiane, which promotes the National Ecotourism Strategy (National Tourism Authority of Lao PDR, 2004), and the work of the newly formed inter-ministerial Ecotourism Technical Cooperation Group, which also receives technical assistance and further support from SNV and Mekong Tourism Development Project advisers. SNV also helped to establish the Lao Sustainable Tourism Network, and in May 2006 launched a 3-year programme, funded by the European Union, to improve the marketing and promotion capabilities of the Lao Association of Travel Agents (LATA), strengthen the organization's management, and institute mechanisms for information-exchange between tour operators and the LNTA. Such examples, along with its cooperation with UNWTO (through the STEP Programme), in disseminating the new Lao Tourism Law (approved by the Lao Parliament on 9 November 2005), show how SNV assists the LNTA in cooperating more closely with the private sector.

Except for its involvement in Savannakhet and Houaphanh, most activities carried out by SNV's international and national advisers in Lao PDR are funded by ADB through the Lao government, an arrangement that emerged during the first phase of the ADB-financed Mekong Tourism Development Project.

SNV and ADB are not the only international organizations working in Lao tourism. Several other ecotourism projects, based on similar principles and with similar aims, are promoted by other aid agencies. These include a CUSO initiative (a Canadian volunteer organization), working in Attapeu Province, DED (the German Development Service) in Phou Khao Khouay National Park, near Vientiane, and also in Udomxay Province, GTZ (German Development Agency) and Vientiane Travel and Tour, its private sector partner, developing an ecotourism programme for eight Akha villages in the Muang Sing area of Luang Namtha, and small European Union projects in Vieng Phoukha District, Luang Namtha and in Pongsali Province (www. ecotourismlaos. com, accessed 3 March 2006). However, most of these are very small in scale and still at the development stage. At the time of writing, ADB and SNV have leading roles in the development and trajectory of tourism in Lao PDR – in so far as it is oriented towards rural, pro-poor community-based tourism.

The Lao National Tourism Administration

In 2005, the National Tourism Authority of Lao PDR was renamed the Lao National Tourism Administration (LNTA) and upgraded to ministerial level within the Prime Minister's Office. It is the main government agency responsible for regulating tourism in the Lao PDR, and formally cooperates with numerous other stakeholders, many of which are also in the public sector (Adams, 1990). It has branch offices in Vientiane and all 16 provinces, and has several important roles: first, tourism planning and co-operation; second, licensing and legal affairs; third, marketing and promotion and finally, training.

First, in coordinating tourism planning, with assistance from the aid agencies, the LNTA develops the National Tourism Strategy and produces master plans for tourism development in the provinces and at specific sites. SNV's part in preparing the National Ecotourism Strategy and Action Plan has already been acknowledged, and the current National Tourism Strategy and Action Plan was formulated with national and international technical assistance from ADB's Mekong Tourism Development Project.

Second, the LNTA licenses tour companies, tour guides, tourist accommodation and restaurants, and sets appropriate standards, guidelines and codes of conduct for them. It publishes a compulsory code of conduct for tour guides and is in the process of setting up a hotel rating system based on good practice elsewhere in the ASEAN region.

Third, most marketing and general promotion of the Lao tourism industry is carried out by LNTA, which produces informational materials and participates in conferences and exhibitions. It also maintains tourist information centres across the country and two websites (www.tourismlaos.gov.la and www.eco-tourismlaos.com). More specific tourism products and services are marketed and promoted directly by the private sector, that is tour companies, hotels and restaurants, which produce their own advertisements and brochures and develop and maintain their own websites.

Finally, although some hoteliers and inbound tour operators train their staff, most training is carried out by government and its NGO partners. The latter are normally responsible for training guides at provincial and village levels, but LNTA trains and registers national tour guides, running an annual course in Vientiane. It also periodically conducts hotel and restaurant management training sessions for the private sector, and runs short tourism management courses for government employees and the private sector. At local level, it assists disadvantaged and poor groups (particularly women) in obtaining employment in the hospitality, guiding and handicraft sub-sectors.

The Private Sector

It was stated earlier that there seems an ambiguity concerning the type of tourists the country's authorities wish to attract. Furthermore, while ADB's support for infrastructural development and sub-regional cooperation is clearly designed to facilitate the movement of all kinds of tourist, and takes up the bulk of its funds (Asian Development Bank (ADB), 2002a), much of its technical support is explicitly directed towards community-based tourism in rural areas. Similarly, the publications and promotional literature emanating from ADB, SNV and the LNTA also have a strong focus on rural tourism, reflecting the fact that its technical advisers are usually specialists in community-based tourism. And while one of the two official websites focuses specifically on the ethnic minorities and cultural attractions of Lao PDR, the other is devoted exclusively to ecotourism (www.tourismlaos.gov.la and www.ecotourismlaos.com).

It is arguable, then, that those who organize and promote Lao tourism are concentrating on what has come to be known as pro-poor, community-based tourism. However, when attention turns to what is happening in the Lao economy, a very different picture emerges. It transpires that one of the key features of economic change since 1986, when a market-oriented economy was introduced, has been the rapid expansion of small, locally-owned businesses. Indeed, by 1996,

> ... (T)here were 146,000 micro/small enterprises employing the equivalent of 259,000 full-time workers and accounting for 6 to 9% of GDP. This is over ten times the 22,000 that were employed by larger enterprises. Indeed, the micro/small enterprise sector accounted for 86% of rural and 13% of urban employment ...
>
> 90% of these are family businesses which tend to be multiple enterprises. However, they provide supplementary rather than principal household income. 63% are female-owned and account for 56% of total employment in this sector. (Enterprise Development Consultants Co. Ltd, 2002)

A decade later, the importance of small/micro business is even more pronounced, and the tourism sector provides a specific example of the more general phenomenon, at least in the provision of accommodation, food and beverages. In 1998, for example, there were only 307 accommodation establishments in the country, whereas by 2005 there were 1088 (Table 12.2). And we are certainly not talking about 'big business', for the average number of rooms was a mere 15, and exceeded 20 only in Khammouane and Vientiane Municipality. Even in Luang Prabang, with the second biggest concentration of establishments in the country, the average was only 11 rooms (Lao National Tourism Administration, 2006, pp. 18–19). The message these figures convey is evident: in all of Lao PDR's main tourism centres, in the accommodation sector (and in restaurants), small/micro businesses are the norm.

Despite criticisms to the contrary (Guttal, 2006), this private sector expansion has been achieved with relatively little support, either from government or from the international aid agencies. Indeed, although SNV, ADB and the government are formally committed to working with the private sector, LNTA's linkages with tourism businesses remain tenuous, and it might be argued that, at least initially, aid agencies and government have looked at the private sector with a degree of suspicion. For their part, private sector tour operators and tourism-related businesses used to complain at a lack of visible outputs and tangible support from LNTA, especially in marketing and promotion, and regulatory or training activities. It is indeed the case that, while the ADB's Mekong Tourism Development Project is primarily focused on infrastructure projects, product and human resource development, it is also committed to developing tourism through the establishment of a Lao Tourism Promotion and Marketing Board, which is intended to increase participation by the private sector and give it a stronger voice on issues related to tourism policy. However, by the time of writing, such a Board had not materialized in Lao PDR (or Viet Nam or Cambodia).

It is remarkable, too, that despite the lack of formal mechanisms for obtaining credit, poor infrastructure and relatively untrained human resources, the vast majority of small and micro enterprises in the accommodation and restaurant sector are financed by *local* capital. Foreign investors are conspicuous by their absence. This might be explained by the relatively late conversion of government to a market orientation, but the ILO study quoted above also indicates that, in 1996, 'the existing legal and policy framework favours large enterprises' (Duffy, 2002). More recently, through legislation enacted by the government in 2004, international investors were offered even more favourable terms, including the possibility of 100% foreign ownership and tax holidays up to 7 years, followed by very low profit taxes thereafter (Lao PDR, 2004, 2006).

Lao Tourism and Poverty Alleviation

At the beginning of this paper, note was made of the emergence in the 1970s of an interest in tourism as a development tool, and the move, at the end of the 1990s, to a closer examination of ways in which it could be specifically targeted to alleviate poverty. This led to an account of how tourism has developed in Lao PDR, and a brief description of the main institutional agencies involved in it. Ideally, what should now follow is a detailed comparison of the role of rural, aid-led, community-based, pro-poor tourism (assuming for the present the last two categories are synonymous) with the relatively unplanned expansion of small tourism-oriented businesses in the private sector. Unfortunately, though, direct comparisons are extremely difficult to make, primarily because detailed data are not available. If the aid agencies *do* carry out full economic costing of their projects, they are not usually accessible to the general public, and in Lao PDR, at any rate, no sustained research on private sector involvement has been carried out. However, enough data are available in sufficient quantity and quality to raise several pertinent issues about the development of tourism in Lao PDR, to inform the debate about the nature of pro-poor tourism more generally, and to suggest priorities for future research.

The Nam Ha Ecotourism Project

In October 2002, 3 years after the NHEP started, its external reviewers were unequivocal that it had been 'a tremendous success in providing a model of how tourism might be used for development in rural and largely subsistent villages and as a mechanism for promoting forest conservation' (Lyttleton and Allcock, 2002, p. 5), and also claimed it 'established a first-class working model for ecotourism activities in areas of great cultural and natural richness' (Lyttleton and Allcock, 2002, p. 6). In its first 3 years, after extensive preparation, a total of 89 guides had been trained, of whom 69 were from the town of Luang Namtha and 20 from the villages. The Nam Ha Ecoguide Service (NHEGS), a local inbound tour service created with technical assistance from the project, was operating three treks and a 1-day boat trip, involving eight ethnic minority villages (with a combined population of about 2000), and had provided accommodation, food and other services to more than 2800 tourists from 38 countries.

Over the same period, the trekking operations earned the NHEGS a gross revenue of US $70,000, of which about 18% went directly into the villages, which also earned money from selling handicrafts and other services, for example, massages (Lyttleton and Allcock, 2002, pp. 17–21, 41–45). As the authors of the external review point out, 'the per cent of income relative to alternative forms of income varies but in one Akha village it is as high as 40%' (Lyttleton and Allcock, 2002, p. 42). There is a similar variation in how (and to whom) the funds going to the villages are distributed but, during the first phase, 8% of all income from treks was placed in Village Development Funds, designed to benefit the community as a whole (Lyttleton and Allcock, 2002, pp. 44–45).

Over the following years, the treks continued, as indicated in Table 12.3, and the NHEP has undoubtedly been recognized in Lao PDR, and indeed throughout South-east Asia, as a successful example of community-based, pro-poor tourism. It also seems, from the little evidence available, that villagers assisted by the project are equally satisfied with its performance, and the reviewers of the Project's first phase note that the Akha villagers they visited 'felt that the tourists dramatically improved their income' (Lyttleton and Allcock, 2002, p. 42). Indeed:

> In each of the Nam Ha target villages, cash income has been increased markedly by the visits of tourists and the established framework

Table 12.3. Income earned by the Nam Ha Ecoguide Service 2001–2005 (US dollars)[a].

Year	No. tours	No. tourists	Gross revenues	Direct tour expenses	PTO 5% tourism promotion fund	Trek permit	Tax	Village fund	Overheads and profit
2001	201	944	17,795	11,655	889	984	427	1626	3125
2002	430	2366	47,837	31,093	2362	2647	1185	4850	7244
2003	165	718	14,755	11,308	748	789	212	306	2855
2004	270	1366	25,662	18,769	1171	1148	291	870	2482
2005	265	1407	31,745	21,927	1326	864	545	1833	3573
Total	1331	6801	137,794	94,752	6496	6432	2660	9485	19,279

Source: Lao National Tourism Administration, 2006.
[a]Discrepancies in the columns are in the original.

> ensures that this income is, to date, reasonably well distributed among the villagers. The villagers are enormously happy to have the tourists visit and feel little in the way of negative impact of their presence. At the same time, the tourist treks have contributed positively to an increased awareness of forest conservation and the means to enact improved conservation practices (Lyttleton and Allcock, 2002, p. 5).

This sounds impressive. However, it *also* has to be emphasized that it has been extremely well resourced. As the external reviewers noted, at the end of the first phase:

> The project has had at its disposal the (virtually) full time presence of 3 foreign TAs, an additional presence of 3 part-time foreign volunteers, and the full-time secondment of 3 NTA [National Tourism Authority] personnel with substantial tourism management skills. If one considers the limited number of target villages (no more than eight in Luang Namtha) it almost equates to 1 TA per village. Clearly this is simplifying the job requirements, but it is unlikely there is a development project in the world that has maintained this level of expertise relative to the number of target communities. While this has directly led to the high level of project achievements, it is impossible that this level of assistance (per target site) be sustained (Lyttleton and Allcock, 2002, p. 47).

The reviewers' conclusions seem to be supported by the data. From 2001 until the end of 2005, as indicated in Table 12.3, the NHEGS operated 1331 tours, catered for 6801 tourists and brought in gross revenue of US$137,794. Of this, $9485 went into village funds, and benefits also accrued to the Provincial Tourism Office and to central government through taxes and payments for trekking permits. However, as indicated in Table 12.4, donor support has been and continues to be considerable. From 1999, when the treks first started, until 2002, the Project received nearly US$500,000 and from 2005 through to 2008, it is to receive more than $100,000 a year from NZAID. Indeed, by 2008 the NHEP will have received more than US $860,000 from international aid agencies.

When examined on a strict profit and loss basis, there is a clear discrepancy in the amount received from donors and that obtained from the treks. However, simply comparing the amounts provided by donors with the net profits received from the treks (even if such detailed figures were consistently available) is perhaps too simplistic. Other factors need also to be considered. First, the villages targeted as partners in NHEP were especially poor, and *any* economic benefits made an immediate difference. Second, one of the reasons participating villages were selected was to reduce the degree of opium consumption in the villages, and their reliance on opium as a source of income. In so far as the Project succeeded in these objectives (and there is evidence that it did), the *economic* significance is difficult to quantify.

Third, it is also necessary to consider how donors' funds are invested in other activities, for example: equipment and training (tour guides, managers and villagers); constructing infrastructure in the villages to contain fresh water and accommodate tourists, and building the Nam Ha Tourist Information Centre in Luang Namtha; monitoring the impacts of the treks in the villages; promoting conservation in the region; and producing advertising materials. Such activities (and there are many others) are carried out with a view to the longer term, and while they are necessary in the accumulation of financial and cultural capital, it is not possible to really incorporate them in a simple balance sheet.

Fourth, the NHEP also contributes to the prosperity of Luang Namtha Province. In 1999, for example, visitors to Luang Namtha simply stayed in the town overnight and then moved on. However, with the development of NHEP, they are now likely to go trekking for up to 3 days, also spending an additional day or two in the town and bringing further income to the town's guest houses and restaurants.

The Nam Ha Project and the Asian Development Bank

In 2002, when the ADB was considering further support for community-based tourism in Lao PDR, the Nam Ha Project was seen as a model for future tourism CBT projects in the rest of Lao PDR and further afield, including Cambodia and Viet Nam. Consequently, as part

Table 12.4. Donor support for the Nam Ha Ecotourism Project 1999–2002 and 2005–2008 (US dollars).

Year	NZODA (NZAID)[a]	Japan IFC	Other agencies[b] WCS, GTZ, EC, UNDCP, SDRI, WTBD; SNV	Financial return from treks, etc.	Achievements
1999–2000	158,400		26,700 (plus technical advisers for specific short-term projects and sundry items).	2001–end of 2005: 137,794 (gross), including 9485 to villages. Also income from handicraft sales, massages, etc.	1331 tours and 6801 tourists Treks and boat trips Income to poor villages 90 + guides trained Investment in infrastructure and cultural capital for future
2000–2001	93,159	106,698			
2001–2002	96,884				
Sub-total	348,442	106,698	26,700		
2005–2006	128,481				
2006–2007	152,686				
2007–2008	127,826				
Total	$757,346	$106,698	$26,700		

Source:Lao National Tourism Administration, 2006.

[a]Figures include 13% to UNESCO in support costs.

[b]Wildlife Conservation Society (WCS), GTZ (German Aid Agency), European Commission (EC), UNDCP (United Nations Drug Control Programme), Sustainable Development Research institute (SDR), Where There Be Dragons (WTBD); SNV (Netherlands Development Organization).

Source: Lyttleton and Allcock (2002, pp. 53–56), UNESCO (2005).

Table 12.5. Financial summary for LNTA/ADB new CBT products: March 2005–February 2006 (US dollars).

	No. villages/estimated no. of significant beneficiaries	No. tours	No. tourists	Total revenue	Village revenue	Permits
Luang **Namtha**						
(1)Pu **Sam Yord Journey**	4/65	89	549	21,119	9406	247
(2) **Nam Ha Camp**	2/40	15	76	3965	1346	228
(3)Akha Trail-Nam Mye **Caves**	4/65	11	37	1821	901	109
Luang **Prabang**						
(4)Chomphet **Trek**	4/65	25	92	4235	420	0
(5)Kiew Kan **Trek**	4/65	23	96	2991	1844	0
(6)Kwang **Si Walk**	2/40	47	215	4098	1864	68
Khammouane						
(7)Phou Hin Poun **2 day Trek**	3/45	14	78	3502	1092	78
(8)Buddha **Cave 1 day Trek**	1/10	23	74	1120	312	5
(9)Kong Lor-Natan Homestay	2/40	24	172	2313	2615	121
(10)Konglor **Boat Trip**	1/20		343	3430	3430	0
Champassak						
(11)Don Daeng Island	1/16	29	171	1300	1207	0
(12)Xe **Pian Forest Excursion**	2/40	15	78	2677	1094	121
(13)Kiet Ngong **Elephant Rides**	1/20	1476	2525	14,397	14,177	0
(14) **Pu Khong Mountain-Don Ko**	1/20	13	50	1227	841	38
Total 14 **products**	32/551	1804	4556	68,194	40,548	1014

Source: Lao National Tourism Administration, 2006.

of the Bank's Mekong Tourism Development Programme's pro-poor, community-based tourism component, up to US$400,000 per province is being made available to Luang Namtha, Luang Prabang, Khammouane and Champassak over the period 2003–2007, an average of $100,000 a year. These projects are now starting to be operational. For the year March 2005 until February 2006, for instance, as indicated in Table 12.5, the 14 new village-based tourism products in the provinces of Luang Namtha, Luang Prabang, Khammouane and Champassak, developed by the Lao National Tourism Administration with ADB financial support, brought in a gross revenue of US$68,194 and contributed US$40,548 to the 32 villages directly involved, with an estimated 551 people deriving

Table 12.6. Financial summary for destinations and tour operators receiving support from the LNTA-ADB Mekong Tourism Development Project. March 2005–February 2006.

Destination and tour operator support	No. tours	No. tourists	Total revenue	Village revenue	Permits/ tickets
Luang Namtha					
(1)Green Discovery Co.	191	756	18,108	10,184	366
(2)Nam Ha Ecoguide Service	270	1366	25,662	18,796	1148
(3)Muang Sing Ecoguide Service	160	567	12,941	8195	0
(4)Vieng Phoukha Ecoguides & Caves	79	238	6885	2272	125
Luang Prabang					
(5)Muang Ngoi Kao Village	0	7800	195,000	97,500	0
Khammouane					
(6)Buddha Cave in Na Kang Xang Village	0	40,000	48,000	40,000	8000
Champassak					
(7)Ban Mai Singsampanh Market	0	14,000	168,000	84,000	0
Total	700	64,727	474,596	260,947	9639

Source: Lao National Tourism Administration, 2006.

a significant portion of their livelihood from tourism-related activities. These figures are the result of the first year of operation for most of the tour products, many of which are located in remote areas far from the main tourism circuits and thus require sustained marketing and promotion before they can become established destinations.

When the examination of benefits brought by these projects is widened to include those accruing to private sector tour operators operating with technical and financial support from the ADB and the NHEP, not only in ADB-supported villages but also on other sites, other economic benefits become apparent. As indicated in Table 12.6, in only 1 year, the four provinces generated some US$474,596 in gross revenue, of which nearly half was realised at the village level, and a considerable (but unquantifiable) percentage of this income might be attributed to direct and indirect support from the Mekong Tourism Development Project.

Before concluding this section, a caveat should perhaps be mentioned. While commercial banks are likely to expect returns on their investments after a period of around 10 years, the ADB allows for loan repayments over 30 years, at a low rate of interest and with a significant interest-free period. This suggests that different criteria of success are being used by the ADB and in the private banking system.

It is also the case that much of the training and HR development in Lao tourism is supported by NHEP and other ADB-supported projects. Not only do they bring immediate economic benefits to the villages where they operate, but also they generate income and employment in the region, raise awareness of cultural and environmental conservation among local people and policy makers, put management systems in place to protect the value and integrity of destinations, and offer training and capacity-building to many people who, without the projects, would be measurably poorer. They also develop models

Table 12.7. Relationship of private sector to selected community-based tourism projects in Lao PDR.

Institution	Area of operation	Partnerships with tour operators	Private sector
LNTA	Throughout Lao PDR		Small-scale, locally owned guest houses and accommodation
NZAID	NHEP: Luang Namtha & Muang Sing; Xieng Khoang	Green Discovery and Diethelm Travel	Non-specialist owners often also with rice lands
ADB & SNV	MTDP: Luang Namtha, Luang Prabang, Khammouane, Champassak	Lao Youth partners with MTDP in Meuang Ngoi, Luang Prabang	Subject to government regulation and licensing
GTZ	Muang Sing	Exotissimo	Profit oriented
CUSO	Attapeu		Major driver of Lao tourism
DED	Phou Khao Khoay National Park (Vientiane Province) & Udomxay Province		
EC	Vieng Phoukha (Luang Namtha) & Pongsali Province		

LNTA = Lao National Tourism Administration; NZAID = New Zealand Aid Agency; ADB = Asian Development Bank; SNV = Netherlands Development Organization; GTZ = German Aid Agency; CUSO = Canadian aid NGO; DED = German Development Service; EC = European Commission.

that can then be adopted by private sector players, several of whom are now cooperating with donor-supported projects in Lao PDR, as indicated in Table 12.7. Such activities may generally be regarded as laudable, but they are almost impossible to quantify.

Despite these caveats, however, when the sums provided by international donors to CBT products in Lao PDR are compared with the short-term financial returns from these projects, the former clearly outweigh the latter. And even if the medium- and longer-term benefits are also taken into account, it could be argued that the contribution of the projects to the financial and cultural capital of the Lao tourism industry, while potentially great, is (again) likely to remain unquantifiable.

Two other important questions remain. The first concerns whether or not international funds invested in these projects could have been spent more profitably to achieve the same aims and objectives, and the second the extent to which the long-term potential of these projects, effectively permeating and transforming the rest of Lao tourism, will actually be realized. In response to the first question, it has to be said that we simply do not know. And the response to the second is that we do not *yet* know. When we *do*, it will be with the benefit of hindsight.

The private sector

It was indicated earlier that since 1986 there has been a rapid expansion of small, locally owned businesses in Lao PDR, especially in tourism, to the extent that Lao tourism is characterized by the emergence of many small-scale, locally owned hotels, guest houses and restaurants, often in urban areas, with an average of only 15 rooms (Lao National Tourism Administration, 2006, pp. 18–19). Furthermore, this expansion has been achieved with little support, either from government or from the international aid agencies, and with very little planning. Whereas CBT ventures in Lao PDR resulted from a deliberate targeting of communities by international agencies, private sector development has occurred because individuals saw entrepreneurial opportunities in tourism and took them.

In this section, a brief account will be given of the development of tourism on the island of Don Det, in the Siphandon ('Four

Thousand Island') region of Champassak, near to the Cambodian border.[2] As indicated earlier (Table 12.2), Champassak is one of the provinces most visited by tourists, and many independent tourists visiting the province pass through the Siphandon region en route to or from Cambodia, in couples or singly, often staying a few days in the peaceful Mekong villages, accurately stated by the *Lonely Planet*, 'to offer fascinating glimpses of tranquil river-oriented village life' in communities reported to be 'self-sufficient, growing most of their own rice, sugarcane, coconut and vegetables, harvesting fish from the Mekong and weaving textiles as needed' (Cummings, 2002, p. 305).

In July 2004, the island of Don Det had a total population of about 880 people, living in two villages, namely Don Det Ok (East) and Don Det Tok (West). The island's tourism started in 1997, when an outsider from the District Centre, Khone Island, built the first guest house. Since then, tourism increased rapidly. In 2002, the island had 19 guest houses, and by 2004 there were 34. Most were concentrated in the village of Don Det Tok, which had a population of 404 people, living in 83 households, as indicated in Table 12.8.

Whether or not villagers of Don Det Tok are considered 'poor' depends largely on definition. According to the Head Man, who used his own assessment criteria rather than the official income criteria (which until recently defined any rural household with an average income of less than US$8.20 per person per month as poor), 17 households (c. 20%) were poor. However, many of the village houses would be considered by the authorities to be 'temporary' (another criterion of poverty), 75 households (90%) had no toilets and few houses were supplied with fresh water. And when interviewed, more than two-thirds of the respondents described *themselves* as poor, usually because they were not self-sufficient in rice (which applied to nearly half the sample), but also for a wide variety of other reasons, ranging from being alone to not having a secure or predictable income.

From interviews with guest house owners of Don Det, it was clear that sources of finance for guest houses and restaurants vary, but inheritance or the sale of livestock or other assets (perhaps themselves inherited) were commonly mentioned. In 2002, one young guest house

Table 12.8. Don Det Tok village, on the island of Don Det, Champassak: July, 2004.

	Don Det Tok
Total households	83 (404 pop)
Sample number of households	55 (66%)
Sample household population	292
Average household size	5.3
Nuclear family	23
Extended family	20
Other	12
Total adults born outside Don Khon/Don Det	27 (9.2%)[a]
Household with at least one in tourism	21 (38%)
At least one person involved in handicrafts	–
Total people in handicrafts	–
Income source	
Agriculture 1	36 (65%)
Agriculture 2	12
Agriculture 3	1
Fishing 1	5 (9%)
Fishing 2	32
Fishing 3	7
Livestock 1	1
Livestock 2	1
Livestock 3	24
Tourism 1	12 (22%)
Tourism 2	6 (11%)
Tourism 3	1
Tourism 4	2
Other 1	3
Other 2	1
Other 3	–
Have own land? No (excluding garden)	13
Sharecropping?	8
Yes	42
Paddy	42
Other	–
Enough rice: yes	29 (52.7%)
No	26 (47.3%)

Continued

Table 12.8. Continued

	Don Det Tok
Self-assessed as poor: yes	37 (67.3%)
No	17
Don't know	1

[a]In 21 households.
[b]One household did not produce enough rice but ran a shop and did not claim to be poor.
Source: Author's survey, July, 2004.

Table 12.9. Summary of sample survey of guest houses of Don Det (July 2004).

	Don Det[a] guest houses
Total	34[b]
Size of sample	19
Number of rooms	105
Average number of rooms	5.5
Restaurant: Yes	14
No	5
All food obtained locally	–
Some food obtained locally	12
No food obtained locally	2
Owner born in Don Det?	10
Owner born outside Don Det?[a]	9[a]
Total with some family labour	1
Number with no family labour	2
Number with only family labour	16
Total workers born in Don Det island	35
Total workers born outside Don Det Island[a]	–
Total workers	20[a]
Average number of staff?	3

[a]Of these owners and workers, all but one had come from nearby islands.
[b]In July 2004, nine guest houses in Don Det were closed.
Source: Author's survey, July 2004.

operator reported that his father had financed the construction of a five-room guest house and restaurant by selling a buffalo and three cows (at the time of writing, equivalent to US$2000), while another owner said the money that financed his guest house and restaurant came from the sale of three buffalo. Both were well on the way to recouping the initial cost, and other guest house and restaurant owners similarly reported that initial investments were recouped after only a few years.[3]

With few exceptions, buildings were made from local materials, with a basic frame of wood and/or bamboo, walls of woven reeds, and thatched roofs (occasionally on a wooden or corrugated metal base). Nearly all the guest houses in the sample were locally owned and owner operated. Although almost half the owners were not from Don Det, all but one of these were from adjacent islands, with family or affinal kin in Don Det. The remaining owner was from Vientiane Province. Most guest houses in the sample were small, with an average of 5.5 rooms, and nearly all had restaurants (as indicated in Table 12.9), invariably run by women in the family, who reportedly had independent control of the profits from the enterprise.

Details of the households sampled in Don Det Tok, and their major sources of income, are provided in Table 12.8. For present purposes, the most relevant points relate to the role of tourism in the village. While 65% of the sampled households indicated that their main source of income was agriculture, in 38% of the sample at least one adult member was involved in catering for tourists, in 22% tourism was their main source of income, and another 11% reported that tourism (and not fishing, as might have been expected) was their second source of income. No household reported any earning from handicrafts, and in fact no weaving was noted in the village.

Within the village, earnings from tourism came from a variety of sources, including accommodation and restaurants, boat trips to other islands (and to view the Irawaddy dolphins of the Mekong River, along the Lao and Cambodia border), tubing, where large inflated inner tubes are provided for tourists to use to float along the river, and the rental of bicycles and small motor bikes. Tourists could also visit a waterfall, travelling by a small shuttle bus, which charged about 60 cents for the journey.

How important was tourism to the village of Don Det Tok? The estimates may be crude but are nevertheless instructive. According to one owner of a six-room guest house, in a good

Table 12.10. Estimate of annual tourist expenditure on the island of Don Det.

Annual tourist expenditure	Total ($US)
34 guest houses with average of 5.5 rooms and 55% occupancy: 37,540 tourists at $1.00 a day	37,540
Restaurants: food and beverages at $8.00 a day for 37,540 tourist days[a]	300,320
Sundries at $3.00 a day for 37,540 tourist days (bicycle hire, boat trips, tubing, etc.)	112,630
Estimated total expenditure	450,950

[a]Lunch or dinner at a restaurant in Don Det Tok costs about $3.00, and with the addition of Lao beer the estimate of $8.00 a day for food and beverages is likely to be an under-estimate.

year he might make US$1200 on accommodation, with much more than double that amount in the restaurant (which was agreed to be the main source of profit). At the normal nightly rate of US$1 a room, this would give him an occupancy rate of 55%, which compares with that of 58% given for Champassak province in 2005 (Lao National Tourism Administration, 2006, p. 17).

From interviews with tourists, and examinations of visitors' books in the guest houses, it was clear that most tourists stayed for at least three nights in the village. The data in Table 12.10 are likely to be an underestimate, but indicate that tourists visiting Don Det in 2004 or 2005, nearly all relatively 'high spenders' from outside South-east Asia, probably spent about US$450,000. A high percentage of this expenditure would have remained in the village, as 12 of the 14 restaurants indicated that they obtained at least some of their supplies locally (e.g. vegetables, fish and some poultry) in Don Det, either from their own kitchen gardens or neighbours, and most of the remainder came from Ban Nakasang, a market town a short journey by boat from the village. And while 36% of the workers in the guest houses were born outside Don Det, as indicated in Table 12.8, they were living in Don Det with other kin, including family members who owned and ran guest houses.

Other details can be mentioned only in passing. The boat owners of Don Det have formed a boat Association, and operate boats for tourists on a rota basis. Part of the Association's monthly proceeds is paid to Don Khong District in tax ($12.00), while $18.00 stays in the Village Fund, and is used for community projects (for example, painting the local primary school and providing it with fans). Indeed, possession of a large boat and an outboard motor is a great asset, as on a good day in the high season a boatman might earn more than $45.00 from tourism. It will be less in the low season, but is still considerably more than the average monthly earnings of many people. Guest Houses must also pay a monthly tax ($1.80) to the District, as must the restaurants ($1.50).

From this brief account of one small island in the Siphandon region of Champassak, it should be clear that tourism is a valuable source of income to many of the 880 residents, and will contribute about US$460,000 annually to the economy of Don Det. Such success was built on previous prosperity, for those who invested in guest houses and restaurants possessed the capital necessary to make the investment, and the research prompted some discussion in the villages as to whether or not tourism benefited the poor. Most villagers interviewed had no doubt that even though the poor were not specifically targeted, they benefited from having a bigger market for their crops and, less often, employment opportunities in the guest houses, as well as opportunities (if they cared to take them) to be sharecroppers on land owned by more well-to-do farmers.

Some redistribution mechanisms were in place, for example the Village Fund, and there were reportedly plans to institute a rice bank to assist the poor when they were short of food. In July 2004, this had yet to materialize, and when carrying out the household sample it was clear that some households were distinctly poorer than others, usually because they had no land of their own (or land that was unproductive).

Issues concerning the distribution of income from tourism also arose when comparing villages. Don Det clearly benefited from tourism, but other islands adjacent to it did not. In July 2004, for instance, Don Det's two villages had a population of 880 in 149 households, and boasted a total of 34 guest houses. By contrast, the village of Ban Hang Khon, on the adjacent island of Don Khon, had a population of 250 people in 45 households, but no guest houses or restaurants, even though tourists would occasionally pass through the village to view the Irawaddy dolphins. It was recognizably poorer than the villages of Don Det, which was probably both a cause and a consequence of its failure to develop tourism.

Conclusion

In the first part of this paper, a brief history of approaches to tourism as a tool for 'development' was given, and it was shown that, in recent years, the focus has moved from mainstream tourism (which was often subjected to severe criticism) to such 'alternative' forms of tourism as ecotourism and pro-poor community-based tourism (CBT). Then followed a discussion of international tourism in Lao PDR, which in recent years has come to rely on tourism as a tool for development, and where government, the Asian Development Bank and international aid agencies have together placed the emphasis (and considerable funds) on pro-poor CBT, with the Nam Ha Ecotourism Project (NHEP) in Luang Namtha as the prototype for projects elsewhere, while at the same time leaving the private sector, essentially small-scale, locally owned guest houses and hotels, to cater for most international visitors to Lao PDR.

In the third section of the paper, the financing and impact of the NHEP was briefly examined, before attention was shifted to the role of the private sector in developing tourism in the Siphandon region of Champassak. It now remains to briefly contrast these two forms of tourism development, as found in Lao PDR, arrive at some tentative conclusions, and suggest future research possibilities. And while, clearly, the comparison arises from the Lao PDR context, the findings may well be more generalizable. The NHEP indeed provides a model for the development of ecotourism in Lao PDR and other parts of South-east Asia, but in its aims, objectives, organization and reliance on donor assistance it has much in common with donor-assisted, community-based tourism projects elsewhere.

As indicated in Table 12.11, it is possible to compare the two models of tourism development under 12 related categories:

1. The NHEP model is said to specifically target the poorest members of the community, frequently ethnic minorities, and a key overall aim is to improve their standard of living. By contrast, the private sector does not specifically target anyone. This difference may be less stark than appears at first sight, for poverty is not the only criterion for developing tourism. The targeted community, or its environment, must be perceived to have some kind of value as an attraction and, in any case, tourists are unlikely to be taken to totally inaccessible, dangerous or unhealthy destinations. By contrast, while the private sector does not specifically target the poor, tour operators are aware of the tourist value of ethnic minorities and their cultures and attempt, perhaps insensitively, to meet tourist demand. In this respect, as in others, the private sector could learn from CBT.
2. Both forms of tourism might be described as 'protocapitalist'. In the NHEP model, initial development costs are born by international donors, who become, in effect, surrogate entrepreneurs. However, unless at some point they involve 'responsible' partners from the private sector, projects are unlikely to survive. For their part, at any early stage of their business development, as in the Siphandon region, guest house owners are also protocapitalists. They may realize capital by selling livestock, which play a major role in the subsistence sector, but they mostly employ unwaged family labour and have yet to engage in product development.
3. In such CBT projects as the NHEP, activities are deliberately planned in consultation with community members, with the aim of 'empowering' them. The extent to which such a process is really 'bottom up' rather than 'top down' must be an empirical question, but 'targeting' communities may (at least)

Table 12.11. Characteristics of Nam Ha ecotourism model and the private sector.

	NHEP model	Private sector
(1)	Targeted at the poor, but not necessarily the poorest, especially ethnic minorities	Not specifically targeted at the poor
(2)	Protocapitalist: initial risks born by aid agencies (surrogate entrepreneurs?)	Protocapitalist. Non-specialist entrepreneurs with local capital and family labour
(3)	Planned activities in 'consultation' with and to 'empower' villagers. Bottom up?	Laissez-faire, without formal planning or consultation re 'empowerment'
(4)	Contributes to spread of cash economy in subsistence (usually rural) areas	Expands cash economy, especially in urban areas
(5)	Attempts to contribute to spread benefits through Village Development Funds	No planned distribution of benefits but may usually occur through village funds
(6)	Variable relationship to market demand	Develops in specific response to market demand
(7)	Benefits from infrastructural improvements (roads, airports, etc.)	Also benefits from infrastructural improvements
(8)	Training from outside agencies with long-term impact (cultural capital, etc.)	Little or no training received from outside agencies or from within private sector
(9)	Linked to conservation of biodiversity in region	No *specific* links with conservation of biodiversity (but depends on indigenous technical knowledge)
(10)	Relies largely on private sector to provide tourists	Supplies tourists to treks but also benefits from extended stays
(11)	Unquantified multiplier effects in locality	Unquantified multiplier effects in locality
(12)	Caters for a minority of tourists	Caters for the majority of tourists

involve a degree of persuasion, which may continue well into the project, and there is frequently a marked element of 'consciousness raising'. By contrast, in the private sector there may be little planning and community consultation, but tourism development does not occur in a social or political vacuum, and formal and/or informal sanctions may be brought to bear on those straying too far from communal norms and values. In fact, to suggest that such projects as NHEP are 'community-based', and that small-scale tourism, as encountered in the Siphandon region, is not, is to draw an artificial and ultimately misleading distinction. The guest houses of Don Det and their owners are an integral part of the community, and the community in which they operate is no more or less cohesive a social entity than villages in Luang Namtha. As a consequence, rather then polarizing CBT (as promoted by international aid agencies) and private sector tourism, it might be more accurate to refer to donor-assisted, community-based tourism (DACBT) and to acknowledge that many private sector enterprises (and even such institutions as guest house associations) may be strongly rooted in communities.

4. DACBT projects undoubtedly contribute to the spread of the cash economy in rural areas, which may hitherto have been characterized by subsistence economies, though tourism is unlikely to be the first source of cash in these communities. Private sector development is also likely to expand the cash economy, though (as in Lao PDR, e.g. in Luang Prabang and Vang Vieng) it is likely to be more active in urban areas and is thus operating in an established cash economy.

5. In DACBT, sustained efforts are usually made to ensure the economic benefits from tourism are spread more or less equally through communities. The *success* of such efforts is not always apparent, and much depends on the nature of the communities being developed. This is also the case in the private sector. In Lao PDR, for instance, both DACBT and the private sector may use (or at least *attempt* to use) Village Funds as a redistribution mechanism, but the state can also become involved in redistribution through taxation and expenditure. Private sector development need not automatically lead to stark inequalities, though (arguably) some inequality will be inevitable if entrepreneurs are to make a profit (and thus sustain their businesses).
6. As the NHEP and ADB-assisted projects demonstrate, communities may initially be involved in tourism without much reference to market demand. Sooner rather than later, however, destinations and treks have to be marketed. Even the most worthy project is unsustainable if it is too far from the main tourist routes, not marketed, or charged at too high a price. By contrast, the private sector – the guest houses in Don Det, for example – develops to *meet* market demand, though it, too, will in turn seek to influence demand through marketing (if the necessary funds can be found). This means that quite early on in the development of a DACBT product, the private sector will have to be involved. This seems to have been one of the lessons emerging from the NHEP experience.
7. Both DACBT and the private sector benefit from infrastructural development. In this respect, ADB's loans to governments to construct and upgrade roads, airports and river jetties favours all tourism sectors (and commercial and industrial development generally). They may also facilitate unforeseen and less desirable activities, e.g. rural–urban migration, trafficking and prostitution.
8. As the NHEP approach indicates, advocates of DACBT emphasize the value of capacity-building and human resource development, and there has already been some cooperation with the private sector, for example inbound tour operators. By contrast, the guest house owners in Siphandon seem to have learned from experience and (as far as is known) receive little training in operating tourism enterprises or developing their products. In this sphere of activity, as in others, there is scope for further collaboration across sectors.
9. In matters of conservation and environmental awareness generally, the private sector might learn from DACBT, which is often developed with a keen awareness of the value of the natural context and the damage that tourism might do. For instance, guest house owners of Don Det could improve their tourist product by offering guided boat tours, focusing on the flora and fauna of the region, and learning how to explain their culture to visitors. Nevertheless, one should not neglect the role of indigenous technical knowledge. In the past, for instance, villages in the Siphandon region have cooperated to prevent over-exploitation of fish stocks (Baird, 1999) and there is no reason to suppose they will be less sensitive to the impacts of tourism.
10. DACBT and the private sector need each other. Most tourists to Lao PDR do not visit with the sole purpose of trekking in Luang Natha with the NHEP, which relies on mainstream tourism to provide it with clients. It is part of their vacation experience, which in Lao PDR is also likely to include visits to Vientiane and Luang Prabang. However, the guest house owners of Luang Namtha also need the NHEP. It makes their destination more attractive, encourages visitors to extend their stay and thus brings them additional business.
11. There is probably more information on the NHEP, its historical development and its impacts, than on similar projects elsewhere, but as yet there is little information on the direct, indirect and induced economic effects it has had on its targeted villages, or the surrounding region. This is also the case for small-scale, locally financed private sector tourism development of the kind found in the Siphandon region of Champassak. More broadly, in tourism research generally there is considerable

discussion of the tourist income multiplier, but few efforts, if any, have been made to calculate it for either DACBT projects or locally owned enterprises. What *can* be said, however, is that such projects as the NHEP rely greatly on technical advice from outside consultants, a substantial proportion of whose income is likely to leave the area, as will also a high percentage of donor funds spent on computers, four-wheel drives and other expensive imports.

12. Finally, despite the burgeoning literature on 'alternative' tourism, ecotourism and community-based tourism, such projects cater for a minority of tourists, and will continue to do so. By contrast, the private sector in Lao PDR, even though relatively undeveloped, undoubtedly caters for the tourist majority – and will continue to do so.

All these issues require further research. This examination of two quite specific forms of tourism development in Lao PDR highlights similarities and differences in a particular tourism context, some of which may well be found elsewhere. The intention is *not* to suggest that 'development' is achieved more by one type of tourism than another, but rather to indicate that the 'pro-poor' tourism label need not be applied exclusively to such donor-assisted, community-based projects as the Nam Ha Ecotourism Project. As has been expressed elsewhere (Harrison, 2001), instead of automatically assuming that tourism enterprises in the private sector are unwelcome and inferior competitors of 'alternative' donor-assisted, community-based tourism projects, they might be considered as potential partners in tourism development, with their own expertise and links to the community, and with an entitlement to at least some of the financial and technical support provided, on a regular basis, by international aid agencies.

It is even possible that, in some circumstances, a 100-bed hotel employing 250 people and directly supporting hundreds of households may be the most appropriate option. At other times it may not. The general point is that dollar for dollar, pound for pound, we need a more realistic, research-based consideration of the role of the private sector in alleviating poverty and (to reintroduce a forgotten term) in bringing 'development' to tourism destination areas.

Notes

[1] Partners are Caroline Ashley (Overseas Development Institute, London), Harold Goodwin (Centre for Responsible Tourism, University of Greenwich, London) and Dilys Roe (Institute for Environment and Development, London).

[2] Research in this region, and elsewhere in Lao PDR, was carried out in two phases by David Harrison, a co-author of this paper, first in June 2002 (as a consultant for the Asian Development Bank) and then in July and August of 2004, as part of an ASEAN-EU Universities Network Programme, 'Building Research Capacity for Pro-poor Tourism'. Funded from 2003 until 2005 by the European Commission, this project involved five partner organizations from Europe and South-east Asia: the International Institute for Culture, Tourism and Development, at London Metropolitan University, the University of Liege (Belgium), the University of Social Science (Viet Nam), Udayana University (Bali, Indonesia) and the National University of Laos. Data on tourism and poverty alleviation were obtained in Lao PDR in July and August 2004, and special thanks are due to Dr Sengdeuane Wayakone, of the National University of Laos, and Dr I Nyoman Adiputra, of Udayana University, who also participated in the Lao part of the research project. Further information on the project can be obtained from www.iictd.org.

[3] The owner of a five-room guest house with an attached restaurant, and 55% occupancy over a year, could expect to receive a gross income of about $9000 ($1004 for the accommodation, at $1.00 a night per room) and about $8000 for food. The initial investment of about $2000 should be recouped within a few years.

Acknowledgements

Steven Schipani wishes to acknowledge the support he has received from the Asian Development Bank and the Lao National Tourism Administration, and David Harrison that of his colleagues at the International Institute for Culture, Tourism and Development at London Metropolitan University and at the Overseas Development Institute in London.

References

Adams, W.M. (1990) *Green Development: Environment and Sustainability in the Third World*. Allen and Unwin, London.

Andronicou, A. (1979) Tourism in Cyprus. In: de Kadt, E. (ed.) *Tourism: Passport to Development*. Oxford University Press, Oxford, pp. 237–264.

Archer, B. (1996) Economic impact analysis. *Annals of Tourism Research* 23(3), 704–707. DOI: 10.1016/0160-7383(95)00097-6.

Ashley, C. (2000) Pro-poor tourism: putting poverty at the heart of the tourism agenda. Overseas Development Institute Paper 51. London.

Ashley, C., Roe, D. and Goodwin, H. (2001) *Pro-Poor Tourism Strategies: Making Tourism Work for the Poor – A Review of Experience*. Overseas Development Institute, International Institute for Environment and Development, and Centre for Responsible Tourism, London.

Asian Development Bank (ADB) (1999a) *Social Sector Profiles: Lao PDR. Strategic Forward Look*. ADB, Manila.

Asian Development Bank (ADB) (1999b) *Regional Technical Assistance 5893: Mekong/Lancang River Tourism Infrastructure Development Project*. ADB, Manila.

Asian Development Bank (ADB) (2002a) *Report and Recommendation of the President to the Board of Directors on Proposed Loans to the Kingdom of Cambodia, Lao People's Democratic Republic, and Socialist Republic of Viet Nam for the Greater Mekong Subregion: Mekong Tourism Development Project*. ADB, Manila.

Asian Development Bank (ADB) (2002b) *Building on Success: A Strategic Framework for the Next Ten Years of the Greater Mekong Subregion Economic Cooperation Program*. ADB, Manila.

Asian Development Bank (ADB) (2002c) *Memorandum of Understanding Between the Government of Lao PDR and the Asian Development Bank for the Mekong Tourism Development Project*. Vientiane.

Asian Development Bank (ADB) (2005) *The Greater Mekong Subregion Tourism Sector Strategy*. ADB, Manila.

Baird, I. (1999) Environmental protection and community development in Siphandone wetland, Champassak Province, Lao PDR: towards sustainable co-management of Mekong river inland aquatic resources, including fisheries, in southern Lao PDR. Champassak Province: CESVI/Centre for Protected Areas and Watershed Management, Department of Forestry.

Boo, E. (1990) *Ecotourism: The Potentials and Pitfalls: (Vols 1 and 2)*. World Wildlife Fund/USAID, Washington, D.C.

Butcher, J. (2003) *The Moralisation of Tourism: Sun, Sand … And Saving the World?* Routledge, London.

Butler, R. (1992) Alternative tourism: the thin end of the wedge. In: Smith, V.L. and Eadington, W.R. (eds) *Tourism Alternatives: Potentials and Problems in the Development of Tourism*. University of Pennsylvania Press, Philadelphia, pp. 31–46.

Butler, R.W. (1999) Sustainable tourism: a state-of-the-art review. *Tourism Geographies* 1(1), 7–25. DOI: 10.1080/14616689908721291.

Cater, E. and Lowman, G. (1994) *Ecotourism: A Sustainable Option*. Wiley, Chichester.

Chambers, E. (2000) *Native Tours: The Anthropology of Travel and Tourism*. Waveland Press, Illinois.

Chazée, L. (2002) *The Peoples of Laos: Rural and Ethnic Diversities*. White Lotus Press, Bangkok.

Cohen, E. (1989) Alternative tourism: a critique. In: Singh, T.V., Theuns, H.L. and Go, G.M. (eds) *Towards Appropriate Tourism: The Case of Developing Countries*. Peter Lang, Frankfurt, pp. 251–269.

Cummings, J. (2002) *Laos*. Lonely Planet, London.

de Kadt, E. (ed.). (1979a) *Tourism: Passport to Development?* Oxford University Press, Oxford.

de Kadt, E. (1979b) The issues addressed. In: de Kadt, E. (ed.) *Tourism: Passport to Development?* Oxford University Press, Oxford, pp. 3–67.

de Kadt, E. (1979c) Preface. In: de Kadt, E. (ed.) *Tourism: Passport to Development?* Oxford University Press, Oxford, pp. ix–xvi.

Denman, R. and Denman, J. (2004) *Tourism and Poverty Alleviation: Recommendations for Action*. WTO, Madrid.

Department for International Development (DFID) (1999) *Tourism and Poverty Elimination: Untapped Potential*. DFID, UK.

Duffy, R. (2002) *A Trip too Far: Ecotourism Politics and Exploitation*. Earthscan, London.

Enterprise Development Consultants Co. Ltd (2002) *Generating Employment through Micro and Small Enterprise and Co-operatives in Lao, PDR*. International Labour Organisation, Geneva.

Fennell, D.A. (1999) *Ecotourism: An Introduction*. Routledge, London.

Fennell, D.A. (2001) A content analysis of ecotourism definitions. *Current Issues in Tourism* 4(5), 403–421. DOI: 10.1080/13683500108667896.

Freitag, T.G. (1994) Enclave tourism development for whom the benefits roll? *Annals of Tourism Research* 21(3), 538–554. DOI: 10.1016/0160-7383(94)90119-8.

Gerosa, V. (2003) Tourism: A viable option for pro-poor growth in Africa? *Expert Group Meeting, Munyonyo Speke Resort*, Kampala, Uganda, 23–24 June.

Groupe Huit (1979) The sociocultural effects of tourism in Tunisia: a case study of Sousse. In: de Kadt, E. (ed.) *Tourism: Passport to Development?* Oxford University Press, Oxford, pp. 285–304.

Guttal, S. (2006) Marketing the Mekong: the Asian development bank and the greater Mekong Sub-region economic cooperation program. Jubilee South. Available at: www.jubileesouth.org/news/EpZyVyEAZFESZsvoiN.shtml (accessed 1 January 2006).

Hall, C.M. (2000) Tourism in Cambodia, Laos and Myanmar: from terrorism to tourism? In: Hall, C.M. and Page, S. (eds) *Tourism in South and South East Asia: Issues and Cases*. Butterworth-Heinemann, Oxford, pp. 178–194.

Hall, C.M., Jenkins, J. and Kearsley, G. (1997) *Tourism Planning and Policy in Australia and New Zealand*. Irwin, Sydney.

Hamilton-Merritt, J. (1999) *Tragic Mountains: The Hmong, the Americans, and the Secret Wars for Laos, 1942–1992*. Indiana University Press, Bloomington.

Harrison, D. (1996) Sustainability and tourism: reflections from a muddy pool. In: Briguglio, B., Butler, R., Harrison, D. and Leal Filho, W. (eds) *Sustainable Tourism in Islands and Small States: Issues and Policies*. Pinter, London, pp. 69–89.

Harrison, D. (ed.) (2001) *Tourism and the Less Developed World: Issues and Case Studies*. CAB International, Wallingford.

Honey, M. (1999) *Ecotourism and Sustainable Development: Who Owns Paradise?* Island Press, Washington.

Jafari, J. (1989) Sociocultural dimensions of tourism: an English language literature review. In: Bystrzanowski, J.. (ed.) *Tourism as a Factor in Social Change, Vol. I: A Sociocultural Study*. International Social Science Council, European Coordination Centre for Research and Documentation in Social Sciences, Vienna, pp. 17–60.

Jamieson, W. (2003) *Poverty Alleviation Through Sustainable Tourism Development*. United Nations, Bangkok.

Johnston, A.M. (2006) *Is the Sacred for Sale?* Earthscan, London and Sterling, VA.

Kiss, A. (2004) Is community-based ecotourism a good use of biodiversity conservation funds? *Trends in Ecology & Evolution* 19(5), 232–237. DOI: 10.1016/j.tree.2004.03.010.

Lao National Tourism Administration (2006) *2005 Statistical Report on Tourism in Laos*. Planning and Cooperation Department Statistics Unit, Vientiane.

Lao PDR, UNDP and WTO (1998) *National Tourism Development Plan for Lao PDR*. Lao PDR, Vientiane.

Lao PDR, UNDP and WTO (1999) *National Tourism Marketing Plan for Lao PDR*. Lao PDR, Vientiane.

Lao PDR (1990) *National Plan for the Development of Tourism in the Lao People's Democratic Republic*. United Nations Development Programme, Vientiane.

Lao PDR (2004) *National Growth and Poverty Eradication Strategy*. Lao PDR, Vientiane.

Lao PDR (2006) Presentation on Lao Investment Promotion at Tourism Round Table, Mekong Tourism Investment Summit, 'Expanding "Win-Win" Business Opportunities in the GMS'. Luang Prabang. 28–30 March.

Lindberg, K., Enriquez, J. and Sproule, K. (1996) Ecotourism questioned: case studies from Belize. *Annals of Tourism Research* 23(3), 543–562.

Lyttleton, C. and Allcock, A. (2002) *Tourism as a Tool for Development: UNESCO-Lao National Tourism Authority Nam Ha Ecotourism Project–External Review July 6–18 2002*. NTAL and UNESCO, Vientiane.

Ministry of Foreign Affairs and Trade (1995) *NZODA support for eco-tourism in Fiji: A report of a study*. Appraisal, Evaluation and Analytical Support Unit, Development Co-operation Division, Wellington.

Momsen, J. (1972) *Report on Vegetable Production and the Tourist Industry in St. Lucia*. Dept. of Geography, University of Calgary, Canada.

Momsen, J.H. (1986) Linkages between tourism and agriculture: problems for smaller Caribbean economies. *Seminar Paper 45, Dept. of Geography*, University of Newcastle upon Tyne.

Mowforth, M. and Munt, I. (2003) *Tourism and Sustainability: Development and New Tourism in the Third World*, 2nd edn. Routledge, London.

Nash, D. (1996) *Anthropology of Tourism*. Pergamon, Oxford.

Nash, D. (1977) Tourism as a form of imperialism. In: Smith, V.L. (ed.) *Hosts and Guests: The Anthropology of Tourism*. Blackwell, Oxford, pp. 33–47.

National Tourism Authority of Lao PDR (2004) *2003 Statistical Report on Tourism in Laos*. NTAL, Statistics, Planning and Co-operation Division, Vientiane.

National Tourism Authority of Lao PDR (2005) *National Ecotourism Strategy and Action Plan: 2004–2010*. NTAL, Vientiane.

Page, S. (1999) *Tourism and Development: The Evidence from Mauritius, South Africa, and Zimbabwe*. Overseas Development Institute, London.

Peters, M. (1969) *International Tourism: The Economics and Development of the International Tourism Trade*. Hutchinson, London.

Pro-Poor Tourism Partnership (2006). Available at: http://www.propoortourism.org.uk/ ppt_pubs_workingpapers.html (accessed 18 April 2006).

Rodenburg, E.E. (1980) The effects of scale in economic development: tourism in Bali. *Annals of Tourism Research* 7(2), 177–196.

Scheyvens, R. (2002) *Tourism for Development: Empowering Communities*. Longman, London.

Scheyvens, R. (2004) Tourism, globalisation and poverty-alleviation: Compatible goals? Paper presented at DevNet Conference: 'Development on the Edge'. Auckland University. 3–5 December.

Schipani, S. (2002) *Ecotourism Status Report: Lao PDR*. SNV Lao PDR, Vientiane.

Smith, V. (1978b) Introduction. In: Smith, V.L. (ed.) *Hosts and Guests: The Anthropology of Tourism*. Basil Blackwell, Oxford, pp. 1–14.

Smith, S. (1983) Class analysis versus world system: critique of Samir Amin's typology of development. In: Limqueco, P. and McFarlane, B. (eds) *Neo-Marxist Theories of Development*. Croom Helm, London, pp. 73–86.

Smith, V. . (ed.) (1978a) *Hosts and Guests: The Anthropology of Tourism*. Blackwell, Oxford.

SNV (2006) SnV: Lao PDR. Available at: http://www.snv.org.la (accessed 2 January 2006).

Telfer, D. (2002) Evolution of tourism and development theory. In: Sharpley, R. and Telfer, D.J. (eds) *Tourism and Development: Concepts and Issues*. Channel View Publications, Clevedon, pp. 35–79.

Timothy, D.J. and Wall, G. (1997) Selling to tourists: Indonesian street vendors. *Annals of Tourism Research* 24(2), 322–340.

Torres, R. and Momsen, J.H. (2004) Challenges and potential for linking tourism and agriculture to achieve pro-poor tourism objectives. *Progress in Development Studies* 4(4), 294–318. DOI: 10.1191/1464993404ps092oa.

UN Economic and Social Commission for Asia and the Pacific (ESCAP) (2005) The contribution of tourism to poverty alleviation. ESCAP Tourism Review No. 25. United Nations.

UNESCO (2005) *Nam Ha Ecotourism Project Phase II, Project Document*. Office of the UNESCO Regional Advisor for Culture in Asia and the Pacific, Bangkok.

United Nations Development Programme (UNDP) (2005) *Human Development Report 2005*. UNDP/ Oxford University Press, Oxford.

van den Berghe, P.L. (1994) *The Quest for the Other: Ethnic Tourism in San Cristobal, Mexico*. University of Washington Press, Seattle and London.

Walpole, M.J. and Goodwin, H.J. (2000) Local economic impacts of dragon tourism in Indonesia. *Annals of Tourism Research* 27(3), 559–576. DOI: 10.1016/S0160-7383(99)00088-2.

Warner, R. (1997) *Shooting at the Moon: Story of America's Clandestine War in Laos*. Steerforth Press, Vermont.
Weaver, D.B. (1998) *Ecotourism in the Less Developed World*. CAB International, Wallingford.
Wheeller, B. (1993) Sustaining the ego. *Journal of Sustainable Tourism* 1(2), 121–129. DOI: 10.1080/09669589309450710.
Wilson, D. (1979) The early effects of tourism in the Seychelles. In: de Kadt, E. (ed.) *Tourism: Passport to Development?* Oxford University Press, Oxford, pp. 205–236.
World Bank (2005) Country Assistance Strategy for the Lao People's Democratic Republic (PDR). *Report No. 31758 LA, 10 March*, East Asia and Pacific Region: Southeast Asia Country Unit.
World Tourism Organization (WTO) (1996) *Compendium of Tourism Statistics: 1990–1994*. WTO, Madrid.
World Tourism Organization (WTO) (1999) *Compendium of Tourism Statistics: 1993–1996*. WTO, Madrid.
World Tourism Organization (WTO) (2001) *Compendium of Tourism Statistics: 2001 Edition*. WTO, Madrid.
World Tourism Organization (WTO) (2002) *Tourism and Poverty Alleviation*. WTO, Madrid.
World Tourism Organization (WTO) (2005) *Compendium of Tourism Statistics: Data 1999–2003*, 2005 edn. WTO, Madrid.

13

Pro-poor Tourism: A Critique

Pro-poor Tourism in Context

After the Second World War economic growth and 'development' were considered vital for the Third World, and debates over the most appropriate theoretical frameworks, ideologies and policies continued for three decades. In the 1970s, for instance, the satisfaction of 'basic needs' was linked to the recognition that economic growth alone did not solve the problems of the poor (Willis, 2005), and Seers argued strongly that development was best achieved by alleviating poverty, reducing unemployment and inequality and, later, by increasing cultural independence and self-reliance (Seers, 1969, 1977).

Such ideas were incorporated into the programmes of many international organizations, including the International Labour Organization and the World Bank, both of which advocated 'bottom-up' approaches to development (Leftwich, 2000; Willis, 2005) while both the 1980 Independent Commission on International Development Issues (Brandt Report) and the 1987 World Commission on Environment and Development (the Brundtland Report) considered poverty alleviation a major development objective (Independent Commission on International Development, 1980; WCED, 1987).

It was soon recognized that international tourism, which had increased rapidly between 1950 and 1970, as indicated in Table 13.1, could play a key role in poverty alleviation. The World Bank led the way in the 1970s, financing infrastructural projects and providing credit for foreign investment, only later moderating its involvement through concerns about tourism's social and environmental impacts (Hawkins and Mann, 2007). And Agenda 21, formulated by the Earth Council in 1996 and adopted by the United Nations in 1999, also led to an increased focus on tourism, especially as a means of involving community members as participants in local development projects.

The adoption in 2000 of the Millennium Development Goals, the first of which is the eradication of extreme poverty and hunger, further reinforced the programmes of numerous multilateral and bilateral institutions in promoting tourism as a way of alleviating poverty. Among the former are regional development banks, including the Asian Development Bank (ADB), which devotes considerable funds to tourism development, especially in the Greater Mekong region (ADB, 1999, ADB, 2002) and such UN agencies as the UN World Tourism Organisation (UNWTO), while the latter include many national aid agencies.

In addition, despite the closure of its Tourism Projects Department in 1979, the World Bank continues to have a crucial role in advising on (some would say imposing)

Table 13.1. International tourist arrivals and receipts, selected years, and percentage share by region.

Year	Total arrivals (million)	Change over the period	Africa	Americas	Asia/Pacific	Europe	Middle East
			Percentage share of international tourist arrivals				
1950	25		2	30	0.1	66	–
1960	69	176%	1.5	24	0.3	72	–
1970	160	132%	1.8	23	0.6	70	–
1980	285	78	2.5	19	7.8	69	2
1990	439	54%	3.5	21.1	12.8	60.5	2.2
1995	540	23%	3.8	20.2	15.3	58.3	2.5
2000	687	27%	4.1	18.6	16.1	57.6	3.5
2001	680	–1%	4.3	18.0	17.8	56.4	3.7
2002	700	3%	4.2	16.7	18.7	56.3	4.2
2003	694	–0.1%	4.5	16.2	16.3	58.7	4.3
2004	765	10%	4.4	16.5	18.9	55.5	4.8
2005	807	5%	4.6	16.6	19.2	54.8	4.9
2005	Total receipts (US$billion)		Percentage share of international tourism receipts				
	680		3.1	21.3	20.4	51.2	4.1
	Average receipt per arrival (US$)		Regional average receipt per arrival				
	840		590	1080	890	790	710

Source: World Tourism Organization (WTO) (1991), *Current Travel and Tourism Indicators*, WTO, Madrid, p. 11; WTO (2006), *Tourism Highlights*, WTO, Madrid, p. 2; and United Nations World Tourism Organization (UNWTO) (2007), *Compendium of Tourism Statistics, Data 2001–2005*, UNWTO, Madrid, pp. 2–3.

structural adjustment programmes (SAPs), now *Poverty* Reduction Strategy Papers (PRSPs), in developing countries and, with the International Monetary Fund, in promoting tourism as one feature of private sector development, an approach which is largely laissez-faire (Mowforth and Munt, 2003).

Academic inputs

When de Kadt's path-breaking collection of papers was published, with financial assistance from the World Bank, he was a colleague of Seers at the Institute of Development Studies, Sussex, where basic needs approaches were much discussed. His 1979 summary, as valid now as three decades ago, lays out the pros and cons of tourism as a development tool (de Kadt, 1979, p. xii). Recognizing tourism's contribution to employment, economic growth, entrepreneurship and inter-sectoral linkages, he also noted it could exacerbate inequalities at international and local levels. And, anticipating later debates, he queried how far mass tourism would really contribute to poverty alleviation:

> More than ever before, the development community is searching for means that will enable the poor to provide for their basic needs through more productive work, more widely available social services, and increased participation in political decision making. It needs to be considered whether the deliberate and large-scale development of tourism, conceived as a major net earner of foreign exchange, leads to results consistent with this newly identified goal of development (de Kadt, 1979).

Contributors to his volume were divided over how far the poor benefited from tourism, as were several of those involved in an equally important collection of papers on the anthropology of tourism (Smith, 1978). In fact, many academics continue to doubt tourism's role as a development tool (Harrison, 2007) with their views often coloured by their commitment to a range of theoretical (and fashionably changing) perspectives, including modernization and dependency theory, liberalism, statism and globalization. A linked tendency, also apparent since the early 1980s, has been to emphasize the contribution to sustainable development of 'alternative', small-scale, community-based or eco-tourist ventures, characterized by high levels of local participation, which reportedly epitomize genuinely 'sustainable' tourism development (Rodenburg, 1980).

While such approaches have undoubtedly broadened the debate, they have also tended to deflect attention away from the more specific question of how far tourism *does* alleviate poverty. Indeed, empirical evidence on this topic is rare and, where assessments have been made, conclusions are contradictory (Harrison and Schipani, 2007). It is in this context, at the end of the 1990s, that the movement for 'pro-poor tourism' (PPT) emerged.

The emergence of pro-poor tourism

Since the end of the 1990s a loose alliance of several individuals, largely outside the academic mainstream of tourism studies, has successfully refocused attention in some quarters on the need to consider tourism as a means of alleviating poverty. There are several accounts of how this association emerged but the outlines are clear (Goodwin, 2005; Roe, 2006; Harrison and Schipani, 2007; DFID, 1999). In 1998 the UK Department for International Development (DFID) commissioned Deloitte and Touche, along with Dilys Roe of the International Institute for Environment and Development (IIED) and Caroline Ashley of the Overseas Development Institute (ODI), to report on the extent to which outbound tourism from the UK could contribute to poverty alleviation in destination societies. The ensuing report suggested that, while tourism already helped poor countries, more could be done with a specific focus on the poor; the criteria for the success of pro-poor tourism were clearly specified:

> Pro-poor tourism strategies are concerned specifically with impacts on poor people, though the non-poor may also benefit. Strategies focus less on expanding the overall size of tourism, and more on unlocking opportunities for specific groups within it (on tilting the cake, not expanding it). However, these strategies have to be integrated within general tourism

> development for two reasons: mainstream activities (such as tourism planning) need to be influenced by pro-poor perspectives; and pro-poor tourism cannot succeed without successful development of the whole tourism destination DFID (1999, p. 1).

After the 1999 Report Roe and Ashley were joined by Harold Goodwin, of the International Centre for Responsible Tourism, and the Pro-poor Tourism Partnership (PPTP) was formed. In the ensuing years they and associates in their respective institutions, along with collaborators elsewhere, produced a range of case studies, often funded by DFID, designed to show how (largely small-scale) tourism could alleviate poverty by stimulating local involvement, partnerships and procurement in numerous destinations, including Southern Africa, Uganda, The Gambia, Nepal, the Caribbean and the Czech Republic. In particular, the case studies focused on partnerships of the private sector, communities and community-based tourism enterprises, including NGOs, and showed how companies, including package tour operators in The Gambia, and local communities might benefit when the latter were brought into direct trading relationships with the former. These were followed by other Working Papers that reviewed methods followed in PPT studies and a range of other topics relevant to PPT, including coping mechanisms of Kenyans when tourism was in decline, international statistics, literature reviews, international codes of conduct and standards, and local sourcing and enterprise development in the Caribbean (Ashley, 2000; Ashley *et al.*, 2001; PPT, 2007).

PPTP personnel also made important contributions to other initiatives, including the high-profile Sustainable Tourism–End Poverty (ST-EP) campaign of the then World Tourism Organization (the UN appellation was added later) (WTO, 2002). Launched at the World Summit for Sustainable Development, Johannesburg, in 2002, the campaign has three dimensions: first, the incorporation of a poverty component in the UNWTO's technical assistance programme; second, the ST-EP Foundation, which finances research and cooperation with other organizations to promote poverty alleviation through tourism; and, third, the ST-EP Trust Fund, which finances technical assistance projects specifically geared to poverty alleviation (UNWTO, 2007).

In line with the focus on poverty, UNWTO followed up the 2002 publication with a series of recommendations on how major stakeholders in tourism could work together to ensure that tourism development was increasingly focused on benefits for the poor (Denman and Denman, 2004).

Publications (with very similar recommendations) arising from a conference organized by the United Nations Economic and Social Commission for Asia and the Pacific (ESCAP) in 2003 also reveal PPT influence. The initial background paper, a resumé of key issues in sustainable tourism with a 'pro-poor' gloss, and practical questions for workshop participants (Jamieson, 2003) was supported by country case studies purportedly showing how (mainly small-scale) tourism alleviated poverty (ESCAP, 2005). More formal collaboration is evident in another summary of key ideas and discussion of possible initiatives and indicators for poverty alleviation through tourism (Jamieson *et al.*, 2004).

Clearly, the focus on tourism as an alleviator of poverty is not new. It was implicit in the early emphasis on tourism as a stimulator of jobs and economic growth, in the work of the World Bank and other agencies, including in the development of strategies and policies centred around the concept of basic needs, and has never been entirely absent from the academic literature on tourism and development. Nevertheless, specifically PPT approaches, promoted in the UK by a small group of researchers and consultants since the end of the 1990s, to some extent by others elsewhere (Gerosa, 2003) and implemented by numerous aid agencies (Rossetto *et al.*, 2007) seem to have led to a popular, simple, sharper and more appealing moral focus on the links poorer residents in destinations have with tourism enterprises.

Key Features of Pro-poor Tourism

Using numerous publications of PPT practitioners as a basis (Ashley, 2000; Ashley *et al.*, 2001) and the various Working Papers produced by the Pro-Poor Tourism Partnership (PPT, 2007) the characteristics of PPT can be summarized fairly succinctly, as indicated in Table 13.2.

Table 13.2. The characteristics of pro-poor tourism.

PPT is not	PPT does
anti-capitalist	focus on incorporating the poor into capitalist markets by increasing job and entrepreneurial opportunities and collective benefits. Like fair trade, it is a form of market intervention, which relies heavily on the private sector
separate from wider tourism systems	depend on existing tourism structures and markets
a theory or model	orientate research to the net benefits from tourism that can or could accrue to the poor
a niche type of tourism, e.g. CBT	apply to *any* kind or type of tourism, including large- and small-scale tourism, *even if the non-poor also benefit*. Can be from regional or national policies or private sector involvement
a specific method	use numerous methods, none of which are specific to PPT, including value chain analysis, to collect data and show how the poor are and can be further involved in tourism
only about 'the poor'	recognize that the non-poor may also benefit from tourism, *even* disproportionately. It is less concerned with the relative than the absolute (net) benefits received by the poor
just about hunger and no/ inadequate incomes	have a broad definition of 'poverty', including lack of freedom, opportunity, power, skills and education. It is about 'development'
only about individual benefits	focus on community benefits – e.g. water, sanitation, health, education, infrastructure, etc.
oonly for those occupying the 'moral high ground'	require wide stakeholder cooperation and commitment, including national and local authorities, planners, the private sector, etc., ideally combining to ensure the poor benefit from tourism

First, and perhaps most importantly, proponents of PPT are not anticapitalist. Rather, strategies derived from a PPT perspective are formulated to incorporate the poor into capitalist markets by increasing the employment and entrepreneurial opportunities, and more collective benefits, available to them. In this respect PPT resembles 'fair trade', in that it is a form of market intervention. Whereas the latter attempts to build in 'added value' to the product on sale, thus creating a qualitatively 'better' product (Mowforth and Munt, 2003, pp. 168–170, pp. 161–162 the former attempts to change market conditions and influence the demand for labour, services and goods provided by the poor.

Second — and crucially — PPT relies on, *and must be integrated into,* wider tourism systems. It is *not* a stand-alone option. As Ashley *et al.* indicate, when reviewing the results of case studies of PPT financed by the UK's DFID, successful PPT depends on the poor having access to markets, on the commercial viability of PPT projects, on a policy framework that provides a secure investment climate (including access to land), and on effective stakeholder cooperation and strategy implementation, categorized as 'implementation issues' (Ashley *et al.*, 2001, pp. 28–34).

Third, PPT is not a specific theory or model, and is not tied to any theories or models. It is not reliant on any such perspective as modernization or underdevelopment, statism or neo-liberalism. Rather, it is an orientation, an approach to any form of tourism which focuses on the net benefits accruing to poor people in tourist destination areas.

Fourth, and as a consequence, the issue is not so much the kind of tourism involved, but the extent to which the poor do or could receive net benefits from it. PPT is not a niche type of tourism, like, for example, community-based tourism (CBT) (where, in any case, the benefits from tourism accruing to the poor may be secondary to conservation or other priorities). And there is no reason why PPT need be associated only (or even at all) with small-scale tourism. Mass tourism, even in its most extreme forms, can potentially be considered a form of PPT.

Fifth, no methods of data collection, targeting or analysis are appropriate only to PPT. As Goodwin notes, statistics of tourist arrivals and expenditure, and tourism satellite accounts, tell little, in themselves, about tourism's impacts at destination level, and he advocates measuring remittances from tourist areas and household income from tourism enterprises, and assessing the results of improved market access (Goodwin, 2006, p. 1). Such methods are clearly relevant to PPT, but they, too, are no more specific to it than examining leakages from destination areas, comparing the relationship of tourism to pre-existing or new small and medium-size enterprises (SMEs), detailing the new employment opportunities tourism creates, assessing the impacts of capacity building and improvements in welfare, or outlining more general changes, for good or ill, in culture and values (Jamieson *et al.*, 2004, pp. 20–23). All these approaches have long existed in mainstream tourism.

Similarly, while value chains (known *inter alia* as global commodity chains, value systems, production networks and value networks (Gereffi *et al.*, 2001, p. 2) have sometimes been analysed by proponents of PPT, value chain analysis (VCA) is well established in examinations of differential market relationships of developing country firms in other economic sectors (Gereffi and Kaplinsky, 2001; Kaplinsky and Morris, 2001; Ashley, 2006). When specifically used in collecting and categorizing tourism-related economic activities, probably because of funding and time constraints, VCA has been applied somewhat simplistically. In particular, attention has focused primarily on linkages *within* tourism destinations, rather than on the wider global context (Mosedale, 2006; Mitchell and Faal, 2007; Mitchell and Phuc, 2007). This is not to deny its usefulness as a framework for organizing data collection, but VCA is not a specific feature of PPT, as indicated by papers on VCA and tourism that make no reference to poverty at all (Yilmaz and Bititci, 2006a, b).

It is also worth noting that VCA in tourism was preceded by numerous less comprehensive but nevertheless relevant attempts to assess tourism's relationship with other market sectors, for example agriculture in the Caribbean (Momsen, 1972, 1973; Britton, 1975) and that some VCA in tourism contexts has tended to cover much the same ground as standard tourism master plans (Mitchell and Phuc, 2007; pp 29–50).

While none of the above methods and concerns apply only to PPT, together they can be brought to bear in developing an understanding of how far tourism does or could contribute to poverty alleviation. The same can be said of other well established tourism policies and planning approaches, including coordination of ministries, development zones, special assistance to SMEs, demonstration projects, investment, capacity building, and so on, all of which have been claimed as essential to PPT (Jamieson *et al.*, 2004, p. 10–14).

Sixth, PPT may not be focused on the very poorest of society. As Ashley notes, 'PPT is not an appropriate tool for reaching the poorest – those with fewest assets and skills who are least able to engage in the commercial economy' (Ashley, 2006, p. 10). Indeed, it is also accepted that non-poor residents of destination areas too may benefit from tourism, even to a disproportionately greater extent than the poor: 'Only occasionally is PPT a matter of transferring gains ... from one group to a poorer group. Thus helping the poor can mean directly helping the better-off clients or employers. Trying to avoid benefiting the non-poor is usually counter-productive' . And again: 'as long as poor people reap net benefits, tourism can be classified as "pro-poor" (even if richer people benefit more than poorer people)'(Ashley *et al.*, 2001).

Seventh, PPT practitioners define poverty in such a way as to include non-economic benefits. PPT strategies are thus considered to include

> infrastructure gains, capacity building and training, and empowerment. While the main focus is generally on direct employment, and indirect employment through micro-enterprise development, it is important to look at the range of potential benefits (and possible negative impacts) and to look at livelihood diversification—the focus on pro-poor tourism is on additional and supplementary livelihoods at the individual and household level. It is far less risky for communities to engage in tourism if that engagement complements their existing livelihood strategies rather than competes with or replaces them. (Goodwin, 2005, p. 5)

Eighth, and despite the above reference to individual and household levels, PPT also focuses on wider collective benefits. For this reason the ADB, as well as having a US$13.5 million 'pro-poor and equitable tourism' component in its Greater Mekong Sub-region Tourism Sector Strategy, primarily to develop community-based tourism, also has a budget of more than $373 million, aimed at specific priority development zones, to improve tourism-related infrastructure, 'feeder roads, piers, and jetties leading to tourist attractions and poor communities, to spread the benefits of tourism more widely *and to reduce poverty*' (ADB, 2002; pp, x, 121–125). Such infrastructural development, along with provision of potable water and efficient sewage systems, is a way of spreading benefits as widely as possible.

Finally, and following from the previous point, optimism, high-mindedness and concern for the welfare of the poor are not enough. PPT is not a case of NGOs and other representatives of civil society working *against* exploitative capitalism, or of the promotion of small-scale, community-based tourism *against* mass tourism. Rather, it is an orientation, covering nearly all forms of tourism, that requires commitment to assisting the poor, commercial viability of projects, and cooperation across all stakeholders – national and local authorities, public and private sectors, government, international organizations and NGOs – to achieve the aim of PPT, namely, to ensure that tourism brings net benefits to the poor.

Pro-poor Tourism: The Problems

At first sight, criticizing PPT might appear somewhat perverse. It seems so evidently morally correct and, after all, who could oppose the interests of the poor? However, some criticisms that can be levelled at PPT have already been made and, indeed, even its major proponents are careful to emphasize its limitations. More specifically, however, criticisms can be made on both conceptual and substantive grounds.

Conceptual objections

Three basic theoretical objections can be made to PPT. The first, and most mentioned, is that, in effect, its advocates are too comfortable with the status quo and thus miss 'the big picture'. Rather than attending to the need for structural change, redistribution of wealth and resources, and addressing international and national power structures, they tacitly accept a neo-liberal approach to development and tinker with the capitalistic international tourism system at the edges, eking out a few resources for small, selected groups of the poor (or relatively poor) in destination areas (Mowforth and Munt, 2003; Scheyvens, 2007).

This criticism is correct in so far as advocates of PPT accept that the approach works best in development destinations where, first, a business case can be made, second, there is already a formal sector involving other tourism stakeholders and, third, conditions are right for capitalist development. However, this is less a theoretical objection than a statement of fact and, as Harrison and Schipani have indicated elsewhere, amounts to little more than 'a crude repackaging' of 1970s and 1980s underdevelopment perspectives.

That PPT advocates operate within a capitalist context is valid as a statement but not as a criticism. They make no claim to be theorists, and see themselves rather as practitioners attempting to improve the benefits that tourism brings to the poor, irrespective of their own beliefs, supportive or otherwise, about varieties of capitalism as political systems. Indeed, a similarly flawed argument could be directed at critics of PPT who voice their objections in books and journals which are part of profit-oriented enterprises.

Second, it can be argued that PPT approaches are conceptually blurred and/or ideologically or even morally indiscriminate, in that any kind of tourism (including sex tourism) that demonstrably increases net incomes of the poor or relatively poor would qualify as PPT.

Logically this does seem to be the case, as the definition of pro-poor tourism focuses on the results of tourism as they have an impact upon the poor, rather than on the type of tourism under investigation. However, current practice is rather different. PPT has no obvious record of supporting sex tourism and – with echoes of debates from the 1960s and 1970s about the extent social scientists should be partisan and, if so, whose side they should take (Becker, 1967;

Gouldner, 1973) – its implicit commitment seems to be the development of business opportunities for the poor or the nearly poor, with little attention paid to, and even some hostility towards, big business, from which (unsurprisingly) PPT seems to attract little but token support. There are exceptions – for example the cooperation of some large UK companies with the UK's Travel Foundation (www.thetravelfoundation.org.uk), an independent charity, and the positive findings of GTZ, the German Aid Agency, concerning PPT features of Sandals Resorts, Jamaica (Karammel and Lengefeld, 2006) – but they are rare and miniscule in scale. That said, the extent to which support for local (as opposed to international) capital should be an explicit priority for PPT adherents might well be a subject for serious consideration

Third, as with sustainable tourism development, the situation of PPT within debates over sustainability is theoretically imprecise and, generally, little discussed. As Chok *et al.* indicate, the focus 'on poor people in the South reflects a strong anthropocentric view ... and ... environmental benefits are secondary to poor peoples' benefits' (Chok *et al.*, 2007, p. 153). Echoing the critique that PPT favours capitalism, they suggest that it is 'reformist' rather than 'radical', and epitomizes a 'weak sustainability position' (Chok *et al.*, 2007, p. 153). This is largely correct but (again) it is irrelevant. As indicated earlier, PPT practitioners make no claims for theoretical expertise or sophistication. Their concerns are more practical.

Fourth, PPT proponents tend to develop close links with NGOs and international NGOs (INGOs) and, like consultants generally, expend much time and effort obtaining external funding, largely from national and international aid agencies. Unfortunately, as a consequence, the theoretical marginality of PPT as an orientation and a series of practices is reinforced and reflected in the personal marginality of its practitioners, both to the business world of tourism and to academia. Lacking permanent academic posts and (relative) financial security, they rarely submit their work to academic peer review and recognized academic journals, thus making future advancement in the academic sphere unlikely. They also tend to remain outside (and largely ignorant of) academic debates, with the added result that a huge body of academic writing is more or less ignored by PPT, for example on enterprises and entrepreneurship, and the insights of adherents of PPT are similarly ignored or underrated in academic circles.

Substantive issues

Some substantive problems arise from the conceptual issues outlined above, and are summarized in Table 13.3. First, PPT's focus on residents of destination areas is conceivably too narrow and parochial. By treating destinations as bounded communities, only the 'poor' or 'near poor' within the boundaries are considered suitable beneficiaries. By contrast, incoming migrants working in tourism and others living outside the boundaries are, by implication, unintended and undeserving beneficiaries, even though it is accepted that the economic position of non-poor locals may also be improved as a result of PPT initiatives.

There is accuracy in this criticism, but it arises more from practical constraints resulting from shortages of funds and from the need to limit the time frame and scope of projects, including value chain analysis, than from any conceptual or ideological orientation.

Second, there are several arguments to the effect that PPT projects have simply not delivered benefits, or *adequate* benefits, to the poor, and that the term 'pro-poor' tourism is a misnomer. Scheyvens, for example, is one such critic (Scheyvens, 2007). But, while her sources suggest tourism benefits have been distributed unevenly, they do not argue that the poor have received no benefits at all (Desforges, 2000; Konadu-Agyemang, 2001). In any case, as indicated earlier, it is accepted by some PPT practitioners that, although the poorest in society may not benefit at all from PPT initiatives, some of the non-poor may do so (Ashley, 2006, p. 10).

Similarly, but more strongly, it has been argued that, by supporting capitalism, PPT initiatives 'undercut "sustainable livelihoods" and exacerbate, rather than alleviate, poverty' (Mowforth and Munt, 2003, p 273). Again, however, there is no empirical evidence for this view, which seems to rest on the (unfounded) assumption that, if the plight of some of the poor is relieved, those who remain are somehow subjected to further deprivation. Even more

Table 13.3. Problems relating to pro-poor tourism.

Conceptual or substantive	Problem	Comment
Conceptual	Tacitly accepts neo-liberal status quo	True but irrelevant; not a theory
	Morally indiscriminate – any tourism can be PPT	Conceptually so but not in practice
	Theoretically imprecise, 'reformist', holding position of 'weak sustainability'	True but irrelevant; not a theory
	Academically and commercially marginal	True
Substantive	Narrow and parochial; focuses only on bounded destinations	True but is more a financial than conceptual constraint
	Does not deliver (enough?) benefits to the poor	Little evidence available, but not *shown* to be more effective than non-PPT private sector
	No clear link of PPT with poverty alleviation; 'normal' tourism may also be PPT	Probably the case?
	Fails to consider equity or to try to change the overall system	True – but is quite explicit about this
	Ignores markets and commercial viability	False; this is more so for CBT
	Ignores problems and PPT features of mass tourism	Does not set out to analyse such problems and has limited involvement with mass tourism

radically (and cynically) it could be argued that the revolutionary cause would be advanced by the continued impoverishment of the poor, and thus *any* attempt to spread the benefits of tourism more widely serves capitalism and delays the onset of revolution. In response to such arguments, however, proponents of PPT can quite explicitly indicate that they are interested more in the absolute than the relative benefits to the poor of PPT, and that their interest is in the local picture and relief of local poverty, rather than in grandiose schemes for the overthrow of capitalism.

That said, it is often impossible to calculate the benefits PPT initiatives do bring to communities. As indicated earlier, rigorous assessments of incomes and expenditures in PPT projects (and, indeed, in many other forms of aid) are difficult to find; by definition intangible benefits are hard to calculate (Ashley *et al.*, 2001; Ashley, 2006). Systematic comparative monitoring of revenues against costs is rare, and where an attempt has been made to do this, as for a landmark ecotourism project in Lao PDR, it is clear that, according to strictly financial criteria, the private sector gives better value for money (Harrison and Schipani, 2007). In such circumstances it is perhaps unsurprising that official recipients of international aid programmes, and sometimes donors, adopt PPT terminology and rhetoric but actually assess the success of aid programmes according to increased numbers of tourists rather than more formal PPT criteria.

Third, and in line with the above, it can be argued that no clear links exist between PPT initiatives and poverty reduction. Many countries most dependent on tourism for their GDP, especially islands and small states, have moved out of the 'least developed' category, as indicated in Table 13.4, even though they were not specifically targeted by PPT strategies. Similarly, surveys in destination areas consistently indicate that residents believe they have benefited economically from tourism (Ap, 1990), although it is also recognized that enthusiasm sometimes wanes as dependency on tourism increases (Smith and Krannich, 1998, pp. 783–786, 798–800; Tosun, 2002), and many claims are made for the role of conventional tourism in alleviating poverty, even when specific categories of

Table 13.4. Tourism expenditure as % of GDP and export of goods and services for the top 25 countries, 2005 or nearest year.

Country	Tourism expenditure as % of GDP	Rank	Human development status of country as ranked by **UNDP for** HDI	Tourism as % of	
				Export of goods	Export of services
Aruba[a]	43.5	1	n/a (High?)	55.8	83.4
St Lucia	41.8	2	Medium	500.0	88.5
Seychelles	38.8	3	High	75.6	73.1
Antigua and Barbuda	36.1	4	High	563.8	70.9
Maldives[b]	35.1	5	Medium	177.2	90.5
Bahamas[a]	32.2	6	High	414.5	86.1
Barbados	30.4	7	High	238.8	62.1
Vanuatu	27.3	8	Medium	244.7	72.7
Lebanon	26.4	9	Medium	257.6	54.6
St Vincent & Grenadines	24.5	10	Medium	244.2	66.9
Sao Tome & Principe	23.9	11	Medium	453.3	75.6
Fiji	23.8	12	Medium	n/a	n/a
St Kitts & Nevis	23.6	13	High	184.5	74.8
Croatia	20.4	14	High	84.8	76.9
Dominica	20.1	15	Medium	133.3	66.7
Samoa	19.5	15	Medium	650.0	69.6
Belize	18.5	17	Medium	63.4	69.6
Jamaica	18.4	18	Medium	107.2	76.5
Mauritius	18.4	18	High	55.5	73.5
Cape Verde	17.3	20	Medium	198.9	63.9
Cambodia	17.2	21	Medium	31.9	83.7
Malta	16.6	22	High	37.2	58.3
Cyprus	15.8	23	High	181.8	40.8
Grenada	15.6	24	Medium	5141.0	55.5
Jordan	13.7	25	Medium	40.9	77.0
For comparison					
France	2.0	102	High	9.6	36.4
Spain	4.7	61	High	27.2	56.6
USA	1.0	128	High	13.7	32.6

Notes: [a]Aruba 2002 and the Bahamas 2003; [b]This was much reduced because of the tsunami at the end of 2004. In 2004 tourism was 58.9% of GDP.
Sources: UNWTO (2007), *Yearbook of Tourism Statistics*, UNWTO, Madrid; and UNDP (2007), *Human Development Report*, Palgrave Macmillan/UNDP, New York, p. 413.

poor are not targeted (Jamieson, 2003; Denman and Denman, 2004).

Clearly, initiatives not normally associated with PPT may perform pro-poor functions. It has been claimed, for instance, that, because they open up areas previously inaccessible to tourists, low-cost airlines are a major form of PPT (Travel Impact Newswire, 2004). Unpopular though this argument might be, especially among environmentalists, it clearly has some merit but, equally clearly, there are offsetting factors, including air travel's increased contribution to global warming. At local levels, too, tourism expansion may be reflected in increased tourist expenditure (which is actually or potentially directed to alleviating poverty) but might also cause damage to environmentally fragile sites, such as Angkor Wat in Cambodia. Such examples provide an additional indication of the need to consider wider, non-economic factors when implementing PPT projects.

Fourth, some critics contend that, to focus on those who are genuinely poor, PPT should focus more on equity. There is clearly some point to this criticism, as unease is understandable if the relatively wealthy benefit as much, or even more, from PPT projects as the poor, who are the prime targets. However, PPT practitioners recognise the need for cooperation among numerous stakeholders and a wide range of policy and planning initiatives (Jamieson *et al.*, 2004, pp. 8–15), and there is nothing to stop governments committed to poverty reduction from using redistributive fiscal or other mechanisms to transform virtually any form of tourism into PPT. At the local level it is a feature of 'pro-poor tourism' to encourage government, aid agencies and the private sector to develop entrepreneurial linkages for the poor in the tourism supply chain (Ashley *et al.*, 2001; Ashley, 2006; Ashley and Mitchell, 2007).

For redistribution really to be the focal point of PPT, closer attention would have to be paid to the role of the state, as well as to the wider world system. As in development matters generally, the impact of any PPT projects, even if on a large scale, is likely to be limited unless a state's entire tourism strategy is constructed around the aim of poverty alleviation. In effect, PPT requires a *developmental* state (Lockwood, 2005, pp. 1–5.).

Even more generally, international equity necessitates both ideological and systemic change ... Such ideological change would need to be accompanied by a change in the international system so that developing countries are granted greater decision making power in institutions such as the World Trade Organization (Schilcher, 2007, p. 182).

This is surely the case, but returns the debate to the issue of how far PPT initiatives can, on their own, overthrow capitalism. In the absence of an ideal developmental state and a more egalitarian global economic system, it is surely legitimate for advocates of PPT to note that they have to work in the real world.

Fifth, it can be argued that PPT has often failed to take sufficient account of commercial viability and access to markets, with NGO and INGO representatives preferring instead to seek aid money for projects which have been subject to little or no prior costing. Again, there is substance to this criticism but, as PPT advocates have noted, it applies most to CBT projects, with which PPT has often (but quite wrongly) been associated. These are not the same and, as Goodwin suggests, 'few [CBT] projects have generated sufficient benefits to either provide incentives for conservation – the objective of ecotourism – or contribute to local poverty reduction' (Ashley, 2006, p 17; Gereffi *et al.*, 2001, p 7.).

Sixth, as indicated earlier, PPT is accused of failing to confront mass tourism, either its 'problems' or its PPT features. The former include poor labour conditions in tourism destinations, 'anti-poor practices' of low wages, and so on (Chok *et al.*, 2007, pp. 157–158). Again, while such conditions clearly exist, PPT advocates can legitimately argue that they are not a campaigning pressure group. Rather, their role is to assist the poor to obtain more of the tourism 'cake'. It is also worth noting, however, that the existence of poor and exploitative working conditions does not invalidate the existence of an entire tourism industry, any more than abolishing the use of child labour in making carpets requires the cessation of carpet manufacture.

From a different perspective PPT advocates are alleged to have ignored the pro-poor characteristics of mass tourism which, as noted above and in Table 13.4, clearly makes a valuable contribution to many national economies. There is substance in this criticism for, while there are critiques from such pressure groups as Tourism

Concern (Beddoe, 2004), there are few detailed, methodologically sound empirical studies of mass tourism and its impacts. However, their absence is related less to a reluctance on the part of potential researchers than to the shortage of research funds or other support to analyse such linkages. When opportunities to study mass tourism and poverty alleviation have emerged, they have been taken, often in cooperation with major hoteliers and tour operators (; Karammel and Lengefeld, 2006; Mitchell and Faal, 2007; Rosetto *et al.*, 20072007, pp. 50–53, 57).

Nevertheless, such cooperation is both minimal and exceptional. In general, transnational company commitment is largely rhetorical, and hoteliers and tour operators have been even less prepared to offer funding assistance than have government and other aid agencies, on which PPT relies for relatively small handouts.

Conclusion

In this paper an attempt has been made to examine the emergence of pro-poor tourism initiatives in the wider context of development studies, and to ask whether or not PPT is, in fact, a distinctive approach to tourism as a development tool. The conclusion is that it is not distinctive and, indeed, that it is far easier to discuss what PPT is not than what it is! It is neither anti-capitalist nor hostile to mainstream tourism, on which it relies; it is neither a theory nor a model, and is not a niche form of tourism; it has no distinctive method and is not only (and sometimes not at all) about the poor. Indeed, by incorporating individual and collective non-economic criteria in its definition of poverty, it returns us to the development debates of the 1970s and 1980s.

Critics of PPT have focused on several conceptual and substantive issues, and an attempt has been made in this paper to confront these objections and consider the extent to which they have some validity, as indicated in Table 13.3. In summary, however, critics have argued that PPT is based on an acceptance of the status quo of existing capitalism, that it is morally indiscriminate and theoretically imprecise, and that its practitioners are academically and commercially marginal. More substantive criticisms include accusations of narrowness and parochialism, a failure to 'deliver' benefits from tourism to the poor, and even to demonstrate clear links between PPT and poverty alleviation; critics also point to a similar failure to focus on equitable distribution of benefits or to make any attempt to change the overall system. It is also claimed that, in developing projects, PPT tends to ignore the crucial role of markets and fails to consider the need for commercial viability and, finally, that its adherents ignore both the 'problems' and the existing PPT features and potential of mass tourism.

The extent to which such criticisms are valid has been discussed in the paper. It has been concluded, in brief, that criticisms focusing on the theoretical inadequacies of PPT are misplaced, simply because PPT advocates make no claim to be making any theoretical contribution, and that, in addition, they accept most of the more substantive criticisms, recognizing that they are working within a context of international capitalism and that their initiatives may bring little benefit to the poorest members of communities.

The main question, however, is how far pro-poor tourism can be considered truly distinctive in its approach or methods, and the conclusion reached here is that, on both counts, it is found wanting. Rather, as it is currently framed and understood, PPT is a stated concern with the poor – a moral injunction – which is sometimes reflected in the application of fairly standard social scientific techniques of data collection and analysis by practitioners (as opposed to theorists) who have obtained funds to assess actual and/or future linkages of poverty alleviation and tourism, normally with a view to improving the individual or collective lot of 'the poor' (broadly defined). At the same time, it could also be described as a movement, an incipient pressure group, which consistently runs the risk of being hijacked by those who seek to claim the high moral ground, the more 'responsible' (and less capitalistic) tourism, and the more 'moral' product.

Does this mean that PPT has no future? That it is a perspective or orientation which should be abandoned? The answer to both questions, it is suggested, is negative. PPT practitioners have been remarkably effective in getting their message across and increasing the focus on poverty. In concentrating on a very simple – and incontrovertibly moral – idea, namely, the net benefits of tourism to the poor, they seem to have cut through much of the development debate and have appealed to many NGOs and INGOs, and numerous national and international aid agencies, whose officials have often incorporated PPT in their own missions. The fact that they have remained academically and commercially marginal is, at least in part, because of a lack of funding.

It is suggested here that the way forward for PPT proponents is to become part of the academic mainstream, to engage with the academic community, especially those involved in development studies, to submit their work to critical academic scrutiny, and to participate in the major debates. There is no future on the academic fringe. In return, the commitment of PPT adherents, their insistence on examining the most basic impacts of tourism *and* on bringing about direct and quantifiable change, will revive in tourism studies a moral dimension which was there in the 1970s and 1980s, but which has too often been subsumed under a welter of theorizing.

Such an association would then allow those in the academic community with an active interest in tourism as a development tool to seek closer associations with national and international development agencies, and with major sections of international tourism. This would undoubtedly take time, but there is little future in eking out small grants and looking at the impacts of ecolodges on small communities. International tourism is big business, its impacts are big and so too, arguably, are its contributions to poverty alleviation. And they could be greater.

The alternatives? PPT can stay on the academic and commercial margins of tourism, while its active practitioners continue to seek both profitable (and undoubtedly useful) consultancies. However, in such circumstances, their practical effects will be minimal and, quite probably, financially wasteful. When looking at tourism and its contributions, if any, to 'development', there is no quick fix, and no easy moral short cut. As in the 1970s tourism can be regarded as a blessing or a blight. Much depends on the observer, stakeholder, context, and on understanding the wider issues. The earlier theoretical debates have not gone away; rather, as globalization has accelerated, they have become more complex. What continues to be needed is a balanced approach to, and research *over time* on, the development of tourism in its various forms, how it is articulated, and whom it benefits. This must be the basis of 'action research', targeted action and policy formulation. There is no future, or logic, in fundamentalism.

References

ADB (1999) *Regional Technical Assistance 5893: Mekong/Lancang River Tourism Infrastructure Development Project*. ADB, Manila.
ADB (1999) *Regional Technical Assistance 5893: Mekong/Lancang River Tourism Infrastructure Development Project*. ADB, Manila.
ADB (2002) *Building on Success: A Strategic Framework for the Next Ten Years of the Greater Mekong Subregion Economic Cooperation Program*. ADB, Manila.

Ap, J. (1990) Residents' perceptions research on the social impacts of tourism. *Annals of Tourism Research* 17(4), 610–616. DOI: 10.1016/0160-7383(90)90032-M.

Ashley, C. and Mitchell, J. (2007) Assessing how tourism revenues reach the poor. Briefing paper, 21 June 2007. Overseas Development Institute, London.
Ashley, C. and Mitchell, J. (2007) Assessing how tourism revenues reach the poor. Briefing paper, 21 June 2007. Overseas Development Institute, London.
Ashley, C. (2000) *Pro-poor tourism: putting poverty at the heart of the tourism agenda, Paper 51*. Overseas Development Institute, London.
Ashley, C. (2000) *Pro-poor tourism: putting poverty at the heart of the tourism agenda, Paper 51*. Overseas Development Institute, London.
Ashley, C. (2006) Participation by the poor in Luang Prabang tourism economy: current earnings and opportunities for expansion, working paper 273. Overseas Development Institute, London.

Ashley, C., Roe, D. and Goodwin, H. (2001) *Pro-poor Tourism Strategies: Making Tourism Work for the Poor—A Review of Experience*. Overseas Development Institute, International Institute for Responsible Tourism and Centre for Responsible Tourism, London.

Bah, A. and Goodwin, H. (2003) Improving access for the informal sector to tourism in the Gambia, PPT working paper 15. Pro-Poor Tourism Partnership, London.

Becker, H.S. (1967) Whose side are we on? *Social Problems* 14(3), 239–247. DOI: 10.2307/799147.

Beddoe, C. (2004) Labour Standards, social responsibility and tourism: a report by tourism concern. Tourism Concern, London.

Britton, R.A. (1975) *Making tourism more supportive of development: Caribbean tourism and St Vincent*. Department of Geography, University of Minnesota.

Chok, S., Macbeth, J. and Warren, C. (2007) Tourism as a Tool for Poverty Alleviation: A Critical Analysis of 'Pro-Poor Tourism' and Implications for Sustainability. *Current Issues in Tourism* 10(2-3), 144–165. DOI: 10.2167/cit303.

Chok, S., Macbeth, J. and Warren, C. (2007) Tourism as a Tool for Poverty Alleviation: A Critical Analysis of 'Pro-Poor Tourism' and Implications for Sustainability. *Current Issues in Tourism* 10(2-3), 144–165. DOI: 10.2167/cit303.

de Kadt, E. (1979) *Tourism: Passport to Development? New York (for the World Bank and UNESCO*. Oxford University Press, New York (for the World Bank and UNESCO).

Denman, J. and Denman, R. (2004) *Tourism and Poverty Alleviation: Recommendations for Action*. WTO, Madrid.

Desforges, L. (2000) State tourism institutions and neo-liberal development: a case study of Peru. *Tourism Geographies* 2(2), 177–192. DOI: 10.1080/14616680050027888.

DFID (1999) *Tourism and Poverty Elimination: Untapped Potential*. DFID, London.

DFID (1999) *Tourism and Poverty Elimination: Untapped Potential*. DFID, London.

ESCAP (2005) *The contribution of tourism to poverty alleviation*. 25. ESCAP Tourism Review, Bangkok.

Gereffi, G. and Kaplinsky, R. (2001) The value of value chains: spreading the gains from globalisation. *IDS Bulletin* 32(3).

Gereffi, G., Humphrey, J., Kaplinsky, R. and Sturgeon*, T.J. (2001) Introduction: globalisation, value chains and development. *IDS Bulletin* 32(3), 1–8. DOI: 10.1111/j.1759-5436.2001.mp32003001.x.

Gerosa, V. (2003) Tourism: a viable option for Pro-poor growth in Africa? 23–24 June 2003. Expert Group Meeting, Munyonyo Speke Resort, Kampala.

Goodwin, H. (2005) *Pro-poor tourism: principles, methodologies and mainstreaming, keynote address to the international conference on Pro-poor Tourism: Mechanisms and Mainstreaming*, Universiti Teknologi, Malaysia, 4–6 May 2005.

Goodwin, H. (2006) *Community-based tourism: failing to deliver? id21insights, 62*. Institute of Development Studies, Brighton.

Gouldner, A. (1973) *For Sociology: Renewal and Critique in Sociology Today*. Allen lane, London.

Harrison, D. (2007) Towards developing a framework for analysing tourism phenomena: a discussion. *Current Issues in Tourism* 10(1), 61–86. DOI: 10.2167/cit300.0.

Harrison, D. (2007) Towards developing a framework for analysing tourism phenomena: a discussion. *Current Issues in Tourism* 10(1), 61–86. DOI: 10.2167/cit300.0.

Harrison, D. and Schipani, S. (2007) Lao tourism and poverty alleviation: community-based tourism and the private sector. *Current Issues in Tourism* 10(2-3), 194–230. DOI: 10.2167/cit310.0.

Harrison, D. and Schipani, S. (2007) Lao tourism and poverty alleviation. *Current Issues in Tourism* 10(2), 196–198.

Hawkins, D.E. and Mann, S. (2007) The world bank's role in tourism development. *Annals of Tourism Research* 34(2), 348–363. DOI: 10.1016/j.annals.2006.10.004.

Independent Commission on International Development (1980) North–South: a programme for Survival—The report of the independent Commission on international development issues under the Chairmanship of Willy brandt. Pan Books, London.

Jamieson, W. (2003) *Poverty Alleviation through Sustainable Tourism Development*. ESCAP, Bangkok.

Jamieson, W., Goodwin, H. and Edmunds, C. (2004) Contribution of tourism to poverty alleviation: pro-poor tourism and the challenge of measuring impacts. Report prepared for the Transport Policy and Tourism Section, Transport and Tourism Division, ESCAP, Bangkok.

Kaplinsky, R. and Morris, M. (2001) *A Handbook for Value Chain Research*. Institute of Development Studies, Brighton.

Karammel, S. and Lengefeld, K. (2006) Can all-inclusive tourism be pro-poor? *id21insights* 62, 6.

Konadu-Agyemang, K. (2001) Structural adjustment programmes and the International tourism trade in Ghana, 1983-99: some socio-spatial implications. *Tourism Geographies* 3(2), 187–206. DOI: 10.1080/14616680010034711a.

Leftwich, A. (2000) *States of Development: On the Primacy of Politics in Development*. Polity Press, Cambridge.

Lockwood, M. (2005) *The State They're In: An Agenda for International Action on Poverty in Africa*. ITDG Publishing, Bourton-on-Dunsmore.

Mitchell, J. and Faal, J. (2007) Holiday package tourism and the poor in the Gambia. *Development Southern Africa* 24(3), 445–464. DOI: 10.1080/03768350701445541.

Mitchell, J. and Phuc, L.C. (2007) Final report on participatory tourism value chain analysis in dA Nang, central Vietnam. MCG Management Consulting and Overseas Development Institute, London.

Momsen, J. (1972) Report on vegetable production and the tourist industry in ST Lucia. University of Calgary, Alberta.

Momsen, J. (1973) Report on food production and the tourist industry in Montserrat. University of Calgary, Alberta.

Mosedale, J. (2006) Tourism commodity chains: market entry and its effects on ST Lucia. *Current Issues in Tourism* 9(4-5), 436–458. DOI: 10.2167/cit271.0.

Mowforth, M. and Munt, I. (2003) *Tourism and Sustainability: Development and New Tourism in the Third World*. Routledge, London, p. 263.

PPT (2007) Pro-poor tourism partnership. Available at: www.propoortourism.org.uk/ppt_pubs (accessed 19 November 2007).

PPT (2007) Pro-poor tourism partnership. Available at: www.propoortourism.org.uk/ppt_pubs (accessed 19 November 2007).

Rodenburg, E. (1980) The effects of scale in economic development: tourism in Bali. *Annals of tourism research* 7(2), 177–196.

Roe, D. (2006) How pro-poor is tourism? *id21 insights* 62.

Rosetto, A., Li, S. and Sofield, T. (2007) Harnessing tourism as a means of poverty alleviation. *Tourism Recreation Research* 32(1), 49–58.

Rossetto, A., Li, S. Sofield, T. and Li, F.M.S. (2007) Harnessing tourism as a means of poverty alleviation: using the right language or achieving outcomes? *Tourism Recreation Research* 32(1), 49–58. DOI: 10.1080/02508281.2007.11081523.

Scheyvens, R. (2007) Exploring the Tourism-Poverty nexus. *Current Issues in Tourism* 10(2-3), 231–254. DOI: 10.2167/cit318.0.

Schilcher, D. (2007) Growth versus equity: the continuum of Pro-Poor tourism and neoliberal governance. *Current Issues in Tourism* 10(2-3), 166–193. DOI: 10.2167/cit304.0.

Schmitz, H. (2005) Value chain analysis for policy makers and practitioners. International Labour Organisation, Geneva.

Seers, D. (1969) The meaning of development. *International development review* 11(4), 2–6.

Seers, D. (1977) The new meaning of development. *International development review* 19(3), 2–7.

Smith, M.D. and Krannich, R.S. (1998) Tourism dependence and resident attitudes. *Annals of Tourism Research* 25(4), 783–802. DOI: 10.1016/S0160-7383(98)00040-1.

Smith, V.L. (1978) *Hosts and Guests: The Anthropology of Tourism*. Basil Blackwell, Oxford.

Tosun, C. (2002) Host perceptions of impacts: a comparative tourism study. *Annals of tourism research* 29(1), 231–234.

Travel Impact Newswire (2004) Low-cost airlines as poverty alleviators. Travel Impact Newswire, 6 July.

UNWTO (2007) ST-EP.. Available at: www.unwto.org/step/index (accessed 19 November 2007).

WCED (1987) *Our Common Future*. Oxford University Press, Oxford, pp. 3–8.

Willis, K. (2005) *Theories and Practices of Development*. Routledge, London, pp. 93–96.

WTO (2002) *Tourism and Poverty Alleviation*. WTO, Madrid, p. 93.

Yilmaz, Y. and Bititci, U. (2006a) Performance measurement in the value chain: manufacturing V. tourism. *International Journal of Productivity and Performance Management* 55(5), 371–389. DOI: 10.1108/17410400610671417.

Yılmaz, Y. and Bititci, U.S. (2006b) Performance measurement in tourism: a value chain model. *International Journal of Contemporary Hospitality Management* 18(4), 341–349. DOI: 10.1108/09596110610665348.

14

Cocoa, Conservation and Tourism: Grande Riviere, Trinidad

Introduction

International tourism was already established in parts of the Caribbean before the Second World War, most notably in Cuba and Jamaica (Schwartz, 1997, pp. 68–69; Taylor, 1973, 1975). However, it was really only after 1945 that it became important throughout the Caribbean. By the beginning of the 21st century it was making a major contribution to the economy of the wider Caribbean and in 2004 accounted for some 14.8% of the region's GDP and for 2.4 million jobs (15.5% of total employment). By 2014 it is expected to account for 16.5% GDP and 17.1% of employment in the region (WTTC, 2004, p. 4).

Tourism in the Caribbean is reportedly the most intensive in the world (Wilkinson, 1997). However, in Trinidad and Tobago, when compared with other Caribbean island states, it is relatively unimportant. In 2002 it contributed only 4.3% of its GDP and in 2003 the country attracted fewer than 6% of all arrivals in the wider Caribbean (Wilkinson, 1997; WTO, 2005). That said, it is the most important industry in Tobago and, as indicated below, in some parts of Trinidad.

Social anthropologists have produced a wealth of data on tourism's impacts (Kousis, 1989; Peake, 1989; Tsartas, 1992; O'Reilly, 2000; McGibbon, 2000; van den Berghe and Flores Ochoa, 2000; UNDP, 2004), but scholars remain divided on its impact in small communities. Debates on whether or not it contributes to 'development', which underpinned the earliest studies of international tourism (Smith, 1978; de Kadt, 1979), remain relevant and, with the added dimensions of 'sustainability', the issues have become even more complex (Harrison, 2001).

While residents in many destination areas have coped with influxes of tourists without 'losing' their own culture (La Flamme, 1979; Connelly-Kirch, 1982; Brandes, 1988; Baumgartner, 1992; Boissevain, 1996; Rogers and Aitchison, 1998; UNVE, 2004), there are real problems, as Pearce, Moscardo and Ross indicate, with '*post hoc* descriptive accounts based on a tradition of single ethnographic case studies' (Pearce *et al.*, 1996, p. 18). Intensive focus on the ethnographic 'moment' makes comparison of destinations difficult. Little comparative work has been done and few longitudinal studies carried out. Indeed, such studies over the years by van den Berghe and his co-authors of Cuzco, Peru (van den Berghe, 1980; Tsartas, 1992; UNVE, 2004) and San Cristóbal, Mexico (UNDP, 2004) are exceptional. Despite the tendency of many academics to denigrate tourism's social impacts, the overall conclusions were that in both Cuzco (UNVE, 2004, pp. 22–23) and (especially) San Cristóbal (UNDP, 2004), it had been largely beneficial, though the 'trickle-down' effects were reportedly greater in the latter than the former. Similar conclusions have been reluctantly reached by others. When Greenwood and Pi-Sunyer, for instance, revisited destinations of their initial research, they revised their initial impressions, swallowed hard, and entered pleas of *mea culpa* (Greenwood, 1989; Pi-Sunyer, 1989).

In the English-speaking Caribbean, international tourism has undoubtedly exacerbated tensions deeply rooted in societies with a history of slavery, and reservations about its economic and sociocultural impacts have been voiced for decades (Zinder, 1969; Levitt and Gulati, 1970; Bryden, 1973; Taylor, 1973, 1975; Cole, 1974; Gray, 1974; Kaufman, 1985, p. 36; Thomas, 1988, pp. 145–167). They were especially prominent during the 'black power' disturbances of the early 1970s (NJAC, 1971; Oxaal, 1971; McKay, 1987; Thomas, 1988; Durant, 1990) and, a decade later, in Grenada (Ferguson, n.d; Payne and Sutton, 1984; Sandford and Vigilante, 1984). Indeed, policies towards tourism have often vacillated according to the ideological hue of the government in power (Mitchell, 1972; Chambers and Airey, 2001). Academics and other commentators added their criticisms and, while some have been circumspect and analytical in their approach (Brereton, 1990; Dann, 1990; Waldren, 1996; Vidich and Lyman, 2000), others have been more critical (Lewis, 1971; McKay, 1987; Kincaid, 1988, pp. 14–19; Antrobus, 1990; Pattullo, 1996; Kempadoo, 1999, pp. 20–22; Perez, 1973–1974). Even now, when the Caribbean region might be considered a 'mature' destination, problems and sensitivities continue (Duperly-Pinks, 2002; Bolles, 2005, p. 140; Getfield, 2005, p. 118; Joseph, 2005).

There is a rich history of ethnographic research in the English-speaking Caribbean, dating from the 1930s (Herskovits and Herskovits, 1947; Henriques, 1953; Smith, 1956, Smith, 1962a, b; Clarke, 1957; Blake, 1961; Klass, 1961; Greenfield, 1968; Rodman, 1971; Waylen, 2005), and there were also many studies on the 'new nations' of the Caribbean (Bell, 1967; Depres, 1967; Oxaal, 1968; Singham, 1968). However, although tourism's importance in these societies is recognized, and its impacts debated, few have researched what actually happens when tourists visit local communities (Durant, 1990). Price's study of Lower Bay, Bequia, St. Vincent, is an exception. Echoing several conclusions of Harrison, 1975a, pp. iii–viii, he suggests that tourism increased employment opportunities, especially for women, and was generally favoured by the local population, but also exacerbated pre-existing stratification, gave prominence to foreign capital, which was resented, and inflated prices for land and foodstuffs (Harrison, 1998, pp. 225, 206–244). In addition, it led to 'friendship/patronage relationships with drifter tourists' for some men in the community, especially the young and marginalized, resulting in clashes with elders and other authorities (Price, 1988, p. 232).

More recently, some brief, small-scale studies have been carried out (Boxill *et al.*, 2002). However, despite four decades of tourism in the region, Boxill notes that 'Caribbean social scientists have been on the periphery of tourism research and development', and adds, 'Academic research on tourism in the Caribbean is in its infancy' (Boxill, 2003, pp. 335, 337).

Context and methods of the study

This paper focuses on Grande Riviere, in North-east Trinidad. After a brief summary of its history, and its socio-economic characteristics in the early 1970s, the main focus is the village as it was three decades later, especially tourism and its impact on community life. The concluding section summarizes the key issues arising from the study.

The author first undertook anthropological fieldwork in Grande Riviere from April 1971 until July 1972 (Harrison, 1975b). During this 15-month residence, an array of investigative methods was used, including collecting pedigrees, formal and informal interviews, the compilation of 32 'personal sets' (elements of wider social networks), and a careful analysis of extended case studies of disputes involving villagers. All these methods were followed within a comprehensive immersion into village culture that might classically be described as participant observation (1975b, pp. 14–36).

Three decades later, after learning that tourists were going to Grande Riviere, the author made two short visits to the village: ten days in April 2004 and 5 days in June 2006. Data were collected on the history of tourism here, arrival numbers and characteristics, and employment in the hospitality sector and other occupations. Informal interviews were held with hoteliers and villagers, and supplemented by written communications with others involved in tourism development and conservation in the area, and with other researchers who had written on Grande Riviere.

During the 1970s, strong and lasting relations were forged with many in Grande Riviere. They were maintained over the intervening years, and the later visits were warmly welcomed by villagers. As a result, these visits yielded unexpectedly rich data and numerous insights into the role tourism was playing in Grande Riviere. Indeed, the availability of a comprehensive historical record, and considerable first-hand knowledge of the village *before* tourists appeared on the scene, makes this longitudinal account of the study of tourism's impacts in Grande Riviere rare, if not unique, in the literature of the English-speaking Caribbean.

A brief history

Grande Riviere is on the north-east coast of Trinidad, in the County of St. David, 60 kilometres from Sangre Grande and 100 kilometres from Port of Spain (Fig. 14.1). There was no slavery in the area, as the mountainous terrain and wet climate were unsuitable for sugar cultivation, and the first settlers were Venezuelan

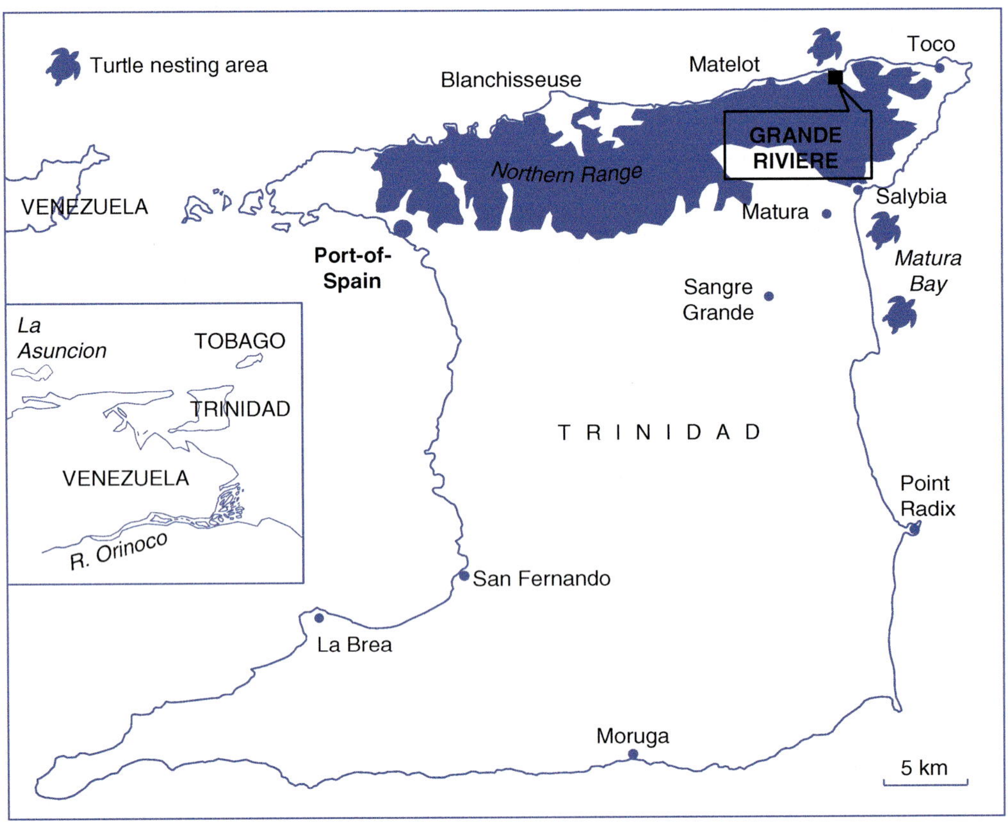

Fig. 14.1. Map of Trinidad.

peons, who first went to North-east Trinidad in the 1860s to escape civil unrest. Settling on the narrow coastal plain, they cultivated subsistence crops and cocoa, which they sold for cash, and were soon joined and outnumbered by arrivals from the island of Tobago, 35 kilometres north-east of Grande Riviere. The development of the village coincided with the period, approximately 1870 to 1920, when cocoa became more important than sugar in the Trinidadian economy (Shephard, 1932, p. 2), and by 1900 the community was thriving.

As Grande Riviere was becoming established, land for cocoa production was a good investment, and outsiders from Port of Spain started to obtain land in the area. The firm of Gordon Grant became especially prominent, not only in North-east Trinidad, but also elsewhere in the country (Harrison, 1979) and, with land obtained by a variety of means, the 650-acre Gordon Grant estate was a crucial feature of Grande Riviere's development. The second decade of the 20th century was undoubtedly the heyday of the village. By 1920, it had developed close links to the rest of Trinidad, especially Port of Spain. Economic life centred on the cocoa estate, which employed 40–70 labourers, depending on the season, and villagers also grew cocoa on their own account, which they sold to the estate or in Port of Spain. In line with its regional importance, the community acquired a police station and a two-storey Freemason's Lodge, in which dances were held every Saturday.

The boom was short-lived. In 1921, the price of Trinidadian cocoa slumped (Shephard, 1932, p. 2) and in the late 1930s the onset of Witchbroom and Black Pod diseases ravaged the cocoa crop and made matters worse. The cocoa estate was run down, and by 1936 wage labourers were employed for only 2 days a week. Attempts to introduce sharecropping failed, and while a new road improved access to the Port of Spain, it also facilitated emigration from the village. Although Trinidad's population was increasing, throughout the County of St. David the population declined, and Grande Rivere's population fell from 718 in 1931 to 550 in 1946. In the early 1930s, the headquarters of the masonic lodge was moved to another village. Later, the police station was also removed and visits from the local magistrate ceased. Most importantly, in 1967 Gordon Grant's cocoa estate was sold. Mohammed, the buyer, soon recouped his outlay by denuding the hillsides of timber, and continued to work the estate with a skeleton staff. To all intents and purposes, Grande Riviere was in irreversible decline.

The early 1970s

In 1972 there were 134 households and only 504 residents, mostly descendants of the peons from Venezuela and the Tobagonians. Clearly, the population decline had continued, even though half had been born outside the village, some elsewhere on the north-east coast, some further afield in Trinidad, and a few on other islands in the Eastern Caribbean.

The most impressive buildings in the village were the Community Centre, a school run by the Anglican Church, and the Roman Catholic Church. However, social life centred on the main shop and its adjacent rum shop, and shopping also satisfied the need to catch up with local news. Also prominent, near the shop and backing on to the beach, were the headquarters of the cocoa estate, which included the overseer's house and several large cocoa 'boxes', where cocoa was dried before being bagged and taken to Port of Spain.

Agriculture provided the main income for more than one-third of the village men. Cocoa remained the most important cash crop, with bananas second, and although the estate itself provided work for less than a dozen men, cocoa was also cultivated on small plots and several farmers had their own cocoa boxes. Subsistence crops from 'gardens' – small plots of perhaps two or three acres – and from domestic kitchen gardens provided most food, along with fish from the local fishermen, all of which was supplemented by wild meat from the forest and from the occasional *Dermochelys coriacea* (leatherback turtle), which had come to nest on the beach. Such turtles were invariably killed, sometimes after suffering considerable abuse (Fig. 14.2; Fig. 14.3) and, unlike meat from the forest, which was usually sold in Sangre Grande some 60 miles away, their meat was distributed widely throughout the village. Except for a few Seventh Adventists (for whom such meat was prohibited by their

Fig. 14.2. Turtle abuse in 1972.

Fig. 14.3. Turtle meat in 1972.

religion), it was for villagers a welcome and rare change from the salt beef sold at the local shop.

Cocoa was an important source of cash but work on government road programmes was more vital, and involved nearly a third of the village men. Such work was preferred to agriculture, as it paid more and income was regular and predictable; and because permanent employment on the road programmes was allocated on the basis of accumulated experience, men born in Grande Riviere were more likely than newcomers to occupy these valued positions. However, most had several sources of income: their garden, for ground provisions and such crops as peppers, fruit, and tomatoes; a plot of several acres for cocoa, perhaps inter-planted with bananas and, if they were fortunate, casual or permanent work on the government road programme.

Social life in the village was structured by criteria relating to age and gender. Generally, men's social life was carried out in public, while the women were home-centred, and while relations among men were usually amicable, there were obvious tensions between women born in the community (who could draw on their female kin for support) and women born elsewhere (who lacked such support), normally arising from accusations of male infidelity. Social life for most, though, was largely restricted to Grande Riviere. Buses passed through the village three or four times a day, but other motorized transport was rarely seen. Along with poor roads, the absence of telephones and electricity meant that Grande Riviere was quite isolated from the rest of Trinidad. Even relationships with nearby villages were clouded by mutual suspicion. Visitors from outside were rare indeed and, unless they were well known or related to the villagers, they received a decidedly frosty welcome.

While the isolation of Grande Riviere deterred visitors, it was an incentive for young people to leave. Their older relatives generally encouraged the emigration, noting the poor infrastructure and limited employment prospects and asserting that the past had been 'plenty better than now'. Aware Grande Riviere was in decline, when they joked they lived 'behind God's back' they were simply recognizing the realities of the situation.

The new millennium

It was clear during visits made in 2004 and 2006 that, in some respects, the village's decline had continued. The main shop, formerly the social centre of the community, had been destroyed by fire and had not been replaced, while the Masonic lodge, a relic of Grande Riviere's central place in North-east Trinidad in the 1920s, had fallen into disrepair and had been dismantled. The population had also declined. In 2000 it was 334, nearly 25% less than in 1972 (McKay, 1987, p. 92). Several houses present in 1972 had been removed and more than 20, some quite handsome, had clearly been abandoned. Further evidence of decline was in the roll of the Anglican school. In 1972 it had 132 pupils; in 2004 there were 37.

By contrast, though, the road to Sangre Grande was much improved, and was used by many more vehicles, even (irregularly) by taxis. Electricity had been installed in the mid-1970s and 30% of households had TV sets, including several with satellite. A similar percentage had telephones (and from 2005 many villagers acquired mobile phones) and most houses had piped water. While in the early 1970s there were no medical facilities, and only irregular visits from a nurse, one now lived in the village, and there was a smart health centre, opened in 1981, with a weekly doctor's surgery, regular visits from a nutritionist, and a weekly schedule of exercise classes, enthusiastically supported by the women.

By far the biggest changes, though, were in the village economy. At the end of the 80s, the government had purchased most of the Mohammed estate, which had completely ceased to function. All but one of its cocoa boxes were removed, and some land was leased to villagers in five-acre plots. Informants suggested that most lessees cultivated the land for only a short period before abandoning the attempt. In fact, cocoa production throughout the village had declined. In 2006, only six men spent much time harvesting cocoa, and only one processed the crop to the point at which it was bagged. The others sold the cocoa 'wet' (in the white and sticky state it is removed from the cocoa pods), leaving the remaining processes to the purchaser, who thus benefited from the annual bonus paid to producers at the end of the season.

Public sector employment continued to be important, and accounted for about 75 jobs, mainly for village men. Several worked for the Public Works Department to maintain the main roads, ten or so in Regional Cooperation (formerly the County Council) to maintain access roads and traces, and a dozen on the Community-based Environmental Protection and Enhancement Programme, introduced by government in 2002. About 20–40 men and women obtained irregular, short-term employment on the Unemployed Relief Programme, the successor to the 70s Prime Minister's Special Project and, finally, of longer term importance, there was the National Reforestation and Watershed Programme, which was started in Trinidad in 2004. Aiming at poverty alleviation, this involves a 10-year programme of reforestation and biodiversity conservation, and employs about 20 villagers to establish and maintain fire traces in the forest, and re-open long-neglected access roads into the Northern Range.

Tourism in Grande Riviere

The change which has had the greatest impact on village life, however, is the introduction of tourism. When the cocoa estate was broken up, the old estate headquarters was for a short time rented to a local resident, who occasionally operated it as a hunting lodge. The picture was transformed in 1993 with the arrival of Piero Guerrini, an Italian photographer. He first rented and then (in 1997) purchased the property, which he developed as a 12-room, beachfront hotel. His arrival coincided with national and international interest in the leatherback turtles increasingly nesting on Trinidad's north-east coast. Local conservation measures had recently been encouraged by the Wildlife section of the Department of Forestry (Livingstone, 2003), and in 1990 in Matura, in the north-east of Trinidad, Nature Seekers Incorporated, a group aiming to conserve the area's natural resources, was formed (Sankar-Oyan, 2004). Interest soon spread to Grande Riviere and, in 1992, the Grande Riviere Environmental Awareness Trust (GREAT) was formed, similarly aimed at protecting the turtles.

Guerrini's 12-room beachfront hotel, the Mt. Plaisir, was ideally situated to benefit from these developments. By 2001 it was receiving glowing reviews in travel guides: it was 'the best hotel in Trinidad' (Guide, 2003), with food that 'could easily hold its own in the swankiest of Port of Spain's eateries' (De-Light and Thomas, 2001, p. 176). In fact, with the emerging focus on turtle conservation and the success of the hotel, a new industry was introduced into Grande Riviere. By 2000, the ten-room Le Grande Almandier hotel and a six-room guest house, McEachnie's Haven, both operated by members of prominent Grande Riviere families, were established. In 2004, they were joined by the French-owned Acajou, another small but relatively upmarket establishment, and this increased the number of permanent hotel rooms available in Grande Riviere to 38. In addition, other villagers began to offer rooms, often on the recommendation of the hotels and guest house at peak periods.

Additional services were soon offered to tourists. Assisted by the United Nations Development Programme and the Department of Forestry, GREAT members started to train as nature guides. In 1999, following some ill-feeling, most left to form the Grande Riviere Nature Tour Guides Association, eventually taking major responsibility for the nesting turtles, while those remaining in GREAT focused primarily on assisting the hatchlings. Despite such tensions, by April 2004 there were 14 registered guides in the village (including one woman), and several unregistered guides, who occasionally accompanied tourist groups in the forest. Most in the guides association are also active members of the Grande Riviere Tourism Development Organization, which was formed, at the suggestion of the Trinidad Industrial Development Corporation, to monitor and develop tourism for the benefit of the village, receiving another small United Nations Development Programme grant in 2003 to purchase some land and construct the Tourist Information Centre. Importantly, the association is the conduit government uses to employ workers on the National Reforestation and Watershed Rehabilitation Programme in this part of the Matura National Park.

By 2006, Grande Riviere was an established holiday destination. The three hotels catered for wealthier international and some regional tourists, and expatriates working in Trinidad, while Trinidadians and some budget-conscious international and regional tourists went to McEachnie's Haven. Accurate statistics of arrivals are not available, but estimates are possible from figures supplied by hoteliers, the beach permits sold at hotels in 2002 and 2003, the permits sold by the Tourist Information Office during 2004 and 2005, or simply by calculating the tourists accommodated in hotels and guest houses over a year, assuming 55% occupancy. From such sources, it seems clear that from 1995, when the first hotel opened its doors, arrivals have increased from a few hundred a year until, by 2006, they were in the region of 10,000, some 65% of which were from overseas, to which one would need to add an unquantified (but possibly larger) number of day-trippers from other parts of Trinidad. Clearly, the isolation of the village is over.

Tourism's Impact

Economically, its impact is considerable and positive, directly accounting for about 80 jobs for villagers, of whom about 50 (mainly women) worked in the hotels. The remainder includes guides, souvenir sellers and other service providers. Other employment in the village is indirectly attributable to tourism. Workers on the Unemployed Relief Programme, for example, prepared the beach carpark, while part of the Community-based Environmental Protection and Enhancement Programme involved enhancing the environment around the beach and the Big River. There is also a tourism component to the National Reforestation and Watershed Rehabilitation Programme, as abandoned access roads in the Northern Range are being revived as trekking routes for tourists. Finally, although linkages of the agricultural sector might be developed further, the hotels and guest house clearly attempt to source their foodstuffs locally, often from their own farming land.

Men, in particular, continue to have a variety of occupations (an advantage, as tourism develops in different stages with different employment requirements), but most aged 15–64 obtain some income from tourism, as do about 40% of the women in the same age range (who now have a wider choice of work compared to the 70s, when they were largely restricted to being water carriers). In the hotels, they learn new skills, such as preparing new dishes, while men use traditional skills in building and horticulture, and learn new skills in carpentry and building construction. In fact, tourism has taken over from cocoa as the major source of employment and no village informant considered work in the hospitality sector demeaning.

More generally, tourism's impacts can be analysed at the systemic, institutional and interactional levels. First, the systemic level. Even in the 1870s, Grande Riviere was economically incorporated in the capitalist world system, situated in a region radically changed by slavery and colonialism, and providing cocoa for the chocolate and sweet drinks of children in the UK (Harrison, 1979). The arrival of electricity, TV and telephones in the 1970s and 1980s (and mobile phones in 2005) further reduced the villagers' geographical and personal isolation. By 2006, cocoa had declined, but tourists came from all over Trinidad, the Caribbean, and elsewhere, to experience the turtle nesting season. In catering for such tourists, and through their involvement in turtle conservation, residents became involved with other outsiders and their organizations: United Nations Development Programme volunteers who helped them obtain grants (Thomas, 1988; Troëng and Drews, 2004); students from the University of Glasgow, who studied marine turtle conservation and ecotourism on Trinidad's north coast (Godley *et al.*, 2001; Livingstone, 2001); and the Caribbean Regional Environment Programme, which brings together community-oriented conservation groups in North-east Trinidad (CREP 2003 Caribbean Regional Environmental Program, 21 April 2002). At the national level, too, through intermittent participation in Matura to Matelot, an environmental umbrella organization in North-east Trinidad (Onwuka, 2004, pp. 174–175) and through contact with the Trinidad Industrial Development Corporation, Grande Riviere is further linked to the outside world in ways which, a few decades ago, would have seemed unthinkable.

Within the village itself, institutional changes clearly occurred. The new economic institutions – hotels, guest houses and guide associations – are of long-term importance. Their influence is comparable to that of the first Venezuelan settlers, who settled on the coastal plain and at the mouth of the Big River – the sites, in fact, of the new hotels (1975b, pp. 38–46). Indeed, Mt. Plaisir hotel incorporates part of the Gordon Grant headquarters and, like its predecessor, is a major employer of village labour. Further continuity is found in the willingness of the hotelier to advance funds to staff, for example to purchase materials for house building and repair.

The members of the Grande Riviere Nature Tour Guides Association undoubtedly obtain political capital from their close involvement with hoteliers and outside organizations. It is comprised of relatively young people (ten of the guides are under 40), who meet regularly, and whose interest in obtaining funds for development is personal, collective, and financial. As a consequence, although the organization, like others elsewhere in North-east Trinidad, has incurred some resentment (Onwuka, 2004, pp. 135, 157–159), it effectively and easily eclipses the Village Council which, in theory, exists to coordinate community activities and represent it to the outside world but which, in practice (as in the 1970s) meets irregularly and is split by damaging gossip and internal disputes (Harrison, 1975b). Relations between the two institutions are undoubtedly strained. As expressed by one of the older guides, in a characteristic turn of phrase: 'They [the Village Council] respect us for what we have done, and we respect them for what they have not done'.

Price suggests that, in Bequia, the development of tourism favoured those with ready access to financial and social capital (1988, p. 217) and, with an important caveat, a similar process occurred in Grande Riviere. The introduction of foreign capital through two of the village's three hotels clearly established a new and upper economic stratum (which is also distinguished ethnically and culturally from others). However, the local owner of Le Almandier drew on previous experience of hotel work in Tobago, and greatly benefited from a prior business relationship with the Italian hotelier. Similarly, the female partner of McEachnie's Haven once worked at Mt. Plaisir, and it is noteworthy that she and her partner are from established and relatively wealthy Grande Riviere families. As in Lower Bay, Bequia, new players entered the stratification system and existing positions have been reinforced.

However, whereas in Bequia tourism reportedly increased the price of land, this has not occurred in Grande Riviere, probably for two reasons. First, prime land is not yet in short supply and, second, the many abandoned houses in the village, on sites apparently ripe for exploitation, indicate they are likely to be on 'family land' and thus inalienable unless all kin agree to the transaction, a consensus difficult to obtain (Harrison, 1975b). In Negril, Jamaica, extensive tourism development reportedly placed a premium on family land, which carries potential both for development and, simultaneously, for family disputes (McKay, 1987, pp. 145–147). This could occur in Grande Riviere if tourism in the village continues its upward trajectory.

At the interactional level, tourism has brought villagers into direct personal contact with people from all over the world. McEachnie's Haven, for instance, which caters primarily for Caribbean tourists, was also visited over a 12-month period by people from Europe, Australia, South Africa and North America. Thirty years ago, when any outsider was greeted with reserve, suspicion, and even hostility, and any white male was assumed to be a priest, such a situation would have seemed impossible.

How close is the interaction? At the very least, villagers and tourists coexist in comfortable symbiosis, but closer associations and friendships reportedly develop. One resident recounted 'how the communication skills of his children improved as they learned to talk to visitors, and how the interaction with visitors increased his self-esteem' (Onwuka, 2004). Onwuka goes on to suggest, somewhat primly, that 'some individuals in Grande Riviere even have a (negative) reputation of making these social relations a part of their livelihood strategy' (2004, p. 153), a clear reference to a 'beach boy' at one of the hotels who prides himself on his 'success' with single women tourists.

In 2004 and 2006, villagers held positive views of tourists. One elderly man, a Seventh-Day Adventist, who informed the author in 1972 that there were three types of people in

Grande Riviere, 'the bad, the worse, and the worsest', said tourists always behaved respectfully and he invariably exchanged greetings with them. Several had visited the homes of guides, or other villagers, and one German reportedly stayed several weeks, establishing friendships with several residents. Another informant, an 80-year-old man, who to the author epitomized traditional, conservative Grande Riviere, said he enjoyed telling tourists about the village's history, and had taken some of them home to show them his 'souvenirs'.

Turtles, Tourism and Conservation

Grande Riviere is an example of conservation-driven tourism. The Grande Riviere Environmental Awareness Trust, later largely superseded by the Grande Riviere Tour Guides Association (GRTGA), was initially formed in 1992 to conserve turtles, not develop tourism, and it was because turtles nested on Grande Riviere beach that Guerrini developed his hotel and first marketed it as an ecotourism centre, a series of decisions which radically affected Grande Riviere.

Precisely why the conservation movement developed in Grande Riviere is difficult to say. The formation of Nature Seekers in Matura was clearly one factor, but so, too, was the drive of the individual who founded GREAT, a Seventh-Day Adventist (and thus a non-consumer of turtle meat). He 'fell in love' with turtles in 1989 after meeting an official from the Forestry Department who visited the village to check on turtle nesting sites. Despite the disputes with other members in GREAT, and its later demise, it is commonly recognized that turtle conservation owes much to him. A third factor was the growing importance throughout Trinidad of environmental awareness, and it figured prominently in the 1991 general election campaign.

It is also unclear why turtles began to nest at Grande Riviere in such numbers. In the early 1970s they came ashore very infrequently and though even then they were legally protected, no stigma was attached to eating their meat. It was not because of an increase in conservation in the village, as this began only in the 90s, and yet it is reported that in 1991 'approximately 300 leatherback nests were deposited in Trinidad by the end of July', and by 2001, 'during peak season, beaches at both Matura and Grande Riviere had approximately 300 leatherbacks in one night' (Maharaj and Ehrhart, 2003). In Grande Riviere, at least, numbers remain at this level, and its one-kilometre-long beach is regularly the scene of some of the most concentrated nesting of leatherback turtles in the world.

Villagers speculate that turtles increasingly nested in Grande Riviere because their food supply, especially jelly fish, had increased, or because beaches elsewhere had become too developed. By contrast, Livingstone suggests three possible reasons (Livingstone, 2006, pp. 19–20, 71–73). The first is that earlier nesting might have been underestimated (which is not the case for Grande Riviere, in the author's view). The second is that turtles sometimes nest at beaches near where they hatched, and because the conservation movement began in Matura before Grande Riviere, the latter might have benefited from these early conservation attempts. As Livingstone recognizes, this might account for some increase since the mid-1990s, but not before. Finally, during the 1950s and 1960s, the government actively promoted commercial fishing, and consequently many larger fish, especially sharks, which were threats to the hatchlings, were removed from the north and north-east coast of Trinidad. Livingstone concludes that 'the increase is most likely a combination of these three factors, although an unequivocal explanation remains unclear' (2006, p. 20).

In any event, the commitment of many villagers, especially the guides, to turtle protection and conservation is quite remarkable, as is their self-confidence and knowledge in explaining the nesting habits of this threatened species to tourists, as well as the wide variety of flora and fauna (including the endemic but threatened *Pawi)* in the forest environment of Grande Riviere. This conclusion is supported by other research, which confirms that conservation has widespread support throughout the village (van den Berghe and Primov, 1977). In short, there are clear indications that, at least for the present, tourism and conservation have come together to the mutual benefit of the leatherback turtle and villagers. Once readily consumed as a valued source of fresh meat, the leatherback

Fig. 14.4. Turtles as a tourist attraction in 2004.

turtle is now protected and a major attraction (Fig. 14.4).

Conclusion

Tourism is relatively unimportant in the overall economy of Trinidad, but has been crucial in Grande Riviere's revival, replacing the failing cocoa business and providing a major source of village employment. Consequently, the standard of living of many has been improved, as seen, for example, in the increased number of children's toys, including bicycles and consumer durables. It further incorporated Grande Riviere into the global economic and cultural network, bringing villagers into contact with numerous national and international organizations, and introduced a new stratum of foreign capital into the village, reinforcing previous social divisions. Above all, it opened this community to the outside world, enabling its members to interact with people from other parts of Trinidad and from overseas, and to develop new skills and expertise, and provided women, especially, with new employment opportunities. With the convergence of turtle conservation and tourism, many in the village (and elsewhere) have been given an interest and education in their natural environment and in protecting their scarce environmental resources. So far, the combination of conservation with tourism in Grande Riviere has been mutually beneficial.

Tourism is not credited here with all the positive changes that have taken place since the early 1970s. Government used the benefits brought by petroleum and natural gas, albeit after strong local agitation, to provide electricity, road improvements, telephones and piped water. Furthermore, out-migration has occurred for more than six decades and most villagers have kin in Port of Spain, North America and the UK. The extent of remittances to the community is unknown but at Christmas, especially, 'barrels' of consumer goods, including TV sets and other electrical items, are sent from overseas to Grande Riviere, and one villager reportedly received a car from the USA. Positive or otherwise, too, the arrival of television may have reduced men's *liming* (relaxing with others, usually in public) and made them more home-centred. Such factors have undoubtedly been important, but tourism must be regarded as the major cause of Grande Riviere's further incorporation into the world system.

This paper is unashamedly presented as a case study of tourism development in one Caribbean village. However, while ethnographic case studies are the basis of much traditional social anthropology, it is recognized that studies have been subjected to some criticism as a genre, primarily because they have been considered too subjective and a-theoretical (Pearce and Moscardo, 1999). Such criticisms, though, have focused largely on the absence of comparative data and the rarity of any longitudinal study (Harrison, 2007), a problem which this study has sought to address.

Ironically, some recent (postmodern) approaches to ethnography allocate more rather than less weight to individual researchers' subjective (and reported) experiences in the field. Even at the intrinsic level, such a study is justifiable if 'in all its particularity *and* ordinariness, this case itself is of interest' (Stake, 2000; emphasis in original), and the anthropologist-observer is viewed differently – as one who presents 'an integrated synthesis of experience and theory' (Denzin, quoted in van den Berghe, 1994, p. 61). Nevertheless, the ethnography is also of wider relevance, in that the researcher is a member of a community of scholars, and to be meaningful to others, the uniqueness of our own research experience gains significance when it is related to the theories of our predecessors and the research of contemporaries (van den Berghe, 1994).

There are other, more established grounds for claiming that such a case study is theoretically relevant. One need not accept Merton's approach to functional analysis to recognize that social science still needs empirical data to inform what he referred to, years ago, as 'theories of the middle range', that is, 'theories intermediate to the minor working hypotheses evolved in abundance during the day-by-day routine of research, and the all-inclusive speculations comprising a master conceptual scheme' (Merton, 1957, p. 5). Put differently, this study of Grande Riviere can be considered an 'instrumental case study' in that it potentially provides insights and furthers an understanding of other issues and contexts, and is thus theoretically relevant (Stake, 2000). Such a claim is hardly radical. Few would deny, for example, the major impact of Greenwood's short case study, in the mid-1970s, of developments in the Spanish town of Fuentarrabia. His initial conclusion, that tourism led to commoditization and a loss of cultural authenticity, became accepted wisdom for a generation of tourism scholars and even now carries considerable appeal, despite his later revisions and other evidence to the contrary (Greenwood, 1989, p. 181); Harrison, 2001). Indeed, material from Grande Riviere clearly indicates that commoditization predated tourism; by the end of the 19th century, more than 80 years before tourists arrived, the village was firmly incorporated into the world economy.

Grande Riviere is not unique. In Tortuguero, Costa Rica, for instance, the conservation of *Chelonia mydas* (the Green turtle) has been promoted by government and international organizations and advertised as an attraction (Lefever, 1992, pp. 128–143; Troëng and Drews, 2004). However, as indicated earlier, with few exceptions (McKay, 1987; Price, 1988), reliable case studies of development in the English-speaking Caribbean are in short supply. If we are to understand tourism's role in development, comparative data are necessary. Without them, any discussion of its impacts is the modern equivalent of medieval Scholastic debate over how many angels can dance on the head of a pin.

Case study material also throws light on established conceptual frameworks of tourism development. Doxey's Index of Irritation, for instance, based partly on a study of Barbados (Doxey, 1971), suggests that antagonism towards tourists (who were usually white) emerged because tourism-led social change had prompted a fear of identity loss (Doxey, 1975, p. 196). However, Doxey ignored the racial tensions then pervading the English-speaking Caribbean, including Grande Riviere which, as already indicated, then received no tourists at all. By contrast, when they did start to arrive in the village, the welcome they received led to and reflected a new openness to outsiders.

Similarly, the trajectory of development in Grande Riviere can be measured against the stages of Butler's tourism destination area cycle of evolution (Butler, 1980). Rather than starting with exploration, followed by local involvement and only later by the arrival of outside interests, tourism in this community was first promoted by an outsider, though his example was quickly followed by local people. In this respect, what happened in Grande Riviere conformed to a postcolonial mode of tourism development, as previously noted in studies of Swaziland, other parts of Africa and the South Pacific (Harrison, 1995, 2000).

Ultimately, 'development' and 'progress' are matters of personal evaluation (Harrison, 1988, pp. 153–159), but the main finding of this research, that tourism did benefit the village, supports the theoretical perspective that sometimes it is indeed a form of development. To date, few of the problems associated with tourism in the Caribbean, as outlined in the opening of this paper, have occurred. Foreign capital has entered the economy, but villagers are appreciative of its arrival and credit Guerrini, owner of the first hotel, with bringing relief to a community which, since the 1920s, has been in continuous decline. Pre-existing stratification has also been reinforced, but there was no widespread criticism of the new entrepreneurs (beyond the assertion, long established, that people with money tend to be mean). Many obtained work in tourism, and no one suggested such employment was demeaning. Indeed, when asked how this business benefited the poor, informants pointed to such opportunities and added (as occurs elsewhere in Trinidad and Tobago) that no one prepared to work should be poor. Has tourism 'developed' Grande Riviere? The villagers, at any rate, believe this to be so.

Nevertheless, some issues are debatable. In the early 1970s, several men made a regular income from fishing. This is no longer so, although residents still fish for subsistence purposes. In the turtle nesting season, they cannot spread their nets in Grande Riviere Bay, as they would entrap turtles (and damage their nets). Consequently, at least during this period, fish are provided by fishermen from Matelot or Toco. Here, conservation rather than tourism might explain the decline in local fishing, but even this is disputed. As informants noted, men can still fish outside the turtle nesting season, but rarely do so, which can be attributed more to local attitudes towards fishing than constraints arising from tourism or turtle conservation.

Grande Riviere is not paradise. In the early 1970s, the community was characterized by several kinds of disputes, and claims to have respect, to be different, were always subjected to trial by gossip and often found wanting (Harrison, 1975b). This has not changed. In addition, new elements of competition occasionally surface, when hotels and guest houses vie with one another for experienced staff, or registered and non-registered guides compete for tourist custom. In the latter instance, the claim of privilege in the accusatory statement, 'you come here; I born and grow here', seems to have emerged among village men, whereas in the past it was largely restricted to the women (Harrison, 1975b, pp. 191–194). This may be part of a wider pattern, in which some feel they are not receiving a 'fair' share of the economic benefits from tourism and turtle conservation (Onwuka, 2004), and some GRTGA members undoubtedly want to professionalize their service, which may be (or may be seen as) a tendency towards exclusivity. It is certainly a trend that should be monitored, and the subject of future research.

Also questionable, perhaps, are the activities of one young man who pointedly befriends single, white women tourists. However, they seemed content with his attention. Beach boys are common throughout the Caribbean, and there was no evidence that such behaviour scandalized others, whether tourist or villager. The same might be said for some young tourists, usually from elsewhere in Trinidad, who seek to purchase marijuana. This causes concern, but recreational use of the drug certainly predates tourism, and most visitors come in couples or small family groups and are primarily interested in turtle-watching and other aspects of the natural environment.

Finally, while turtle protection and tourism have developed in tandem, there are tensions between them. Hoteliers are not always diligent in keeping beachfront lighting to the minimum (Livingstone, 2003), and one informant suggested that construction of the Acajou hotel, on the bank of Big River, might adversely affect wildlife and affect drainage and the river's flow. In addition, the route taken by land crabs to the sea for the annual 'washing' is likely to be blocked by newly installed fences around some of the hotels.

At the time of writing, tourism in Grande Riviere corresponds closely to Honey's definition of ecotourism, in that it involves responsible travel to a natural area, is small scale, has minimal negative impact, encourages environmental awareness, provides direct benefits for conservation, financially benefits and empowers local people, and fosters respect for local culture (Honey, 1999, pp. 22–25). However, sustainable development does not occur automatically, and the future of tourism in Grande Riviere will depend on several factors, including continued nesting by Leatherback turtles, continued demand from repeat and first-time tourists for the environmental attractions of Grande Riviere, the growth of facilities in the village, including secondary attractions, and the ability of members of the Grande Riviere Tourism Development Organization and the Grande Riviere Tour Guides Association to understand, monitor, and – with outside help – control tourism development in the interests of conservation and all members of the community.

Case studies are snapshots of an area at any one time. In this sense, no single case study is too early or too late. Rather, the more there are, and the longer the period covered, the more comprehensive the picture of development that emerges, and the more early changes are situated within a wider historical context. So far, considerable data exist on Grande Riviere's pre-tourism development, and this paper, focusing on tourism's impacts over the last decade, should be seen as but the first analysis of the village as a destination. However, to date, this business has brought numerous economic and other benefits and few undesirable consequences. Indeed, in this village, and by extension in the North-east of Trinidad, it could be instrumental in halting and reversing a process of economic and social decline which started in the 1920s and continued at least until the early 1990s. If this does happen, tourism in this part of Trinidad will indeed prove to be a blessing, not a blight.

References

Antrobus, P. (1990) Gender Issues in Caribbean Tourism. *Presented at the first Caribbean Conference on Tourism and Sociocultural Change in the Caribbean*, Trinidad and Tobago.

Baumgartner, R. (1992) Tourism and socioeconomic change: the case of the Rolwaling Valley in Eastern Nepal. In: Singh, T., Fish, M., Smith, V. and Richter, L. (eds) *Tourism Environment: Nature, Culture, Economy*. Inter-India Publications, New Delhi, pp. 76–93.

Bell, W. (ed.) (1967) *The Democratic Revolution in the West Indies*. Schenkman, Cambridge.

Blake, J. (1961) *Family Structures in Jamaica*. Free Press, New York.

Boissevain, J. (1996) Introduction. In: Boissevain, J. (ed.) *Coping with Tourists: European Reactions to Mass Tourism*. Berghahn Books, Providence, pp. 1–26.

Bolles, A. (2005) One love, one heart and skin teeth: women workers in Negril, Jamaica. In: Jayawardena, C. (ed.) *Caribbean Tourism: People, Service and Hospitality*. Ian Randle, Kingston, pp. 136–150.

Boxill, I. (2003) Challenges for academic research on tourism in the Caribbean. In: Jayawardena, C. (ed.) *Tourism and Hospitality, Education and Training in the Caribbean*. University of the West Indies Press, Mona, pp. 329–337.

Boxill, I., Taylor, O. and Maerk, J. (eds) (2002) *Tourism and Change in the Caribbean and Latin America*. Arawak Publications, Kingston.

Brandes, S. (1988) *Power and Persuasion: Fiestas and Social Control in Rural Mexico*. University of Pennsylvania Press, Philadelphia.

Brereton, V. (1990) Tourism Awareness Programmes and their Content. *Presented at the first Caribbean Conference on Tourism and Sociocultural Change in the Caribbean*, Trinidad and Tobago.

Bryden, J. (1973) *Tourism and Development: A Case Study of the Commonwealth Caribbean*. Cambridge University Press, Cambridge.

Butler, R.W. (1980) The concept of a tourist area cycle of evolution: implications for management of resources. *The Canadian Geographer/Le Géographe canadien* 24(1), 5–12. DOI: 10.1111/j.1541-0064.1980.tb00970.x.

Chambers, D. and Airey, D. (2001) Tourism policy in Jamaica: a tale of two governments. *Current Issues in Tourism* 4(2–4), 94–120. DOI: 10.1080/13683500108667884.

Clarke, E. (1957) *My Mother who Fathered Me*. George Allen and Unwin, London.

Cole, M. (1974) Introduction. In: Abdulah, N. (ed.) *The Social Impact of Tourism on Tobago*. 13. University of the West Indies, St. Augustine, 3. 3–4. DOI: 10.2753/RPO1061-040513013.

Connelly-Kirch, D. (1982) Economic and social correlates of handicraft selling in Tonga. *Annals of Tourism Research* 9(3), 383–402. DOI: 10.1016/0160-7383(82)90020-2.

CREP 2003 Caribbean Regional Environmental Program (21 April 2002) M2M Matura to Matelot: NGO/CBO community consultations: selection of demonstration activities, NE Trinidad segment of the T and T No. 1 priority Amenity area, CREP consultation. Toco Anglican School. Available at: www.toco.interconnection. org/documents/CREP

Dann, G. (1990) Sociocultural Impacts of Caribbean Tourism: Theoretical and Methodological Issues. *Presented at the first Caribbean Conference on Tourism and Sociocultural Change in the Caribbean*, Trinidad and Tobago.

de Kadt, E. (ed.) (1979) *Tourism: Passport to Development?* Oxford University Press, Oxford.

De-Light, D. and Thomas, P. (2001) *The Rough Guide to Trinidad and Tobago*. Rough Guides, London.

Depres, L. (1967) *Cultural Pluralism and Nationalist Politics in British Guiana*. Rand McNally, Chicago.

Doxey, G. (1971) *The Tourist Industry in Barbados: A Socioeconomic Assessment*. Dusco Graphics, Kitchener.

Doxey, G. (1975) A Causation Theory of Visitor-Resident Irritants: Methodology and Research Inferences. *The Travel Research Association, Sixth Annual Conference Proceedings, The Impact of Tourism*, pp. 195–198.

Duperly-Pinks, D. (2002) Community Tourism: "Style and Fashion" or Facilitating Empowerment? In: Boxill, I., Taylor, O. and Maerk, J. (eds) *Tourism and Change in the Caribbean and Latin America*. Arawak Publications, Kingston, pp. 137–161.

Durant, O. (1990) Tourism, Crime and Deviance in the Caribbean. *Presented at First Caribbean Conference on Tourism and Social Change in the Caribbean*, Trinidad and Tobago.

Ferguson, J. (n.d) *Grenada: Revolution in Reverse*. Latin American Bureau, London.

Getfield, J. (2005) Tourism harassment in Jamaica: crime or career? In: Jayawardena, C. (ed.) *Caribbean Tourism: People, Service and Hospitality*. Ian Randle, Miami, pp. 106–118.

Godley, B., Broderick, A., Blackwood, S., Collins, L., Glover, K. *et al.* (2001). *Marine Turtle Newsletter* 61, 15–18.

Gray, H. (1974) Towards an economic analysis of tourism policy. *Social and Economic Studies* 23, 386–397.

Greenfield, S. (1968) *English Rustics in Black Skin*. College and University Press, New Haven.

Greenwood, D. (1989) Culture by the pound: an anthropological perspective on tourism as cultural commoditization. In: Smith, V. (ed.) *Hosts and Guests: The Anthropology of Tourism*, 2nd edn. University of Pennsylvania Press, Philadelphia, pp. 171–185.

Guide, R. (2003) Destination Guides. Available at: http://www.hotelus.com/guides/travelguide/cid_4372-/action_/viewLocation-formId_/106202

Harrison, D. (1975a) *Report on the Pilot Survey of Tourism in St Vincent*. Overseas Development Administration, Ministry of Overseas Development, London.

Harrison, D. (1975b) Social relations in a Trinidadian village. Unpublished PhD thesis. Department of Anthropology, University College London.

Harrison, D. (1979) The changing fortunes of a Trinidadian Peasantry. In: Cross, M. and Marks, A. (eds) *Peasants, Plantations and Rural Communities in the Caribbean*. Department of Sociology, Surrey, UK, and the Department of Caribbean Studies, Royal Institute of Linguistics and Anthropology, Leiden, the Netherlands, pp. 54–85.

Harrison, D. (1995) Tourism in Swaziland. *Annals of tourism research* 22, 135–156.

Harrison, D. (1998) *Sociology of Modernization and Development*. Routledge, London.

Harrison, D. (2000) Tourism in Africa: the social and cultural framework. In: Dieke, P. (ed.) *The Political Economy of Tourism Development in Africa*. Cognizant, New York, pp. 37–51.

Harrison, D. (2007) Towards developing a framework for analysing tourism phenomena: a discussion. *Current Issues in Tourism* 10(1), 61–86. DOI: 10.2167/cit300.0.

Harrison, D. (ed.) (2001) *Tourism and the Less Developed World: Issues and Case Studies*. CAB International, Wallingford.

Henriques, F. (1953) *Family and Colour in Jamaica*. McGibbon and Kee, London.

Herskovits, M. and Herskovits, F. (1947) *Trinidad Village*. Alfred A. Knopf, New York.
Honey, M. (1999) *Ecotourism and Sustainable Development: Who Owns Paradise?* Island Press, Washington, DC.
Joseph, N. (2005) Service vs Servility: creating a service attitude in the Caribbean hospitality industry. In: Jayawardena, C. (ed.) *Caribbean Tourism: People, Service and Hospitality*. Ian Randle, Miami, pp. 182–199.
Kaufman, M. (1985) *Jamaica under Manley: Dilemmas of Socialism and Democracy*. Zed Books, London.
Kempadoo, K. (1999) *Sun, Sex and Gold: Tourism and Sex Work in the Caribbean*. Rowman and Littlefield, Lanham.
Kincaid, J. (1988) *A Small Place*. Virago, London.
Klass, M. (1961) *East Indians in Trinidad: A Study of Cultural Persistence*. Columbia University, New York.
Kousis, M. (1989) Tourism and the family in a rural Cretan community. *Annals of Tourism Research* 16(3), 318–332. DOI: 10.1016/0160-7383(89)90047-9.
La Flamme, A. (1979) The impact of tourism: a case from the Bahama islands. *Annals of Tourism Research* 6, 137–148.
Lefever, H. (1992) *Turtle Bogue: Afro-Caribbean Life and Culture in a Costa Rican Village*. Susquehanna University Press, Selinsgrove.
Levitt, K. and Gulati, I. (1970) Income effect of tourist spending: Mystification multiplied: a critical comment on the Zinder report. *Social and Economic Studies* 19, 326–343.
Lewis, G. (1971) The U.S. virgin islands: prototype of the Caribbean tourist economy. In: Mathews, T. and Andic, F. (eds) *Politics and Economics in the Caribbean*. Institute of Caribbean Studies, University of Puerto Rico, Rio Pedras, pp. 217–243.
Livingstone, S. (2003) Trinidad marine turtle conservation conference 2003 TMTCC community conference. Organized by University of Glasgow Darwin Initiative Project and Trinidad government wildlife section, forestry division.
Livingstone, S. (2006) Sea turtle ecology and conservation on the North coast of Trinidad, West Indies. Unpublished PhD thesis. University of Glasgow, Division of Environmental and Evolutionary Biology, Institute of Biomedical and Life Sciences.
Livingstone, S. (ed.) (2001) *University of Glasgow Trinidad Expedition 2000: Final Report*. University of Glasgow, Glasgow.
Maharaj, A. and Ehrhart, L. (2003) Nesting Physiology and Hatching Success of Leatherbacks at two Nesting Beaches in Trinidad. *Proceedings of the 22nd annual symposium on sea turtle biology and conservation*.
McGibbon, J. (2000) *The Business of Alpine Tourism in a Globalising World*. Vetterling Druck, Rosenheim.
McKay, L. (1987) Tourism and changing attitudes to land in Negril, Jamaica. In: Besson, J. and Momsen, J. (eds) *Land and Development in the Caribbean*. Macmillan, London, pp. 132–152.
Merton, R. (1957) *Social Theory and Social Structure*, 2nd edn. Free Press, New York.
Mitchell, J. (1972) To Hell with Paradise: A New Concept in Caribbean Tourism. *Address at the Caribbean Travel Association Press Conference*.
NJAC (1971) *Slavery to Slavery*. National Joint Action Committee, Port of Spain.
Onwuka, O. (2004) Welcome. During your visit, take nothing but pictures, kill nothing but time, leave nothing but footprints. *Unpublished manuscript, University of Amsterdam*.
Oxaal, I. (1968) *Black Intellectuals come to Power*. Schenkman, Cambridge.
Oxaal, I. (1971) *Race and Revolutionary Consciousness: A Documentary Interpretation of the1970 Black Power Revolt in Trinidad*. Schenkman, Cambridge, MA.
O'Reilly, K. (2000) *The British on the Costa del Sol*. Routledge, London.
Pattullo, P. (1996) *Last Resorts: The Cost of Tourism in the Caribbean*. Cassell, London.
Payne, A. and Sutton, P. (1984) *Grenada: Revolution and Invasion*. Croom Helm, London.
Peake, R. (1989) Tourism and Swahili identity in Malindi old town, Kenyan coast. *Africa* 59, 209–220.
Pearce, P. and Moscardo, G. (1999) Tourism community analysis: asking the right questions. In: Pearce, D. and Butler, R. (eds) *Contemporary Issues in Tourism Development*. Routledge, London, pp. 31–51.
Pearce, P., Moscardo, G. and Ross, G. (1996) *Tourism Community Relationships*. Pergamon, Oxford.
Perez, L. (1973–1974) Aspects of underdevelopment: tourism in the West Indies. *Science and Society* 37, 473–480.
Pi-Sunyer, O. (1989) Changing perceptions of tourism and tourists in a Catalan resort town. In: Smith, V. (ed.) *Hosts and Guests: The Anthropology of Tourism*, 2nd edn. University of Pennsylvania Press, Philadelphia, pp. 187–199.

Price, N. (1988) *Behind the Planter's Back*. Macmillan, London.
Rodman, H. (1971) *Lower-class Families: The Culture of Poverty in Negro Trinidad*. Oxford University Press, New York.
Rogers, P. and Aitchison, J. (1998) *Towards Sustainable Tourism in the Everest Region of Nepal*. World Conservation Union and International Centre for Protected Landscapes, Nepal.
Sandford, G. and Vigilante, R. (1984) *Grenada: The Untold Story*. Madison Books, New York.
Sankar-Oyan, R. (2004) Nature seekers Inc
Schwartz, R. (1997) *Pleasure Island: Tourism and Temptation in Cuba*. University of Nebraska Press, Lincoln.
Shephard, C. (1932) *The Cacao Industry of Trinidad: Some Economic Aspects (Part V)*. Government Printer, Port of Spain.
Singham, A. (1968) *The Hero and the Crowd in a Colonial Polity*. Yale University Press, New York.
Smith, M. (1962a) *Kinship and Community in Carriacou*. Yale University Press, New Haven.
Smith, M. (1962b) *West Indian Family Structure*. University of California Press, Seattle.
Smith, R. (1956) *The Negro Family in British Guiana*. Routledge, London.
Smith, V. (ed.) (1978) *Hosts and Guests: The Anthropology of Tourism*. Basil Blackwell, Oxford.
Stake, R. (2000) Case studies. In: Denzin, N. and Lincoln, Y. (eds) *Handbook of Qualitative Research*, 2nd edn. Sage, Thousand Oaks, pp. 435–454.
Taylor, F. (1973) Tourist industry in Jamaica, 1919–1939. *Social and Economic Studies* 22, 205–227.
Taylor, F. (1975) *Jamaica: The Welcoming Society: Myths and Reality. Working Paper 8*. Mona, Jamaica.
Thomas, C. (1988) *The Poor and the Powerless: Economic Policy and Change in the Caribbean*. Latin American Bureau, London.
Troëng, S. and Drews, S. (2004) Money talks: economic aspects of marine turtle use and conservation. Available at: http://assets.panda.org/downloads/moneytalks.pdf
Tsartas, P. (1992) Socioeconomic impacts of tourism on two Greek Isles. *Annals of Tourism Research* 19(3), 516–533. DOI: 10.1016/0160-7383(92)90134-B.
UNDP (2004) UNDP small grants program: an integrated community development approach to biodiversity conservation and a viable ecotourism enterprise in grande Riviere – project fact sheet. Available at: http://www.undp.org/sgp/cty/LATIN_AMERICA_CARIBBEAN/TRINIDAD_AND
UNVE (2004) United nations volunteers for the environment 28 October 2004. Available at: http://unvcaribbean.undp.org.tt/tto.htm
van den Berghe, P.L. (1980) Tourism as ethnic relations: a case study of Cuzco, Peru. *Ethnic and Racial Studies* 3(4), 375–392. DOI: 10.1080/01419870.1980.9993312.
van den Berghe, P. (1994) *The Quest for the Other: Ethnic Tourism in San Cristóbal. Mexico*. University of Washington Press, Seattle.
van den Berghe, P.L. and Flores Ochoa, J. (2000) Tourism and nativistic ideology in Cuzco, Peru. *Annals of Tourism Research* 27(1), 7–26. DOI: 10.1016/S0160-7383(99)00043-2.
van den Berghe, P. and Primov, G. (1977) *Inequality in the Peruvian Andes: Class and Ethnicity in Cuzco*. University of Missouri Press, Columbia.
Vidich, A. and Lyman, S. (2000) Qualitative methods: their history in sociology and anthropology. In: Denzin, N. and Lincoln, Y. (eds) *Handbook of Qualitative Research*, 2nd edn. Sage, Thousand Oaks, pp. 37–84.
Waldren, J. (1996) *Insiders and Outsiders: Paradise and Reality in Mallorca*. Berghahn Books, Oxford.
Waylen, K. (2005) Turtles and tourism, perceptions and Pawi: perceptions of natural resources and the effect of ecotourism in Grande Riviere, Trinidad. Unpublished Master thesis. Imperial College of Science, Technology and Medicine, Faculty of Life Sciences, University of London.
Wilkinson, P. (1997) *Tourism Policy and Planning: Case Studies from the Commonwealth Caribbean*. Cognizant, New York.
Wilson, P. (1973) *Crab Antics*. Yale University Press, New Haven.
WTO (2005) *Compendium of Tourism Statistics*. World Tourism Organization.
WTTC (2004) *The Caribbean: The Impact of Travel and Tourism on Jobs and the Economy*. World Travel and Tourism Council, London.
Zinder, H., Associates (1969) *The Future of Tourism in the Eastern Caribbean*. Agency for International Development USAID, Washington, DC.

15

Tourism Culture(s): The Hospitality Dimension (with Peter Lugosi)

Introduction

In the broad anthropological sense, culture may be defined as everything we learn about taking our place in human society and underpins the framing of our world views, our norms and values, and conditions our negotiated path through everyday social reality. It is almost a truism to note that cultures vary considerably across ethnic and national boundaries and affect the performance of organizations (Hofstede *et al.*, 2010) and prioritizing culture as an important factor in tourism is hardly new. It runs like a thread through the entire fabric of travel and tourism and the movement of tourists (and other migrants) and the movement of cultures are intimately related. It is indeed the case that 'cultures travel, too' (Rojek and Urry, 1997, p. 10). Culture is a vital conditioning factor in the relationships of tourists and residents of destination areas (Pearce, 1995, 2005, pp. 113–134; Hottola, 2004; Reisinger, 2009), in the formation and re-formation of patterns of everyday life in tourist destinations (Pi-Sunyer, 1989; Grünewald, 2002; Fisher, 2004), or in the way policies are determined for tourism development (Schroeder and Sproule-Jones, 2012). Culture is also at the heart of the tourism 'product', whether formally sold as 'heritage' or 'cultural tourism' (Harrison and Hitchcock, 2005) or presented as an element in the attraction of 'difference' and 'the other'. As such, when marketed to tourists, 'culture' or 'heritage' may bring groups or categories of people together in a common identity (Soper, 2008, p. 55) or, perhaps simultaneously, be the locus of hotly contested identities and perceptions of place (Harrison, 2005a, b).

This paper examines the interaction between hospitality and culture. More specifically, it is argued that hospitality spaces (hotels, resorts, cruise ships and service spaces within them, including restaurants, bars and cafes) are focal points in the production of tourism culture(s). The paper suggests that they are particular organizational contexts that facilitate specific embodied experiences, performances of self, modes of interaction and power relationships among the staff and the tourists. In offering a thematic framework for examining the interaction of tourism cultures and hospitality, the paper draws on three conceptual themes: 'third spaces' (Bhabha, 1994), 'socio-technical systems' (Ball *et al.*, 2003) and 'total institutions' (Goffman, 1968) and thus furthers an appreciation of the role of hospitality spaces within tourism experiences and of a framework for understanding both the *processes* in hospitality spaces and the *outcomes* of those processes.

Conceptualizing Tourism Culture in Hospitality Spaces

It is possible to talk about tourism *cultures* and refer, for example, to the tourism culture of the beach (Urbain, 2003; Metusela and Waitt, 2012), of the British in Majorca (Andrews, 2009), of sports tourists, of mountain tourism, and so on. However, more specifically, 'tourism culture' is used here to denote a way of life that characterizes the physical and social interface at which tourists and residents meet one another in settings where commercial hospitality is provided. Following Ryan (2003, p. 280), tourism cultures have a number of characteristics, involving a temporary spatial-temporal distinction from everyday norms and cultural practices, where leisured, hedonistic consumption is central to the consumer experience. Within service encounters, cultural symbols draw on various stereotypes and caricatures; interaction between tourists and frontline staff involves shifting superior/inferior relationships; frontline staff and tourists do not fully conform to the habits or norms of their usual peer groups; and communication between tourists and staff may be through intermediaries and partly-spoken languages. In such circumstances, where interaction is often based on relatively basic concepts that are linked to limited domains of behaviour, the potential for misunderstanding is always present.

This conceptualization of tourism culture does not assume it is a static or objective entity; rather, it recognizes that it exists as social practice, performed and reproduced within specific organizational settings. It takes a myriad of forms, depending on the type of tourism and tourist involved and the nature of the destination. Different (sometimes quite radically different) cultures come to meet in a bounded physical space, or even a *series* of bounded physical spaces, operated by commercial hospitality providers, sharing views (to some extent, at least) as to the nature of the institution and their own roles within it. However, tourism culture is essentially fluid, reflecting the different types of tourist and resident occupying the space, who carry with them their own cultures and expectations but who may yet learn attitudes and behaviour patterns from one another via various processes of inner and outer acculturation. In what follows, the emphasis is on tourism culture as it is learned, performed and encountered in situations where *commercial hospitality* is on offer and, in particular, at hotels and resorts in developing countries. While recognizing that hotels and resorts come in many different forms, we consider how they provide organizational spaces that differentially facilitate the ongoing negotiation and (re)construction of tourism cultures.

Hotels and resorts vary in terms of their size, the configuration of the physical service environment or 'servicescape' (Bitner, 1992), experiential propositions, service levels and complexity and, importantly for this paper, in terms of the level of social separation of tourists and staff from the everyday norms and cultures that exist outside these spaces. However, academics have begun to conceptualize these as particular bounded spaces of interaction in which established cultural norms and behaviour may be challenged and replaced with alternatives (Weaver, 2005; Pritchard and Morgan, 2006; Berdychevsky *et al.*, 2013). Various dimensions of experiences within such hospitality spaces have been considered, and from a number of different perspectives. For example, researchers have paid considerable attention to service cultures and the experiences of service workers (Mars and Nicod, 1984; Adler and Adler, 2004; Sherman, 2007). Studies have also examined various aspects of consumer experiences within such hospitality contexts (Knutson *et al.*, 2007, 2009; Walls *et al.*, 2011), and related issues such as stereotyping in service encounters (Èivre *et al.*, 2013). However, in the context of commercial hospitality, the emergence of tourism culture remains under-examined and under-conceptualized. In response to this gap in knowledge, this paper suggests a thematic framework for understanding and thus researching tourism cultures.

Several conceptual frameworks could be used to assist in understanding the nature of tourism culture, as schematized in Fig. 15.1. First, as signalled by Hollinshead (1998), the interface could be described as a *third space*, a term derived from Bhabha (1994), which is characterized by a shifting, negotiable and contestable hybrid culture which is sourced from aspects of tourist and resident cultures but is yet different from both of them.

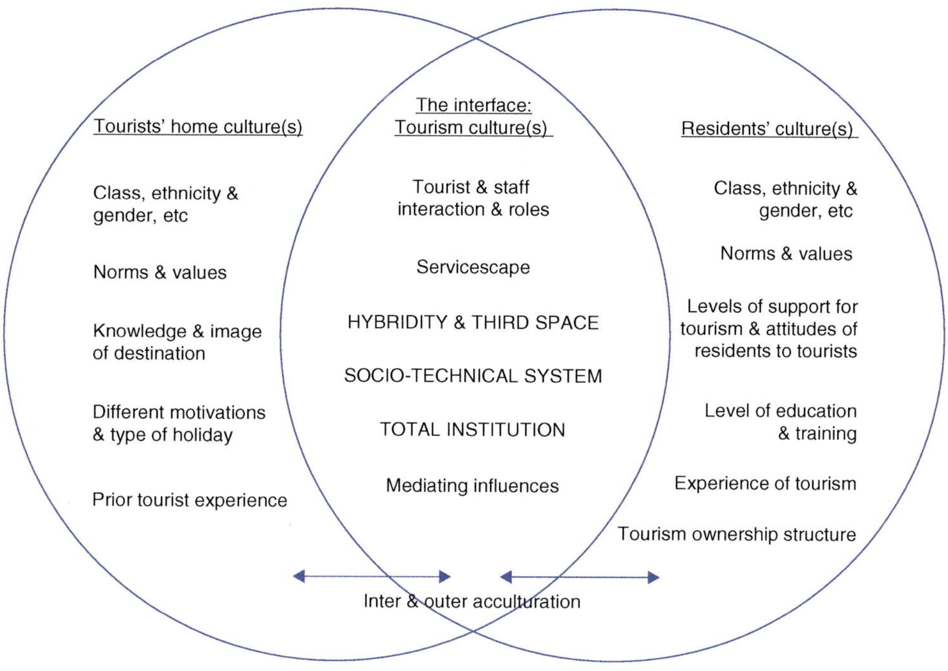

Fig. 15.1. Tourism culture(s). Source: Authors.

Second, and in tandem with the first perspective, tourism culture could be considered both a key component and outcome of a *socio-technical system* (Trist, 1981; Pasmore *et al.*, 1982), which incorporates both technical and social sub-systems. According to this view, what operators, workers and consumers bring to the business – their attitudes, cultures and orientations – is essential to the way the business operates.

Third, it is possible to envisage hotels and resorts as being total institutions, defined by Goffman as a place 'of residence and work where a large number of like-situated individuals, cut off from the wider society for an appreciable period of time, together lead an enclosed, formally administered round of life' (1968, p. 11).

These conceptualizations of hotels and resorts do not exhaust the possible frameworks of analysis. A strong case might also be made, for example, albeit at greater length than this short piece permits, for seeing them as locations of a specific *habitus* (Bourdieu, 1984, pp. 466–484), and it has been argued elsewhere that at least some forms of tourism culture might be considered a distinctive form of quasi-spiritual community or *communitas* (Turner and Turner, 1978, p. 250; Graburn, 2001, pp. 47–48; Jack and Phipps, 2005, pp. 103–105). Similarly, while a substantial literature exists on processes of knowledge transfer and creation that characterize work-related migration across international boundaries (Williams, 2006), such processes might *also* operate within national boundaries when workers move to occupational contexts that are markedly different from those of their home culture. Elements of such approaches can be found in the three frameworks discussed below which, it is argued, enhance our understanding of the role of hospitality space in developing tourism culture(s) (Fig. 15.1).

The framework of analysis adopted for any in-depth study of tourism cultures in hotels and resorts will affect the nature of the data obtained. A third space approach, for instance, directs attention to the uncertainties and negotiated aspects of the tourism culture, to issues of difference and otherness, while analysis from a socio-technical perspective is likely to emphasize the systemic interaction of workers' cultures and its overlaps, differences and contradictions with the technical requirements for operating the business. Finally, analysis of the hotel or resort as a total institution will involve examining the physical and social boundaries that separate the institution from the outside world and the cultural practices within it that serve to further differentiate (and/or protect) it – and its population – from the outside. In the remaining part of this paper, elements of tourism culture will be discussed to illustrate how these frameworks might be utilized.

Examining Tourism Culture(s) in Hospitality Spaces

Hospitality spaces and/as third spaces

Third space reflects a particular conception of culture and identity as fundamentally hybrid; constructions or articulations of both react to draw on and integrate others or notions of otherness. Bhabha, 1994, 1996 argued that notions of culture and self should not be seen as static entities or the fixed property of individuals; instead, they are constantly (re)produced through social engagement and re-interpretation. This perspective suggests that cultures and identities are fluid and subject to change as they are performed in different social and cultural spaces. Identities and cultures are constructed in and through spatial relations (Hall and Du Gay, 1996; Hetherington, 1998), including in hospitality contexts (Lugosi, 2009, 2014). The concept of third spaces thus points to particular ways of being, but also reflects the creation of social and cultural spaces in which notions of self are negotiated and articulated.

The extent to which the *'third space'*, is distinct from the cultural space(s) occupied by tourists and residents before they enter that of the hotel or resort will depend on the specifics of the situation, but the daily content of any tourism culture is highly fluid. In Hollinshead's terms, 'difference' and 'otherness' (1998, pp. 126–128) converge to be reshaped into a hybrid tourism culture mediated by the physical and social structure of the hotel or resort.

What this means in practical terms can be briefly illustrated by reference to some common scenarios. Focusing primarily on East Africa, Blanton notes several 'barriers to host–guest communication' (1981, pp. 120–125), which, for present purposes, can be seen as issues that emerge in the third space. First, in many developing societies, only a minority of local residents can afford to patronize hotels and resorts, and many will not have visited them. As a consequence, for new employees, the physical environment of the hotel may often be quite intimidating. Second, attitudes to time vary. Life in many non-industrialized developing societies is less dominated by 'clock time' than in Western societies, a factor that may be marketed as one of the positives of the holiday destination but is less attractive when reflected in slow and (apparently) desultory service or in the non-appearance of staff when they are supposed to be on duty. Finally, whereas social relations in developing societies are based on strong and established kinship ties, those in hotels and resorts are short-term and impersonal (Crick, 1989). Tourists and locals encounter each other across a 'mutual gaze' (Maoz, 2006), and both must also negotiate relations within their own categories. And where regional or ethnic differences are present, among staff or tourists, social 'reality' can indeed be stressful (Nicholls, 2011).

Within hotels and resorts, massive adaptation by staff members might be required. The entire world of food and beverages can be considered a foreign culture and while there is rich material on cross-cultural (and even intra-cultural) social differences in food production, consumption and taste (Douglas, 2003), and numerous guides on cultural differences in etiquette (e.g. Foster, 2000), little attention has been given by tourism researchers to complexities of the gastronomic interface of tourist and residents within hospitality spaces (See e.g. Pratt, 2013). In Fiji, for example, owners of backpacker resorts accustomed to subsistence agriculture reportedly do not understand why tourists expect a choice of meals, and the food they offer – or are prohibited from offering – to guests is much influenced by their totemic relationship to specific plants, fish and animals (Gibson, 2013). Indeed, even a cursory examination of a standard operational manual for bar staff reveals that basic training is far more than a technical matter: good sanitation and safety procedures and proper food storage seem neutral enough, but cultural factors impinge on the need to know the *range* of alcoholic drinks sold and their characteristics (including which drinks should be served chilled), understanding how to approach different types of guests and their needs, including those who are drunk, and the projection of a *favourable* image of self, including making eye contact. And once in the restaurant, not only might the food served to tourists be totally foreign to the staff, but also the array of table cutlery will be daunting, especially if people normally eat with their fingers (Blanton, 1981, p. 124).

Similar cultural issues are present in other areas of hotel and resort life, for example, front office and housekeeping sections. Blanton reports the advice of one hotelier in Indonesia that 'one should start housekeeping training with the *concept* of a bed, not how to make it' (1981, p. 124), and suggests that 'industry workers are put into a new cultural environment where their familiar assumptions, values, behaviour and ideas are constantly challenged' (Blanton, 1981, p. 125).

The above examples, involving what might has been described as 'culture shock' (Pearce, 1995) or 'culture confusion' (Hottola, 2004), centre primarily on possible adaptation required by local residents when becoming involved in tourism culture (not the least of which may be to learn a new language). However, adaptation is also required of tourists, especially holidaymakers, and this has been little studied. They *may* have travelled more than local residents and appear more knowledgeable about other parts of the world but, in some respects, their successful participation in tourism culture requires more adaptation. The spectacular faint by Shirley Valentine's friend Jane, in the film Shirley (1989) set in the Greek Island of Mykonos, on realizing she was eating squid, is but one example of a tourist reaction to 'funny food'. Tourists' taste can be highly conservative (Rogerson, 2012) and encouragement that they should 'eat local' often falls on deaf ears.

Considerable differences also exist in what is 'acceptable' public behaviour (e.g. public displays of affection between couples, especially if they are gay, and in standards of dress or undress, especially for women), which is why so much consideration has been given to 'demonstration effects' (Fisher, 2004). Staff from highly conservative societies may find it difficult to tolerate alcohol consumption and the behavioural excesses to which it might lead (Andrews, 2009). In this context, it is noteworthy that hotels and resorts are often considered commendable areas of freedom (or Western dens of iniquity) relative to the rest of society.

In fact, learning how *not* to work, and for only a few intensive days or weeks, might be considered more difficult than learning to be a member of staff at a hotel or resort. Away from home and a work-structured environment, for 24×7 (alone or with family or friends), in what is for most a relatively strange place, tourists must negotiate a *modus vivendi* to take them through the duration of their stay. They may already be in relationships with their travelling companions, including those which are highly intimate (and which over a holiday period can become highly stressful), but even these will be further mediated through a sense of place (Trauer and Ryan, 2005). Just as staff members, from the CEO down to the cleaner, must learn their role in the tourist culture, so too must the tourist. And whereas the former can draw on the organizational structure for support, the latter is more or less on his or her own.

Hospitality spaces as socio-technical systems

The notion of socio-technical systems refers to the purposive mobilization of technologies, materials, embodied performances of self, organizational codes, norms and practices in order to maintain efficient operation (Trist, 1981; Pasmore *et al.*, 1982) and – within hospitality contexts –to ensure a quality (service or consumption) experience (Ball *et al.*, 2003). To facilitate this, businesses of any size will be organized hierarchically, headed by a CEO, comprised of departments (e.g. food and beverages, front office, housekeeping and human resources) with specific operational rules and procedures, and department heads and supervisors entrusted to ensure conformity to the rules and regulations – to the 'corporate culture' – and to further the hotel or resort's mission. That, any rate, will be the formal structure of the organization. However, the business does not operate in a sociocultural vacuum and will be affected by numerous cultural factors introduced by staff and tourists from 'outside,' as well as forms of behaviour that arise specifically from within.

National and regional cultures play their part. As indicated earlier, businesses need punctual and efficient attendance and frequently expect staff to clock on and off, whereas in many developing societies this apparent regimentation of life may be quite foreign (Blanton, 1981, p. 122; Hope, 2004, p. 50). McDonald's, for example, must adapt to different cultural expectations in Germany and the UK, which *inter alia* affects its hiring practices (Royle, 1995), as well as the food it serves, while management styles are also subject to considerable cross-cultural variation. Alternatively, expatriate managers at international hotels in Russia must adapt to a hierarchical and paternalistic system descended from the Soviet system, where empowering employees is innovatory and a service ethic is little in evidence (D'Annunzio-Green, 2002). Attempts to introduce Western management patterns in Chinese hotels have experienced similar problems (Mwaura *et al.*, 1998).

Other cultural factors are also important. For example, work–family conflicts are common in the hospitality sector, especially because of the need for shift work and its effects on the non-work roles of women (Harrison, 1997, p. 181; Karatepe, 2010; Jung Choi and Tae Kim, 2012). In developing societies, too, staff members may absent themselves from work to attend to kinship or other social obligations, often at short notice, which is often not acceptable to management, especially if expatriate managers have no formal induction into the culture of the developing society (D'Annunzio-Green, 2002).

The hotel/resort hierarchy is likely to be based on achievement but in many developing societies status may be based on age. Consequently, if department heads or supervisors are younger than junior staff, conflicts arise, especially if requests from older staff members to younger but more senior colleagues involve contravening the operational procedures of the hotel/resort. Similarly, staff may be discouraged from being over-familiar with and borrowing money from each other or from guests, whereas (in Parsonian terms) local norms might encourage inter-personal behaviour that is particularistic, ascriptive, affective and collective in orientation (Parsons and Shils, 1951, pp. 80–88).

Another common rule is for social distance to be maintained between staff members and guests, but guests might seek to reduce the divide by claiming or seeking familiarity with staff members (or vice versa). Where this happens, especially if the guest is drunk, such overtures are likely to be unwelcome; in other circumstances, friendships, romantic attachments and even marriage can result, but in a tourism culture there are also huge risks of misunderstanding and conflict, especially in an alcohol-fuelled environment where inhibitions are reduced and culturally induced misunderstandings correspondingly exacerbated.

Intoxication, in particular, is not an issue only among guests. It is also an occupational hazard for staff. As O'Neill notes, with specific reference to managers: 'one of the defining characteristics of the hotel industry macroculture may be an emphasis on partying' (O'Neill, 2012, p. 84).

Within tourism culture, the behaviour of some holidaymakers can have a negative impact on the guest experience of others. This can include 'unsupervised children running around, staring, smoking and noticeably drunk [sic], which are all behaviours which vary in their prevalence and acceptability from culture to culture' (Nicholls, 2011, p. 210). Differences may be further exacerbated when tourists are from different social classes or national cultures, and thus follow different standards of public behaviour (Nicholls, 2011, p. 214).

At other times, behaviour that is normal in the resident society may intrude on the guest experience. The exuberance of housekeeping staff, accustomed to rising early in the morning, will be unappreciated by hotel guests wanting to 'lie in'. Gibson notes a similar lack of awareness among staff that guests might wish to be quiet (or make love) in the afternoon (2013).

All the above instances may operate against the business's mission, as it is formally expressed, while some features of local or tourists' cultures may be positive. In many developing societies, for example, the 'natural' hospitality and friendliness of local residents is marketed as one of the key attractions. The key point to note, however, is that the variety of cultural orientations brought into the work context by staff and tourists inevitably interact with the *formal* structure of the organization to produce a distinctive 'mix' of organizational features which will vary across both time and place.

Hospitality spaces as total institutions

For Goffman, total institutions are separated, often by physical barriers, from the rest of society, and fall into one of five broad categories: (i) those caring for incapable and harmless people (e.g. homes for the blind); (ii) those looking after people both incapable and a threat to the community (e.g. mental hospitals); (iii) those that isolate people who apparently deliberately threaten the community (e.g. prisons); (iv) instrumental institutions (e.g. army barracks, ships and boarding schools), and (v) retreats or training stations for the religious (e.g. monasteries) (1968, p. 16).

Having defined total institutions and suggested possible types, some of which are clearly mechanisms of social control, Goffman then lists several key features likely to be present, to a greater or lesser extent, in any such organization. Together, they might be said to constitute an 'ideal type' of total institution (Weber, 1949, pp. 90–102) and comprise the following (Goffman, 1968, p. 17): individuals sleep, play and work in the same place and under the same single authority. Daily activity 'is carried on in the immediate company of a large batch of others, all of whom are treated alike and required to do the same thing together'. 'All phases of the day's activities are tightly scheduled, with one activity leading at a prearranged time into the next, the whole sequence of activities being imposed from above by a system of explicit formal rulings and a body of officials.' Finally, 'the various enforced activities are brought together into a single rational plan purportedly designed to fulfil the official aims of the institution.'

Goffman (1968) did not focus on institutions specializing in providing hospitality but others have recognized the prevalence of social control in many aspects of hospitality and tourism (Cheong and Miller, 2000; Peacock and Kübler, 2001; Lynch *et al.*, 2011). The highly organized holiday camp, especially popular in the UK in the years following the Second World War, perhaps conforms most closely to the ideal type total institution (Read, 1986) but Ritzer and Liska note a 'softer' form of control in cruise ships and hotels (Ritzer and Liska, 1997, p. 106). Dann, too, has argued 'whether one is referring to historical or contemporary tourism, the overriding message of order and control is both omnipresent and perennial (Dann, 1996, p. 100) and, more relevant to present purposes, he discusses at some length the characteristics of total institutions in holiday hotels organized for the elderly (Dann, 2000).

That said, precisely which types of tourism institutions fall into which of the above categories is arguable. However, hotels have been placed on islands far from established population centres to protect the population from 'negative' impacts (e.g. the Maldives) or, alternatively, to protect holidaymakers from everyday life outside the resorts (as in the case of many 'boutique' resorts of the South Pacific and the Caribbean, including all-in resorts). Cruise ships could easily be placed in the fourth category, and while hotels and resorts are not formally religious institutions, the similarities of tourism with pilgrimage and the achievement of *communitas* are sufficiently established to allow that some hotels or resorts, at least, may have a quasi-religious function (Turner and Turner, 1978, pp. 20, 240; Harrison, 1997, pp. 178–182).

Other features mentioned by Goffman, which tend to follow from the features common to all total institutions, are also important (1968, pp. 20–22) and can be briefly mentioned in tabular form, along with an indication of their possible relevance to hotels and resorts (Table 15.1).

Much more could be said about the above features, but further reference here will be made to only a few points. First, in discussing the role of hotels as agents of social control, Wood draws attention to two processes of change: domestication, where hotels increasingly provide 'home comforts' in guests' rooms (to which they are increasingly restricted), and proletarianization, in that hotels are no longer the preserve of an elite (1994, p. 74). With reference to the latter process, involving increased guest numbers across a wider range of social classes, he also notes, in passing, that the 'interactional constraints presented by the hotel, as well as more mechanical ones, raise questions about the compatibility of the 'industrialization of service' with the provision of hospitality' (Wood, 1994, p. 78).

Table 15.1. Features of total institutions and their relevance to hotels and resorts.

Goffman's characteristics	Relevance to hotels and resorts
1. A split between the managed group and the staff.	Found in hotels/resorts. Staff and guest quarters totally separate; different entrances/exits; staff/guest social relations are clearly delineated.
2. Restricted conversation across boundaries and little mobility across stratum.	Operational manuals clearly note the need to maintain social distance from guests and spell out disciplinary measures if rules are not followed. In addition, restricted access to hotel/resort via dress codes etc.
3. 'Inmates' are excluded from decisions about own lives.	Implicit in hospitality sector. Can operate efficiently only if staff and guests know their place. Measures designed to ensure health and safety (e.g. well-lit areas, controlled access of locals, physical borders, etc.) all serve to control guest behaviour.
4. Mismatch with society's work structure (either much more or much less).	Staff work – sometimes long hours on shift work – but guests do not. Less relevant in hotels catering primarily to business tourists.
5. Incompatibility with the family.	In hotels/resorts, family is often subservient to providing service to guests.

Source: Authors.

Second, when discussing the mismatch of the work structure in a total institution (e.g. a prison or a hospital) to that of the wider society, Goffman remarks: 'Sometimes so little work is required that inmates, often untrained in leisurely pursuits, suffer extremes of boredom' (1968, pp. 20–21). No great leap of imagination is required to relate this to hotels and resorts, and to consider, in this context, the role of the daily events and activities organized by many hotels and resorts for their guests, including children. The continuous supply of piped or live music, fish-feeding, water sports, horse riding, coconut husking, competitions, dancing, village visits, floor shows and all the other forms of entertainment (or edutainment) found at the world's hotels and resorts serve a double function: they are simultaneously part of the attraction (authentic or otherwise) *and* a form of social control to allay guest boredom. Arguably, excess alcohol and food consumption may also be more likely when such functions are not being met.

Conclusion

It has been suggested that distinct 'tourism cultures' are entangled with and emerge from the hospitality interface of resident and guest cultures: they are hybrid cultures which contain elements of both but are yet different, influenced as they are by the operational nature of the hotel or resort. It is also suggested that little research has been carried out on tourism culture in such contexts, and that such conceptual frameworks as third space, socio-technical systems or total institutions might productively be applied to orientate research and further our understanding of the nature of tourism at different kinds of destinations. They might also assist our understanding of the extent to which different kinds of hospitality organization have different impacts, on both staff and holidaymaker.

Other issues also emerge. First, although it has been argued that 'industry training is not a "conversion" to another culture, but a specialized form of education which will facilitate service and communication' (Blanton, 1981, p. 120), tourism culture in developing countries is markedly different from the cultures of many (if not most) employees. Exactly *how* different will vary and can be determined only by research but, even if it is not a 'conversion', to participate in the tourism culture of hotels, resorts and similar institutions may require radical adaptation by staff members.

Second, such research is not merely a theoretical exercise. Understanding the varieties of tourism culture is of clear value to anyone wishing to plan hospitality and tourism development, whether in the private or the public sector. And if service quality is really important, it has to be based on understanding of the tourism culture at specific destinations. There is no one size that fits all.

Third, the pervasiveness of tourism culture, in all its variety, belies the notion that one can dispense with education for hospitality and tourism and focus, instead, on merely training functionaries to slot into an industrialized hotel sector which allegedly, but probably wrongly, requires 'job-ready applicants' (Zaitouni *et al.*, 2012). As Blanton noted three decades ago, 'training in social and cultural skills ... should build upon what is already known: it should be a "problem-posing education" which makes [students] critical thinkers' (1981, p. 125).

In a context where the relevance of universities, in particular, to tourism and hospitality education, is increasingly being questioned (Tribe, 2000; Airey, 2004; Holden and Wickens, 2007; Ayikoru *et al.*, 2009), and where universities in some developing countries are dispensing with any attempt at producing reflective and critical thinkers, it is a timely reminder that working in hotels and resorts is about more than the ability to fold a table napkin or produce the perfect soufflé.

References

Adler, P. and Adler, P. (2004) *Paradise Laborers: Hotel Work in the Global Economy*. Cornell University Press, Ithaca, New York.

Airey, D. (2004) From here to uncertainty. In: Tribe, J. and Wickens, E. (eds) *Critical Issues in Tourism Education: Proceedings of the 2004 Conference of the Association for Tourism in Higher Education*. Missenden Abbey, Buckinghamshire, UK.

Andrews, H. (2009) 'Tits out for the boys and no back chat': gendered space on holiday. *Space and Culture* 12(2), 166–182.

Ayikoru, M., Airey, D. and Tribe, J. (2009) Reading tourism education: Neoliberalism unveiled. *Annals of Tourism Research* 36(2), 191–221.

Ball, S., Jones, P., Kirk, D. and Lockwood, A. (2003) *Hospitality Operations: A Systems Approach*. Thomson, London.

Berdychevsky, L., Poria, Y. and Uriely, N. (2013) Hospitality accommodations and women's consensual sex. *International Journal of Hospitality Management* 34, 169–171. DOI: 10.1016/j.ijhm.2013.03.006.

Bhabha, H. (1994) *The Location of Culture*. Routledge, London and New York.

Bhabha, H. (1996) Culture's in-between. In: Hall, S. and Du Gay, P. (eds) *Questions of Cultural Identity*. Sage, London, pp. 53–60.

Bitner, M.J. (1992) Servicescapes: the impact of physical surroundings on customers and employees. *Journal of Marketing* 56(2), 57–71. DOI: 10.1177/002224299205600205.

Blanton, D. (1981) Tourism training in developing countries the social and cultural dimension. *Annals of Tourism Research* 8(1), 116–133. DOI: 10.1016/0160-7383(81)90072-4.

Bourdieu, P. (1984) *Distinction: A Social Critique of the Judgement of Taste*. Routledge, London.

Cheong, S.-M. and Miller, M. (2000) Power and tourism: a Foucauldian observation. *Annals of Tourism Research* 27(2), 371–390.

Crick, M. (1989) Representations of international tourism in the social sciences: sun, sex, sights, savings, and servility. *Annual Review of Anthropology* 18(1), 307–344. DOI: 10.1146/annurev.an.18.100189.001515.

Dann, G. (1996) *The Language of Tourism: A Sociolinguistic Perspective*. CAB International, Wallingford.

Dann, G. (2000) Overseas holiday Hhtels for the elderly: total bliss or total institution? In: Robinson, M., Long, P., Evans, N., Sharpley, R. and Swarbrooke, J. (eds) *Reflections on International Tourism: Motivations, Behaviour and Tourist Types*. Centre for Travel and Tourism in Association with Business Education Publishers, Sunderland, pp. 83–94.

Douglas, M. (2003) Standard uses of food: introduction. In: Douglas, M. (ed.) *Mary Douglas: Collected Works*. Routledge, Oxon, pp. 1–39.

D'Annunzio-Green, N. (2002) An examination of the organizational and cross-cultural challenges facing international hotel managers in Russia. *International Journal of Contemporary Hospitality Management* 14(6), 266–273. DOI: 10.1108/09596110210436805.
Èivre, Z., Kneževiæ, M., Baruca, P. and Fabjan, D. (2013) Facial attractiveness and stereotypes of hotel guests: an experimental research. *Tourism Management* 36(1), 57–65.
Fisher, D. (2004) The demonstration effect revisited. *Annals of Tourism Research* 31(2), 428–446. DOI: 10.1016/j.annals.2004.01.001.
Foster, D. (2000) *The Global Etiquette Guide to Asia: Everything you Need to Know for Business and Travel Success*. Wiley, New York.
Gibson, D. (2013) Living in the moment: cultural challenges faced by Indigenous-owned budget resorts in Fiji. Case studies from Wayalai, Yasawa island group. PhD work in progress. University of the South Pacific, Suva.
Goffman, E. (1968) *Asylums: Essays on the Social Situation of Mental Patients and Other Inmates*. Penguin Books, Harmondsworth.
Graburn, N. (2001) Secular ritual: a general theory of tourism. In: Smith, V.L. and Brent, M. (eds) *Hosts and Guests Revisited: Tourism Issues of the 21st Century*. Cognizant, New York, pp. 42–50.
Grünewald, R. (2002) Tourism and cultural revival. *Annals of Tourism Research* 29(4), 1004–1021. DOI: 10.1016/S0160-7383(02)00005-1.
Hall, S. and Du Gay, P. (eds) (1996) *Questions of Cultural Identity*. Sage, London.
Harrison, D. (1997) Globalization and tourism: some themes from Fiji. In: Oppermann, M. (ed.) *Pacific Rim Tourism*. CAB International, Wallingford, pp. 167–183.
Harrison, D. (2005a) Introduction: contested narratives in the domain of world heritage. In: Harrison, D. and Hitchcock, M. (eds) *The Politics of World Heritage: Negotiating Tourism and Conservation*. Channel View Publications, Clevedon, pp. 1–10.
Harrison, D. (2005b) Levuka, Fiji: contested heritage. In: Harrison, D. and Hitchcock, M. (eds) *The Politics of World Heritage: Negotiating Tourism and Conservation*. Channel View Publications, Clevedon, pp. 66–89.
Harrison, D. and Hitchcock, M. (eds) (2005) *The Politics of World Heritage: Negotiating Tourism and Conservation*. Channel View Publications, Clevedon.
Hetherington, K. (1998) *Expressions of Identity: Space, Performance, Politics*. Sage, London.
Hofstede, G., Hofstede, G. and Minkov, M. (2010) *Cultures and Organizations*, 3rd edn. McGraw-Hill, New York.
Holden, A. and Wickens, E. (2007) 'Citizen of the world' or 'management puppet'? The place of tourism studies in higher education. *Tourism Recreation Research* 32(2), 41–47. DOI: 10.1080/02508281.2007.11081275.
Hollinshead, K. (1998) Tourism, hybridity, and ambiguity: the relevance of Bhabhas 'third space' cultures. *Journal of Leisure Research* 30(1), 121–156. DOI: 10.1080/00222216.1998.11949822.
Hope, C.A. (2004) The impact of national culture on the transfer of "best practice operations management" in hotels in St. Lucia. *Tourism Management* 25(1), 45–59. DOI: 10.1016/S0261-5177(03)00059-1.
Hottola, P. (2004) Culture confusion: intercultural adaptation in tourism. *Annals of Tourism Research* 31(2), 447–466.
Jack, G. and Phipps, A. (2005) *Tourism and Intercultural Exchange: Why Tourism Matters*. Channel View Publications, Clevedon, UK.
Jung Choi, H. and Tae Kim, Y. (2012) Work-family conflict, work-family facilitation, and job outcomes in the Korean hotel industry. *International Journal of Contemporary Hospitality Management* 24(7), 1011–1028. DOI: 10.1108/09596111211258892.
Karatepe, O. (2010) The effect of positive and negative work-family interaction on exhaustion: does work social support make a difference? *International Journal of Contemporary Hospitality Management* 22(6), 836–856.
Knutson, B.J., Beck, J.A., Kim, S.H. and Cha, J. (2007) Identifying the dimensions of the experience construct. *Journal of Hospitality & Leisure Marketing* 15(3), 31–47. DOI: 10.1300/J150v15n03_03.
Knutson, B.J., Beck, J.A., Kim, S. and Cha, J. (2009) Identifying the dimensions of the guest's hotel experience. *Cornell Hospitality Quarterly* 50(1), 44–55. DOI: 10.1177/1938965508326305.
Lugosi, P. (2009) The production of hospitable space: commercial propositions and consumer co-creation in a bar operation. *Space and Culture* 12(4), 396–411. DOI: 10.1177/1206331209348083.
Lugosi, P. (2014) Mobilising identity and culture in experience co-creation and venue operation. *Tourism Management* 40, 165–179. DOI: 10.1016/j.tourman.2013.06.005.

Lynch, P., Molz, J.G., Mcintosh, A., Lugosi, P. and Lashley, C. (2011) Theorizing hospitality. *Hospitality & Society* 1(1), 3–24. DOI: 10.1386/hosp.1.1.3_2.

Maoz, D. (2006) The mutual gaze. *Annals of Tourism Research* 33(1), 221–239. DOI: 10.1016/j.annals.2005.10.010.

Mars, G. and Nicod, M. (1984) *The World of Waiters*. George Allen and Unwin, London.

Metusela, C. and Waitt, G. (2012) *Tourism and Australian Beach Cultures: Revealing Bodies*. Channel View Publications, Bristol.

Mwaura, G., Sutton, J. and Roberts, D. (1998) Corporate and national culture - an irreconcilable dilemma for the hospitality manager? *International Journal of Contemporary Hospitality Management* 10(6), 212–220. DOI: 10.1108/09596119810232211.

Nicholls, R. (2011) Customer-to-customer interaction (CCI): a cross-cultural perspective. *International Journal of Contemporary Hospitality Management* 23(2), 209–223. DOI: 10.1108/09596111111119338.

O'Neill, J.W. (2012) The determinants of a culture of partying among managers in the hotel industry. *International Journal of Contemporary Hospitality Management* 24(1), 81–96. DOI: 10.1108/09596111211197818.

Parsons, T. and Shils, E. (1951) Values, motives and systems of action. In: Parsons, T. and Shils, E. (eds) *Toward a General Theory of Action: Theoretical Foundations for the Social Sciences*. Harper and Row, New York, pp. 47–278.

Pasmore, W., Francis, C., Haldeman, J. and Shani, A. (1982) Sociotechnical systems: a North American reflection on empirical studies of the seventies. *Human Relations* 35(12), 1179–1204. DOI: 10.1177/001872678203501207.

Peacock, M. and Kübler, M. (2001) The failure of 'control' in the hospitality industry. *International Journal of Hospitality Management* 20(4), 353–365. DOI: 10.1016/S0278-4319(01)00022-6.

Pearce, P.L. (1995) From culture shock and culture arrogance to culture exchange: ideas towards sustainable socio-cultural tourism1. *Journal of Sustainable Tourism* 3(3), 143–154. DOI: 10.1080/09669589509510719.

Pearce, P. (2005) *Tourist Behaviour*. Channel View Publications, Clevedon.

Pi-Sunyer, O. (1989) Changing perceptions of tourism and tourists in a Catalan resort town. In: Smith, V. (ed.) *Hosts and Guests: The Anthropology of Tourism*. University of Pennsylvania Press, Philadelphia, pp. 187–199.

Pratt, S. (2013) Minimising food miles: issues and outcomes in an ecotourism venture in Fiji. *Journal of Sustainable Tourism* 21(8), 1148–1165. DOI: 10.1080/09669582.2013.776060.

Pritchard, A. and Morgan, N. (2006) Hotel Babylon? Exploring hotels as liminal sites of transition and transgression. *Tourism Management* 27(5), 762–772. DOI: 10.1016/j.tourman.2005.05.015.

Read, S. (1986) *Hello Campers*. Transworld Publishers, London.

Reisinger, Y. (2009) *International Tourism – Cultures and Behaviour*. Butterworth-Heinemann, Oxford.

Ritzer, G. and Liska, A. (1997) McDisneyization and 'post-tourism': contemporary perspectives on contemporary tourism. In: Rojek, C. and Urry, J. (eds) *Touring Cultures: Transformations of Travel and Theory*. Routledge, London, pp. 96–109.

Rogerson, C.M. (2012) Strengthening agriculture-tourism linkages in the developing world: opportunities, barriers and current initiatives. *African Journal of Agricultural Research* 7(4), 616–623.

Rojek, C. and Urry, J. (1997) Transformations of travel and theory. In: Rojek, C. and Urry, J. (eds) *Touring Cultures: Transformations of Travel and Theory*. Routledge, London, pp. 1–19.

Royle, T. (1995) Corporate versus societal culture: a comparative study of McDonald's in Europe. *International Journal of Contemporary Hospitality Management* 7(2), 52–56.

Ryan, C. (2003) *Recreational Tourism: Demand and Impacts*. Channel View Publications, Bristol.

Schroeder, K. and Sproule-Jones, M. (2012) Culture and policies for sustainable tourism: a South Asian comparison. *Journal of Comparative Policy Analysis: Research and Practice* 14(4), 330–351. DOI: 10.1080/13876988.2012.698579.

Sherman, R. (2007) *Class Acts: Service and Inequality in Luxury Hotels*. University of California Press, Berkeley.

Shirley, V. (1989) *Dir. Lewis Gilbert, with Screenplay by Willy Russell*. Paramount Pictures.

Soper, A.K. (2008) Mauritian landscapes of culture, identity, and tourism. In: Knudsen, D.C., Metro-Roland, M.M., Soper, A.K. and Greer, C.E. (eds) *Landscape, Tourism and Meaning*. Ashgate, Aldershot, pp. 51–64.

Trauer, B. and Ryan, C. (2005) Destination image, romance and place experience—an application of intimacy theory in tourism. *Tourism Management* 26(4), 481–491. DOI: 10.1016/j.tourman.2004.02.014.

Tribe, J. (2000) Balancing the vocational: the theory and practice of liberal education in tourism. *Tourism and Hospitality Research* 2(1), 8–25. DOI: 10.1177/146735840000200106.

Trist, E. (1981) *The Evolution of Socio-technical Systems*, Occasional Paper No.2. Ontario Quality of Working Life Centre, Toronto.

Turner, V. and Turner, E. (1978) *Image and Pilgrimage in Christian Culture*. Columbia University Press, New York.

Urbain, J.-D. (2003) *At the Beach*, Translated by Porter, C. University of Minnesota Press, Minneapolis.

Walls, A., Okumus, F., Wang, Y. and Kwun, D.J.-W. (2011) Understanding the consumer experience: an exploratory study of luxury hotels. *Journal of Hospitality Marketing & Management* 20(2), 166–197. DOI: 10.1080/19368623.2011.536074.

Weaver, A. (2005) Spaces of containment and revenue capture: 'super-sized' cruise ships as mobile tourism enclaves. *Tourism Geographies* 7(2), 165–184. DOI: 10.1080/14616680500072398.

Weber, M. (1949) *The Methodology of the Social Sciences*. Free Press, New York.

Williams, A.M. (2006) Lost in translation? International migration, learning and knowledge. *Progress in Human Geography* 30(5), 588–607. DOI: 10.1177/0309132506070169.

Wood, R.C. (1994) Hotel culture and social control. *Annals of Tourism Research* 21(1), 65–80. DOI: 10.1016/0160-7383(94)90005-1.

Zaitouni, M., Breakey, N. and Robinson, R. (2012) Do We Know What the Hotel Industry Expects From Graduates? *Paper presented to the panel 'Changing Times in Tourism and Hospitality Education'. THE-ICE 6th International Panel of Experts Forum*, The Emirates Academy of Hospitality Management, Dubai, UAE, 18–20 November.

16

Towards Developing a Framework for Analysing Tourism Phenomena: A Discussion

Introduction

Since the middle of the 19th century and the construction of the railways, mass tourism has been a significant social phenomenon. It first developed in western Europe and North America but, since the Second World War, tourists have increasingly crossed international boundaries, boosting the contribution of travel and tourism to the economies of both developed and less developed countries. Even in the 19th century, however, tourism provoked vociferous and hostile opposition (Gilbert, 1975, pp. 192–196; Ousby, 1990, pp. 191–194), and international tourism similarly brought with it critics of 'the golden hordes', especially as they affected less developed countries (Turner and Ash, 1975). Was tourism a blessing or a blight? For some observers, undoubtedly, it was the latter (Young, 1973; Wolfers, 1974, pp. 60–72).

Despite an early warning from MacCannell not to engage in a 'rhetoric of moral superiority' towards tourists (MacCannell, 1976, p. 9), he later branded tourists 'an expeditionary force without guns' (MacCannell, 1989, p. xviii), and academics studying tourism have often emphasized tourism's perceived negative impacts. Most contributors to Smith's (1978a) first edition of *Hosts and Guests* would fall into this category. Contributors to an equally seminal collection edited by de Kadt (1979) were somewhat more circumspect, but a consistent indictment of tourism as a negative influence soon emanated from 'dependency' theorists of the 1980s (Kent, 1975; Britton, 1982, 1989; Harrison, 2001a). The trend continued into the 1990s and, as Brown was to note, there was an 'enormous amount of academic material devoted to criticizing tourism and tourists' (Brown, 1998). Indeed, the level of antagonism towards tourism has led one established tourism academic to wonder why 'so many colleagues are bent on turning their backs on the (un?) real world? Why are the fiercest critics of tourism to be found in that part of academia that provides mostly teaching jobs?' (Aramberri, 2001, p. 757).

Despite the hostility of some academics to tourism, however, in recent years scholarly studies of tourism have ranged widely. Some contributors focused on images of destination areas apparently conveyed to potential tourists (Cohen, 1993a; Selwyn, 1993; Dann, 1996). Others examined tourism's role in modern or postmodern societies (MacCannell, 1976; Krippendorf, 1987; Urry, 1988, 1990, 1995; Munt, 1994; Franklin, 2003) and, as indicated later in this paper, interest in tourism's relationship to sustainable development has recently become a major preoccupation (Mowforth and Munt, 2003). At a less abstract level of analysis, much of Cohen's early work was on tourist typologies and the dynamics of social interaction involving tourists and local residents (Cohen, 1972, 1979, 1985). More generally, over the years ethnographic studies of destination areas in the less developed world have mushroomed.

Clearly, debates over the nature and importance of tourism have not lessened, but wider academic coverage now makes it well-nigh impossible to keep track of every contribution to them. In the early 1970s, for example, *Annals of Tourism Research* was one of very few journals devoted entirely to the academic study of tourism, whereas by 2004 there were 'some 75 core academic journals in tourism, hospitality and leisure' (Hall, 2005), including *Current Issues in Tourism*, which was first published in 1998. Further evidence of expansion, if it were needed, is in the number of doctoral dissertations produced on tourism. It has been calculated that, in North America alone, from 1951 to 2000 there were 526, with 72% produced since 1986 (Meyer-Arendt and Justice, 2002, pp. 1172–1173).

Interestingly, the above categorization of PhDs was based on their disciplinary origins, with departments of recreation (including tourism studies) accounting for more than a quarter of the post-1986 doctorates. This highlights another debate, notably the extent to which tourism can be considered a discipline in its own right. It is a discussion which, at times, has bordered on the acrimonious (Tribe, 1997, 2000; Leiper, 2000), and is based, at least to some extent, on a comparison (and defence) of what is taught under university curricula with the more rigorous epistemological demands made of a subject, and its methods of enquiry, for it to be considered a 'scientific' discipline. The situation is compounded, at least in parts of Europe and north America, by the fact that while tourism is taught as a subject in many 'new' universities and institutions of higher education, it is noticeably absent from the curricula of most older, more established and (in general) more prestigious universities. At the time of writing, then, while tourism has been approached from many disciplinary perspectives, in what might almost be regarded as a 'mix and match' manner (Aramberri, 2001, p. 739), it remains a moot point as to whether or not 'tourism studies' is itself a discipline.

These three conditioning factors – the continuing tendency of some academics to evaluate tourism (especially mass tourism) negatively, the burgeoning tourism literature, and the unresolved debate as to how far 'tourism studies' is a discipline in its own right – have led to a situation in which it has become increasingly difficult to incorporate the wide range of contributions to the literature of tourism into a coherent framework. Three decades ago, when tourism was little studied, it was relatively easy to arrive at a comprehensive knowledge of the tourism literature. Now, quite simply, the problem for newcomers to tourism studies, as well as for those with some experience in the field, is how to make sense of all that is on offer. In the following pages, first, some problems involved in assessing tourism's social and cultural role in society will be discussed. Second, a methodological basis for a conceptual framework to structure an analysis of tourism will be presented and, third, it will be then applied more specifically to tourism phenomena.

Problems in Assessing Tourism's Social and Cultural Role in Society

Contributions to the tourism literature are not easy to classify. In a major review of publications on tourism's social impact, for instance, Pearce and Moscardo (1999) suggest that most fall into one of two basic categories: ethnographic case studies, mainly of less developed countries and regions, and social survey research. They conclude that there are basic problems in defining and measuring tourists, tourism and 'community', that those conducting attitude surveys have too often imposed their own perceptions of tourism's impacts on respondents, and that – above all – 'the atheoretical nature of much tourism research is a major limiting factor in its development' (Pearce and Moscardo, 1999, pp. 41–46). According to them, the result is confusion.

There is much to recommend this analysis. Social anthropological examinations of tourism's social impact have indeed produced a wealth of data but, despite the richness of the ethnography, intensive focus on the ethnographic moment makes comparison of tourist destinations difficult. Few comparative studies have been carried out and, on the rare occasions study areas have been revisited, some researchers had to revise their initial impressions, swallow hard, and enter pleas of *mea culpa* (Greenwood, 1989, pp. 181–185; Pi-Sunyer, 1989, pp. 196–199). Furthermore, perhaps because of characteristics they share with some tourists (Nash, 2001), there are still traces of academic snobbery among those who carry out research on tourism. As Nash notes, anthropologists 'still have a way to go towards scientific maturity' (Nash, 1996).

In addition, although many social surveys of residents of destination areas have been carried out (Pearce *et al.*, 1996, pp. 21–23), most have been after tourism development has taken place and attempts to ascertain local attitudes in advance are rare (Prentice, 1993). Where contrary examples can be found, largely in developed societies, the planning process has been both extensive and expensive – as in Flathead County, Montana (Culbertson *et al.*, 1996). In less developed countries, political and other constraints make it even more difficult for extensive community participation to occur, except perhaps in small-scale 'ecotourist' projects (Tosun, 2000, p. 627), and even then much depends on the extent to which local communities are internally cohesive, have access to local resources and can draw on the support of relatively powerful, external, interest groups (Harrison and Price, 1996, pp. 8–18).

There are severe limitations, then, in both ethnographic studies and social surveys. However, while the analysis of Pearce *et al.* has considerable merit, their solution – the application of what they term 'social representations theory' (Pearce *et al.*, 1996, pp. 31–107) – is problematic. According to them, studies of tourism should be situated in 'complex meta-systems of everyday knowledge' that 'help define and organize reality, ... allow for communication and interaction', and enable 'new and abstract concepts/events' to be incorporated into existing frameworks (Pearce *et al.*, 1996; Pearce and Moscardo, 1999). However, while such an approach is an established *sine qua non* for competent ethnography, it is neither a theory nor a conceptual framework.

Their suggestion that we examine tourism holistically makes it no easier to compare ethnographic accounts of tourism in different destinations. Tourism's role and effects vary according to place and time and are greatly determined by structural factors and types of tourism (Dogan, 1989; Brunt and Courtney, 1999). While general lists of tourism's impacts are common (Craik, 1995; Smith, 2001, p. 110; Puppim de Oliveira, 2003, pp. 100–101), they function primarily and only as useful prompts for researchers. That is all they can be, for the interpretation placed even on agreed impacts may vary widely among researchers and residents of destination areas. With reference to Table 16.1, for example, whereas Smith suggests that upward mobility is a positive effect of tourism development (2001, p. 110), others consider increased inequality a negative consequence of tourism (Stonich, 2000, pp. 141–143). And while some allege that tourism leads to a 'phoney folk culture' and commoditization (or commodification) (Jordan, 1980), others suggest that commoditization need not lead to a loss of culture or a diminution of standards (Harrison, 1992).

Debates over the issue of 'demonstration effects' have been similarly confused (Fisher, 2004), as have those focusing on the role of tourism in the dynamics and perceptions of group and category formation, including ethnicity. Indeed, it is increasingly recognized that individual and social 'identities' are resources that are both manipulable and flexible according to circumstance and the definitions of social actors (Hitchcock, 1999).

The key point here is that although much valuable research has been carried out, based on numerous disciplines, and there is a mass of empirical data 'out there', we are no nearer now than three decades ago to a clear understanding of tourism's sociocultural impacts on destination areas in less developed regions and countries or, indeed, in societies generally considered to be 'developed'. This is partly because academic and lay assessments are inevitably subjective (Harrison, 1992), but it is also because, across the social sciences, there is no generally accepted framework in which to incorporate the various types of tourism studies. Instead, refuge is taken in describing tourism studies as 'interdisciplinary' or 'multidisciplinary' when, in fact, they too frequently lack disciplinary rigour of any kind.

Table 16.1. Selected cost-benefit factors of tourism.

	Positive aspects	*Negative aspects*
Economic	Hard currency	Economic leakages and inflation
	Multiplier effect	Seasonality
	Labour-intensive services	In-migration of managers/ labour
	Better infrastructure	Costs of offsetting crime
	Income from residents' use of amenities	Loss of receipts from external crises
Sociocultural	Widening social perspective	Loss of local identity
	Preservation of family ties	Coca-cola-isation
	Upward mobility	Commoditization of tourists as 'things'
	Appreciation of heritage and ethnic identity	Fear of terrorism and crime
	Stimulus for folklore and museums	Deterioration of historic sites through over-use
		Commoditization of culture and loss of tradition
		Misuse of intellectual property rights

Source: Based on Smith, 2001, p. 110.

The Methodological Basis for a Conceptual Framework

A conceptual framework for analysing tourism's sociocultural role in society must be based on existing social theory. As a consequence, at the risk of oversimplifying complex issues, it is necessary to consider the major levels of analysis currently applied in mainstream social science before examining their implications for studying tourism's impacts. This is easier said than done, for there is no clear agreement among social scientists about how many actually exist. However, when referring to 'research strategies', Pawson (1999) suggests there are four, namely: 'ethnographers', who focus on intentional action; 'surveyors', who aim to discover social regularities through the use of quantitative methods; 'comparative researchers', who carry out cross-cultural research; and 'evaluation researchers', who are primarily interested in policy-related matters (Pawson, 1999, pp. 23–29).

When these are narrowed down, however, they relate to three quite basic levels of analysis, or abstraction, which focus first on social action and interaction, second, on social institutions and the structures within which they occur and, third, on the wider social structures and sociocultural systems which (for methodological purposes) such structures are said to constitute.

This conceptualization owes much to the view of Berger and Luckmann (1966) that 'society' can be seen as subjective reality or objective reality, and that the subjective becomes objectified through the process of institutionalization (Berger and Luckmann, 1966, pp. 65–109). Such levels of conceptualization represent alternative but (arguably) complementary routes towards understanding social phenomena. They are not incompatible with one another, and no one perspective can claim superiority. Rather, they are entry points selected according to the individual researcher's interests, methodological propensities and definition of the task at hand. A diagrammatic representation of this (very) basic model, with a small selection of its proponents, is provided in Fig. 16.1.

While there is inevitably much blurring across boundaries between levels, major contributors to social science are commonly linked to one or more of them. For instance, those grounding their approach on the interactionism associated with Weber (Gerth and Mills, 1948; Weber, 1964), and developed by Schutz, Mead, Goffman and Garfinkel (Morris, 1977; Wolff, 1978; Craib, 1984, pp. 71–97), examine direct

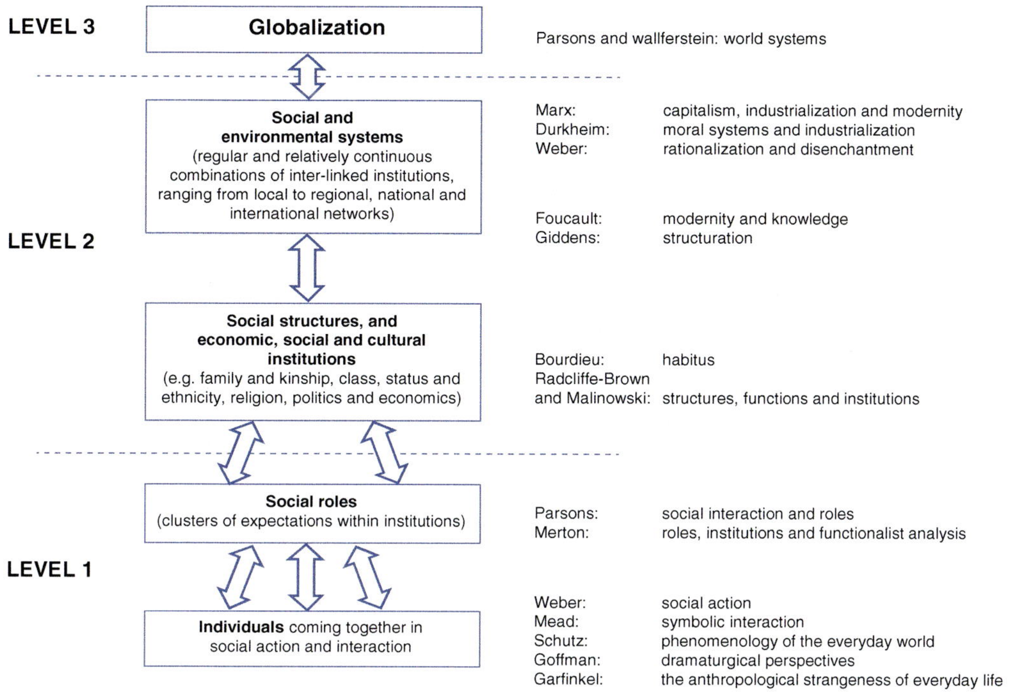

Fig. 16.1. Three levels of social analysis.

and mediated social action and interaction. Operating at the most basic levels of interaction (level one in Fig. 16.1), they stress the intersubjective, negotiated and dynamic nature of social life, building on the 'first order' constructs of social actors to produce what phenomenologists refer to as second-order 'objectifications' or 'constructs' of social experience (Morris, 1977, pp. 19–20; Rogers, 1983, p. 90).

Other social scientists prefer to bracket the ongoing and contingent nature of social life and speak in terms of social roles, institutions and structures. Operating what Merton depicts as 'middle range theory' (Merton, 1968, p. 39), social interaction is concretized in roles and institutions, indicated as level two in Fig. 16.1. From this perspective, the ordered nature of social life is possible because individuals take on social roles (clusters of expectations) which provide the limits of how they should behave. Such roles operate within specific groupings of social activity which, for convenience, are described as social institutions, for example families, clans, tribes or ethnic groups, religious groups and organizations, political parties, businesses, and social classes or status groups. These vary considerably over time and place, but all are characterized by (at least) two key features: first, they endure across generations and, second, social actors are constrained (to a greater or lesser extent) in the performance of their roles by social sanctions, formal or informal, and positive or negative. When taken together, these social institutions link dynamic and face-to-face social interaction with social structures which, for the purpose of the analysis, are deemed relatively stable over time, even though they are inevitably subject to exogamous and endogenous pressures to change.

Institutional analysis was prominent in the seminal work of such social anthropologists and sociologists as Radcliffe-Brown (1952, p. 10), Malinowski (Parsons, 1957, pp. 58–61; Richards, 1957, pp. 24–26), Parsons (Parsons and Shils, 1951, pp. 53–109; Dawe, 1979, pp. 400–408) and Merton (1968) and, shorn of its functionalist overtones, and despite attempts to develop social network analyses (Mitchell, 1969; Gulliver, 1971), continues to play a key role in social scientific thinking

(Seymour-Smith, 1986, p. 153). Indeed, defined as 'practices which have the greatest time-space extension within such (societal) totalities', social institutions figure prominently in the theory of structuration provided by Craib (1992, pp. 51–55); Giddens (1994, p. 80), and in the work of Bourdieu, for whom *habitus* is 'the system of structured, structuring dispositions' (1994, p. 95) that enables institutions to 'attain full realization' (1994, p. 101).

Broader and more abstract theorizing about societies (level three in Fig. 16.1) is evident in what C. Wright Mills once described as 'grand theory', where the aim of social science is to produce 'systematic theory' of 'the nature of man and society' (quoted in Skinner, 1985, p. 3). Examples of this genre can be found in the work of structural functionalists and structuralists. The latter, in particular, were especially prominent in France, through Saint-Simon, Comte and Durkheim (then via Radcliffe-Brown to the USA), Levi-Strauss, Piaget and Althusser, and in Germany, most notably through Hegel and Marx (Bottomore and Nisbet, 1979, pp. 559–594). In essence, structuralism is based on the notion that 'the relation is more important than the parts' (Bottomore and Nisbet, 1979, p. 558) or, putting it slightly differently, that social phenomena are to be understood only if underlying patterns and structures can be isolated. This necessarily presupposes that the parts are inter-related in an orderly, systemic manner. Within such schema, individual actions, and the outcomes of their actions, are conditioned by underlying social forces. Put somewhat crudely by Engels, 'the course of history is governed by inner general laws' (Engels, 1968, p. 612), whereas Marx, although insistent that 'sum of productive forces' conditions human history, reflects that 'circumstances make men just as much as men make circumstances' (Marx, 1977).

The extent to which human agency can be reconciled to the apparent structured or systemic order of everyday social life has been a consistent topic in social science, and preoccupied systems theorists and interactionists during the second half of the 20th century (Buckley, 1967; Gouldner, 1973; Dawe, 1979; Craib, 1984, pp. 37–58). Indeed, when postmodernists proclaimed the end of the 'grand narrative' (Skinner, 1985, p. 12; Smart, 1993, pp. 28–39). they were (arguably) stating little more than the obvious, especially after the decline of Marxism that followed the fall of the Berlin Wall. Nevertheless, despite the apparent certainty of such assertions, in retrospect it is difficult to isolate a period when any special narrative held sway!

More recently, all these debates have re-emerged and have been incorporated into globalization theory (Waters, 1995), an umbrella term denoting a wide range of perspectives examining the nature and inter-relations of such global actors as societies, nation states, social classes, political parties, transnational economic institutions and other national and international formal organizations. Linked by little more than the idea – as often implicit as explicit – of system, the ensuing melange brings together established elements of Marxism, neo-Marxism and 'bourgeois' theory. They include the world systems theory of Wallerstein, with its primary focus on economic linkages (Harrison, 1992; Waters, 1995, pp. 22–26), the issues of cultural convergence, homogenization and discontinuities arising from the work of Giddens and Robertson, and the reiteration of capitalism as a global force by Sklair and Harvey (Waters, 1995, pp. 47–52; Sklair, 1999, pp. 139–150). They also include at least some contributions from postmodernists, who reject 'meta-narratives', claim 'that the promise of modernity to deliver order, certainty and security will remain unfulfilled' (Smart, 1993, p. 27), and yet – in providing their own illustrated accounts of contemporary society – go some way to undermining their own relativism (Swingewood, 1999, p. 71).

Like other social theorists, globalization theorists disagree on whether or not the system takes analytical precedence over the individual actor, and over how far economic and social structures, social and cultural processes, institutions and human agency can be integrated (if at all) in any single perspective (Scott, 1997, pp. 18–20; Germain, 2000; Hay and Marsh, 2001). It is generally agreed, however, that there is still no commonly agreed single vision, no single theory or perspective, that can be used exclusively to analyse or explain social phenomena. In this respect, perhaps, postmodernists are correct, but they peddle yesterday's news, for no such perspective ever existed. Instead, as Weber recognized, research methodologies arise from within communities of social scientists, whose

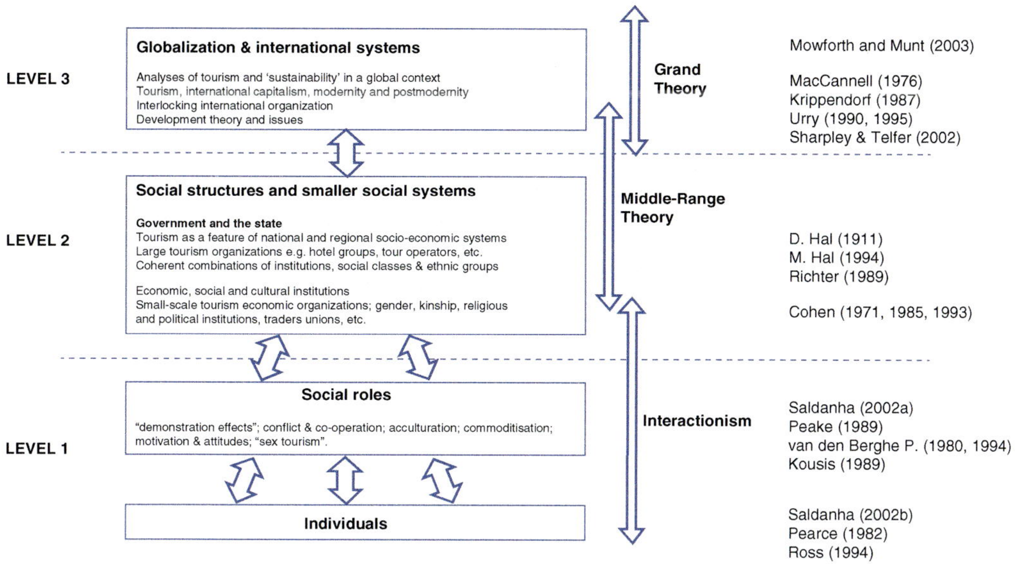

Fig. 16.2. Levels of analysis in tourism studies.

research interests and various attempts to penetrate 'the infinite causal web are determined by the evaluative ideas which dominate the investigator and his age' (Weber, 1949, p. 84).

Towards a Conceptual Framework for Analysing Tourism Phenomena[1]

The conceptual framework outlined in the previous section can be applied to the tourism literature, as indicated in Fig. 16.2, but requires tourism scholars to recognize, first, that levels of analysis do differ and, second, that all theoretical and empirical contributions to the debate can be clearly situated within and/or across these levels, so that comparison of similar social phenomena can occur.

Elsewhere, an attempt has been made to relate types of formal organizations to these levels of analysis (Harrison, 2001b, p. 258), but the aim here is rather to suggest examples of where specific texts in tourism studies can be situated within the wider methodological framework, and indicate how awareness of the three analytical levels might help assist in framing research questions and comparing research results with material others have obtained in other contexts.

From the outset, though, four caveats are necessary. First, it is not intended here to provide a new or revised 'paradigm' into which all studies of tourism can be inserted. Inevitably, perspectives differ across and within disciplines, and it has already been suggested that, in social science generally, no such paradigm, in the sense used by Thomas Kuhn, actually exists (Kuhn, 1970a, b). However, it does seem sensible to try to increase awareness within tourism studies of the nature and importance of different methodological approaches, and provide some criteria as to how they can be comparatively assessed. As a result, those confronted with an often bewildering array of tourism writing, ranging from small-scale empirical studies to meta-theoretical narratives, can situate contributions within one or more of these analytical levels.

Second, as even a cursory scan of tourism literature indicates, boundaries between levels of abstraction are blurred. The work of any one author is unlikely to be restricted to a single level of analysis, and substantial variation may be found even within texts. MacCannell's early study of tourism (1976), for example, may be interpreted as a theoretical but untestable review of tourism's role in modern societies, but it can also be used to derive empirically verifiable assertions about tourist motivation and how far 'the tourist' seeks 'authenticity'. In this way, it contributes to an understanding of the characteristics of social systems and of the components of social action.

Third, adherents of different methodological approaches may disagree, for example, on whether or not conclusions reached by some scholars can or should be empirically verifiable and subject to 'scientific' proof or falsification. Because in tourism studies, as in social science generally, there is no dominant paradigm, competing academics choose different points at which to penetrate Weber's 'infinite causal web', not always graciously recognizing contributions informed by other methodological positions. Nevertheless, irrespective of the extent to which positions are 'testable', they can be situated at one of the three levels. How far they are or should be testable is not at issue here.

Finally, for a conceptual framework to be useful in analysing tourism's impacts, it should be flexible and applicable across a wide variety of tourism-related contexts, where similar questions can be asked of the data being supplied, irrespective of the scale of the studies, which may range from small-scale case studies to more comprehensive and non-empirically based theoretical contributions.

Level 3: Globalization and Systems Approaches

If one commences with a global system perspective, all tourism-related activity must ultimately and by definition occur within it. However, 'grand theory' in tourism studies emerged at its very start, in MacCannell's pathbreaking introduction to tourism in modern society, which was influenced by Levi-Strauss's structuralism and Marx's alienation theory (MacCannell, 1976, pp. 2–4, 19–21). His focus is on late capitalism and its corresponding cultural system, where tourism is both a symptom and a result of major shifts in social organization. The tourist, an epitome of 'modern man', lives in a structurally differentiated world and searches for security and authenticity in a context of increasing mass leisure, where the centrality of work is displaced by leisure and where work itself becomes an object of curiosity. Alienated from others, and from the productive system, the tourist embarks on a quasi-religious and ultimately fruitless quest for a 'lost' heritage and social identity (MacCannell, 1976, p. 13).

The work of Urry on the 'tourist gaze' (1990, 1995) and, with his co-writers (Abercrombie and Urry, 1983; Lash and Urry, 1987), on postmodern societies, complements that of MacCannell. For Urry, contemporary capitalist societies have moved from organized to disorganized capitalism, where emphasis has shifted from production to consumption. As a consequence, there has emerged a new service class, which in turn is reflected in a new kind of tourist – the 'post-tourist' – who represents the new service class at play, who wants independent travel and individual care and attention (Urry, 1990, p. 96), and who 'reads' the countryside as landscape, as spectacle, through the 'romantic gaze' of the reflective individual (Urry, 1990, p. 174).

Krippendorf adopts a similar position to that of MacCannell, but focuses on industrial (rather than modern) society, and uses a cruder and more openly systemic approach to mass tourism, which he considers an unsuccessful attempt by tourists to 'escape' the problems of industrial society. As tourists, we simply take our problems with us, and drag along all the evils of residential and industrial areas from which we would like to escape, and bring them as presents to our hosts: traffic problems, air pollution, noise, metropolitan architecture, alienation – to name but a few (Krippendorf, 1987, p. 33).

For present purposes, whether or not changes in modern capitalism are as depicted, or the new service class actually exists, is irrelevant. What does matter is that such writers see tourism largely as a reflection of wider social movements. Modern or postmodern society has impacts on the nature of tourism which, in turn, may lead to consequences in regions visited by tourists from these societies.

Krippendorf is perhaps the most strident of these critics of modern society, but his vision of a brighter, greener, more sustainable form of industrial society (and tourism) can be linked to another important strand of scholarship, which focuses on 'development' issues (Krippendorf, 1987, pp. 86–148). Here, the emphasis is again on the global economic and sociocultural system, but this time on how far tourism can be considered a tool for 'development' or (for later writers) 'sustainable development', in societies considered 'undeveloped', 'less developed', or 'underdeveloped'(Harrison, 1992, 2001a; de Kadt, 1979; Selwyn, 1996; Sharpley and Telfer, 2002).[2] One of the most recent and comprehensive contributions to this debate focuses on such themes as globalization, sustainability and development, and examines the links of tourism in the 'Third World' (especially new or 'alternative' forms of tourism) with changes occurring in Western societies (Mowforth and Munt, 2003). It is a project that necessitates excursions into grand theory but also requires an examination of the global political and economic institutions that characterize international tourism and its various social and economic impacts on 'Third World' peoples. In this sense, it links the most abstract of models of the world system with institutional analysis.

Level 2: Social and Economic Structures and Institutions

Elsewhere, doubt has already been expressed that such perspectives as modernization, dependency, economic neo-liberalism and alternative development can really be considered 'paradigms' (Telfer, 2002, pp. 38–78), but it is perhaps inaccurate to suggest that 'there has been very little work' to connect development theory and tourism studies (Telfer, 2002). The work of Mowforth and Munt (2003), indicated above, is a comprehensive example to the contrary and, as indicated below, numerous studies on tourism's impact on less developed countries and regions have focused on national or regional social structures, institutions and relationships. Some of these have also examined the interactional involvements of residents in tourist destination areas – albeit with little effort to assess tourism's impacts on developed societies.[3] Similarly, social geographers have researched the importance of tourism in national and regional contexts (Hall, 1991; Williams and Shaw, 1991), while others have focused more particularly on the importance of tourism in the sphere of national and international politics (Hall, 1994).

Among those who might be described as tourism theorists of the middle range, Cohen is surely pre-eminent. His anthropological interests effectively start where MacCannell's grand theory ends. Indeed, as he and others have pointed out, the implication (derived from MacCannell and, to a lesser extent, Urry) that there is only one type of tourist, is quite unsustainable (Schudson, 1979; Cohen, 1984, p. 378; Bruner, 1991, p. 241; Franklin, 2003, p. 269), and Cohen is but one of several social scientists to have developed typologies of tourists, whom he categorizes according to the extent they seek security and novelty in more or less institutionalized and organized social contexts (Cohen, 1972, 1973, 1974, 1979, 1973; Smith, 1978b; Yiannakis and Gibson, 1992; Wall, 1993). He has also studied varieties of 'culture broker', including those involved in sexual interaction, accompanied by real or imagined romance, which occurs between Arab boys and Jewish girls (Cohen, 1971) and between *farangs* and locals in Thailand (Cohen, 1982a, b,1993b). Guides, too, are culture brokers, and Cohen has also examined the roles performed by tourist guides in Thailand (Cohen, 1982c) and more generally (Cohen, 1985), categorizing them according to the type of tourism in which they are involved and its level of development. Finally, he looks at the interaction involved when tourists seeking a quick and easy profit are exploited, in turn, by confidence tricksters (Cohen, 1996).

More generally, sociologists and anthropologists have produced invaluable case studies of tourism development. Kousis, for example, provides a relatively early study of the impact of tourism on kinship in a rural Cretan community (Kousis, 1989), as does Peake for Malindi Old Town, in Kenya (Peake, 1989), and more recently McGibbon, in her careful analysis of tourism in an Austrian Alpine village, and its impacts on the economy, kinship and political institutions, situates changes within an overall context of globalization and European Union membership (McGibbon, 2000). Similar studies, with more of an emphasis on second-home and retirement tourism, have been carried out in Mallorca (Waldren, 1996) and the Costa del Sol (O'Reilly, 2000). Destinations in less developed countries have been similarly studied, and van den Berghe's work on tourism in Peru (van den Berghe, 1980) and Mexico (Van den Berghe,

1994), for example, places tourist–resident interaction within wider historical, ethnic and national structures.

Except for Clancy's examination of the role of the state in tourism development in Mexico (Clancy, 2001), and work by Burns, who distinguishes between tourism planning based on allegedly rational economic criteria and that centred more on democratically agreed social goals (Burns, 1999, 2004), most studies have not explicitly focused on 'development', and have not been presented through an explicitly theoretical lens. Nevertheless, development issues have been utterly intrinsic to such research. The extent to which tourism is an effective 'tool' for development underlies all such case studies, and while the central focus is usually on local social, cultural and economic institutions, the level of analysis ranges from wider global political and economic systems to social interaction and role performance.

Level 1: Interactionist Perspectives and Analyses

Analyses at this level are more difficult to characterize than at system or institutional levels. Some relevant data can be found at the systems level (e.g. in MacCannell's discussion of the motivation and practices of 'the tourist' [1976]) but most are found in studies carried out by social psychologists and anthropologists. In addition, for those prepared to excavate less obvious fields, useful and stimulating insights are found in a wide range of travel writing, both modern and historical.

As indicated earlier, social psychologists have administered numerous attitude surveys among residents in tourism destination areas, and most tourism authorities and companies also carry out customer questionnaires on a regular basis. Academic surveys normally occur in established tourist destinations, but despite their ubiquity the information they provide is remarkably banal. They tend to suggest that most residents of destination areas want tourism for the economic benefits it brings, but are ambivalent about other impacts (rightly or wrongly) attributed to tourism. Indeed, after reviewing 50 such surveys, Pearce *et al.* indicate, 'few consistent patterns or relationships were uncovered' (Pearce *et al.*, 1996, p. 25). They suggest this is because definitions and measurements of tourists, tourism and 'community' differ, because researchers tend to confront respondents with set categories and responses, and because the overall exercise has not been founded in any kind of theory (Pearce *et al.*, 1996, pp. 25–27). More particularly, perhaps, it might be argued that these clipboard, whistle-stop surveys neither uncover the interactional bases for the expressed opinions, nor demonstrate how attitudes are likely to be contingent on circumstance over time and place. As a consequence, they reveal relatively little of what is happening and changing in the everyday worlds of tourist and resident in destination areas.

Attempts to attribute motivations to tourists are similarly limited. Much has been written on why people travel for leisure purposes (Pearce, 1982; Ross, 1994, pp. 14–29), but while some efforts have been made to incorporate the dynamic aspects of tourist–resident interaction (especially Pearce, 1982, pp. 68–85), the result is usually the production of yet more typologies. As in the early 1980s (Dann, 1981; Iso-Ahola, 1982), we still do not really know why people travel and how far their attitudes, and those of other tourists and residents, are altered by the interactional contexts in which all become involved. There may be 'demonstration effects', but these have been more asserted than studied, and thus our understanding of what happens, when it happens, and when it does not happen is poor (Fisher, 2004, p. 441).

In his 'reading' of O'Rourke's well-known film documentary *Cannibal Tours*, Bruner suggests that tourist–resident interaction is more likely to lead to changes in attitude and behaviour on the part of the latter rather than the former, primarily because of power differences (Bruner, 1991, p. 241). However, more recently there has been a return to concepts from earlier acculturation theory, and it has been argued that a model of cultural confusion might provide an appropriate conceptual framework against which to assess field data of different types of interaction in inter-cultural situations involving tourists and residents (Hottola, 2004).

Whether or not tourists face 'culture shock' or 'culture confusion', though, there is ample ethnographic evidence that residents of numerous tourism destination areas obtain economic benefits from tourism and cope with repeated influxes of visitors without 'losing' their own culture (Brandes, 1988, pp. 88–109; Connolly-Kirch, 1980; Krippendorf, 1987; Baumgartner, 1992; Rogers and Aitchison, 1998). As Boissevain notes:

> Host communities take specific, active measures to protect their values and customs threatened by outsiders. The means they utilize covert action, hiding from tourists, communal celebrations, fencing them out, organized protests and even overt aggression to protect their interests (Boissevain, 1996, p. 21).

Contributors to Boissevain's volume frequently indicate that interaction with tourists may also directly affect relationships that residents have with one another (Odermatt, 1996), as also shown in an ethnography of an Austrian ski resort which describes the numerous divisions and tensions that emerge when local residents (not, in themselves, a homogeneous category) are joined by 'outsiders' from other parts of Austria, from other Western societies, and by tourists (McGibbon, 2000, pp. 114–163). Similarly, ethnographic research in Goa shows that 'local' attitudes to tourism cannot be understood without reference to the self-interest of those involved and the nature of their connections with institutions and organizations outside the tourism destination area (Saldanha, 2002a). A study by the same author of rave culture in Goa, and the bodily interaction it encourages, also provides a further example of how diligent ethnography can contribute to the interactional level of analysis (Saldanha, 2002b).

Careful ethnography is also seen in the work of Edensor (1998, p. 2001), who follows Goffman in focusing on tourist performance of specific roles on tourist 'stages', which vary across place, for example, in organized 'enclaves' or more heterogeneous public 'stages' (Edensor, 2000, pp. 328–333), as well as over time. As he makes clear, performances may vary from repetition of well established parts to reflexivity and innovation when faced with situations of confusion. When norms are transcended, and the Schutzian 'cookbook menus' no longer apply, 'we are able to see how the material and sensual qualities of particular spaces interrupt the equanimity of disciplined and regulated embodied dispositions' (Edensor, 2001).

When expectations are confounded, or when none exist, possibilities of closer interaction, for good or ill, become possible. Two areas of research that do focus, at least to some extent, on interactional aspects of tourism are, first, those dealing with sex and tourism and, second, those that analyse relationships between public and private aspects of various types of home-stay tourism. Sex and tourism and, most specifically, kinds of 'sex tourism', have been studied at length, if not in depth, but accounts that avoid sensationalism and expressions of moral outrage and, instead, provide insights into such relationships, and their contingent and negotiated nature, are relatively few (Cohen, 1982a, b; Oppermann, 1998; Montgomery, 2001, p. 69–88; Bauer and McKercher, 2003).

Studies of home-stay tourism normally occupy a much lower key. However, compared to more heavily commercialized settings, they show a different side of holidaymaking and resident–tourist interaction, where the boundaries between 'public' and 'private', 'front stage' and 'back stage', continually have to be negotiated. As in Blackpool boarding houses in the 19th century (Walton, 1978, pp. 84–90), such establishments facilitate closer, more varied, more satisfying, and yet sometimes more stressful forms of social interaction involving hospitality providers, their families and their guests (Stringer, 1981; Bouquet, 1987; Pearce, 1990; Lynch, 1999; Nilsson, 2002).

Finally, insights into the interactional aspects of holidaymaking can be obtained from a plethora of historical and travel writing that is outside the category of social science. Jakle's recognition of the importance to tourists of seemingly unimportant chance encounters (Jakle, 1985, pp. 36–39) is amply demonstrated (to take examples more or less at random) in Mark Twain's account of his visit to Europe and the Holy Land (1954, first published 1869), including time spent railing against the hapless guides he encountered (Twain, 1954, first published 1869, pp. 118, 171–174), or in Steinbeck's journey (with his dog) through the USA (Steinbeck, 1962). Similarly, Theroux's accounts of journeys through the rundown seaside resorts of East Kent, in England, for example, or of his brief meeting with a lone rambler on the cliffs of Dover (Theroux, 1984, pp. 28–45), invest casual encounters with a significance rarely matched in social scientific writing.

The literature discussed in this section is indeed eclectic, but the examples demonstrate possible sources of raw material for understanding the dynamic and emergent properties that arise from social interaction. Unlike the studies discussed for levels two and three, data that facilitate an understanding of the ongoing and contingent nature of tourism phenomena do not come in tidy social-scientific, gift-wrapped packages. In fact, very little social scientific research has focused specifically on the interactional contexts prompted by tourism, and most of what has been done has focused on interaction of 'hosts' with their 'guests' – categories of increasingly doubtful analytic value (Aramberri, 2001). Tourism's influence on the way holidaymakers interact with one another, or how resident relations are changed, are rarely considered. Most surprisingly, no one (to this author's knowledge) has systematically examined the occurrence of tourism-oriented conflicts involving tourists and/or residents. Extended case studies of such instances would provide valuable comparative data that could link ongoing but contingent social interaction with changing role performance and, in turn, to institutional and structural change. Nevertheless, taken together, all this interaction provides a fascinating and sometimes electric 'mix', which has largely been ignored. Equally importantly, it provides the ongoing, negotiated basis for institutional change which, in turn, has ramifications at the social system level.

Conclusions: Further Issues in Developing a Framework for the Social Analysis of Tourism

At the start of this paper it was argued that differing evaluations of mass tourism, especially among academics, a vast increase in tourism literature, and the unresolved debate as to how far 'tourism studies' is discipline, have all contributed to a situation in which it has become increasingly difficult to incorporate the wide range of contributions to the literature of tourism into a coherent framework.

In some respects, the problem is more general. Since the breach of the Berlin Wall in 1989, and the consequent decline in 'grand narratives' in development studies generally, there has been an increased reliance on empirical studies, and the resulting analysis has been at a relatively low conceptual level. As Booth comments: 'For those engaged in local studies, disillusionment with neo-Marxist structural theories has tended to lead in *practice* to abandonment of the terrain of "political economy"' (Booth, 1994, p. 18, his emphasis). In tourism studies, the issue is perhaps even more complex, for while its much-vaunted interdisciplinarity may indeed facilitate a wide range of research, all too frequently the result is a lack of any methodological foundation. Instead, after nearly three decades, we have a mass of empirical data, from numerous and different types of research, but still lack a comprehensive picture of tourism's economic and sociocultural importance in either developed or less developed countries.[4]

There are notable exceptions (Tsartas, 1992; Wall, 1996; Simpson and Wall, 1999), but most tourism research is specific to place and time, and in the literature there are few comparative studies. Given the global economic importance of tourism, and its social and cultural importance in countless communities, lists of negative and positive characteristics are simply not enough. There are indeed established conceptual frameworks which facilitate comparative analysis, for example Doxey's (1975) index of irritation and Butler's (1980) tourist area life cycle, but, despite their relatively ancient pedigree, they have been used but little in any consistent comparative analysis (Agarwal, 2002; Butler, 2005a, Butler, 2005b; Harrison, 1995). Other concepts have also facilitated description and analysis of the cultural processes related to the levels of analysis outlined in this paper. Globalization, commoditization (or commodification), demonstration effects, inward and outward acculturation, the evolution of arts and crafts, the construction and consumption of place, history and heritage, staged authenticity, and so on, all refer to social groups and categories and the different and competing world views, beliefs and attitudes with which they can be characterized. Such concepts have been profitably applied by many social scientists who have studied tourism, but these are not paradigms, and the cultural processes they describe need to be situated in a more coherent conceptual framework.

It should again be emphasized that the aim in this paper is not to introduce a new 'paradigm' or to belittle the valuable empirical and theoretical work on tourism carried out over the last three decades. The objective is more modest: to focus explicitly on three interrelated levels of analysis – the interactional, institutional and systemic – which have long been established in social science, especially sociology, and to argue that together they provide a useful framework for coherently categorizing and comparing contributions from a wide range of literature and disciplines. Put simply: they provide a way of making sense of much of the tourism literature.

The conceptual framework proposed here is not the only possible relating to tourism, but it is of value at both the research stage and when 'reading' any finished text on tourism and society, irrespective of its source or disciplinary basis. At the primary research stage, it serves to clarify the methodological orientation and the nature of questions to be asked. Such questions, indicated in Table 16.2 arise logically from the frame of reference adopted by the writer or researcher. And when research results are presented, or theories (grand or otherwise) articulated, use of this framework enables a more conscious and closer depiction and analysis of interaction, institutions and social systems, helps clarify the level of analysis being applied, and shows how any given contribution might be integrated into the wider corpus of tourism literature. If it can facilitate knowledge of tourism in this way, it contributes to a wider, more coherent and comprehensive methodological framework for social scientific research in tourism.

As has been mooted elsewhere (Harrison, 1996, pp. 80–83), it may be appropriate to reconsider the application of systems theory to tourism

Table 16.2. Typical questions at appropriate levels of tourism analysis.

Level 3: Globalization
What is the extent and nature of the system? Its basis (e.g. capitalism, industrialism, production, consumption)? Inter-relationships of social, cultural, economic and environmental systems? What are their formal characteristics? What are the formal properties of the system (e.g. regions, nations, states, states, ethnic groups)? Are they inter-related in any specific way? Is there a concept of balance or equilibrium? Where does power lie? What processes are involved in maintaining the system and its sub-systems? Formal and informal international organizations? How is change conceptualized and measured? What is the role of tourism in the overall unit of analysis? Can it be seen as a dependent or independent variable? Are there properties, processes and organizations to tourism? How does tourism contribute to or reflect change?
Level 2: Structures and institutions
Basic demography and development description. Economic, cultural and social contexts and main actors. Are they coterminous with societies or cultures? What are the key social, cultural, political, economic and religious institutions that comprise the overall structure? Where does power reside? How is it indicated? Who owns scarce resources? Where are conflicts and tensions within the unit of analysis (country, state, region, town/resort, etc.)? Are stresses and strains apparent? Are there institutions to deal with them? Are there institutions that specifically relate to tourism, and only to tourism? How is tourism organized (accommodation, tour operators and travel agencies, transport, etc.)? How far does tourism influence other institutions? How does tourism fit into the pattern? Tourism markets, demand and supply.
Level 1: Interaction
What are the main units of analysis: town/village/settlement/resort? Who are the key social actors? How are they distinguished? Role and other conflicts? How revealed? Attitudes of different participants in the interaction? Are there social stereotypes? Who is interacting with whom, in what context, and to what effect? Where do tourists and residents meet? In what context and to what effect? How do tourists interact with other tourists? With different kinds of resident? How does tourism affect the interaction of residents with other residents? Who ARE the different tourists and residents? Age, class, gender, ethnic and other divisions? Extended case studies of tourism-related interaction.

phenomena. Indeed, reference to tourism's role in globalization virtually presupposes some kind of systems perspective. A similar point has been made by Farrell and Twining-Ward (2004), who have recently suggested a need to 'reconceptualize' tourism studies. How far such a project requires recourse to the systems theory of the mid-1990s is not the issue here, but there might be some value in considering the extent to which the work of more recent systems theorists, who see social systems as adaptive systems, can be applied to societies where tourism is, or may be, an agent and indicator of rapid social and cultural change (Wilson and Sober, 1994; Wilson, 2002).

For social scientists of 'a certain age', the prospect of a revival in systems theory may be somewhat daunting, and it is not suggested here that any such perspective should or could replace all other orientations in tourism studies, or that a single, comprehensive paradigm is either necessary or possible. Indeed, some tourism scholars may be quite content with the existing situation, with all its theoretical and disciplinary divisions. However, the position taken in this paper is that there is a need for a greater awareness of different levels of analysis, that empirical and/or theoretical work at one level can be related to contributions made at other levels, and that the conceptual framework discussed in this paper could facilitate such an understanding.

Notes

[1] In this paper, no attempt is made to link these levels of analysis to approaches in economics. However, there are clear parallels. Macroeconomic theory concentrates on the role of tourism in international economic systems and at the national level, as measured, for example, in GNP and GDP. Examples of analyses of tourism's role in national economies are Clancy, for Mexico (2001), and Poirier and Wright, for Tunisia (Poirier and Wright, 1993). In the international arena, firms may be large-scale transnational companies involved in tourism, for example airlines, international hotel chains and tour operators (Dunning and McQueen, 1982; Go, 1988; Lundberg *et al.*, 1995, pp. 49–132; Williams, 1995; Hanlon, 1996; Sinclair and Stasler, 1997), and economists have also studied smaller firms in destination areas (Rodenburg, 1980). By contrast, microeconomic theory, which also holds individual behaviour constant, deals with stereotypical behaviour of social actors operating within market settings, as consumers and suppliers of goods and services. Similarly, no reference is made here to the related 'disciplines' of management and planning. Arguably, like tourism studies, they lack methods which are specific to them alone, and might best be regarded as social action by individuals or groups to organize a desired set of future institutional and other outcomes in the face of numerous other possibilities and stakeholder interests. Such a position would conceive them less as technical procedures and more as examples of applied policy initiatives (cf. Hall *et al.*, 1997, p. 11).

[2] For an approach to tourism development specific to economic theory, see Sinclair (1998).

[3] Very little research seems to have been carried out on tourism's impacts in developed societies. Is it because we find it less easy to patronise residents in such destination areas as Blackpool or Brighton? Or because tourism is well established? And what about the problems that occur – as in East Kent, England, for instance – when people cease to visit areas once popular among tourists?

[4] This is not simply a theoretical issue. During the outbreak of foot and mouth disease in the UK in the spring of 2001, for example, the attention of the public and politicians was almost entirely on the farming community, even though, in economic terms, tourism was a far greater source of employment in rural areas (Chrisafis, 2001; Middleton, 2001; Whitehead, 2001).

References

Abercrombie, N. and Urry, J. (1983) *Capital, Labour and the Middle Classes*. Allen and Unwin, London.

Agarwal, S. (2002) Restructuring seaside tourism: the resort lifecycle. *Annals of Tourism Research* 29(1), 25–55.

Aramberri, J. (2001) The host should get lost: paradigms in tourism theory. *Annals of Tourism Research* 28(3), 738–761.
Bauer, T.G. and McKercher, B. (eds) (2003) *Sex and Tourism: Journeys of Romance, Love and Lust*. Haworth Hospitality Press, New York and London.
Baumgartner, R. (1992) Tourism and socio-economic change: the case of the Rolwaling Valley in eastern Nepal. In: Singh, T.V., Fish, M., Smith, V.L. and Richter, L.K. (eds) *Tourism Environment: Nature, Culture, Economy*. Inter-India Publications, New Delhi, pp. 76–93.
Berger, P. and Luckmann, T. (1966) *The Social Construction of Reality*. Penguin, Harmondsworth.
Boissevain, J. (1996) Introduction. In: Boissevain, J. (ed.) *Coping with Tourists: European Reactions to Mass Tourism*. Berghahn Books, Providence and Oxford, pp. 1–26.
Booth, D. (1994) Rethinking social development: an overview. In: Booth, D. (ed.) *Rethinking Social Development: Theory, Research and Practice*. Longman, Harlow, pp. 3–34.
Bottomore, T. and Nisbet, R. (1979) Structuralism. In: Bottomore, T. and Nisbet, R. (eds) *A History of Sociological Analysis*. Heinemann, London, pp. 557–598.
Bouquet, M. (1987) Bed, breakfast and evening meal: commensality in 19th and 20th century farm households in Hartland (Devon). In: Bouquet, M. and Winter, M. (eds) *Who From Their Labours Rest*. Avebury, Aldershot, pp. 93–104.
Bourdieu, P. (1994) Structures, habitus and practices. In: *The Polity Reader in Social Theory*. Polity, Cambridge, pp. 95–110.
Brandes, S. (1988) *Power and Persuasion: Fiestas and Social Control in Rural Mexico*. University of Pennsylvania Press, Philadelphia.
Britton, S.G. (1982) The political economy of tourism in the third world. *Annals of Tourism Research* 9(3), 331–358. DOI: 10.1016/0160-7383(82)90018-4.
Britton, S.G. (1989) Tourism, dependency and development: a mode of analysis. In: Singh, T.V., Theuns, H.L. and Go, F.M. (eds) *Towards Appropriate Tourism: The Case of Developing Countries*. Peter Lang, Frankfurt and Berne, pp. 93–110.
Brown, F. (1998) *Tourism Reassessed: Blight or Blessing?* Butterworth-Heinemann, Oxford.
Bruner, E.M. (1991) Transformation of self in tourism. *Annals of Tourism Research* 18(2), 238–250. DOI: 10.1016/0160-7383(91)90007-X.
Brunt, P. and Courtney, P. (1999) Host perceptions of sociocultural impacts. *Annals of Tourism Research* 26(3), 493–515. DOI: 10.1016/S0160-7383(99)00003-1.
Buckley, W. (1967) *Sociology and Modern Systems Theory*. Englewood Cliffs, Prentice-Hall, New Jersey.
Burns, P. (1999) Paradoxes in planning tourism elitism or brutalism? *Annals of Tourism Research* 26(2), 329–348. DOI: 10.1016/S0160-7383(98)00099-1.
Burns, P. (2004) Tourism planning: a third way? *Annals of Tourism Research* 31(1), 24–43.
Butler, R.W. (1980) The concept of a tourist area cycle of evolution: implications for management of resources. *The Canadian Geographer/Le Géographe canadien* 24(1), 5–12. DOI: 10.1111/j.1541-0064.1980.tb00970.x.
Butler, R. (ed.) (2005a) *The Tourism Area Life Cycle, Vol. 1: Applications and Modifications*. Channel View Publications, Clevedon.
Butler, R. (ed.) (2005b) *The Tourism Area Life Cycle, Vol. 2: Conceptual and Theoretical Issues*. Channel View Publications, Clevedon.
Chrisafis, A. (2001) Brown gets £12bn bill for tourism loss. *Guardian* 8.
Clancy, M. (2001) *Exporting Paradise: Tourism and Development in Mexico*. Pergamon Press, Amsterdam.
Cohen, E. (1971) Arab boys and tourist girls in a mixed Jewish-Arab community. *International Journal of Comparative Sociology* 12(4), 217–233. DOI: 10.1177/002071527101200401.
Cohen, E. (1972) Toward a sociology of international tourism. *Social Research* 39(1), 164–182.
Cohen, E. (1973) Nomads from affluence: notes on the phenomenon of drifter-tourism. *International Journal of Comparative Sociology* 14(1–2), 89–103. DOI: 10.1177/002071527301400107.
Cohen, E. (1974) Who is a tourist?: a conceptual clarification. *The Sociological Review* 22(4), 527–555. DOI: 10.1111/j.1467-954X.1974.tb00507.x.
Cohen, E. (1979) A phenomenology of tourist experiences. *Sociology* 13(2), 179–201. DOI: 10.1177/003803857901300203.
Cohen, E. (1982a) Marginal paradises: bungalow tourism on the islands of southern Thailand. *Annals of Tourism Research* 9(2), 189–228.
Cohen, E. (1982b) Thai girls and Farang men the edge of ambiguity. *Annals of Tourism Research* 9(3), 403–428. DOI: 10.1016/0160-7383(82)90021-4.

Cohen, E. (1982c) Jungle guides in northern Thailand – the dynamics of a marginal occupational role. *The Sociological Review* 30(2), 234–266. DOI: 10.1111/j.1467-954X.1982.tb00756.x.

Cohen, E. (1984) The sociology of tourism: approaches, issues, and findings. *Annual Review of Sociology* 10(1), 373–392. DOI: 10.1146/annurev.so.10.080184.002105.

Cohen, E. (1985) The tourist-guide: the origins, structure and dynamics of a role. *Annals of Tourism Research* 12(1), 5–29.

Cohen, E. (1993a) The study of touristic images of native people: mitigating the stereotype of a stereotype. In: Pearce, D.G. and Butler, R.W. (eds) *Tourism Research: Critiques and Challenges*. Routledge, London, pp. 36–69.

Cohen, E. (1993b) Open-ended prostitution as a skilful game of luck: opportunity, risk and security among tourist-orientated prostitutes in a Bangkok soi. In: Hitchcock, M., King, V.T. and Parnwell, M.G. (eds) *Tourism in South-East Asia*. Routledge, London, pp. 155–178.

Cohen, E. (1996) Touting tourists in Thailand: tourist-orientated crime and social structure. In: Pizam, A. and Mansfield, Y. (eds) *Tourism, Crime and International Security Issues*. Wiley, Chichester, pp. 77–90.

Connolly-Kirch, D. (1980) A comment on boundaries Betwen work andd private worlds. *Annals of tourism research* vii(4), 608–610.

Craib, I. (1984) *Modern Social Theory: From Parsons to Habermas*. Wheatsheaf, Brighton.

Craib, I. (1992) *Anthony Giddens*. Routledge, London.

Craik, J. (1995) Are there cultural limits to tourism? *Journal of Sustainable Tourism* 3(2), 87–98. DOI: 10.1080/09669589509510713.

Culbertson, K., Snyder, D., Mullen, S., Kane, B., Zeller, M. *et al.* (1996) Finding common ground in the last best place: the Flathead County, Montana master plan. In: Price, M. (ed.) *People and Tourism in Fragile Environments*. Wiley, Chichester, pp. 139–158.

Dann, G.M.S. (1981) Tourist motivation an appraisal. *Annals of Tourism Research* 8(2), 187–219. DOI: 10.1016/0160-7383(81)90082-7.

Dann, G.M.S. (1996) *The Language of Tourism*. CAB International, Wallingford.

Dawe, A. (1979) Theories of social action. In: Bottomore, T. and Nisbet, R. (eds) *A History of Sociological Analysis*. Heinemann, London, pp. 362–417.

de Kadt, E. (1979) *Tourism: Passport to Development?* Oxford University Press, Oxford.

Dogan, H.Z. (1989) Forms of adjustment: sociocultural impacts of tourism. *Annals of Tourism Research* 16(2), 216–236.

Doxey, G.V. (1975) A causation theory of visitor-resident irritants: methodology and research inferences. *The Travel Research Association, Sixth Annual Conference Proceedings, 'The Impact of Tourism', San Diego, California* 8–11(September), 195–198.

Dunning, J.H. and McQueen, M. (1982) Multinational corporations in the international hotel industry. *Annals of Tourism Research* 9(1), 69–90. DOI: 10.1016/0160-7383(82)90035-4.

Edensor, T. (1998) *Tourists at the Taj: Performance and Meaning at a Symbolic Site*. Routledge, London.

Edensor, T. (2000) Staging tourism: tourists as performers. *Annals of Tourism Research* 27(2), 322–344.

Edensor, T. (2001) Performing tourism, staging tourism: (re)producing tourist space and practice. *Tourist Studies* 1(1), 59–81. DOI: 10.1177/146879760100100104.

Engels, F. (1968) Ludwig Feurbach and the end of classical German philosophy. In: Marx, K. and Engels, F. (eds) *Selected Works*. Lawrence and Wishart, London, pp. 584–622.

Farrell, B.H. and Twining-Ward, L. (2004) Reconceptualizing tourism. *Annals of Tourism Research* 31(2), 274–295. DOI: 10.1016/j.annals.2003.12.002.

Fisher, D. (2004) The demonstration effect revisited. *Annals of Tourism Research* 31(2), 428–446. DOI: 10.1016/j.annals.2004.01.001.

Franklin, A. (2003) *Tourism: An Introduction*. Sage, London.

Germain, R.D. (2000) Introduction: globalization and its critics. In: Germain, R.D. (ed.) *Globalization and its Critics: Perspectives from Political Economy*. Macmillan, Houndmills, pp. xiii–xx.

Gerth, H. and Mills, C.W. (eds) (1948) *From Max Weber; Essays in Sociology*. Routledge, London.

Giddens, A. (1994) Elements of the theory of structuration. In: Press, Polity. (ed.) *The Polity Reader in Social Theory*. Press, Polity, Cambridge, pp. 79–88.

Gilbert, E.M. (1975) *Brighton: Old Ocean's Bauble*. Harvester, Hassocks.

Go, F. (1988) Key problems and prospects in the international hotel industry. *Travel & Tourism Analysis* 1, 27–48.

Gouldner, A.W. (1973) *For Sociology: Renewal and Critique in Sociology Today*. Allen Lane, London.

Greenwood, D. (1989) Culture by the pound: an anthropological perspective on tourism as cultural commoditization. In: Smith, V.L. (ed.) *Hosts and Guests: The Anthropology of Tourism*, 2nd edn. University of Pennsylvania Press, Philadelphia, pp. 171–185.

Gulliver, P.H. (1971) *Neighbours and Networks*. University of California Press, Berkeley, Los Angeles and London.

Hall, D. (1991) *Tourism and Economic Development in Eastern Europe and the Soviet Union*. Belhaven, London.

Hall, C.M. (1994) *Tourism and Politics: Policy, Power and Place*. Wiley, Chichester and New York.

Hall, M.C. (2005) Systems of surveillance and control: commentary on 'An analysis of institutional contributors to three major academic tourism journals: 1992–2001'. *Tourism Management* 26(5), 653–656. DOI: 10.1016/j.tourman.2004.05.002.

Hall, C., Jenkins, J. and Kearsley, G. (1997) Preface. In: Hall, C.M., Jenkins, J. and Kearsley, G. (eds) *Planning and Policy in Australia and New Zealand*. 12. Irwin, Sydney, pp. vii–0. DOI: 10.1111/j.1527-2001.1997.tb00293.x.

Hanlon, P. (1996) *Global Airlines: Competition in a Transnational Industry*. Butterworth-Heinemann, Oxford.

Harrison, D. (1992) International tourism and the less developed countries: the social consequences. In: *Tourism and the Less Developed Countries*. Belhaven Press, Chichester, pp. 19–34.

Harrison, D. (1995) Development of tourism in Swaziland. *Annals of Tourism Research* 22(1), 135–156. DOI: 10.1016/0160-7383(94)00051-S.

Harrison, D. (1996) Sustainability and tourism: reflections from a muddy pool. In: Briguglio, L., Archer, B., Jafari, J. and Wall, G. (eds) *Sustainable Tourism in Islands and Small States*. Pinter, London, pp. 69–89.

Harrison, D. (2001a) Less developed countries and tourism: the overall pattern. In: Harrison, D. (ed.) *Tourism and the Less Developed World: Issues and Case Studies*. CAB International, Wallingford, pp. 1–22.

Harrison, D. (2001b) Afterword. In: Harrison, D. (ed.) *Tourism and the Less Developed World: Issues and Case Studies*. CAB International, Wallingford, pp. 251–263.

Harrison, D. and Price, M. (1996) Fragile environments, fragile communities? An introduction. In: Price, M. (ed.) *People and Tourism in Fragile Environments*. Wiley, Chichester, pp. 1–18.

Hay, C. and Marsh, D. (2001) Introduction: Demystifying globalization. In: Hay, C. and Marsh, D. (eds) *demystifying Globalization*. Palgrave, Houndmills, pp. 1–17.

Hitchcock, M. (1999) Tourism and ethnicity: situational perspectives. *International Journal of Tourism Research* 1(1), 17–32. DOI: 10.1002/(SICI)1522-1970(199901/02)1:1<17::AID-JTR145>3.0.CO;2-L.

Hottola, P. (2004) Culture confusion: intercultural adaptation in tourism. *Annals of Tourism Research* 31(2), 447–466.

Iso-Ahola, S.E. (1982) Toward a social psychological theory of tourism motivation: a rejoinder. *Annals of Tourism Research* 9(2), 256–262. DOI: 10.1016/0160-7383(82)90049-4.

Jakle, J.A. (1985) *The Tourist: Travel in Twentieth-Century America*. University of Nebraska Press, Lincoln and London.

Jordan, J.W. (1980) The summer people and the natives some effects of tourism in a Vermont vacation village. *Annals of Tourism Research* 7(1), 34–55. DOI: 10.1016/S0160-7383(80)80005-3.

Kent, N. (1975) A new kind of sugar. In: Finney, B.R. and Watson, K.R. (eds) *A New Kind of Sugar: Tourism in the Pacific*. East-West Center, Honolulu, pp. 169–180.

Kousis, M. (1989) Tourism and the family in a rural Cretan community. *Annals of Tourism Research* 16(3), 318–332. DOI: 10.1016/0160-7383(89)90047-9.

Krippendorf, J. (1987) *The Holiday Makers: Understanding the Impact of Leisure and Travel*. Heinemann, London.

Kuhn, T.S. (1970a) *The Structure of Scientific Revolutions*, 2nd edn. University of Chicago Press, Chicago.

Kuhn, T.S. (1970b) Reflections on my critics. In: Lakatos, I. and Musgrave, A. (eds) *Criticism and the Growth of Knowledge*. Cambridge University Press, Cambridge, pp. 231–278.

Lash, S. and Urry, J. (1987) *The End of Organized Capitalism*. Polity Press, Cambridge.

Leiper, N. (2000) An emerging discipline. *Annals of Tourism Research* 27(3), 805–809. DOI: 10.1016/S0160-7383(99)00118-8.

Lundberg, D.E., Krishnamoorthy, M. and Stavenga, M.H. (1995) *Tourism Economics*. Wiley, New York and Chichester.

Lynch, P.A. (1999) Host attitudes towards guests in the homestay sector. *Tourism and Hospitality Research* 1(2), 119–144. DOI: 10.1177/146735849900100203.

MacCannell, D. (1976) *The Tourist: A New Theory of the Leisure Class*. Macmillan, London and Basingstoke.
MacCannell, D. (ed.) (1989) *The Tourist: A New Theory of the Leisure Class*. Macmillan, London and Basingstoke.
Marx, K. (1977) The German ideology. In: McLellan, D. (ed.) *Karl Marx: Selected Writings*. Oxford University Press, Oxford, pp. 157–191.
McGibbon, J. (2000) *The Business of Alpine Tourism in a Globalising World*. Vetterling Druck, Rosenheim.
Merton, R.K. (1968) *Social Theory and Social Structure*, 2nd edn. Free Press, New York.
Meyer-Arendt, K.J. and Justice, C. (2002) Tourism as the subject of North American doctoral dissertations, 1987–2000. *Annals of Tourism Research* 29(4), 1171–1174. DOI: 10.1016/S0160-7383(02)00038-5.
Middleton, V.T.C. (2001) Foot and mouth disease: carry on killing the countryside. *Tourism 109* 10.
Mitchell, J.C. (1969) *Social Networks in Urban Situations*. Manchester University Press, Manchester.
Montgomery, H. (2001) *Modern Babylon? Prostituting Children in Thailand*. Berghahn Books, New York and Oxford.
Morris, M.B. (1977) *An Excursion into Creative Sociology*. Basil Blackwell, Oxford.
Mowforth, M. and Munt, I. (2003) *Tourism and Sustainability: Development and New Tourism in the Third World*, 2nd edn. Routledge, London.
Munt, I. (1994) The 'Other' Postmodern Tourism: Culture, Travel and the New Middle Classes. *Theory, Culture & Society* 11(3), 101–123. DOI: 10.1177/026327694011003005.
Nash, D. (1996) *Anthropology of Tourism*. Pergamon, Oxford.
Nash, D. (2001) On travelers, ethnographers and tourists. *Annals of Tourism Research* 28(2), 493–496. DOI: 10.1016/S0160-7383(00)00053-0.
Nilsson, P.A. (2002) Staying on farms: an ideological background. *Annals of Tourism Research* 29(1), 7–24.
Odermatt, P. (1996) A case study of neglect: the politics of (re)presentation: a Sardinian case. In: Boissevain, J. (ed.) *Coping with Tourists: European Reactions to Mass Tourism*. Berghahn Books, Providence and Oxford, pp. 84–111.
Oppermann, P. (1998) *Sex Tourism and Prostitution: Aspects of Leisure, Recreation and Work*. Cognizant, New York.
Ousby, I. (1990) *The Englishman's England: Taste, Travel and the Rise of Tourism*. Cambridge University Press, Cambridge.
O'Reilly, K. (2000) *The British on the Costa del Sol*. Routledge, London.
Parsons, T. (1957) Malinowski and the theory of social systems. In: Firth, R. (ed.) *Man and Culture: An Evaluation of the Work of Bronislaw Malinowski*. Routledge and Kegan Paul, London, pp. 53–70.
Parsons, T. and Shils, E. (1951) Values, motives, and systems of action. In: Parsons, T. and Shils, E. (eds) *Toward a General Theory of Action: Theoretical Foundations for the Social Sciences*. Harvard University Press, Cambridge, MA, pp. 47–109.
Pawson, R. (1999) Methodology. In: Taylor, S. (ed.) *Sociology: Issues and Debates*. Palgrave, Houndmills, pp. 19–49.
Peake, R. (1989) Tourism and Swahili: identity in Malindi old town, Kenyan coast. *Africa* 59(2), 209–220.
Pearce, P.L. (1982) *The Social Psychology of Tourist Behaviour*. Pergamon Press, Oxford.
Pearce, P.L. (1990) Farm tourism in New Zealand: a social situation analysis. *Annals of Tourism Research* 17(3), 337–352.
Pearce, P.L. and Moscardo, G. (1999) Tourism community analysis: asking the right questions. In: Pearce, D.G. and Butler, R.W. (eds) *Contemporary Issues in Tourism Development*. Routledge, London, pp. 31–51.
Pearce, P.L., Moscardo, G. and Ross, G.F. (1996) *Tourism Community Relationships*. Pergamon, Oxford.
Pi-Sunyer, O. (1989) Changing perceptions of tourism and tourists in a Catalan resort town. In: Smith, V.L. (ed.) *Hosts and Guests: The Anthropology of Tourism*, 2nd edn. University of Pennsylvania Press, Philadelphia, pp. 187–199.
Poirier, R.A. and Wright, S. (1993) The political economy of tourism in Tunisia. *The Journal of Modern African Studies* 31(1), 149–162. DOI: 10.1017/S0022278X0001185X.
Prentice, R. (1993) Community-driven tourism planning and residents' preferences. *Tourism Management* 14(3), 218–227. DOI: 10.1016/0261-5177(93)90023-E.
Puppim de Oliveira, J.A. (2003) Governmental responses to tourism development: three Brazilian case studies. *Tourism Management* 24(1), 97–110. DOI: 10.1016/S0261-5177(02)00046-8.
Radcliffe-Brown, A.R. (1952) *Structure and Function in Primitive Society*. Cohen and West, London.
Richards, A. (1957) The concept of culture in Malinowski's work. In: Firth, R. (ed.) *Man and Culture: An Evaluation of the Work of Bronislaw Malinowski*. Routledge and Kegan Paul, London, pp. 15–32.

Rodenburg, E.E. (1980) The effect of scale in economic development: tourism in Bali. *Annals of Tourism Research* 7(2), 177–196.
Rogers, M.F. (1983) *Sociology, Ethnomethodology, and Experience: A Phenomenological Critique*. Cambridge University Press, Cambridge.
Rogers, P. and Aitchison, J. (1998) *Towards Sustainable Tourism in the Everest Region of Nepal*. World Conservation Union and International Centre for Protected Landscapes, Nepal.
Ross, G.F. (1994) *The Psychology of Tourism*. Hospitality Press, Melbourne.
Saldanha, A. (2002a) Identity, spatiality and post-colonial resistance: geographies of the tourism critique in Goa. *Current Issues in Tourism* 5(2), 94–111. DOI: 10.1080/13683500208667910.
Saldanha, A. (2002b) Music tourism and factions of bodies in Goa. *Tourist Studies* 2(1), 43–62. DOI: 10.1177/1468797602002001096.
Schudson (1979) Review essay: on tourism and modern culture. *The American Journal of Sociology* 84(5), 1249.
Scott, A. (1997) Introduction: globalization – social process or political rhetoric? In: Scott, A. (ed.) *The Limits of Globalization: Cases and Arguments*. Routledge, London, pp. 1–22.
Selwyn, T. (1993) Peter pan in south-east Asia: views from the brochures. In: Hitchcock, M., King, V.T. and Parnwell, M.J.G. (eds) *Tourism in South-East Asia*. Routledge, London, pp. 117–137.
Selwyn, T. (ed.) (1996) *The Tourist Image: Myths and Myth Making in Tourism*. Wiley, Chichester.
Seymour-Smith, C. (1986) *Macmillan Dictionary of Anthropology*. Macmillan Press, London.
Sharpley, R. and Telfer, D.J. (eds) (2002) *Tourism and Development: Concepts and Issues*. Channel View Publications, Clevedon.
Simpson, P. and Wall, G. (1999) Consequences of resort development. A comparative study. *Tourism Management* 20(3), 283–296. DOI: 10.1016/S0261-5177(99)00025-4.
Sinclair, M.T. (1998) Tourism and economic development: a survey. *Journal of Development Studies* 34(5), 1–51. DOI: 10.1080/00220389808422535.
Sinclair, M. and Stasler, M. (1997) *The Economics of Tourism*. Routledge, London.
Skinner, Q. (1985) Introduction: the return of grand theory. In: Skinner, Q. (ed.) *The Return of Grand Theory in the Human Sciences*. Cambridge University Press, Cambridge, pp. 3–20.
Sklair, L. (1999) Globalization. In: Taylor, S. (ed.) *Sociology: Issues and Debates*. Palgrave, Houndmills, pp. 321–345.
Smart, B. (1993) *Postmodernity*. Routledge, London.
Smith, V.L. (ed.) (1978a) *Hosts and Guests: The Anthropology of Tourism*, 1st edn. Blackwell, Oxford.
Smith, V. (1978b) Introduction. In: Smith, V.L. (ed.) *Hosts and Guests: The Anthropology of International Tourism*. Oxford, Blackwell, pp. 1–14.
Smith, V.L. (2001) Tourism change and impacts. In: Smith, V.L. and Brent, M. (eds) *Hosts and Guests Revisited: Tourism Issues of the 21st Century*. Cognizant, New York, pp. 107–121.
Steinbeck, J. (1962) *Travels with Charley: In Search of America*. Odhams, London.
Stonich, S.C. (2000) *The Other Side of Paradise: Tourism, Conservation, and Development in the Bay Islands*. Cognizant, New York.
Stringer, P.F. (1981) Hosts and guests the bed-and-breakfast phenomenon. *Annals of Tourism Research* 8(3), 357–376. DOI: 10.1016/0160-7383(81)90004-9.
Swingewood, A. (1999) Sociological theory. In: Taylor, S. (ed.) *Sociology: Issues and Debates*. Palgrave, Houndmills, pp. 50–72.
Telfer, D.J. (2002) Conclusion: tourism and development. In: Sharpley, R. and Telfer, D.J. (eds) *Tourism and Development: Concepts and Issues*. Channel View Publications, Clevedon, pp. 338–348.
Theroux, P. (1984) *The Kingdom by the Sea*. Penguin, London.
Tosun, C. (2000) Limits to community participation in the tourism development process in developing countries. *Tourism Management* 21(6), 613–633. DOI: 10.1016/S0261-5177(00)00009-1.
Tribe, J. (1997) The indiscipline of tourism. *Annals of Tourism Research* 24(3), 638–657. DOI: 10.1016/S0160-7383(97)00020-0.
Tribe, J. (2000) Indisciplined and unsubstantiated. *Annals of Tourism Research* 27(3), 809–813. DOI: 10.1016/S0160-7383(99)00122-X.
Tsartas, P. (1992) Socioeconomic impacts of tourism on two Greek Isles. *Annals of Tourism Research* 19(3), 516–533. DOI: 10.1016/0160-7383(92)90134-B.
Turner, L. and Ash, J. (1975) *The Golden Hordes: International Tourism and the Pleasure Periphery*. Constable, London.
Twain, M. (1954, first published 1869) *The Innocents Abroad*. Collins, London and Glasgow.

Urry, J. (1988) Cultural change and contemporary holiday-making. *Theory, Culture & Society* 5(1), 35–55. DOI: 10.1177/026327688005001003.
Urry, J. (1990) *The Tourist Gaze: Leisure and Travel in Contemporary Societies*. Sage, London.
Urry, J. (1995) *Consuming Places*. Routledge, London.
van den Berghe, P.L. (1980) Tourism as ethnic relations: a case study of Cuzco, Peru. *Ethnic and Racial Studies* 3(4), 375–392. DOI: 10.1080/01419870.1980.9993312.
Van den Berghe, P.L. (1994) *The Quest for the Other: Ethnic Tourism in San Cristobal, Mexico*. University of Washington Press, Seattle and London.
Waldren, J. (1996) *Insiders and Outsiders: Paradise and Reality in Mallorca*. Berghahn Books, Providence and Oxford.
Wall, G. (1996) Perspectives on tourism in selected Balinese villages. *Annals of Tourism Research* 23(1), 123–137. DOI: 10.1016/0160-7383(95)00056-9.
Wall, G. (1993) Towards a tourism typology. In: Nelson, J.G., Butler, R. and Wall, G. (eds) *Tourism and Sustainable Development: Monitoring, Planning, Managing*. University of Waterloo, Waterloo, pp. 45–58.
Walton, J. (1978) *The Blackpool Landlady: A Social History*. Manchester University Press, Manchester.
Waters, M. (1995) *Globalization*. Routledge, London.
Weber, M. (1949) *The Methodology of the Social Sciences*. Free Press, New York.
Weber, M. (1964) *The Theory of Social and Economic Organization, Translated and edited by E. Shils and H.A. Finch*. Free Press, New York.
Whitehead, T. (2001) 2001 a tourism odyssey. *Tourism* 111(Winter), 7–8.
Williams, A.M. and Shaw, G. (1991) *Tourism and Economic Development: Western European Experiences*, 2nd edn. Wiley, Chichester.
Williams, A.M. (1995) Capital and the transnationalism of tourism. In: Montanari, A. and Williams, A.M. (eds) *European Tourism: Regions, Spaces & Structures*. Wiley, Chichester, pp. 163–176.
Wilson, D.S. (2002) Reviving functionalism in the human social sciences. Talk given to the 2002–2003 Marschak Colloquium at University College Los Angeles, 4 October. Available at: www.anderson.ucla.edu/research/marschak/2002-2003/04oct02.htm.
Wilson, D.S. and Sober, E. (1994) Reintroducing group selection to the human behavioral sciences. *Behavioral and Brain Sciences* 17(4), 585–608. DOI: 10.1017/S0140525X00036104.
Wolfers, M. (1974) *Black Man's Burden Revisited*. Allen and Busby, London.
Wolff, K.H. (1978) Phenomenology and sociology. In: Bottomore, T. and Nisbet, R. (eds) *A History of Sociological Analysis*. Heinemann, London, pp. 499–556.
Yiannakis, A. and Gibson, H. (1992) Roles tourists play. *Annals of Tourism Research* 19(2), 287–303. DOI: 10.1016/0160-7383(92)90082-Z.
Young, G. (1973) *Tourism: Blessing or Blight?* Penguin, Harmondsworth.

17

Tourism and Development: From Development Theory to Globalization

The Emergence of Development Theory

Neither tourism nor the study of development is especially new. Mass tourism goes back at least to the mid-19th century and the introduction of cheap railway travel, though much earlier if one considers pilgrimage a variant of tourism (Aramberri, 2010, pp. 2–3), while concern over the trajectory of 'modern' societies can be traced back to such classical political economists as Adam Smith, Marx, Ricardo, Mill and Weber, who were greatly exercised by the causes and directions of social 'evolution' (Harrison, 1988, pp. 1–8; Payne and Phillips, 2010, pp. 10–32). That said, it was not really until after the Second World War, with the emergence of numerous 'new nations', that the specific concept of 'development' came to be seen as intensely important by social theorists and politicians, especially those involved in the continuous confrontations of the Cold War. From then on, at least until the end of the 1990s, social scientific theory became intertwined, sometimes murkily, with national and international development policies, and theoretical and ideological debates were often indistinguishable.

In 1977, after some reflection, Seers concluded that development basically involved economic growth, accompanied by reduced inequality, unemployment and poverty, achievable only through a radical distribution of power and increased economic and cultural self-reliance, possibly led by nationalistic elites (Seers, 1977). None of this was nation-specific; instead, it would involve 'working on common worldwide problems, by one and all' and would make it 'almost meaningless' to restrict the concept of development only to the 'Third World' (Seers, 1977, p. 6). The development project was for everyone.

Arguably, nearly four decades later, this definition of development, or at least its aims, has not been bettered, despite numerous theoretical perspectives applied to what are variously known as developing or less developed countries or even (still) the Third World. Details may vary, but the broad outline of how perspectives have changed is clear and many have summarized them (Harrison, 1988, 1992a, pp. 10–11; Isbister, 2001; Telfer, 2002; Willis, 2005; Mowforth and Munt, 2009, pp. 31–46; Payne and Phillips, 2010). According to Sharpley, whose outline can be considered representative (Table 17.1), since the 1950s Modernisation and Dependency Theory, Neo-liberalism and Alternative/Sustainable Development have offered various routes to development but over the last decade, there has been something of an

Table 17.1. Development theory from the 1950s.

Timeline	*Development process*	*Key concepts and strategies*
1950s–1960s	Modernization theory	*Dominance of Western economic growth based models:* • Stages of growth • Structural theories • Diffusion: growth poles and trickle down • State intervention: regulation/protectionism
1960s–1970s	Modernization theory/ dependency theory	*Underdevelopment the result of domination/ exploitation by developed countries* • Economic restructuring: import substitution, protectionism; development of domestic markets • Limits to growth: neo-Malthusian theories in response to environmental concerns
1970s–1980s	Neo-liberalism	*Promotion of the free market* • Limits on government intervention in economic activity • Deregulation/privatization • Structural adjustment programmes • New economic order; one world
1980s	Neo-liberalism/alternative development	*Awareness of effects of development on different cultures/ societies* • Grassroots/people-centred development • Basic needs: food, housing, education, health • Local context/indigenous knowledge • Environmental sustainability
1990s	Alternative/ sustainable development	*Dominance of sustainable development paradigm, but emergence of post-development school* • Grassroots/people-centred development • Environmental management • Engagement with globalization • The development 'impasse'
2000s	Beyond the impasse: a new paradigm?	*Post-development: rejection of over-arching development concepts* • Global environmental policies/protocols • Transnational movements • Micro-level strategies • Poverty reduction • State security and development

Source: Sharpley, 2009, p. 39.

'impasse', characterized by considerable disillusion with all major theories of development (Sharpley, 2009, p. 39).

This depiction is largely valid but some caveats should be noted. First, not everyone was impressed with existing paradigms. Writing as early as the mid-1970s, and comparing Brazil and China – '[then] the biggest capitalist and socialist experiments in the Third World,' – Berger was to note that the 'theories are delusional and the sacrifices are indefensible. Rejection of both the Brazilian and the Chinese models is the starting point for any morally acceptable development policy' (Berger, 1976, pp. 13–14).

Second, perspectives vary on when the 'impasse' in development thinking occurred. During the 1960s there was growing disillusionment with Grand Theory and this, along with virulent neo-Marxist critiques, led to reduced support for modernization theory (though development practitioners continue to follow many of its tenets). However, neo-Marxists were also in disarray, though their denouement has been dated as early as the 1970s (Payne and Phillips, 2010, p. 85) and as late as the mid-1980s (Booth, 1985; Kitching, 1985, pp. 144–5).

More generally, throughout the 1980s there was a sense of development issues not being addressed. Even the Brandt Report (Report of the Independent Commission on International Development Issues, 1980), with its proposal of a major transfer of resources to developing countries, and the Brundtland Commission (World Commission on Environment and Development, 1987), which placed environmental sustainability at the heart of the international development agenda, were unable to dispel the sense of development failure.

Third, as Sharpley recognises, the paradigms he describes have long coexisted. Dependency theory actually emerged in the 1950s, at the same time as modernization theory, and when popularized by Frank (1969) it was based on 'a current already old in Latin American thought' (Cardoso, 1977, p. 9). And neo-liberalism, with its insistence on the role of the market, limited government and democracy, had strong links with modernization theory even before the 1970s and remained a powerful force in the 1990s, when it continued to oppose neo-statism (Payne and Phillips, 2010, pp. 98–115).

By the late 1980s it was realised that none of the competing perspectives were entirely satisfactory. The almost exclusive focus by modernization theory and dependency theory on internal structures and external linkages, respectively, was clearly inadequate to describe, let alone explain, global inequalities. Furthermore, the sustainability perspective – despite international conferences in 1992 (Rio de Janeiro), 2002 (Johannesburg) and 2012 (again, in Rio de Janeiro), and increasing evidence of global warming – came to be criticized as vague, idealistic, contradictory, too general, and directed more at the symptoms of the problem, rather than addressing the 'real' causes of the mismatch of non-renewable resources and consumption patterns (Adams, 1990, p. 10; Sharpley, 2009, pp. 64–67).

Indeed, 'alternative development' in general (Payne and Phillips, 2010, pp. 118–144), with its focus on bottom-up, local initiatives, and its many worthy aims (such as basic needs and the Millennium Development Goals) is neither a theory nor a paradigm.

However ideologically attractive, it remains a catch-all category that includes everything but conventional development – an issue that reappears when the focus is on alternative tourism.

In development theory, then, 'paradigms' or (better) perspectives continue but none is dominant and there is a general mistrust of grand narratives. Instead, there has been renewed emphasis on empirical research and a decreased tendency to make grandiose statements about the state of the world. We have not (yet) experienced the death of theory, but its advocates are now more modest and their claims reduced.

In fact, many of the features found in the competing perspectives have been incorporated into what can broadly be described as globalization theory (Robertson, 1992; Waters, 1995; Hirst and Thompson, 1999; Hay and Marsh, 2000), which can encompass the global and the local – hence glocalization (Robertson, 2012). Along with the social, economic, cultural, and environmental, globalization enables theorists of various persuasions to operate in a more sober, ideology-free and certainly less exciting environment than that of the heady 1970s and 1980s.

And development studies? According to Payne and Phillips, 'contemporary development studies, although still a sizeable academic enterprise in many Western countries, is at the same time an uncertain and under-confident discipline' (2010, p. 3). Their response to this impasse is to 'relocate the study of development squarely within the intellectual project of political economy and the diverse theoretical traditions associated with it' (2010, p. 181). We have come full circle with a return to the classical political economists and Seers (1977): development is the concern of everyone.

Tourism Through the Lens of Development Theory

The rapid growth of tourism during the post Second World War period coincided with the emergence of development theory (Telfer, 2002, pp. 50–51). Not surprisingly, tourism soon came to be considered by many countries and development agencies (including the World Bank) as a valuable tool for development. Economists, especially, welcomed this new phenomenon as an industry 'without chimneys'. The unquestioned welcome did not last, however, and over the years perspectives vacillated from the initial 'advocacy' of tourism to caution, adaptation and then a more objective 'knowledge-based' approach (Jafari, 1989, pp. 19–25). Even then, such platforms co-existed, much as they do now, and many academic and non-academic remain highly critical of international tourism's role in development.

Two publications, in particular, were instrumental in setting the pattern for future studies of tourism and development. The first was Valene Smith's first edited collection (Smith, 1977) of case studies of tourism in both developed and developing societies, which was to be of lasting influence for anthropologists and others with an interest in tourism's impacts in less developed regions (Scott and Selwyn, 2010). The second, especially germane to this paper, was a similar collection examining the role of tourism in developing societies. Edited by Emanuel de Kadt (1979), a colleague of Seers at the Institute of Development Studies at Sussex in the UK, it focused on the problematic role of tourism as a possible 'passport to development', an issue as valid now as it was then. His emphases on national policies and planning, types of tourism, the social, economic and cultural impacts of tourism, and the nature of the tourism encounter (de Kadt, 1979, pp. 3–33) have been continued by a generation of scholars who have looked at the global, national and local implications of international tourism (Harrison, 1992b, pp. 19–34; 2007, p. 70).

Much has changed since these early contributions. Development economists and sociologists have been joined by a range of academics from other disciplines, including marketing and management, all of whom might consider themselves scholars 'of tourism development' (Harrison, 2010, p. 41). This diversity is reflected in the range and volume of publications. The third edition of Mowforth and Munt's popular text on tourism and sustainability (Mowforth and Munt, 2009), for instance, lists nearly 700 references, while a recent text on tourism's role in poverty reduction listed more than 200, most of which were from non-academic sources (Mitchell and Ashley, 2010)!

Despite the volume of publications on tourism and development, however, some generalisations can be made. First, while development theory is undoubtedly relevant to tourism (Harrison, 2001a, pp. 7–8; Mowforth and Munt, 2009, pp. 52–60), their relationship is often implicit. In this respect, Telfer's work (2002) is exceptional. After reviewing the literature, he suggests that the four 'paradigms' of modernization, dependency, neo-liberalism and alternative tourism can be related to tourism under two broad categories: first, the scale and control of development and, secondly, linkages to the local community and the environment (Telfer, 2002, pp. 52–78). However, as he recognises, this involves considerable generalisation; for instance, 'it is not correct to assume automatically that small-scale alternative tourism is sustainable and large-scale tourism is not' (p. 62). In fact, many advocates of tourism, especially those who emphasize its economic benefits, are unaware of the modernization perspective that underpins their work and the countless development plans that remain the standard approach to tourism as a development tool.

Among academics, too, a modernization approach is often implicit. MacCannell's depiction of the tourist as 'modern man' (MacCannell, 1976), debates over tourism's impacts on indigenous arts and crafts, authenticity, tradition and social structures in tourist destinations, tourism entrepreneurship, and tourism as an agent of commoditisation and social change generally, can all be subsumed under the modernization umbrella (Harrison, 2001a, pp. 6–7). One need not believe that 'the modern' is good to recognize that modernization is occurring. Among the few more explicit applications of modernisation theory to tourism, Macnaught's early defence of tourism (and Pacific cultures) in Pacific islands is noteworthy (Macnaught, 1982), while studies of tourism in China have focused on processes of modernization (Oakes, 1998; Sofield and Li, 1998). More recently, Andriotis (2003) and Sharpley (2003) have analysed tourism as a mode of modernization in the islands of Crete and Cyprus, respectively, while Aramberri has taken the bold step of defending modern mass tourism as a welcome feature of modernity (2010).

Studies of tourism from the perspective of underdevelopment, or dependency, have been more numerous and emphasize the alleged downsides of tourism, especially leakages from income received from tourism (Wood, 1993; Telfer, 2002, pp. 53–56). A much quoted example is that of Britton (1982) who, like MacNaught, focused on Pacific islands, especially Fiji, and argued that their peripheral economic status vis-à-vis the metropolitan countries from which tourists originated led to a situation of dependency, where metropolitan countries both defined the tourist experience and dominated the island economy through transnational hotels. Most other examples of tourism development utilising an underdevelopment perspective were also published some time ago, for example, Oglethorpe (1984) and Palmer on the Bahamas (1994). A more recent contribution examines leakages in tourism expenditure from Thai villages, but recognises the need also to examine regional and intra-village structures (Geoffrey Lacher and Nepal, 2010, pp. 965–966).

As Bianchi has noted, a cluster of global institutions, such as the WTO, WTTC, World Trade Organization, EU and NAFTA, have provided a powerful institutional impetus to the expansion of neo-liberal modes of tourism development in different regional contexts' (Bianchi, 2002, p. 297). His concerns have been echoed by others (Telfer, 2002, pp. 56–58; Mowforth and Munt, 2009, pp. 308–310). A particularly good example is from Southeast Asia, where between 2003 and 2011 alone the Asian Development Bank 'provided US 58.7 million in loan and grant assistance to the GMS [Greater Mekong Subregion] tourism industry' (Asian Development Bank, 2012, p. 11), much of which went to improved road and river facilities. Indeed, the key priorities of the ADB's regional cooperation strategy and programme for the GMS have a strong flavour of neo-liberalism, and include making cross-border travel easier, integrating national markets, and developing the private sector (Asian Development Bank, 2012, p. 14).

Finally, it is necessary to address the tourism equivalent of alternative development (Sharpley, 2009, p. 39). Like its parent concept, the notion of 'alternative tourism' is vague. It can refer to almost anything other than mass tourism, including 'nature based', 'special interest', 'niche', 'small scale', backpacking', 'new' and 'pro-poor' tourism, none of which is necessarily sustainable. In fact, the only two features they have in common is that they are not mass tourism and will never replace it (Aramberri, 2010, pp. 311–352). Ecotourism? Definitions vary, but few businesses could meet the purist criteria listed by Honey (1999) that includes minimal impact, conservation and support for human rights and democratic movements. In fact, evidence suggests ecotourism and community-based projects do little for conservation (Goodwin, 2006, p. 7) and that their environmental and social benefits are frequently overestimated (Butler, 1999, pp. 12–13). For Telfer (2002, pp. 68–78), the preferred version is 'appropriate' or 'sustainable' tourism, which is characterized by a mix of large- and small-scale enterprises. However, it is unclear for whom or what it is appropriate and this version of alternative tourism seems to be an uneasy compromise between mass and small-scale tourism.

'Sustainable tourism' is often put forward as another variant of alternative tourism and – even more than sustainable development – is a problematic concept (Butler, 1999). Sustaining the physical environment is difficult enough, as 'nature' is always subject to trade-offs for human comfort and it is even harder to apply to social structures and cultures. Socio-cultural phenomena are changeable and adaptive, and identifying when they cease to be sustainable is virtually impossible (Harrison, 1996). As defined by Butler (1999, p. 12), sustainable tourism development that is viable without degrading the human or physical environment or prohibiting successful development elsewhere, is indeed a worthy aim. However, even its advocates are pessimistic. As Mowforth and Munt note, large and small tour operators 'will increasingly deploy links with conservation, ecology and matters ethical, to their own ends' (2009, p. 376). For them, at least, the future of international tourism, in general, is likely to be 'more of the same' (2009, p. 377).

Theoretical and Methodological Reflections: Issues and Challenges

There is every indication that tourism will continue to be promoted as a 'passport' to development in both developed and developing countries for the foreseeable future, and that both domestic and international tourism will continue to grow. Current forecasts are that international arrivals alone will reach 1.8 billion by 2030 (United Nations World Tourism Organisation, 2013, p. 14) and, if current proportions remain the same, domestic tourist arrivals, currently 4 billion, are likely to reach more than 8.5 billion over the same period (Pierret, 2011). As with development generally, the focus is moving East. China's impact, as a destination and a source of tourists, is only beginning, the potential of India is similarly enormous (Aramberri, 2010, pp. 313–314), and trends all point to continued growth in other regions, especially South-east Asia.

Even more than now, holidaymaking will be the norm and few regions will be unaffected: niche markets and 'alternative' tourism (including space tourism) will multiply, but mass tourism to 'natural' areas and to urban 'heritage' sites will undoubtedly continue to constitute the bulk of international arrivals. At the same time, uncertainties will remain. There will be more global and regional financial crises, along with religious fundamentalism, political unrest and related terrorism and violence, all of which will have local impacts on the global movement of tourists.

Under such circumstances, simply charting the movement of tourists is a key role for anyone with a stake in studies of travel and tourism. Beyond this, though, several major issues emerge from this brief analysis of the relationship of tourism to development. First, there is a continued need for theoretically informed perspectives on tourism's relationship to development. In some respects, this goes against the trend, inherited from development debates, to mistrust theory and marginalise theorists, a trend reinforced by the incorporation of such disciplines as marketing and management into academic units focusing on tourism (even to the extent of having 'management' in their titles or, perhaps, operating within business schools, rather than traditional academic disciplines). The theories should not be as ideologically-driven as the development perspectives of the last four decades, that are no longer adequate, but they do need to recognize global realities. For example, with the implosion of the 'second world' in 1989, the 'third world' and 'emerging economies' became part of one world.

In like fashion, international tourism should be seen in a global context. Questions raised in the established 'paradigms' remain but now need to be more globally situated and to be supplemented by approaches from disciplines other than development economics. Studies of tourism's impacts in small states and islands, for example, remain relevant, but so too are those relating to tourism in other developing and in developed societies. What goes on in Greece, Cyprus or Italy, Egypt and Tunisia, or the UK and France, is often inter-related and the processes are comparable. This is where globalization theory is appropriate. It is already being applied to tourism (Cartier and Lew, 2005; Hjalager, 2007; Cooper, 2008; Telfer and Sharpley, 2008, pp. 58–79, pp. 58–79; Theuns, 2008; Wood, 2008; Aramberri, 2010, pp. 56–76) and, it is argued here, is an appropriate (and suitably broad) basis for future studies of tourism and development.

Second, when tourism is conceived as a comprehensive system, with globalization as its major feature (Fig. 17.1), it is not the province of any one academic discipline or subject to any one methodology (defined here as the underlying logic or philosophy behind the perspective). Rather, the cross-fertilisation of tourism and development is reflected in the systemic linkages of societies that supply and receive tourists, among the various industries that make up global tourism, in the economic, socio-cultural and environmental characteristics and impacts of different types of tourism, and at the level of social interaction. These various levels have been discussed at length elsewhere (Harrison, 2007) but Butler's well-known tourism area life cycle approach (TALC) (Butler, 2006a, b), Sharpley's focus on destination capitals, where sustainability is a local concern (Sharpley, 2009, pp. 175–198), and the application by Hamzah and Hampton of resilience theory (2013) are a few examples of more specific, lower level concepts than can be easily accommodated in a globalization perspective. At the same time, wider issues relating to tourism's intricate relationship with the environment will remain of central concern, including the sustainability of mass tourism, which should surely become a central preoccupation of future theorists and practitioners (Weaver, 2000).

Third, a globalization orientation also accommodates the kind of political economy advocated by Bianchi (2002), itself a version of earlier Marxist approaches, where attention is directed to the power of transnational tourism corporations (TTCs) in both developing and developed countries. Like political economy approaches generally, globalization perspectives focus both on cross-boundary processes (thus linking with dependency theory) and on what goes on within destinations (for long the implicit or explicit province of modernization theory).

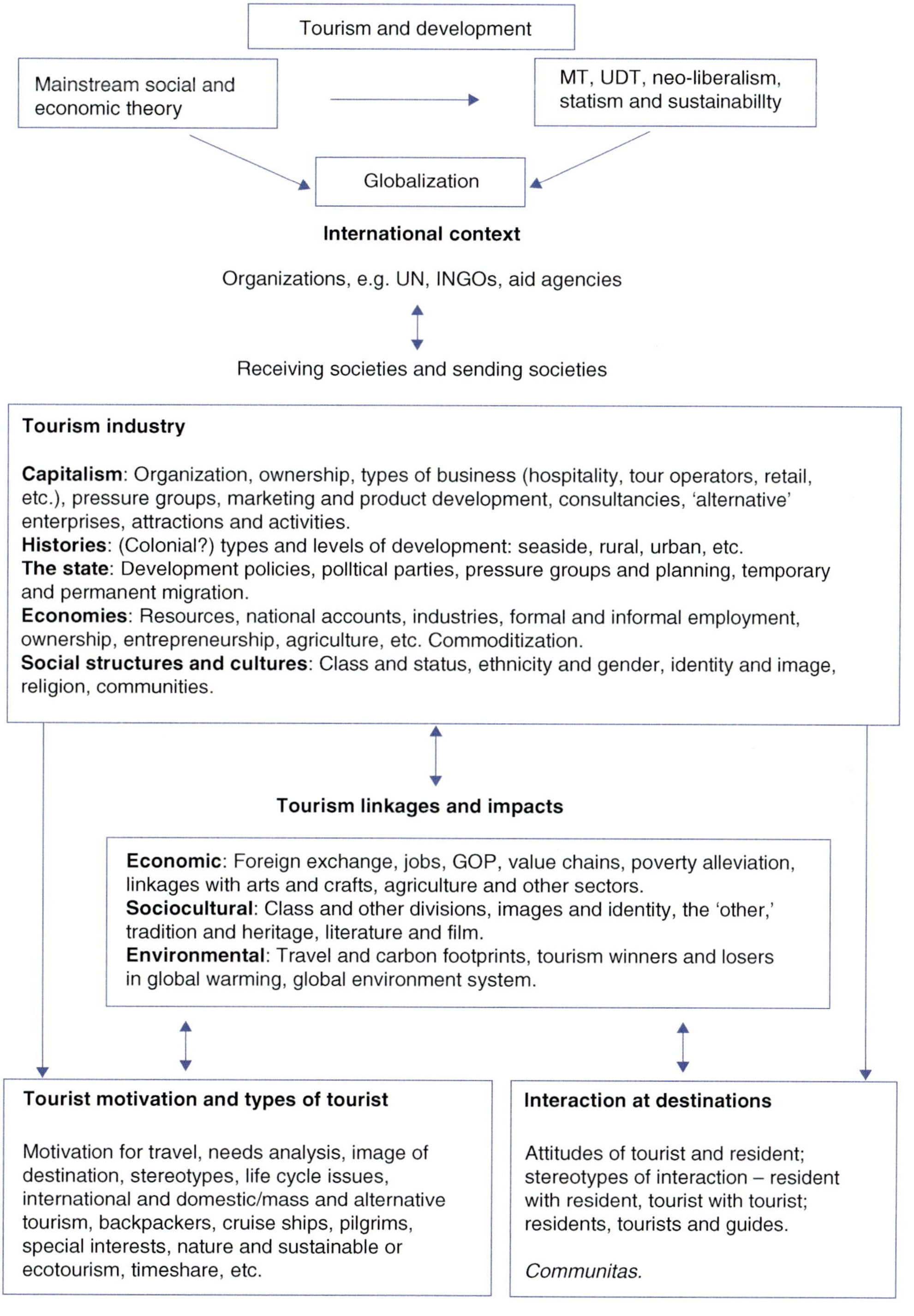

Fig. 17.1. Theoretical perspectives informing approaches to tourism and development.

Source: Harrison, 2010, p. 42.

Fourth, whether the emphasis is on developed or developing societies, globalization perspectives facilitate analysis of the role of the state (Harrison, 2001b, pp.33–39). In this respect, neo-liberalism and neo-statism are not exclusive 'paradigms', whose advocates are unable to engage in debate; they are more akin to policy orientations, the validity of which can be determined by empirical research, by evidence arrived at by comparing the state's role in tourism in different societies. And as Lockwood bluntly notes, if a state – irrespective of its ideological colours – is not committed to furthering the welfare its of its citizens, development will not occur (Lockwood, 2005).

Finally, and astonishingly, with notable exceptions (Andriotis, 2003; Sharpley, 2003; Aramberri, 2010), conventional mass tourism has been largely neglected by tourism scholars, except to berate it for its failings, in favour of (minority) forms of 'alternative tourism'. And yet, it is the major form of domestic and international tourism and is capable of transforming communities, for better or worse, over very short periods. In particular, the role of transnational tourism corporations has often been criticized, but the extent to which they occur, the actual reach of the global corporations, their inter-relations and their different characteristics have all been little researched (Allan, 1995; Bianchi, 2002; Meyer, 2003). For many, these global corporations are the epitome of mass tourism and are thus highly problematic. Indeed, as Aramberri remarks, mass tourism in general 'has thus become a usual suspect for every crime and misdemeanour with the sole exception of patricide and good taste' (Aramberri, 2010, p. 55). In querying further if tourism is indeed responsible for so many ills, he suggests 'most of those flights of imagination vanish into thin air' (2010, p 55). This may seem an over-statement but the truth is that we simply do not know. The research has yet to be done.

References

Adams, T. (1990) *Green Development: Environment and Sustainability in the Third World*. Routledge, London.

Allan, W. (1995) Capital and the Transnationalisation of tourism. In: Montanari, A. and Williams, A. (eds) *European Regions, Spaces and Restructuring*. Wiley, Chichester, pp. 163–176.

Andriotis, K. (2003) Tourism and Modernisation in Crete: a form of modernisation. *Current Issues in Tourism* 6(1), 23–53.

Aramberri, J. (2010) *Modern Mass Tourism Bingley*. Emerald Group Publishing, Bingley, UK.

Asian Development Bank (2012) *Greater Mekong Sub Region: Tourism Sector Assessment, Strategy and Road Map*. Asian Development Bank, Mandaluyong City, Philippines.

Berger, P. (1976) *Pyramids of Sacrifice: Political Ethics and Social Change*. Allen Lane, London.

Bianchi, R. (2002) Towards a new political economy of global tourism. In: Sharpley, R. and Telfer, D. (eds) *Tourism and Development: Concepts and Issues*. Channel View, Clevedon.

Booth, D. (1985) Marxism and development sociology: interpreting the impasse. *World Development* 13(7), 761–787. DOI: 10.1016/0305-750X(85)90107-X.

Britton, S.G. (1982) The political economy of tourism in the third World. *Annals of Tourism Research* 9(3), 331–358. DOI: 10.1016/0160-7383(82)90018-4.

Butler, R.W. (1999) Sustainable tourism: a state-of-the-art review. *Tourism Geographies* 1(1), 7–25. DOI: 10.1080/14616689908721291.

Butler, R. (ed.) (2006a) *The Tourism Area Life Cycle, Vol.1: Applications and Modifications*. Channel View, Clevedon.

Butler, Richard. (ed.) (2006b) *The Tourism Area Life Cycle, Vol.2:Conceptual and Theoretical Issues*. Channel View, Clevedon.

Cardoso, F. (1977) The consumption of dependency theory in the United States. *Latin American Research Review* 12(3), 7–24.

Cartier, C. and Lew, A. (eds) (2005) *Seductions of Place: Geographical Perspectives on Globalization and Touristed Landscapes*. Routledge, Oxon, UK.

Cooper, C. (2008) Globalization is more than an economic phenomenon. *Tourism Recreation Research* 33(1), 109–111. DOI: 10.1080/02508281.2008.11081296.

de Kadt, E. (1979) The issues addressed. In: Kadt, Emanuelde. (ed.) *Tourism: Passport to Development?* Oxford University Press, Oxford, pp. 3–76.

Frank, A.G. (1969) *Latin America: Underdevelopment or Revolution?* Monthly Review Press, New York and London.

Geoffrey Lacher, R. and Nepal, S.K. (2010) Dependency and development in northern Thailand. *Annals of Tourism Research* 37(4), 947–968. DOI: 10.1016/j.annals.2010.03.005.

Goodwin, H. (2006) *Community-based tourism: failing to deliver? id21 Insights 62*. Institute of Development Studies, Brighton.

Hamzah, A. and Hampton, M.P. (2013) Resilience and non-linear change in island tourism. *Tourism Geographies* 15(1), 43–67. DOI: 10.1080/14616688.2012.675582.

Harrison, D. (2001b) Tourism and less developed countries: key issues. In: Harrison, David. (ed.) *Tourism and the Less Developed World: Issues and Case Studies*. CAB International, Wallingford, pp. 23–46.

Harrison, D. (1988) *The Sociology of Modernization and Development*. Unwin and Hyman, London.

Harrison, D. (1992a) International tourism and the less developed countries: the background. In: Harrison, D. (ed.) *Tourism and the Less Developed Countries*. Belhaven, London, pp. 1–18.

Harrison, D. (1992b) Tourism in less developed countries: the social consequences. In: Harrison, D. (ed.) *Tourism and the Less Developed Countries*. Belhaven, London, pp. 19–34.

Harrison, D. (1996) Sustainability and tourism: reflections from a muddy pool. In: Briguglio, L., Archer, B., Jafari, J. and Wall, G. (eds) *Sustainable Tourism in Islands and Small States: Issues and Policies*. Pinter Publishers, London, pp. 69–89.

Harrison, D. (2001a) Less developed countries and tourism: the overall pattern. In: Harrison, D. (ed.) *Tourism and the Less Developed World: Issues and Case Studies*. CAB International, Wallingford, pp. 1–22.

Harrison, D. (2007) Towards developing a framework for analysing tourism phenomena: a discussion. *Current Issues in Tourism* 10(1), 61–86. DOI: 10.2167/cit300.0.

Harrison, D. (2010) Tourism and development: looking back and looking ahead – more of the same? In: Pearce, D. and Butler, R. (eds) *Tourism Research: a 20-20 Vision*. Goodfellow Publishing, Oxford, pp. 40–52.

Hay, C. and Marsh, D. (2000) Introduction Demystifying globalization. In: Hay, C. and Marsh, D. (eds) *Demystifying Globalization*. Palgrave, Basingstoke, pp. 1–17.

Hirst, P. and Thompson, G. (1999) *Globalization in Question*, 2nd edn. Polity Press, Cambridge.

Hjalager, A.-M. (2007) Stages in the economic globalization of tourism. *Annals of Tourism Research* 34(2), 437–457. DOI: 10.1016/j.annals.2006.10.006.

Honey, M. (1999) *Ecotourism and Sustainable Development: Who owns Paradise?* Island Press, Washington D.C.

Isbister, J. (2001) *Promises not Kept: The Betrayal of Change in the Third World*, 5th edn. Kumarian, Connecticut.

Jafari, J. (1989) Sociocultural dimensions of tourism. In: Bystrzanowski, J. (ed.) *Tourism as a Factor of Change: A Sociocultural Study*. European Coordination Centre for Research and Documentation in Social Sciences, Vienna, pp. 7–16.

Kitching, G. (1985) Politics, Method and evidence in the 'Kenya debate'. In: Henry, Bernstein. and Campbell, Bonnie. (eds) *Contradictions of Accumulation in Africa: Studies in Economy and State*. Sage, Beverly Hill, CA, pp. 115–151.

Lockwood, M. (2005) *The State They're in*. Barnes and Noble, New York.

MacCannell, D. (1976) *The Tourist: A New Theory of the Leisure Class*. Macmillan Press, London.

Macnaught, T.J. (1982) Mass tourism and the dilemmas of modernization in Pacific island communities. *Annals of Tourism Research* 9(3), 359–381. DOI: 10.1016/0160-7383(82)90019-6.

Meyer, D. (2003) *The U.K. Outbound Tour Operating Industry and implications for Pro-poor Tourism. PPT Working Paper 17*. Overseas Development Institute, London.

Mitchell, J. and Ashley, C. (2010) *Tourism and Poverty Reduction: Pathways to Prosperity*. Earthscan, London.

Mowforth, M. and Munt, I. (2009) *Tourism and Sustainability: Development, globalisation and new tourism in the Third World*, 3rd edn. Routledge, London.

Oakes, T. (1998) *Tourism and Modernity in China*. Routledge, London.

Oglethorpe, M.K. (1984) Tourism in Malta: a crisis of dependence. *Leisure Studies* 3(2), 147–161. DOI: 10.1080/02614368400390131.

Palmer, C. (1994) Tourism and colonialism: the experience of the Bahamas. *Annals of Tourism Research* 21(4), 792–811.

Payne, A. and Phillips, N. (2010) *Development*. Polity Press, Cambridge.

Pierret, F. (2011) Some points on domestic tourism. Available at: http://www2.unwto.org/en/agora/ (accessed 12th May 2013).

Report of the Independent Commission on International Development Issues (1980) *North-South: A Programme for Survival*. Pan Books, London.

Robertson, R. (1992) *Globalisation: Social Theory and Culture*. Sage Publications, London.

Robertson, R. (2012) Globalisation or glocalisation? *Journal of International Communication* 18(2), 191–208 DOI: 10.1080/13216597.2012.709925.

Scott, J. and Selwyn, Teds. (2010) *Thinking Through Tourism*. Berg Publishers, Oxford.

Seers, D. (1977) The new meaning of development. *International Development Review* 19(3), 2–7.

Sharpley, R. (2003) Tourism, Modernisation and development on the island of Cyprus: challenges and policy responses. *Journal of Sustainable Tourism* 11(2-3), 246–265. DOI: 10.1080/09669580308667205.

Sharpley, R. (2009) *Tourism Development and the Environment: Beyond Sustainability*. Earthscan, London.

Smith, Ved. (1977) *Hosts and Guests: The Anthropology of Tourism*. University of Pennsylvania Press, Philadedelphia.

Sofield, T. and Li, S. (1998) China: tourism development and cultural policies. *Annals of Tourism Research* 25.

Telfer, D. (2002) The evolution of tourism and development theory. In: Sharpley, Richard. and Telfer, David. (eds) *Tourism and Development: Concepts and Issues*. Channel View, Clevedon, pp. 35–50.

Telfer, D. and Sharpley, R. (2008) *Tourism and Development in the Developing World*. Routledge, London.

Theuns, H.L. (2008) Globalization and tourism: pros and cons. *Tourism Recreation Research* 33(1), 99–105. DOI: 10.1080/02508281.2008.11081294.

United Nations World Tourism Organisation (2013) *UNWTO Tourism Highlights*, 2012 edition. UNWTO, Madrid.

Waters, M. (1995) *Globalization*. Routledge, London.

Weaver, D. (2000) A broad context model of different destination scenarios. *Tourism Management* 21(3), 217–224.

Willis, K. (2005) *Theories and Practices of Development*. Routledge, London.

Wood, R. (1993) Tourism, culture and the sociology of development. In: Hitchcock, M., King, V. and Parnwell, M. (eds) *Tourism in South-East Asia*. Routledge, London, pp. 48–70.

Wood, R.E. (2008) Globalization and tourism: mapping the terrain. *Tourism Recreation Research* 33(1), 106–108. DOI: 10.1080/02508281.2008.11081295.

World Commission on Environment and Development (1987) *Our Common Future*. Oxford University Press, Oxford.

18

Looking East but Learning from the West: Mass Tourism and Emerging Nations

Introduction

There is much in modern tourism that replicates the past. Visitors to ancient Egypt scratched their names on the walls of the Egyptian pyramids, and drunken tourists at religious festivals in ancient Egypt, Greece and Rome scandalized local residents by their 'immoral' behaviour (Casson, 1974, pp. 31–32, 142–143; Harrison and Sharpley, 2017). Whether on the 17th century Grand Tour of Europe, visiting spas in the 18th and 19th century, or on the first package tours in the England of the mid-19th century, travellers were generally welcomed for the economic benefits they brought, greeted with ambivalence for the different cultures they represented, and criticized for their behaviour when contravening the norms and values of their 'hosts.'

This reference to history is no simple diversion. When examining the current development of mass tourism in emerging countries, an historical perspective is vital. Consequently, this paper compares the emergence of domestic mass tourism in 19th century England and international mass tourism in the post-1945 period with the later rise of mass tourism in East Asia, notably Japan and China, and suggests that while there were some differences, the processes were broadly similar.

Domestic Mass Tourism in England in the 19th Century

Relatively large-scale tourism first developed in the mid-1700s, when physicians popularized the medical benefits of sea bathing (Gilbert, 1954, pp. 56–86), thus providing an alternative to inland spas, which had long been visited for health and recreational purposes (van Tubergen and van der Linden, 2002). The resulting move to the sea was greatly facilitated by advances in stage coach design and improvements in roads, a 'golden age of coaching' which lasted until the 1830s (Briggs, 1994).

Possibly delayed by the Napoleonic Wars (1792–1815), which were highly costly and led to considerable poverty, also (arguably) crowding out civilian investment (Englander, 1995, pp. 534–535; Esdaile, 2014, pp. 159–162), the British economy soon went (quite literally) full steam ahead. The impact of the railways on the British landscape and economy was enormous. They revolutionized the distribution of consumer goods, especially perishable products (Briggs, 1994), and similarly facilitated the movement of passengers from one part of the country to another in 'monster trains' (Bradley, 2015, p. 89), including those who participated in Thomas Cook's early package tours. As Briggs notes,

Cook called for 'railways for the millions.... And he got them' (1994, p. 236).

Mass travel by train became unstoppable: excursions were run to executions, to (illegal) bare-knuckle prize fights, and to horse races, with hundreds of trains taking spectators to the most prestigious events. However, purposes of travel varied considerably. Women valued Cook's package tours for the security they provided and made up most of his early clients, and his tours always had a strongly educational focus. Theatre and concert audiences increased enormously, as did attendance at special exhibitions. Cook himself ran trips taking a total of some 165,000 passengers to the 1851 Great Exhibition in the Crystal Palace in London's Hyde Park.

Not everyone rejoiced at this unprecedented movement of the largely urban population. In the mid-1840s, the poet William Wordsworth, having previously written a guide to the (English) Lake District, did his utmost to stop railways penetrating the rural idyll, anticipating crowds of lower-class visitors, 'wrestling matches, horses and boat races without number,' and an increase in 'pothouses and beershops' run 'by the lower class of innkeepers' (in Ousby, 1990, p. 192). Similarly, although resorts such as Brighton, in the south, and Blackpool, in the north, became holiday destinations for workers from nearby cities (London, on the one hand, and northern cotton towns, on the other), seaside resort residents (and their more refined middle-class visitors) often objected to the behaviour of largely working-class holidaymakers. Drunkenness, boisterousness and vulgarity topped the list and many resorts legislated against noisy behaviour (Gilbert, 1954, p. 187; Walton, 1978, pp. 138–139), and naked sea bathing came to scandalize some resorts (Walvin, 1978, p. 78; Travis, 1993; Walton, 2000, p. 135). On the 'supply side', commercialization and competition for the visitors' money was increasingly considered intrusive (Walton, 1978, pp. 119–122; Travis, 1993), while visitors were frequently said to bring little economic benefit to a resort because they brought their own food and drink with them (Gilbert, 1954, pp. 205–206).

The railways facilitated large-scale movements of people and a wide range of accommodation was developed to cater for their various needs but none of this could have occurred had the *demand* not been present. In fact, as the railways flourished, so did the productivity, living standards and material expectations of many workers, in England, Western Europe and North America (Walvin, 1978, pp. 61–65; Soane, 1993, pp. 7–23). There were also a lot more of them. Following advances in public health, more (and more nutritious) food and a falling death rate, the English population rose from 16.9 million in 1851 to 30.8 million in 1901 (Briggs, 1994), while in Europe and North America generally, between 1850 and 1900, residents in towns of more than 100,000 increased from nearly 15 million to 81 million (Soane, 1993).

All these factors, coupled with a growing recognition of the benefits of holidays, as reflected in the British 1870 Bank Holiday Act (and, much later, the 1938 Holidays with Pay Act), meant that English seaside resorts, in particular, had entered their most popular period, which lasted until the outbreak of the First World War in 1914. At the same time, following a pattern begun by such religious organizations as the Boy's Brigade, in 1894 the first British holiday camp, for men who pledged 'not to touch alcohol or use improper language', was formed (Drower, 1982, p. 7). Here, at least, Thomas Cook's legacy continued. An annual excursion to the seaside or, better still, a week's stay, was increasingly becoming the norm.

Attractions ranged from the garish, noisy and vulgar to the more refined, from dance bands on newly constructed piers, blacked-up 'nigger' minstrels, music halls and freak shows for some, to manicured gardens, museums, plays and classical music concerts for others. Many seaside towns had resident orchestras, and others encouraged touring companies to visit, and throughout the summer season visitors at popular resorts listened and danced to music at seafront band stands, on piers, or in especially designed Winter Gardens and theatres, performed by orchestras and bands made up of well-known professional musicians, sometimes supplemented by amateurs, who played a range of popular and light classical European and British music, often giving premieres to local and nationally known composers (Pimlott, 1947, p. 132; Walvin, 1978, p. 75; Anon, 1993; Johnson, 2001).

Clearly, from the outset, the 'mass tourist' was not a homogeneous category. True, most visitors were attracted to the sea and the beach, but away from the beach they took 'different routes, for different periods and to sharply contrasting resorts and enjoyments (Walvin, 1978, p. 82).

If land was owned by one or only a few owners, or local authorities were diligent in implementing building and health regulations, seaside resorts could cater for different classes of visitor in separate zones (Walton 2000, p. 16-17), leading to clear differences in architecture, facilities and social 'tone' (Walvin, 1978, pp. 77–78). Where such zoning was not possible, adjacent coastal towns emerged to offer 'superior' entertainment and facilities for wealthier visitors (Walvin, 1978, p. 77).

Several key points emerge from this brief but necessary venture into 19th century English history. First, domestic mass tourism resulted from the coincidence of a growing and increasingly prosperous population with major advances in travel and communications – most notably, the railways. Second, as in previous periods, the (mis)behaviour of tourists was often frowned upon. Third, tourists came from different classes and had a range of motivations and interests, which were satisfied by entrepreneurs providing an appropriate range of accommodation and entertainment. Finally, the character of the urban landscape, seaside towns and tourism generally, changed over the decades (Pimlott, 1947, pp. 116–121). Fishing villages became thriving tourist towns, later to decline or be rejuvenated, as in the case of Brighton, on England's south coast, which became a highly popular urban environment with two universities, achieving much-vaunted city status as Brighton and Hove in 2001.

International Mass Tourism in the 20th Century

As Briggs notes, it took more than a century for mass tourism in the UK to be extended across national borders (Briggs, 1994, p. 236). Indeed, international tourism was not new to the 20th century, but its *mass* nature was. While the focus in this section is on travel by air, the emphasis is partial; during the early decades of the 20th century there were undoubtedly major developments in motorized transport and its supporting infrastructure. The mass production of cars, starting with the Model T Ford, brought independent road travel affordable, first to the USA population and then elsewhere. As cheaper, faster, more reliable and fuel-efficient vehicles were developed, the landscapes of vast swathes of North America and (later) Europe were transformed. By 1953, 93% of all recreational trips in the USA were by car (Jakle, 1985, p. 186). Travel and holiday patterns changed; new roads absorbed the increased private and commercial traffic; hotels and motels, increasingly owned or operated by large corporations, provided accommodation; and by the 1960s the 'traffic jam' was a fact of life, especially at holiday periods (Jakle, 1985, pp. 185–198; Jakle *et al.*, 1996; Middleton and Lickorish, 2005, pp. 55–56, 105–106).

The years immediately after 1945 were ones of austerity for former combatants of the Second World War. However, North America soon recovered and, building on pre-war advances (Briggs, 1994), European economies improved steadily during the 1950s. In the UK, Billy Butlin had built his first holiday camp in 1937, and the 1938 Holidays with Pay Act reflected much of what was already happening in many parts of the UK (Briggs, 1994). Prolonged peace had its rewards, and in 1957 Macmillan, the British Prime Minister, announced that the British people 'have never had it so good'.

From the early 1950s, all-in, short-haul charter tourism was developed in the UK by such entrepreneurs as Raitz and Laker, who used war-surplus aircraft and airstrips constructed during the Second World War (Middleton and Lickorish, 2005; Buswell, 2011, 78, 103, 119–124). They and other tour operators later benefited from the development of turbo-prop and jet aircraft (Lyth, 2009), and from hard-won 1960 legislation that enabled them to compete more easily with the national airlines. Consequently, by 1967, five million Britons were holidaying abroad, more than double the number of 1960.

Charter tours also developed in northern Europe, especially Scandinavia and Germany. While Nice, Antibes and Rimini were early beneficiaries, Mallorca, in particular, where outside developers were able to purchase large swathes of the coast, was soon transformed by millions of visitors (Longfren, 1999, p. 173;

Table 18.1. International tourism arrivals and receipts: selected years, 1950–2014.

Year	Arrivals (million)	Receipts (US$bn)	Year	Arrivals (million)	Receipts (US$bn)
1950	25.3	2.1	1999	639.6	464.5
1960	69.3	6.9	2000	687.0	481.6
1965	112.9	11.6	2001	686.7	469.9
1970	165.8	17.9	2002	707.0	488.2
1975	222.3	40.7	2003	694.6	534.6
1980	278.1	104.4	2004	765.1	634.7
1985	320.1	119.1	2005	806.6	682.7
1990	439.5	270.2	2006	847.0	742.0
1991	442.5	283.4	2007	903.0	856.0
1992	479.8	326.6	2008	917.0	939.0
1993	495.7	332.6	2009	882.0	851.0
1994	519.8	362.1	2010	940.0	927.0
1995	540.6	410.7	2011	995.0	1042.0
1996	575.0	446.0	2012	1035.0	1075.0
1997	598.6	450.4	2013	1087.0	1159.0
1998	616.7	451.4	2014	1133.0	1245.0

Source: Adapted from UNWTO data.

Buswell, 2011). From 1960, charter tourism 'remapped the Mediterranean' (Longfren, 1999, p. 184), 'turning old, barren and depopulated peripheries into centres of economic growth' (Longfren, 1999, p. 184). Spain, Italy, Greece, Turkey, Tunisia, Morocco, Egypt and Israel all became sun, sea and sand destinations, and some also winter destinations, as were Yugoslavia, Romania and Bulgaria, whose communist governments wanted much-needed foreign exchange (Allcock, 1991, pp. 244–246; Carter, 1991; Hall, 1991, p. 95-104). Switzerland and Austria, too, became popular venues for winter holidays (Williams and Shaw, 1988, p. 16).

The rapid expansion of international tourism is reflected in figures for international arrivals and receipts (Table 18.1). Arrivals increased ten times from 25 million international trips in 1950 to 278 million in 1980, and receipts by a much greater proportion. The process was repeated in central and eastern Europe after the post-1989 collapse of the Soviet bloc and the opening up of its former members' borders. Along with the ensuing expansion of the European Union (and its 'open skies' policy), new members of the EU, again utilizing former military airfields and again benefiting from the operation of budget airlines (Hall *et al.*, 2006, p. 15; Kiralova, 2006, pp. 112–113), opened up the region to tourism, much of which is centred on the capital cities (Hall *et al.*, 2006).

At first sight, mass tourism involves a standard tourism product (Longfren, 1999, p. 191), reflecting a wider 'McDonaldization' of production and consumption in society (Ritzer, 2014). However, even 'standard' mass tourist Mediterranean resorts cater for many tourist interests and offer a wide range of summer and winter attractions (Williams and Shaw, 1988, pp. 17–21). Comparing British clients of three apparently similar charter tour companies, Wright notes that once at the destination, first-time visitors 'used the package as a springboard to independent activity' (Wright, 2002, p. 199), adding that 'to dismiss package tourism as mass tourism is to ignore the multiplicity of experiences and meanings that package tourism represents to its participants' (2002, p. 199).

More generally, tourist typologies developed over the last three decades show that 'Western tourists' – a highly problematic category – are prompted by a much wider range of motivations than allowed for in MacCannell's somewhat simplistic depiction (Cohen, 1979; Schudson, 1979, p. 1252; Ross, 1994; Ryan, 2003, pp. 73–91). Therefore, 'mass tourism' is best seen as a generic term, an 'ideal type' against which specific examples of tourism development might be assessed (Harrison and Sharpley, 2017; Weber, 1949, pp. 89–102), and neither 'Western' tourists nor their 'African' or 'Asian' counterparts should be lumped together in some kind of meaningless, homogenous category.

Japanese Tourism: 1960s–2000s

Since the early 1950s, international trips have been mainly intra-regional, to Europe (66%) and North America (30%) (Harrison, 1992, p. 4). By 1990, they totalled nearly 440 million (Table 18.1), but the market shares of Europe and the Americas was reduced to 64% and 20% respectively (Harrison, 1992, p. 5). Twenty years later, the growing importance of East Asia, in particular, was obvious. Arrivals and receipts from international tourism had increased but Europe's share of international tourist arrivals of 1.133 billion was now reduced to less than 52%, that of the Americas to about 16%, while that of Asia and the Pacific (of negligible importance in 1950) had risen to more than 23% and accounted for more than 30% of all international tourism receipts (UNWTO, 2015, pp. 4–5). This increase of international arrivals to East and South-east Asia is well documented (Fig. 18.1), and the importance of East Asia, in particular, will be the focus in the remainder of this paper. The development of mass tourism in Japan followed a similar pattern to that of 19th century England and post-1945 Europe. Peace facilitated prosperity, transport and communications improved, and new destinations and new or revamped attractions satisfied the growing demand.

As elsewhere, mass domestic tourism came first. It originated in mass pilgrimages that became popular during the pre-industrial Edo period (Graburn, 1995, p. 47; Thompson, 2009) and benefited greatly from post-1945 improvements in roads, railways and air transport (Funck, 2013, p. 79). Following Ehrentraut (1995), Graburn depicts post-war domestic tourists as seeking 'tradition' and 'Japan-ness' in historical

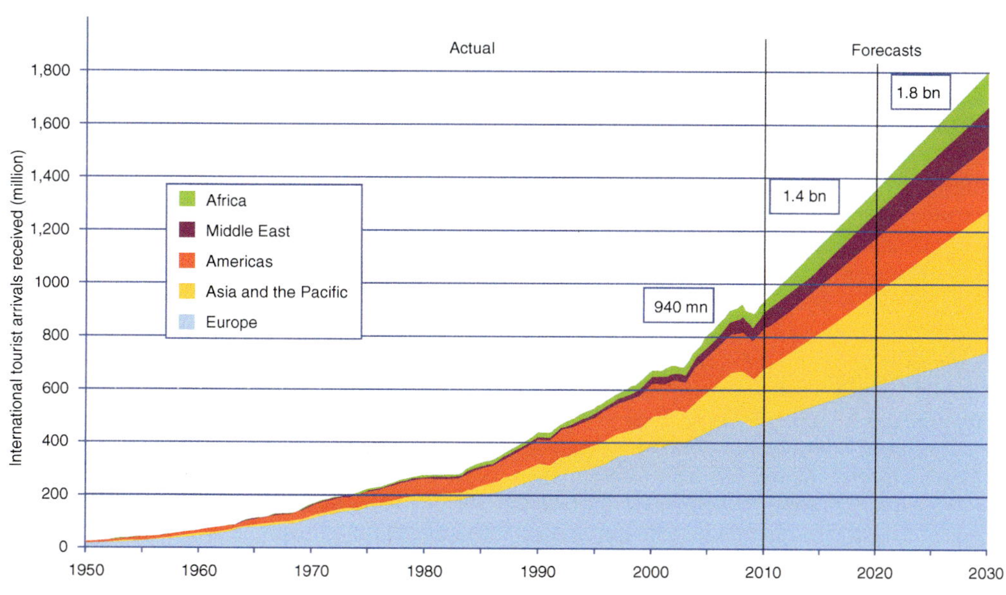

Fig. 18.1. Tourism towards 2030: actual trend and forecast 1950–2030.

shrines, temples and castles, thus responding to government's encouragement to 'rediscover' – even 'reinvent' – the 'rural', as epitomized in collectively experienced traditional hot spring resorts and clean, managed and sexualized nature (Graburn, 1995). Arguably, such nostalgia arose from a deep post-war malaise and collective loss of confidence, as well as an increasingly urban, crowded and frenetic everyday modernity but, as Graburn notes, 'Japanese history is no stranger to the longings of nostalgia' (Graburn, 1995, p. 66). Put differently, the Japanese search for tradition is itself traditional.

Such nostalgia is reflected in a tendency to reproduce an idealized past, albeit on a small scale, as tourist attractions. Indeed, domestic and outbound tourism converge in theme parks based on small-scale reconstructions of a real or fictional national past, or on similarly fantastical Western attractions (Dutch towns, English villages, Shakespeare's house, Anne of Green Gables, Don Quixote, and so on), which are increasingly popular in Japan and some Southeast Asian countries (Hendry, 2008; Surman, 2008, p. 197). As noted by Guichard-Anguis, 'the foreign is well packaged, but in a Japanese parcel' (Guichard-Anguis, 2008, p. 14). It is also nearer, cheaper and safer than 'the real thing'.

Japanese domestic tourism has been more consistent than inbound or outbound tourism. It declined somewhat in the mid-2000s but picked up again in 2011. In 2012, according to the Japan Tourism Agency (JTA), expenditure by domestic visitors (overnight stays) was 68.2% of all tourism expenditure (Japan Tourism Agency (JTA), 2014, p. 75) and, in 2014, nights spent by domestic tourists in all kinds of accommodation were nearly ten times those of inbound tourists (OECD (Organisation for Economic Co-operation and Development), 2016, p. 216). According to the World Travel and Tourism Council (WTTC), in 2014 travel and tourism contributed 7.5% of Japan's GDP, of which 92% came from domestic travel (World Travel and Tourism Council, 2015b, p. 1, 5).

Compared to domestic tourism, Japanese outbound tourism developed very slowly after the Second World War. A reduction in foreign exchange restrictions in 1964, their removal in 1978, and the arrival in the 1970s of wide-bodied jet aircraft (Mackie, 1992) led to limited growth but, in 1979, only 5% of the Japanese public travelled overseas (Williams, 2013). A substantial spurt did occur in the mid-1980s, when government, under pressure to reduce its positive trade balance with the West, incentivized outbound travel, which then became more culturally acceptable and by 1996 reached 16.7 million. From then, though, an under-performing economy, the high cost of education, a preference for long-haul travel, and a reluctance to take vacations (Williams, 2013) led to stagnation (Table 18.2). In 2012, there were only 18.5 million outbound trips, which fell to 16.2 million in 2015 (JTB Tourism Research and Consulting Co, 2016a). For much of this period, the USA was the preferred long-haul destination, while the most favoured short-haul destinations, initially Singapore, Hong Kong, South Korea and Taiwan (Mackie, 1992), have recently become China (the most popular), followed by South Korea, Thailand, Taiwan, Hong Kong and Singapore (Herrero and Iwahara, 2016).

Outbound Japanese tourism reflects the changing demographics of an ageing society and (again) indicates considerable varieties of tourist typologies and motivations. There is now a substantial 'silver market' (Funck and Cooper, 2013, pp. 212–213) and by 2007 nearly 90% of outbound tourists travelled independently, rather than on group tours, as compared to nearly 50% in 1964 (Funck, 2013, pp. 78–79). Japanese *kisaeng* (male sex tourism) of the 1970s and 1980s in South Korea and Thailand has been much discussed (Hall, 1992; Cohen, 1993; Leheny, 1995; Bishop and Robinson, 1997; Moon, 2008, pp. 152–155), and Moon notes several other categories of outbound tourist: the niche market of returnee colonists to South Korea (2008, p. 155); young, single 'office ladies', a distinct category of Japanese tourist, whose *esute* tourism combines an interest in cosmetics, healthcare and shopping (Moon, 2008, pp. 158–164); *hallyu* tourism, where older Japanese women nostalgically seek a long-gone version of Japan in popular Korean TV soap operas; and history tourism, practiced by a more restricted, intellectually inclined group of visitors (Moon, 2008, pp. 164–165).

Other categories of Japanese outbound tourism include 'dark' tourism, where elderly Japanese visit Second World War sites in Palau or the South Pacific, and adventure tourism, where younger tourists (often honeymooners) go diving or simply

Table 18.2. Japanese outbound and inbound tourism: selected years, 1973–2015.

Outbound		Inbound		
Year	Total departure	All arrivals	Holidaymakers	% of arrivals
1973	2,288,066	784,691	447,274	57.0
1975	2,466,326	811,672	446,420	55.0
1985	4,948,366	2,327,000	1,329,000	57.1
1990	10,997,431	3,235,860	1,879,497	58.1
1995	15,298,125	3,345,274	1,731,439	51.8
1999	16,357,572	4,437,863	2,560,343	57.7
2000	17,818,590	4,757,146	2,693,357	56.6
2005	17,403,565	6,727,926	4,368,573	64.9
2010	16,637,224	8,611,175	6,361,974	73.9
2011	16,994,200	6,218,752	4,057,235	65.2
2012	18,490,657	8,358,105	6,041,645	72.3
2013	17,472,748	10,363,904	7,962,517	76.9
2014	16,903,388	13,413,467	10,880,604	81.1
2015	16,213,763	19,737,409	16,969,126	87.2

Sources: Tokuhisa, 1980; Wu *et al.*, 2000 p. 143; WTO (World Tourism Organisation, 1990; JTB Tourism Research and Consulting Co, 2016a and JTB Tourism Research and Consulting Co, 2016b

enjoy the laid-back 'paradise' of tropical islands (Yamashita, 2008). Similar attractions await in Hawai'i, whereas mainland USA is valued more for its elements of 'modernity', such iconic sites as Hollywood and Disneyland, and the availability of 'marker goods' and souvenirs (Moore, 1985, pp. 633–640). By contrast, Japanese who visit the UK, while not eschewing such brands, look for 'authentic' British culture and 'traditions,' explore sites associated with such literary figures as Beatrix Potter and the Brontes, or locations of films made in the UK, for example, the Harry Potter series (Surman, 2008).

Finally, like outbound tourism, inbound tourism to Japan grew slowly, but it was more consistent (Table 18.2) and in 2015 exceeded the former (JTB Tourism Research and Consulting Co, 2016a and JTB Tourism Research and Consulting Co, 2016b). Japan's historic villages, castles, shrines and temples, national parks and UNESCO heritage sites, along with such centres of entertainment as Tokyo Disneyland, have increasingly attracted international as well as domestic visitors. For a long time, South Korea was the most important source of tourists, followed by Taiwan, Hong Kong and the USA. However, from a very low base in 2003, visitors from China have increased, especially since 2013, and in 2015a new, young and aspiring Chinese middle class, benefiting from the depreciation of the yen, overtook South Koreans as the main source of tourists (Fig. 18.2), representing 25% of all tourist arrivals to Japan and accounting for more than 40% of all tourist expenditure (btrax, 2016).

To summarize: in 2015, tourism in Japan reportedly accounted for almost 8% of Japan's GDP and some 7% of the country's employment, and Japan was 'the fourth largest travel and tourism economy in the world' (World Travel and Tourism Council, 2015a). Indeed, it was the emergence of outbound and inbound tourism in Japan that initially made East Asia a force to be reckoned with in global tourism. The centre of gravity in global tourism was starting to shift.

The Development of Chinese Tourism

China's growth generally, and as a recipient and source of international tourists, has been spectacular. It has superseded Japan as the growth pole of global tourism and yet has hardly

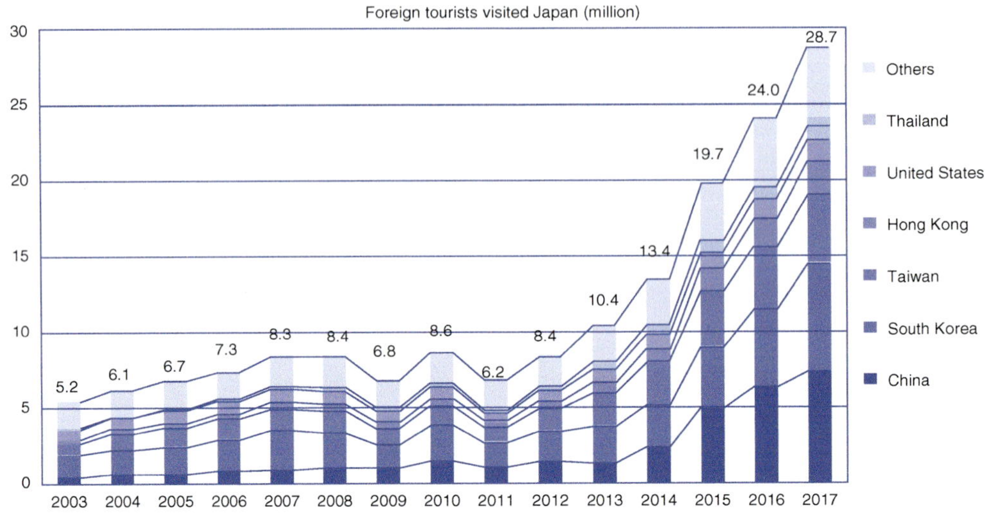

Fig. 18.2. Foreign tourists visited Japan (thousand).

started developing its potential as a destination or (especially) a source of tourists (Herrero and Iwahara, 2016). However, this emerged only after a long period of internal and external unrest. Since the late 1900s, there was periodic conflict with Japan, culminating in the Second Sino-Japanese War (1937–1945) and, after the formation of the Chinese Communist Party in 1921, intermittent civil war occurred between nationalists and communists, ending in 1949 with Mao Zedong's victory and the installation of the Peoples' Republic of China government (PRC).

China's travails did not end in 1949. International tensions – the Korean War (1952–1953), war with India (1962) and fraught relations with the USSR – accompanied by domestic policies initially based on a Soviet model of top-down industrial development, successive five-year plans, and enforced land reform and redistribution of land to the peasants. During the early 1950s, government also promoted a series of radical 'movements' to advance its policies and neutralize opposition, costing an estimated 5–10 million lives (Berger, 1977, first published 1974). However, Kruschchev's 1956 critique of Stalin opened up a rift with Mao, who then initiated several consecutive disruptive (and contradictory) campaigns, including the 1957 Hundred Flowers Campaign, which encouraged criticism of the government and the Communist Party, and the Anti-Rightist Movement of the same year, which effectively ended such licence. With the Great Leap Forward (1958–1961), Mao introduced a new 'model' of self-reliant development, introducing collective farming and industrial decentralization. None of this could prevent (and might have exacerbated) widespread famine, which from 1958 until 1962 led to extensive loss of life, estimates of which range from 23 million to 45 million (Peng, 1987, p. 649; Dickotter, 2010, p. 333).

In 1960, all Soviet aid to China was withdrawn. Mao then launched the Cultural Revolution (1966), aiming to eliminate a Soviet-style bureaucracy by attacking middle class bureaucrats and intellectuals. For months, Mao's student supporters, the Red Guards, went on an ideological rampage and killings and suicides were widespread. Eventually, the movement imploded, its excesses were finally recognized, and in 1967 the Peoples' Liberation Army restored order. By 1968, the radical phase of the Cultural Revolution was over, but officially it ended only in 1976, after having profoundly negative impacts on the economy (Guo, 2010, p. 110, p. 153–154).

By the mid-1970s, albeit at a massive human cost, it seems starvation had been eliminated and most Chinese were better-off than their 1949 counterparts, but poverty remained widespread and standards of living were uniformly low (Berger, 1977, first published 1974, pp. 185–186). In addition, under Mao, any internal travel in China was tightly controlled. From 1956, the PRC government used the established *hukou*, or *huji* system of household registration to ensure that, except for elite Party members, rural to urban migration was almost impossible (Sofield and Li, 1998, p. 380; Walker, 2015, p. 78, 92). Under such constraints, neither domestic nor international tourism was an option for the average Chinese citizen.

In the 1970s, international tensions eased. President Nixon visited China in 1972 and Mao ZeDong died in 1976. When Deng Xiaoping became leader in 1978 he ended self-reliance as a development strategy and announced the 'Four Modernisations' of agriculture, national defence, science and technology. Inbound international tourism was promoted as a source of foreign exchange, foreign investment encouraged, developers were offered incentives to build hotels, and foreigners were accorded privileged access to transport facilities and special shopping centres (Zhong *et al.*, 2015). The race towards modernity had begun.

Since 1978, the average annual growth rate in China has exceeded 10%, more than three times the average of developed nations, double that of India, and higher than Singapore, Hong Kong, Taiwan and South Korea (Guo, 2010, p. 111). This achievement, obtained through a high rate of capital accumulation, containment of wage growth, and integration into the world economy, is not unproblematic, as poverty remains and growth has been accompanied by a rise in regional inequalities (Guo, 2010, pp. 113–127). Nevertheless, a substantial (and increasing) middle class has emerged, which has been both driver and beneficiary of economic development. Indeed, Song and Cui suggest that by 2017 the number in the middle class could be as high as 340 million (Song and Cui, 2009, p. 129), while Shorrocks and Davies note that since 2002 the size of the middle class in China has exceeded that in the USA (Shorrocks and Davies, 2015, p. 129).

Inbound, outbound and domestic tourism have all increased along with economic growth, though at different rates. When 'open door' policies were first introduced, inbound tourism was prioritized but domestic tourism was neglected and controls on internal migration remained (Xu, 1999; Wu *et al.*, 2000; Chiang, 2012). Some restrictions on domestic travel were removed in 1978, more in the 1980s, when identity cards were introduced, and in 2011, when migration to small and medium-sized towns was facilitated (Huang and Pieke, 2003). Nevertheless, for most of the 1980s the official policy towards domestic travel for leisure purposes was 'no encouragement' (Guangrui, 1985, p. 141).

Despite government policies, as disposable incomes increased, so did domestic tourism (Guangrui, 1985, p. 141). In the 1990s the policy changed, and the inauguration of Golden Weeks Holidays in 1999 reflected new awareness of domestic tourism's economic value, and the positive *social* functions of 'heritage' attractions that promoted traditions, including those of ethnic minorities, and the unity-in-diversity ideology government wished to convey to both domestic and international visitors (Sofield and Li, 1998, pp. 372–376; Wu *et al.*, 2000, p. 298; Graburn, 2015, pp. 176–177). Travel to natural or protected areas was also encouraged, along with 'red tourism', featuring visits to areas noted for the birth and exploits of communist heroes (Li and Hu, 2008). Domestic tourism thus became 'a catalyst for the development of inland regions' (Ghimire and Li, 2001, pp. 98–102; Chiang, 2012).

Such statistics are problematic, as they include rural migrants working in coastal cities (Chiang, 2012), but mass domestic tourism is clearly crucial to the Chinese economy. From 1994 to 2014, domestic tourists increased from 524 million to 3.6 billion (Chiang, 2012; www.statistica.com/statistics/277254, accessed 6 August 2016), rising to an estimated four billion in 2015 (Ryan and Shen, 2017). In 2010 domestic tourism in China reportedly accounted for 81% of all tourism revenue and 94% of all tourists, comprising 'the largest domestic tourism market in the world', with the remaining 19% and 6% respectively coming from 'foreign' visitors and 'compatriot' Chinese from Hong Kong, Macau and Taiwan (Chiang, 2012).

Over the years, the nature of domestic tourism changed: whereas in 1994 most (39%) domestic tourists were urbanites, by 2010 it was 75% (Chiang, 2012), and the most favoured destinations were coastal cities, especially Shanghai, Beijing and Chongqing.

The fertile socio-economic conditions that encouraged domestic tourism also prompted outbound tourism (Xie and Li, 2009; Arlt, 2013; Xiang, 2013Arlt, 2013; Zhang, 2016; Dai *et al.*, 2017). While commentators differ on its precise trajectory (Xie and Li, 2009; Zhang, 2016), the overall pattern is clear. From 1983 until the mid-1990s, Chinese citizens could visit friends and relatives (at the latter's cost), initially in Hong Kong, then Macau, and later Singapore, Malaysia, Thailand, the Philippines and other border areas.

Some restrictions were lifted in 1995, and approved destination status (ADS), seen by government as a way of exercising 'soft power' (Arlt, 2016, pp. 8–12), was increasingly granted to other countries, especially after 2001, when China joined the World Trade Organization. Consequently, by 2013, nearly 150 countries had obtained ADS (Xie and Li, 2009, pp. 226–230; Arlt, 2013). Statistics of outbound tourism from China are also problematic, as they exclude journeys of unregistered residents of major cities, yet include multiple business trips by individuals and regular cross-border commutes to Hong Kong, Macau and Russia (Arlt, 2013). Irrespective of such caveats, though, the overall trend is inexorably upwards (Fig. 18.3), and likely to continue. As Arlt notes, 'even the 95million border crossings estimated by the China Outbound Tourism Research Institute for 2013 translate into just 3% of the mainland Chinese population crossing a border in that year' (2013, p. 128), and with current passport possession a mere 3–5% and a potential market of at least 1.2 billion people (95% of the Chinese population) (Li, 2016), virtually every tourist destination worldwide is setting its pitch for the Chinese market (Xie and Li, 2009, pp. 232–234; ; Arlt, 2016, pp. 8–12).

At the same time, the profile of Chinese outbound tourism is changing. Leisure tourism overtook business trips in 2000 and currently most trips overseas (85%) are for leisure purposes (Xie and Li, 2009), mainly for shopping and sightseeing (Arlt, 2013; Dai *et al.*, 2017). Most outbound tourism is intra-regional, to Hong Kong and Macau (70%) and other Asian countries, notably Japan, South Korea and Thailand (20%). By contrast, such long-haul destinations as the USA and Europe take only a small percentage of Chinese visitors, but numbers will increase, through a 'ripple effect', as Chinese tourists become more experienced and adventurous, travel further afield, rely less on tour agency packages, and opt more for self-organized itineraries (Zhang and Heung, 2002, p. 8; Xie and Li, 2009; Arlt, 2013, pp. 130–131). And while younger travellers increasingly seek entertainment and adventure, long-haul tourists to the West focus more on history, and natural and cultural heritage (Xiang, 2013; Andreu *et al.*, 2014).

China is the world's fastest growing source of tourists. In 2014 their expenditure was 13% of global tourism receipts, more than any other nation (UNWTO, 2015). In the process, they

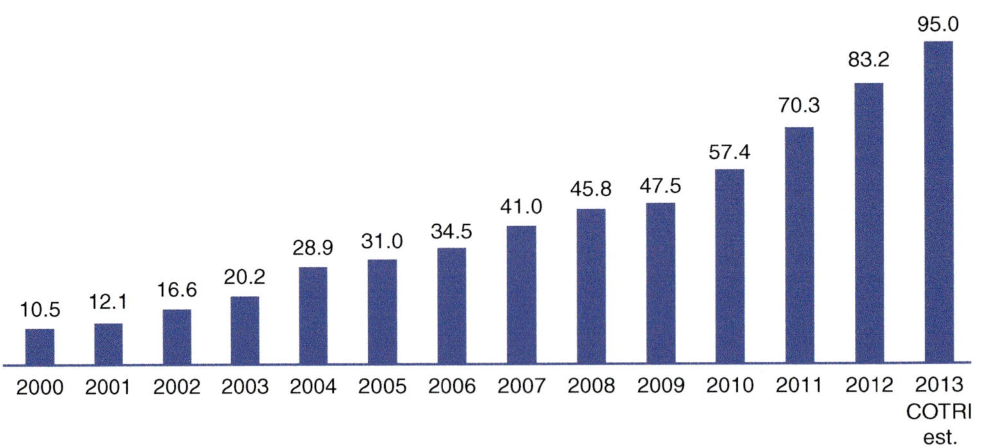

Fig. 18.3. Chinese outbound trips 2000–2013.

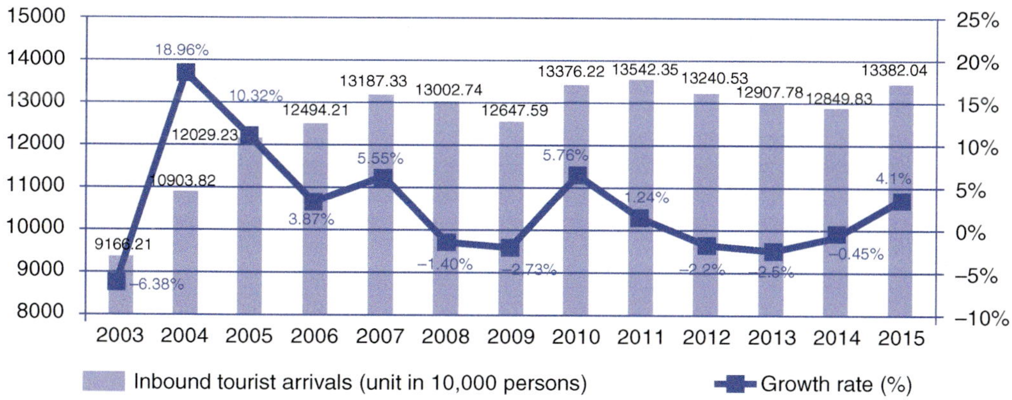

Fig. 18.4. Inbound tourist arrivals and growth rate.

have changed and increased the tourism profile of several destinations, with more to come. As Li notes, this growth may be just an appetizer with the main course yet to come (Li, 2016).

By contrast, the increase in inbound tourism to China has been unspectacular. In 1978, when rapprochement with the West began, China received 229,600 'foreign' (i.e. non-Chinese) tourists (Guangrui and Lew, 2003). Inbound tourism was then increasingly prioritized as a source of foreign exchange and was first included in the national plan for social and economic development in 1986 (Lim and Pan, 2005). From then on, excluding temporary but drastic falls in arrivals following the 1989 Tiananmen Square student protests and the 2003 Severe Acute Respiratory Syndrome (SARS) crisis, arrivals steadily increased (Lim and Pan, 2005, pp. 500–502; Yang *et al.*, 2010, pp. 827–828). By 2015, total arrivals (including excursionists) exceeded 131 million (Fig. 18.4) and nearly 55 million overseas visitors stayed overnight, of whom two thirds were from Hong Kong, Macau and Taiwan, while the remainder were 'foreign' tourists. In all cases, however, numbers have remained relatively static (Fig. 18.4, Table 18.3).

In 2015, including direct and indirect impacts, travel and tourism reportedly accounted for 9.4% of China's GDP and 8.6% of all employment (World Travel and Tourism Council, 2015a 2015, p. 1). As indicated earlier, the level of domestic tourism far outstrips inbound tourists. As disposable incomes continue to rise, and with government

Table 18.3. China inbound tourism: 2006–2015 (millions).

Year	Total arrivals (overnights)	Foreign	Hong Kong	Macau	Taiwan
2006	49.9134	18.1061	27.9668		
2007	54.7198	21.3989	25.3814	3.9183	4.0212
2008	53.0492	19.7041	25.6686	3.8790	3.7975
2009	50.8752	17.6969	25.4979	3.8480	3.8324
2010	55.6645	21.2764	26.0945	3.9289	4.3647
2011	57.5807	21.9410	26.9159	4.2794	4.4444
2012	57.7249	21.9487	26.7100	4.3157	4.7504
2013	55.6859	20.8071	26.0759	4.2297	4.5732
2014	55.6220	20.8127	25.8745	4.2075	4.7274
2015	56.8857	20.2858	27.0899	4.6660	4.8441

Source: https://www.travelchinaguide.com/tourism, accessed 4 August 2016.

planning to introduce a national system of paid holidays by the end of 2020, the number of domestic trips is expected to rise by 16% annually (Hernandez *et al.*, 2015). Outbound tourism in also booming. In 2014, Chinese tourists reportedly took 109 million international trips, and competed with Germans and Americans as the highest spenders on international travel, spending US$155 billion, while at the same time China was the fourth most popular world destination (Hernandez *et al.*, 2015; UNWTO, 2015, p. 6).

In short, whether occurring in or outside China, Chinese mass tourism is big business. However, it is not unproblematic. The transport system is said to be poorly integrated, and investment in infrastructure, construction and maintenance inadequate (Zhou and Szyliowicz, 2006, pp. 12–14). As with air transport, investment in the railways has been similarly constrained by state ownership (Zhang and Chen, 2003) and the capacity of road infrastructure, though increasing, is behind the increase in car ownership (Jie and van Zuylen, 2014). Consequently, transport systems and popular destinations are often choked by the sheer number of visitors (Xu, 1999, p. 209, p. 216; Pearce and Chen, 2012, p. 399). Carrying capacities of protected areas may be similarly exceeded (Li, 2004), and social tensions can arise from excessive concentration of tourism accommodation, income inequalities and the uneven distribution of tourism resources (Wen and Sinha, 2009; Yang *et al.*, 2013), while residents in 'heritage' sites may become (or at least feel) marginalized (Wang *et al.*, 2010; Xu *et al.*, 2013; Guo and Sun, 2016). Alternatively, like rural to urban migrants (and inexperienced Chinese tourists in, for example, Hong Kong, Taiwan or Thailand), visitors from rural areas may receive a cold welcome from urban dwellers, often being seen as uncivilized, unhygienic or criminal (Huang and Pieke, 2003).

Conclusion: Western Past and East Asian Present

Mass tourism developed in the West and in East Asia in broadly similar ways. Domestic mass tourism came first, emerging when the underlying socio-economic and political order was stable, when significant sectors of the population had a disposable income, and when road, rail or air transportation systems could transfer people in bulk from their homes to destinations. Mass international tourism followed the same pattern, as holidays had by then been institutionalized, often with legislative support, though there were significant differences in the control exercised by government. As in Central and Eastern Europe before the collapse of the Soviet bloc, governments of Japan and China, especially, played a major role in facilitating and/or controlling outbound tourism.

Historically and currently, tourists have often been lampooned as national stereotypes and treated as figures of fun or derision, especially when contravening norms that prevail in destination regions. In this respect, Chinese tourists are merely the latest to have caused resentment in destination areas, prompting the Chinese government, and some overseas destinations, to issue guidelines about how they are expected to behave (Sziakov, 2013; Economist, 2013; eTurboNews, 2015; Tatlow, 2015; Teon, 2015).

Such (mis)behaviour is likely to be modified. Generally, as tourists (individually and collectively) gain more travel experience they rely less on pre-paid packages, travel further afield more confidently and seek an increasing variety of tourist experiences. Younger, better educated travellers, especially, are more likely to engage in independent travel, as indicated in figures for both Japan and China (Jaff, 2013; Market Intelligence Group, 2013).

Finally, a theoretical issue: despite obvious commonalities in the emergence of tourism across time and place, it has recently been argued that tourism in 'emergent' nations has been sufficiently intense – and different - to require a different 'paradigm' to that hitherto applied to Western tourism. Following Winter (2009, p. 23), Cohen and Cohen, for instance, claim Western theories of tourism have been dominated by a 'culture bound', ethnocentric and Western-oriented 'discourse of authenticity' (2012, p. 2180; 2015a, p. 11). Chang, too, argues the need to 'Asianise the field' to ensure 'the diversity of Asian identities, perspectives and contributions are fore-grounded in tourism scholarship', and enable 'the anticipated Asian tourism wave [to] be understood on its own terms rather than under the gaze of Western eyes' (Chang, 2015, pp. 96–97).

For their part, Cohen and Cohen suggest the 'New Mobilities Paradigm' (NMP), when applied to tourism in emerging societies, is less ethnocentric than 'the discourse of ethnicity,' 'does not distinguish between a centre and periphery of tourist activity ... does not assume a single point of dissemination of tourism ... and does not prioritise a particular motivation for tourism' (Cohen and Cohen, 2015b, p. 163). However, their account of tourism in 'emerging world regions' (Cohen and Cohen, 2015a, pp. 16–17–27), said to be informed by the NMP, is, in fact, a thoroughly understandable historical description and, indeed, covers very similar ground to this paper.

Moreover, what it means to understand Asian tourism 'on its own terms' is unclear, and although 'the discourse of authenticity' has undoubtedly been important in Western tourism studies, its 'dominance' is questionable, while the NMP, even if it *were* a paradigm, is arguably as 'Western' as 'modernist studies of tourism'. In any case, whereas some Asian scholars have gone so far as to suggest the need for a 'paradigm shift' if research on Chinese tourism is to somehow be freed of 'Western' approaches (Xiao and Smith, 2006; Huang *et al.*, 2014), Cohen and Cohen baulk at supporting calls for 'alternative social sciences in non-Western societies' (Cohen and Cohen, 2015b, p. 159).

At this point, three (brief) responses must suffice. First, neither Western nor Chinese academics have neglected Chinese tourism. Clearly, a time lag between the growth of tourism in a region and academic interest in the topic is normal. International tourism was well under way before MacCannell's pioneering study *The Tourist* (MacCannell, 1976) was published, and only then did tourism studies emerge as a field of study (Jafari and Aaser, 1988). Similarly, in his review of 187 English language papers on Japanese tourism, published from 1980 until 2009, Williams notes that 61% were published from 2000 to 2009 (2013, pp. 47–48).

The Chinese case is even more marked. All literature reviews show that recently there has been a veritable explosion of academic publications on Chinese tourism, in English and Chinese (Kong and Cheung, 2009, p. 343; Hsu *et al.*, 2010, p. 148; Xiao *et al.*, 2010, p. 330; Huang, 2011; Tsang and Hsu, 2011; Liu *et al.*, 2013; Zhong *et al.*, 2013; Andreu *et al.*, 2014, p. 346; Zhong *et al.*, 2015). Indeed, Huang and Chen cite a 2013 survey by Zhang and her colleagues of 16,024 tourism articles published in China from 2003 to 2012 (Huang and Chen, 2016, p. 12). Such a volume of scholarship belies the notion that Chinese tourism has been neglected, and also demonstrates that both Chinese and Western tourism academics cover a wide range of topics from an equally wide range of perspectives.

Secondly, social scientists have long debated the meaning of 'paradigm' (Eckberg and Hill, Jnr., 1979). Indeed, Kuhn, its key exponent, repeatedly redefined it over his career. Ultimately, though, he saw paradigms as bodies of knowledge that reflected the practices, values and lexicons of communities of scientists. These communities (alone) produce and validate scientific knowledge and socialize members into their disciplines, and are distinct from other scientific communities in that their positions are *incommensurable* from other bodies of knowledge (Kuhn, 2012, p. 205). This needs to be emphasized: for Kuhn, incommensurability with other bodies of thought was *always* the criterion of whether or not a 'paradigm' exists.

Advocates of a new 'paradigm' for Asian tourism use the term in a much looser sense. For Winter, it denotes a shift in attention from West to East tourist movement to that from East to West (Winter, 2009, p. 29), though he does not explain why this change in *subject* necessitates a paradigm change. By contrast, Cohen and Cohen variously refer to the status of the NMP as 'a model' (Cohen and Cohen, 2012, p. 2181), a 'theoretical approach', a 'theory' or an 'innovative insight' (2012, p. 2185), whereas Huang, van der Veen and Zhang depict a move from positivistic to qualitative perspectives as a shift in paradigms (Huang *et al.*, 2014, p. 381). In the Kuhnian sense used here, however, none of these positions are incommensurable; rather, they simply reflect the established fact that tourism studies has long accommodated different – but complementary – disciplines and perspectives (Harrison, 2007).

Whether or not the above conceptualization of paradigm is accepted, the possible nature of a *non-Western* paradigm in the Kuhnian sense has rarely been articulated. Quoting Lin Yutan, an earlier American-based Chinese writer, Huang, van der Veen and Zhang contrast a 'Western mind' which is 'generally analytical and thus science-oriented' with a Chinese way of thinking that is 'largely intuitive, dialectic or synthetic, non-conclusive and common-sense based' and then, on the basis of this dubious stereotype, advocate a Chinese paradigm that is 'more appropriate to study Chinese tourists' (Huang *et al.*, 2014, p. 382), though quite what such a 'Chinese perspective' would mean, in practice, is not explained.

Other questions underlie this debate: Is there a separate (social) science for 'Europeans' and 'Asians?' Are Asians and Chinese, or British and Europeans, homogeneous categories? Why not have a 'German' 'Italian' or 'Indian' paradigm? What is so special about 'Asians' and 'Chinese' and their tourism patterns?' Has it been shown that 'Western' scholarship cannot address the differences (if any) between Chinese tourism and its Asian and/or Western equivalents? Put more pithily, is there a tourism studies equivalent of acupuncture?

References

Allcock, J.B. (1991) Yugoslavia. In: Hall, D. (ed.) *Tourism and Economic Development in Eastern Europe and the Soviet Union*. Belhaven Press, London, pp. 236–258.

Andreu, R., Claver, E. and Quer, D. (2014) Destination attributes and Chinese outbound tourism to Europe. *Journal of China Tourism Research* 10(3), 275–291.

Anon (1993) Music matters: municipal and resort Orchestras, 1880s to 1930s. BBC Radio 3. 14 November.

Arlt, W.G. (2013) The second wave of Chinese outbound tourism. *Tourism Planning & Development* 10(2), 126–133. DOI: 10.1080/21568316.2013.800350.

Arlt, W.G. (2016) China's outbound tourism: history, current development and outlook. In: Li, X. (R). (ed.) *Chinese Outbound Tourism 2.0*. Apple Academic Press, Oakville, Canada and Waretown, USA, pp. 3–37.

Berger, P. (1977, first published 1974) *Pyramids of Sacrifice: Political Ethics and Social Change*. Penguin, Harmondsworth.

Bishop, R. and Robinson, L. (1997) *Night market: sexual culture and the Thai economic miracle*. Routledge, London.

Bradley, S. (2015) *The Railways: Nation, Network and People*. Profile Books, London.

Briggs, A. (1994) *A Social History of England*. Weidenfeld and Nicolson, London.

btrax (2016) *Travel Trends: Inbound & Outbound Japanese Tourism* [online]. Available at: http://blog.btrax.com/en/2016/03/09/travel-trends-inbound-outbound-japanese-tourism/ (accessed 24 June 2016).

Buswell, R.J. (2011) *Mallorca and Tourism: History, Economy and Environment*. Channel View Publications, Bristol.

Carter, F.W. (1991) Bulgaria. In: Hall, D. (ed.) *Tourism and Economic Development in Eastern Europe and the Soviet Union*. Belhaven Press, London, pp. 220–236.

Casson, L. (1974) *Travel in the Ancient World*. The John Hopkins University Press, Baltimore and London.

Chang, T.C. (2015) The Asian wave and critical tourism scholarship. *International Journal of Asia Pacific Studies* 11(1), 83–101.

Chiang, M.-H. (2012) The changing role of tourism in China's economy. *Journal of China Tourism Research* 8(2), 207–223. DOI: 10.1080/19388160.2012.677372.

Cohen, E. (1979) Rethinking the sociology of tourism. *Annals of Tourism Research* 6(1), 18–35. DOI: 10.1016/0160-7383(79)90092-6.

Cohen, E. (1993) Open-ended prostitution as a skilful game of luck, opportunity, risk and security among tourist-orientated prostitutes in a Bangkok soi. In: Hitchcock, M., King, V.T. and Parnwell, M. (eds) *Tourism in South-East Asia*. Routledge, London and New York, pp. 155–178.

Cohen, E. and Cohen, S.A. (2012) Current sociological theories and issues in tourism. *Annals of Tourism Research* 39(4), 2177–2202. DOI: 10.1016/j.annals.2012.07.009.

Cohen, E. and Cohen, S.A. (2015a) A mobilities approach to tourism from emerging world regions. *Current Issues in Tourism* 18(1), 11–43. DOI: 10.1080/13683500.2014.898617.

Cohen, E. and Cohen, S.A. (2015b) Beyond Eurocentrism in tourism: a paradigm shift to mobilities. *Tourism Recreation Research* 40(2), 157–168. DOI: 10.1080/02508281.2015.1039331.

Dai, B., Jiang, Y., Yang, L. and Ma, Y. (2017) China's outbound tourism – stages, policies and choices. *Tourism Management* 58, 253–258. DOI: 10.1016/j.tourman.2016.03.009.

Dickotter, F. (2010) *Mao's Great Famine: The History of China's Most Devastating Catastrophe*. Bloomsbury, London.

Drower, J. (1982) *Good Clean Fun: The Story of Britain's First Holiday Camp*. Arcadia Books, London.

Eckberg, D.L. and Hill, Jnr., L.H. (1979) The paradigm concept and sociology. *American Sociological Review* 44(6), 925–937.

Economist (2013) Chinese tourists: mind your manners. 6 November.

Ehrentraut, A. (1995) Heritage authenticity and domestic tourism in Japan. *Annals of Tourism Research* 20(2), 262–278. DOI: 10.1016/0160-7383(93)90054-7.

Englander, D. (1995) Napoleonic wars. In: Gardiner, J. and Wenhorn, N. (eds) *The History Today Companion to British History*. Collins and Brown, London, pp. 533–535.

Esdaile, C.J. (2014) *The Wars of Napoleon*. Routledge, London.

eTurboNews (2015) Thailand issues etiquette manuals for Chinese tourists [online]. Available at: http://www.eturbonews.com (accessed 17 February 2015).

Funck, C. (2013) Domestic tourism and its social background. In: Funck, C. and Cooper, M. (eds) *Japanese Tourism: Spaces, Places and Structures*. Berghahn Books, New York and Oxford, pp. 62–95.

Funck, C. and Cooper, M. (2013) Conclusions: retrospect, challenges and the future. In: Funck, C. and Cooper, M. (eds) *Japanese Tourism: Spaces, Places and Structures*. Berghahn Books, New York and Oxford, pp. 210–215.

Ghimire, K.B. and Li, Z. (2001) The economic role of national tourism in China. In: Ghimire, K.B. (ed.) *The Native Tourist: Mass Tourism Within Developing Countries*. Earthscan, London, pp. 86–108.

Gilbert, E. (1954) *Brighton: Old Ocean's Bauble*. Methuen, London.

Graburn, N. (1995) The past in the present in Japan: Nostalgia and neotraditionalism in contemporary Japanese domestic tourism. In: Butler, R. and Pearce, D. (eds) *Changes in Tourism: Peoples, Places, Processes*. Routledge, London, pp. 47–70.

Graburn, N. (2015) Ethnic tourism in rural China: cultural or economic 'development'? In: Diekmann, A. and Smith, M. (eds) *Ethnic and Minority Communities as Tourist Attractions*. Channel View Publications, Bristol, pp. 176–197.

Guangrui, Z. (1985) China ready for new prospect for tourism development. *Tourism Management* 6(2), 141–143. DOI: 10.1016/0261-5177(85)90023-8.

Guangrui, Z. and Lew, A.A. (2003) Introduction: China's tourism boom. In: Lew, A.A., Yu, L., Ap, J. and Gangrui, Z. (eds) *Tourism in China*. Haworth Press, New York, pp. 3–11.

Guichard-Anguis, S. (2008) Introduction: The culture of travel (*tabi no bunka*) and Japanese tourism. In: Guichard-Anguis, S. and Moon, O. (eds) *Japanese Tourism and Travel Culture*. Routledge, London, pp. 1–17.

Guo, R. (2010) *An Introduction to the Chinese Economy*. Wiley, Chichester.

Guo, Z. and Sun, L. (2016) The planning, development and management of tourism: the case of Dangjia, an ancient village in China. *Tourism Management* 56(1), 52–62. DOI: 10.1016/j.tourman.2016.03.017.

Hall, D. (1991) Evolutionary pattern of tourism development in eastern Europe and the Soviet Union. In: Hall, D. (ed.) *Tourism and Economic Development in Eastern Europe and the Soviet Union*. Belhaven Press, London, pp. 78–115.

Hall, C.M. (1992) Sex tourism in South-east Asia. In: Harrison, D. (ed.) *Tourism and the Less Developed Countries*. Belhaven Press, London, pp. 64–74.

Hall, D., Smith, M. and Marciszewska, B. (2006) Introduction. In: Hall, D., Smith, M. and Marciszewska, B. (eds) *Tourism in the New Europe: The Challenges and Opportunities of EU Enlargement*. CAB International, Wallingford, pp. 3–19.

Harrison, D. (2007) Developing a framework for analysing tourism phenomena. *Current Issues in Tourism* 10(1), 61–86.

Harrison, D. and Sharpley, R.A.J. (2017) Introduction. In: Harrison, D. and Sharpley, R.A.J. (eds) *Mass tourism in a Small World*. CAB International, Wallingford.

Harrison, D. (1992) International tourism and the less developed countries: the background. In: Harrison, D. (ed.) *Tourism and the Less Developed Countries*. Belhaven Press, London, pp. 1–18.

Hendry, J. (2008) Fantasy travel in time and space: a Japanese phenomenon. In: Guichard-Anguis, S. and Moon, O. (eds) *Japanese Tourism and Travel Culture*. Routledge, London, pp. 129–143.

Hernandez, J., Bahut, J., Wang, X. and Garcia, R.G. (2015) Sector report: the tourism market in China. EU SME Centre [online]. Available at: http://www.ccilc.pt/sites/default/files/eusme_centre_report_-_tourism_market_in_china_update_-_sept_2015.pdf (accessed 6 August 2016).

Herrero, A.G. and Iwahara, K. (2016) Special Report: Economic Research – Chinese Outbound Tourism Resembling Japan in the 1980s? [online]. Available at: http://www.iberchina.org/files/2016/tourism_chinese_natixis.pdf (accessed 22 July 2016).

Hsu, C., Huang, J. and Huang, S. (2010) Tourism and hospitality research in mainland China: trends from 2000 to 2008. In: Pearce, D.G. and Butler, R.W. (eds) *Tourism Research: A 20-20 Vision*. Goodfellow, Oxford, pp. 147–160.

Huang, Songshan (Sam). (2011) Tourism as the subject of China's doctoral dissertations. *Annals of Tourism Research* 38(1), 316–319. DOI: 10.1016/j.annals.2010.08.005.

Huang, Songshan (Sam). and Chen, G. (2016) Current state of tourism research in China. *Tourism Management Perspectives* 20(1), 10–18. DOI: 10.1016/j.tmp.2016.06.002.

Huang, P. and Pieke, F.N. (2003) China Migration Country Study. *Conference on Migration, Development and Pro-poor Choices in Asia*, Dhaka, 22–24 June.

Huang, Songshan (Sam)., van der Veen, R. and Zhang, G. (2014) New era of China tourism research. *Journal of China Tourism Research* 10(4), 379–387. DOI: 10.1080/19388160.2014.952909.

Jafari, J. and Aaser, D. (1988) Tourism as the subject of doctoral dissertations. *Annals of Tourism Research* 15(3), 407–429. DOI: 10.1016/0160-7383(88)90030-8.

Jaff, G. (2013) *The Atlantic* 21 October [online]. Available at: http://www.theatlantic.com/ china/ archive/2013/10/how-the-chinese-learned-to-embrace-independent-travel/280737 (accessed 8 August 2016).

Jakle, J.A. (1985) *The Tourist*. University of Nebraska Press, Lincoln and London.

Jakle, J.A., Sculle, K.A. and Jefferson, S.R. (1996) *The Motel in America*. John Hopkins University Press, Baltimore and London.

Japan Tourism Agency (JTA) (2014) *White Paper on Tourism in Japan: The Tourism Situation*. Ministry of Land, Infrastructure, Transport and Tourism, Tokyo.

Jie, L. and van Zuylen, H.J. (2014) Road traffic in China. *Procedia - Social and Behavioral Sciences* 111, 107–116. DOI: 10.1016/j.sbspro.2014.01.043.

Johnson, W.H. (2001) *Seaside Entertainment in Sussex*. S.B. Publications, Seaford.

JTB Tourism Research and Consulting Co (2016a) Japanese outbound tourists statistics [online]. Available at: www.tourism.jp/en/ statistics/outbound (accessed 22 June 2016).

JTB Tourism Research and Consulting Co (2016b) Japan-bound statistics [online]. Available at: www.tourism.jp/en/statistics/inbound/ (accessed 22 June 2016).

Kiralova, A. (2006) Tourism in the Czech Republic. In: Hall, D., Smith, M. and Marcisszweska, B. (eds) *Tourism in the New Europe: The Challenges and Opportunities of EU Enlargement*. CAB International, Wallingford, pp. 104–115.

Kong, H. and Cheung, C. (2009) Hotel development in China: a review of the English language literature. *International Journal of Contemporary Hospitality Management* 21(3), 341–355. DOI: 10.1108/09596110910948332.

Kuhn, T. (2012) *(With an Introductory Essay by Ian Hacking) The Structure of Scientific Revolutions*, 4th edn. The University of Chicago Press, Chicago and London.

Leheny, D. (1995) A political economy of Asian sex tourism. *Annals of Tourism Research* 22(2), 267–384.

Li, W. (2004) Environmental management indicators for ecotourism in China's nature reserves: a case study in Tianmushan nature reserve. *Tourism Management* 25(5), 559–564. DOI: 10.1016/j.tourman.2003.06.001.

Li, X. (2016) Introduction. In: Li, X. (ed.) *Chinese Outbound Tourism 2.0*. Apple Academic Press, Oakville, Canada and Waretown, USA, pp. xxv–0.

Li, Y. and Hu, Z. (2008) Red tourism in China. *Journal of China Tourism Research* 4(2), 156–171.

Lim, C. and Pan, G.W. (2005) Inbound tourism developments and patterns in China. *Mathematics and Computers in Simulation* 68(5–6), 498–506. DOI: 10.1016/j.matcom.2005.02.004.

Liu, C., Li, J. and Pechacek, P. (2013) Current trends of ecotourism in China's nature reserves: a review of the Chinese literature. *Tourism Management Perspectives* 7(1), 16–24. DOI: 10.1016/j.tmp.2013.03.001.

Longfren, O. (1999) *On Holiday: A History of Vacationing*. University of California Press, Berkeley.

Lyth, P. (2009) Flying visits: the growth of the British package tours, 1945-1975. In: Segreto, L., Manera, C. and Pohl, M. (eds) *Europe at the Seaside: The Economic History of Mass Tourism in the Mediterranean*. Berghahn Books, New York and Oxford, pp. 11–30.

MacCannell, D. (1976) *The Tourist: A New Theory of the Middle Class*. Macmillan, London.

Mackie, V. (1992) Japan and South-east Asia: the International division of labour and leisure. In: Harrison, D. (ed.) *Tourism and the Less Developed Countries*. Belhaven Press, London, pp. 75–84.
Market Intelligence Group (2013) *Japan*. European Travel Commission, Brussels.
Middleton, V.T.C. and Lickorish, L.J. (2005) *British Tourism: The remarkable story of growth*. Elsevier, Amsterdam.
Moon, O. (2008) Japanese tourists in Korea: colonial and post-colonial encounters. In: Guichard-Anguis, S. and Moon, O. (eds) *Japanese Tourism and Travel Culture*. Routledge, London, pp. 147–172.
Moore, A. (1985) Rosanzerusu is Los Angeles an anthropological inquiry of Japanese tourists. *Annals of Tourism Research* 12(4), 619–643. DOI: 10.1016/0160-7383(85)90081-7.
OECD (Organisation for Economic Co-operation and Development) (2016) *Tourism Trends and Policies, 2016*. OECD Publishing, Paris.
Ousby, I. (1990) *The Englishman's England*. Cambridge University Press, Cambridge.
Pearce, P.L. and Chen, T. (2012) Citizens' representations of China's golden weeks. *Asia Pacific Journal of Tourism Research* 17(4), 394–415. DOI: 10.1080/10941665.2011.628331.
Peng, X. (1987) Demographic consequences of the great leap forward in China's provinces. *Population and Development Review* 13(4), 639–670. DOI: 10.2307/1973026.
Pimlott, J.A.R. (1947) *The Englishman's Holiday: A Social History*. Faber and Faber, London.
Ritzer, G. (2014) *The McDonaldization of Society*, 8th edn. Sage, London.
Ross, G.F. (1994) *The Psychology of Tourism*. Hospitality Press, Melbourne.
Ryan, C. (2003) *Recreational Tourism: Demand and Impacts*. Channel View Publications, Clevedon.
Ryan, C. and Shen, H. (2017) Mass tourism and China. In: Harrison, D. and Sharpley, S. (eds) *Mass Tourism in a Small World*. CAB International, Wallingford.
Schudson, M.S. (1979) On tourism and modern culture. *American Journal of Sociology* 84(5), 1249–1258.
Shorrocks, A.S. and Davies, G. (2015) *Global Wealth Databook, 2015*. Research Institute, Credit Suisse A.G, Zurich.
Soane, J.V.N. (1993) *Fashionable Resort Regions: their Evolution and Transformation, with Particular Reference to Bournemouth, Nice, Los Angeles and Wiebaden*. CAB International, Wallingford.
Sofield, T.H.B. and Li, F.M.S. (1998) Tourism development and cultural policies in China. *Annals of Tourism Research* 25(2), 362–392. DOI: 10.1016/S0160-7383(97)00092-3.
Song, K. and Cui, A. (2009) Targeting key segments of China's diverse and rapidly emerging middle class will be crucial as household incomes rise. China Business Review. 1 January.
Surman, B. (2008) The search for the real thing: Japanese tourism to Britain. In: Guichard-Anguis, S. and Moon, O. (eds) *Japanese Tourism and Travel Culture*. Routledge, London, pp. 193–201.
Sziakov, O. (2013) Chinese tourists misbehaving again [online]. Available at: http://www.eturbonews.com (accessed 12 October 2013).
Tatlow, D.K. (2015) China hopes 'blacklist' will curb its travelers' misbehaviour abroad. *The New York Times* 25 May 2015.
Teon, A. (2015) Now the Japanese are Mad at Chinese Tourists [online]. Available at: https://thenanfang.com/mainland-chinese-tourists-bad-behaviour-angers-japanese (accessed 6 August 2016).
Thompson, C. (2009) Cultural solutions to ecological problems in contemporary Japan: heritage tourism in Asuke. In: Singh, S. (ed.) *Domestic Tourism in Asia: Diversity and Divergence*. Earthscan, London, pp. 51–79.
Tokuhisa, T. (1980) Tourism within, from and to Japan. *International Social Science Journal* 32(1), 129–150.
Travis, J.F. (1993) *The Rise of the Devon Seaside Resorts, 1750-1900*. University of Exeter Press, Exeter.
Tsang, N.K.F. and Hsu, C.H.C. (2011) Thirty years of research on tourism and hospitality management in China: a review and analysis of journal publications. *International Journal of Hospitality Management* 30(4), 886–896. DOI: 10.1016/j.ijhm.2011.01.009.
UNWTO (2015) *UNWTO Tourism Highlights, 2015 Edition*. UNWTO, Madrid.
van Tubergen, A. and van der Linden, S. (2002) A brief history of spa therapy. *Annals of the Rheumatic Diseases* 61(3), 273–275. DOI: 10.1136/ard.61.3.273.
Walker, A.G. (2015) *China under Mao: a Revolution Derailed*. Harvard University Press, Cambridge, MA.
Walton, J. (1978) *The Blackpool Landlady: A Social History*. Manchester University Press, Manchester.
Walton, J.K. (2000) *The British Seaside: Holidays and resorts in the twentieth century*. Manchester University Press, Manchester.
Walvin, J. (1978) *Leisure and Society: 1830-1950*. Longman, London and New York.

Wang, H., Yang, Z., Chen, L., Yang, J. and Li, R. (2010) Minority community participation in tourism: a case of Kanas Tuva villages in Xinjiang, China. *Tourism Management* 31(6), 759–764. DOI: 10.1016/j.tourman.2009.08.002.

Weber, M. (1949) *The Methodology of the Social Sciences (Translated and Edited by Edward A. Shils and Henry A. Finch)*. Free Press, New York.

Wen, J.J. and Sinha, C. (2009) The spatial distribution of tourism in China: trends and impacts. *Asia Pacific Journal of Tourism Research* 14(1), 93–104. DOI: 10.1080/10941660902756776.

Williams, D. (2013) 30 years of tourism research: a bibliography of Japanese tourism (1980-2009). *Josai International Bulletin* 21(7), 43–70.

Williams, A. and Shaw, G. (1988) Western European tourism in perspective. In: Williams, A. and Shaw, G. (eds) *Tourism and Economic Development: Western European Experiences*. Wiley, Chichester, pp. 12–39.

Winter, T. (2009) Asian tourism and the retreat of anglo-western centrism in tourism theory. *Current Issues in Tourism* 12(1), 21–31. DOI: 10.1080/13683500802220695.

World Travel and Tourism Council (2015a) *Travel and Tourism: Economic Impact 2015 – China*. WTTC, London.

World Travel and Tourism Council (2015b) *Travel and Tourism: Economic Impact 2015 – Japan*. WTTC, London.

Wright, S. (2002) Sun, sea, sand and self-expression: mass tourism as an individual experience. In: Berghoff, H., Korte, B., Schneider, R. and Harvie, C. (eds) *The Making of Modern Tourism: The Cultural History of the British Experience, 1600-2000*. Palgrave, Houndmills, pp. 181–202.

WTO (World Tourism Organisation (1990) *Compendium of Tourism Statistics, 1985-1989*. WTO, Madrid.

Wu, B., Zhu, H. and Xu, X. (2000) Trends in China's domestic tourism development at the turn of the century. *International Journal of Contemporary Hospitality Management* 12(5), 296–299. DOI: 10.1108/09596110010339652.

Xiang, Y. (2013) The characteristics of independent Chinese outbound tourists. *Tourism Planning & Development* 10(2), 134–148. DOI: 10.1080/21568316.2013.783740.

Xiao, H. and Smith, S.L.J. (2006) Towards a paradigm shift of knowledge: implications for tourism research in China. *Chinese Tourism Research* 2(4), 402–422.

Xiao, H., Su, D. and Li, M. (2010) Diffusion of knowledge across linguistic communities: the case of using "foreign" sources for tourism research in China. *Journal of China Tourism Research* 6(4), 326–342. DOI: 10.1080/19388160.2010.527569.

Xie, Y. and Li, M. (2009) Development of China's outbound tourism and the characteristics of its tourist flow. *Journal of China Tourism Research* 5(3), 226–242. DOI: 10.1080/19388160903149965.

Xu, G. (1999) Socio-economic impacts of domestic tourism in China: case studies in Guilin, Suzhou and Beidaihe. *Tourism Geographies* 1(2), 204–218. DOI: 10.1080/14616689908721310.

Xu, K., Yan, T. and Zhu, X. (2013) Commodification of Chinese heritage villages. *Annals of Tourism Research* 40, 415–419. DOI: 10.1016/j.annals.2012.08.009.

Yamashita, S. (2008) The Japanese encounter with the South: Japanese tourists in Palau. In: Guichard-Anguis, S. and Moon, O. (eds) *Japanese Tourism and Travel Culture*. Routledge, London, pp. 172–192.

Yang, C.-H., Lin, H.-L. and Han, C.-C. (2010) Analysis of international tourist arrivals in China: the role of World Heritage Sites. *Tourism Management* 31(6), 827–837. DOI: 10.1016/j.tourman.2009.08.008.

Yang, J., Ryan, C. and Zhang, L. (2013) Social conflict in communities impacted by tourism. *Tourism Management* 35, 82–93. DOI: 10.1016/j.tourman.2012.06.002.

Zhang, G. (2016) Evolution of China's policy of outbound tourism. In: Li, X. (R). (ed.) *Chinese Outbound Tourism 2.0*. Apple Academic Press, Oakville, Canada and Waretown, USA, pp. 21–37.

Zhang, A. and Chen, H. (2003) Evolution of China's air transport development and policy towards international liberalization. *Transportation Journal* 42(3), 31–49.

Zhang, H.Q. and Heung, V.C.S. (2002) The emergence of the mainland Chinese outbound travel market and its implications for tourism marketing. *Journal of Vacation Marketing* 8(1), 7–12. DOI: 10.1177/135676670200800102.

Zhong, L., Leung, D., Law, R. and Wu, B. (2013) eTourism in China: a review of the Chinese-Language literature. *Asia Pacific Journal of Tourism Research* 18(5), 464–482. DOI: 10.1080/10941665.2012.680974.

Zhong, L., Wu, B. and Morrison, A.M. (2015) Research on China's Tourism: A 35-Year Review and Authorship Analysis. *International Journal of Tourism Research* 17(1), 25–34. DOI: 10.1002/jtr.1962.

Zhou, W. and Szyliowicz, J. (2006) The development and current status of China's transportation system. *World Transport Policy and Practice* 12(4), 10–16.

19

Mass Tourism in a Small World (Co-authored with Richard Sharpley)

The Emergence of Mass Tourism

Large-scale tourism is not an entirely new phenomenon: its antecedents can be found in the large-scale festivals and games of ancient Egypt, Greece and Rome, in pilgrimage, which has long been a feature of the major world religions, and in spa tourism, which predates Roman times and has long been established in many regions in Europe and North America (Towner, 1996, pp. 53–95; van Tubergen and van der Linden, 2002). The Grand Tour of Europe, popular over much the same period, involved relatively few travellers but nevertheless had a lasting impact on places visited by tourists, on the travellers themselves, and on the culture of their home societies (Hibbert, 1987; Black, 1992; Towner, 1996, pp. 96–138).

However, consistent large-scale, systematic and regular travel for leisure purposes really only emerged in the second half of the 18th century. Impetus came in the mid-1700s from Dr Richard Russell and other physicians who popularized the medical benefits of sea bathing (Gilbert, 1954), thus providing an alternative to inland spas, but it was the railways, from the mid-1800s, that established mass tourism in the UK and elsewhere (Towner, 1996) and which enabled Thomas Cook, almost by accident, to develop (though not start) package tours, later taking tourists overseas (Pimlott, 1947, pp. 168–169, 191–194; Buzard, 1993). Increasingly, for nearly a century, holidaymakers in the developed world went on holiday en masse and small coastal communities accessible to metropolitan centres themselves became urbanized. The prominence of such resorts as Blackpool and Brighton in the UK, for instance, dates back to this time, and others soon followed, in the UK (Pimlott, 1947; Gilbert, 1954; Walton, 1978), in other parts of Europe (Corbin, 1994, pp. 255–281) and also in the USA, as seen in the growth in the early 1800s of Atlantic City on the New Jersey coast and resorts on the California coast (Jakle, 1985).

Road improvements and technical advances in transport led to other developments. In the late 1800s, increasing numbers of enthusiasts took up cycling, though it remained popular for much longer in Europe than in the USA (Lofgren, 1999, p. 69). Better road surfaces, improved technology in automobile production and decreasing prices opened up the countryside to motorized transport, especially after the motor car began to be mass produced. As Jakle (1985, p. 270) notes, in the USA during the period between the two world wars 'automobiling was embraced by and dominated by the masses', leading to the transformation of the countryside, not least by increasing numbers of camping sites and motels that catered to these new tourists, while national parks such as Yosemite and Yellowstone came under increasing pressure to accommodate visitors (Jakle, 1985; Lofgren, 1999).

For more than a century, then, coastal resorts and other destinations in the UK, mainland Europe and the USA catered for the leisure and entertainment needs of vast numbers of their increasingly affluent but highly differentiated populations (Pimlott, 1947; Jakle, 1985; Soane, 1993). It was undoubtedly mass tourism, but it occurred largely within national boundaries. Later, the first shoots of mass international tourism appeared in the 1920s and 1930s, some of which were state sponsored (Baranowski, 2005, p. 130; Dann and Parrinello, 2009), but it was only after 1945, as Europe reaped the dividends arising from peace, increased prosperity, a widespread rise in disposable income, good communications and improvements in aircraft technology, that international leisure travel developed rapidly on a mass scale.

Modern Mass Tourism: The General Pattern

In 1950, when little more than 25 million international tourist arrivals were recorded, most international trips were almost exclusively to Europe (66%) and North America (30%), usually within the same region. By 1990, although arrivals to all regions had substantially increased, totalling nearly 440 million, the market shares of Europe and North America were 64% and 20%, respectively (Harrison, 1992a, p. 5, 2001). If we then fast forward to 2014, it is clear that while both international arrivals and receipts from international tourism had increased, as indicated in Table 19.1, destinations in other parts of the world had grown in importance. As economic development has occurred elsewhere, Europe's share of international tourist arrivals of 1.133 billion has been reduced to less than 52%, that of the Americas to about 16%, while that of Asia and the Pacific (of negligible importance in 1950) is now more than 23%, and accounts for more than 30% of international tourism receipts (United Nations World Tourism Organization (UNWTO), 2015).

Table 19.1. International tourism arrivals and receipts: selected years, 1950–2014. (Compiled by the authors from UNWTO data.)

Year	Arrivals (million)	Receipts (US$ billion)	Year	Arrivals (million)	Receipts (US$ billion)
1950	25.3	2.1	1999	639.6	464.5
1960	69.3	6.9	2000	687.0	481.6
1965	112.9	11.6	2001	686.7	469.9
1970	165.8	17.9	2002	707.0	488.2
1975	222.3	40.7	2003	694.6	534.6
1980	278.1	104.4	2004	765.1	634.7
1985	320.1	119.1	2005	806.6	682.7
1990	439.5	270.2	2006	847.0	742.0
1991	442.5	283.4	2007	903.0	856.0
1992	479.8	326.6	2008	917.0	939.0
1993	495.7	332.6	2009	882.0	851.0
1994	519.8	362.1	2010	940.0	927.0
1995	540.6	410.7	2011	995.0	1042.0
1996	575.0	446.0	2012	1035.0	1075.0
1997	598.6	450.4	2013	1087.0	1159.0
1998	616.7	451.4	2014	1133.0	1245.0

The emergence of China as both a destination and a source of tourists is especially noteworthy. In 1978, the year it opened out to the West, it received 229,600 foreign (i.e. non-Chinese) tourists (Guangrui and Lew, 2003). In 2015, 20.3 million arrived, along with 27.1 million and 4.7 million from Hong Kong and Macau, respectively, and a further 4.8 million from Taiwan (Guide, 2016). China is now one of the world's top destinations, attaining fourth place in the United Nations World Tourism Organization's (UNWTO) league tables of international arrivals and third place in that of international tourism receipts (United Nations World Tourism Organization (UNWTO), 2015). It is also the world's fastest growing source of tourists: since 2012 its citizens have spent more than those of any other nation, in 2014 accounting for more than 13% of global tourism receipts (United Nations World Tourism Organization (UNWTO), 2015), in the process changing and increasing the tourism profile of several destinations visited by Chinese tourists. Currently, Hong Kong, Macau, South Korea, Thailand, Singapore, Malaysia, the USA, Vietnam, Japan and France are preferred destinations, though the Chinese are targeted worldwide as tourists. Moreover, as only about 5% of Chinese citizens possessed a passport in 2014 (Arlt *et al.*, 2016), at the time of writing all the evidence points to the continuation of these trends in the foreseeable future, a situation which has attracted increasing academic attention, from within and outside China, over the last decade (Harrison, 2017). Indeed, the UNWTO estimates that, more generally, by 2030 international arrivals will amount to some 1.8 billion, with tourism in emerging nations growing at twice the rate of that in advanced economies, by which time the former will have 57% of the market share of international arrivals (United Nations World Tourism Organization (UNWTO), 2015).

Finally, it is worth noting that while international mass tourism has increased remarkably since the 1950s, so too has domestic tourism, building on a much wider base since the mid-19th century. Currently, domestic trips are reported by UNWTO to number between 5 billion and 6 billion a year (United Nations World Tourism Organization (UNWTO), 2015), a margin of difference that suggests hugely unreliable statistics (Ghimire, 2001, pp. 11–15; United Nations World Tourism Organization (UNWTO, 2013, pp. 13–14). Nevertheless, globally domestic tourism is of major economic importance, including in many developing countries where domestic arrivals sometimes amount to more than five times those of international tourists (Ghimire, 2001; United Nations World Tourism Organization (UNWTO), 2013; Harrison, 2017). Indeed, as Ryan notes in Chapter 13, this volume, in 2015 the Chinese undertook around 4 billion domestic tourism trips.

Scholarly Approaches to Mass Tourism

Historians have carried out considerable research on mass tourism. Its development has been traced generally by Boyer (2007) while others have charted the more specific history of holidaymaking in English resorts (Pimlott, 1947; Gilbert, 1954; Walton, 1978, 2000; Travis, 1993) and elsewhere (Black, 1992; Hudson, 1993; Towner, 1996). There have also been important comparative studies of resort development in France, the USA and Germany (Soane, 1993; Berghoff *et al.*, 2002; Borsay and Walton, 2011) while other historians and some social scientists have focused more thematically on the sea or the role of the beach – especially as a liminal, pleasure periphery – in popular culture (Shields, 1991, pp. 73–116; Corbin, 1994; Lencek and Bosker, 1999; Urbain, 2003).

Interestingly, numerous historical records exist of reactions to tourists. These include: (i) Seneca's condemnation of drunks in the 1st century AD at Baiae (Casson, 1974, p. 143); (ii) disagreements over the value of the Grand Tour (Hibbert, 1987; Black, 1992; Towner, 1996); (iii) the Reverend Francis Kilvert's 1870 characterization of English tourists as 'vulgar, illbred, offensive and loathsome' (Plomer, 1992); and (iv) Wordsworth's objections that 'his' English Lake District was being threatened by excursionists from the industrial towns of Lancashire and Yorkshire, who wanted 'wrestling matches, horses and boat races without number' (Ousby, 1990, p. 192). Anticipating many modern criticisms of tourists, too, residents of late 19th-century Brighton blamed tourism for drunkenness, licentious behaviour and prostitution, asserting that working-class visitors, in particular, were of little economic benefit to the town because they carried their own food and drink with them (Gilbert, 1954, pp. 204–206). Criticisms of tourists in other parts of Europe took a similar (and equally modern) tone (Buzard, 1993, pp. 37–40).

Arguably, though, there has been little cross-fertilization across the boundary between historians of tourism and social scientists operating within the field of tourism studies. As Walton (2005, p. 2) suggests, this might be the result of stylistic and methodological differences, but other factors are also at work. For example, blessed with the benefit of hindsight, historians have generally adopted a largely neutral position, whereas non-historians considering modern tourism and its immediate impacts, especially in developing societies, have often been more judgemental (and possibly more anxious to distance themselves from other visitors). This was certainly evident in the 1970s when MacCannell, whose pioneering study set the standard for so much of what was to follow, once referred to tourists as 'an expeditionary force without guns' (1999, p. xxiv). Similarly, Krippendorf, an economist and environmentalist, depicted modern tourists as vainly trying to escape alienation from industrial and differentiated society, adding that while their 'carefree and ignorant' behaviour (1987, p. 43) brought them little benefit, it did great damage to their destinations, creating 'a new and devious form of colonialism' (Krippendorf, 1987).

Over the last three decades, however, scholarly attitudes towards tourism have reportedly shifted. According to Jafari (1989), they have moved from acceptance, through caution and adaptation, to a more 'knowledge-based' approach. Though somewhat simplistic – Jafari himself accepts none of these 'platforms' have been superseded – the knowledge-based approach frames our understanding of tourism in 'developed' societies. This seems evident when comparing the work of sociologists and anthropologists in the 1970s and 1980s with later studies of developed country destinations, especially in the Mediterranean. Much of the latter centred on mass tourism's economic impacts (Segreto *et al.*, 2009), its social and cultural features for both tourists (Andrews, 2011) and residents (Waldren, 1996; O'Reilly, 2000), and its relationship to modernity, such as in the case of Cyprus (Sharpley, 2003). Others examined the interactional spaces created by tourism (Boyer, 2007; Obrador Pons *et al.*, 2009) and how residents have reacted and adapted to mass tourism more generally (Boissevain, 1996), while McGibbon has carried out similar research on tourism's impacts on an Austrian Alpine village (McGibbon, 2000). Importantly, too, there have been serious efforts to address the need to manage the impact of tourism in coastal resorts in Southern Europe (Boyer, 2007) and – of especial value, because of its comparative approach – to analyse tourism's problematic impacts in coastal resorts across the world (Agarwal and Shaw, 2007).

Over recent decades, the growth of mass tourism in European cities, some already established as destinations for centuries, has been phenomenal (Berger, 2015), bringing many undoubted economic benefits (Ashworth and Tunbridge, 1990, p. 260). Rapid expansion in city tourism is also occurring elsewhere: Dubai, for instance, is expecting 20 million visitors by 2020, of which it hopes some 500,000 will be medical tourists (eTurboNews, 2015). Inevitably, problems have emerged, especially in some European capitals (Ashworth and Tunbridge, 1990, pp. 254–265; Dumbrovská and Fialová, 2014; Settis, 2016), sometimes leading to a resident backlash (Keeley, 2014; Tjolle, 2014; Jones, 2016, 19 February).

More generally, there are clearly concerns about tourism's environmental impacts and its contribution to climate change (Croall, 1995; Holden, 2007; Becken, 2013). For some time, there have been increasingly urgent warnings about the impacts of uncontrolled mass tourism in Venice (Settis, 2016), and Barcelona is now imposing restrictions on further tourism development, instigating a moratorium on new accommodation facilities and diverting income from taxes raised from tourism away from marketing the city to social and environmental improvement programmes (Blanchar, 2015). However, generally tourism in developed countries is considered to be an economic – even social – necessity: the Lake District welcomes visitors, Brighton is now a highly popular city, and cities across the developed world compete for international tourists. Despite numerous reservations, no one seems to be advocating that London, Paris, Rome, Prague or Budapest – or even Venice and Barcelona – cease being tourist attractions. Instead, one country after another, developed or developing, competes to attract yet more tourists, especially (in view of the economic upsurge of East Asia) the Chinese. Indeed, in France, one of their preferred European destinations, it has been suggested that only tourism has the ability to save the national economy (Dedieu and Mathieu, 2015), while the UK government's 2015 *Five Point Plan* also makes a special point of attracting more Chinese visitors (Department for Culture, Media and Sport (DCMS), 2015). This is not to suggest that mass tourism does not raise serious concerns; however, rather than appealing for tourism to be abolished, the common reaction is to call for controls and more effective planning of tourist numbers.

Arguably, tourism in the developing world has attracted much greater criticism. Again, numerous balanced studies exist, including work in Indonesia by anthropologists such as Picard (1996) and Cole (2008) and on tourism's role in poverty reduction (e.g. Mitchell and Mann, 2010). Nevertheless, in the mid-1970s, tourism was branded 'a corrupter of innocence' in the South Pacific, while in Kenya it allegedly perpetuated 'the myth of the white hunter' and reinforced 'colonial values' (Turner and Ash, 1975, p. 175). Contributors to important early volumes edited by Smith (1977) and de Kadt (1979) were clearly sceptical about tourism's benefits, though some modified their objections over time (Smith and Brent, 2001, p. 7). Indeed, while local residents were largely appreciative, Smith herself was horrified when the island of Boracay in the Philippines became a tourist resort (Smith, 1992) and later developments did little to change her opinion (Smith, 2001).

Others, following development theory, have depicted changes in theoretical orientation from modernization to dependency, neo-liberalism to alternative development and, most recently, to sustainability (Sharpley, 2009; Telfer, 2015; Harrison, 2015). Again, such chronologies can be misleading and apparently dated perceptions of tourism continue to coexist in repackaged form: modernization merged into neo-liberalism and the advocacy of trade liberalization, while dependency theory, along with its critique of capitalist (under) development (Harrison, 2014, pp. 146–148), has contributed to environmentalism and the advocacy of sustainability and 'alternative' tourism.

Critics of mass tourism's impacts in developing societies (but not, it would seem, in most developed societies) point to a series of destructive features: (i) commoditization undercuts tradition and leads to inauthenticity; (ii) jobs in tourism are unskilled and demeaning; (iii) income from large-scale (and highly capitalistic) tourism is frequently reduced by damaging 'leakages'; and (iv) family structures and local economies are unbalanced by the search for the tourist dollar. Such criticisms can be countered (Macnaught, 1982; Harrison, 1992b), but the focus on underdevelopment and advocacy of sustainable tourism development in developing societies remains. As an example, the popular text by Mowforth and Munt on tourism and sustainability in what they continue to refer to as 'The Third World', ends with the somewhat gloomy prediction that, in a world dominated by capitalism and unequal power relations, change is unlikely to occur. There might be some local improvements, but all we can expect is 'more of the same' (Mowforth and Munt, 2016).

The dominance of sustainability perspectives in the study of tourism is, in itself, evidence of major dissatisfaction – at least in academic circles, but less so among practitioners – with mass tourism (which is frequently regarded as synonymous with modern tourism). Emphasis on the desirability of and benefits from 'alternative' tourism (in its many forms), along with sustainable tourism development, really *presupposes* the validity of the critique of mass tourism. Nevertheless, the usefulness of such notions as 'sustainable tourism development' and 'alternative' tourism has not gone uncontested. Some time ago, for example, Butler (1992, p. 44) argued mass tourism was with us to stay and that 'alternative tourism' was less of a panacea than often claimed, while Lanfant (1992) suggested alternative tourism had 'not broken radically with the "other tourism"' and, for Cohen (1992, p. 275), 'criticism of established mass tourism appears too radical'. In fact, there is little academic consensus on the meaning of sustainable tourism development and little empirical evidence that such a concept can be usefully applied (Sharpley, 2009, p. 175), as revealed when focusing on social structures and cultures (Harrison, 1996, 2014, p. 148). This has not prevented Weaver from suggesting (2012, 2013), that any conceptualization of sustainable tourism development must include the possibility that, along with sustainable alternative tourism, there must be a willingness to bring about sustainable mass tourism.

At the same time, powerful defences of mass tourism have also emerged.... Butcher (2003) has challenged portrayals of 'the new tourist' as morally superior to the mass tourist, arguing instead that 'guilt-free' mass tourism brings genuine development and opportunities to developing societies (Butcher, 2003), while Aramberri makes a similar claim, provocatively asserting that, through liberal capitalism, 'mass societies have generally put more cash in their members' pockets, henceforth creating, among other things, an impressive growth in MMT [Modern Mass Tourism]' (Aramberri, 2010).

The role of such writers in urging the need to address the issues of mass tourism, rather than simply subjecting it to blanket criticism, is acknowledged, as are efforts of those who have tried to create typologies of tourists. Special mention should be made here of Cohen's (1972) categorization of tourists as 'institutionalized' or 'non-institutionalized', according to the extent they rely on others to plan their trips, and his later continuum of tourist experience, ranging from secular hedonism to the search for ultimate meaning (Cohen, 1979). His and other key contributions have been summarized by Smith (2000), Ryan (2003) and Chen (2015), who show that tourists can be defined, *inter alia*, by their motivation for travel, their interests and personality types, their age or position in the life cycle, their interests, and the economic, social, cultural and environmental impacts on the places they visit. By contrast, like MacCannell (1999) and Krippendorf (1987), others have seen tourism's changing features as a reflection of their home societies, relating 'new tourism' to 'the burgeoning new middle classes of the First World' and 'new middle classes from the Third World, most particularly from South-east Asia' (Mowforth and Munt, 2016).

Towards an Ideal Type of Mass Tourism

Too often, tourism academics, non-government agencies and, in particular, aid agencies have ignored the 'elephant in the room' and focused on small-scale, 'sustainable' projects. As argued elsewhere (Harrison, 2015, pp. 65–71), however, this position is untenable: (i) capitalism and international tourism are likely to continue into the foreseeable future; (ii) mass tourism will continue to be the norm; and (iii) virtually all forms of 'alternative' tourism will remain linked to and dependent on mass tourism, and none of them will replace it.

We live in a globalized world. Correspondingly, the overall unit of analysis is a global, capitalist system involving mass production and mass consumption, within which, as the pattern of international tourist movements indicates, national, regional and international boundaries are continuously criss-crossed, for most of the time as a matter of course. As in any system, what happens in one part has knock-on effects on others. The 1920s' fashion for a sun tan led relatively wealthy Northern Europeans to seek the sun in the Mediterranean; later, after the Second World War, surplus aircraft, improved technology and increased disposable incomes prompted many more holidaymakers from the colder parts of Europe to leave their traditional seaside resorts (and their unpredictable weather), in favour of (cheap) sun, sea and sand in the Southern Mediterranean. In short, the rise of the Mediterranean resorts led *directly* to the decline of such UK resorts as Blackpool, Eastbourne and Brighton. More recent examples include the impact on tourist demand of terrorist attacks on such popular destinations as Bali and Tunisia, and the general impact of the global financial crisis of 2008–2010.

A further example of globalization is the post-Second World War spread of the transnational tourist corporation (TTC), especially hospitality chains and tour operators. As Bianchi (2015, p. 328) has observed, 'an increasingly complex and differentiated geography of tourism production, distribution and exchange has emerged, underwritten by the forces of economic globalization and market liberalization'. More recently, the influence of TTCs has spread to China, whose renminbi is sought by destinations in developed and developing countries alike. It is but one example of the fact that, irrespective of their rankings in World Bank income tables, nearly all nation states seek tourists, frequently competing with one another for visitors from both old and new markets. Indeed, with the spread of TTCs and the emergence of East and South-east Asia as sources and destinations of international tourists, it no longer makes sense to distinguish artificially between 'development' in the West and elsewhere. Rather, it is imperative they are seen as component and intrinsic parts of a global system (Harrison, 2015).

It then follows that the continuing failure to compare historical studies of Western tourism with studies of tourism development in today's developed and developing countries cannot be justified. As pointed out elsewhere, 'nearly 150 years after the introduction of package tours for large numbers of people in Europe, the debates it engendered and the processes it involved continue to be relevant' (Harrison, 2001, p. 3). In particular, the increased purchasing power of a growing middle class, along with improvements in transport and communications, which characterized the emergence of large-scale tourism in the 19th and 20th centuries in Europe and the Americas, are similarly influential in the growth of modern tourism from and to East Asia and the Pacific. New tourism centres and peripheries are undoubtedly being formed, but the *processes* underlying their formation are not new (Harrison, 2017).

With the above axioms in mind, we move to our analysis of mass tourism. Following Weber, we define it by noting the presence of major features which, taken together, form an ideal type, an heuristic device which facilitates comparison of social phenomena with one another and against a constant yardstick (Weber, 1949). We consider the following features of mass tourism to be critical, though not all need be (or are likely to be) found in any one example of mass tourism; rather, in every case, there will be more or less conformity to the ideal type.

1. There is a regular and systematic movement, involving many industries, of large numbers of people away from their normal places of residence, primarily for holiday purposes.
2. Tourist numbers at destinations are concentrated and seasonal.
3. The major stakeholders in mass tourism – governments and providers of transport, accommodation and attractions – operate in order to make a profit and may benefit from economies of scale.
4. Travel, accommodation and attractions, designed for large numbers of people, are structured and organized by specialist organizations, often with strong links across national boundaries, and often in packages.
5. Control of key elements of the tourism industry rests outside the destinations, possibly with transnational companies.
6. There is concern about the economic, social, cultural or environmental consequences of tourism, which may include commoditization and standardization of production, changing social structures, loss of 'authenticity' and environmental degradation.
7. National and international institutions have emerged to support or oppose tourism and respond to its consequences.
8. Impacts are contingent not only on absolute numbers of tourists, but also on the nature of the destination population, its prior experience of tourism, and cultural and other differences between tourist and destination resident.
9. Interaction between tourist and resident is fleeting and superficial and restricted to commercial contexts involving the provider and recipient of services.
10. Where there are marked economic, cultural or social inequalities between tourist and resident, social interaction might be underpinned by stereotypes and/or power disparities.
11. Distinctions might be made: (i) between tourists who focus primarily on 'nature' (admittedly a broad term) and those whose activities occur predominantly within an urban environment; and (ii) between those interested primarily in hedonistic pursuits and others involved in 'higher' or more 'spiritual' activities. Nevertheless variations may well occur through what Pearce has described as the travel career ladder (TCL), and tourists' motivation can be analysed across many typologies (and combinations of motivation and tourist types) (Pearce, 2005).

In portraying these ideal typical features as continua (Table 19.2), first, it should be noted

Table 19.2. Mass tourism: ideal type continua.

LOW	Numbers of tourists and ratios to resident population	HIGH
LOW	Physical concentration of tourist accommodation	HIGH
LOW	Levels of seasonality	HIGH
LOW	Extent tourism is established in the region	HIGH
LOW	Extent tourism is for profit	HIGH
LOW	Local control of tourism sector	HIGH
LOW	Level of formal organization of tourism sector	HIGH
LOW	Economic impacts	HIGH
LOW	Social impacts	HIGH
LOW	Environmental impacts	HIGH
LOW	Opposition to tourism	HIGH
LOW	Intensity of tourist–local interaction	HIGH
LOW	Levels of social inequality between tourists and residents	HIGH
LOW	Extent to which interaction of tourists and locals is mediated by others (e.g. guides)	HIGH
LOW	Nature–urbanization	HIGH
LOW	Spirituality–hedonism	HIGH

that we have not further specified types of tourist, though other criteria are clearly possible. Indeed, the term 'mass tourism' covers a wide variety of travel contexts, including: (i) travellers' motivations; (ii) the organization and underlying process of their journey; (iii) their choice of destination; and (iv) the activities they pursue on arrival. However, in some sense, tourists in virtually every typology produced by tourism scholars can be regarded as mass tourists, participating within an overall system designed to move people, en masse, cheaply, safely and quickly from one part of the globe to another.

Second, it is recognized that several features of the ideal type might also be applied to other travellers, including the military, traders and pilgrims. Writing at a time when there is a mass movement of displaced people from the war zones of the Middle East and Africa, it should be evident that other forms of 'mobility' often overlap with tourism. Indeed, despite the often tragic differences, economic migrants and refugees have much in common with recreational tourists (Hannam, 2009; Amin, 2016, 4 March; Pechlaner and Nordhorn, 2016).

Third, although not addressed in this foray into mass tourism, if tourism development is to be genuinely comparative across time as well as place, research methods need to be considerably refined and expanded. Tools developed by macroeconomists to assess economic processes underlying modern tourism development have been well honed (Dwyer *et al.*, 2010) but are difficult to apply to tourism development in earlier periods. Department for Culture, Media and Sport (DCMS) (2015) Irritation Index, too, is perhaps less appealing or appropriate when used to assess (if indeed it were possible) local reactions to tourism over, say, even a 50-year period. Butler's model of the Tourism Area Life Cycle (TALC) (Butler, 1980) has proved more versatile, having been used extensively to assess the social impact of increasing tourist numbers across a wide range of destinations in both developed and developing societies (Butler, 2006a, b) and, in at least one instance, has been applied to more distant historical material (Weaver, 2006). More complex concepts do exist, for example 'carrying capacity' and its more recent variant 'limits of acceptable change', or the Organization for Economic Co-operation and Development's (OECD) core indicators for environmental performance, as applied to a Spanish mass tourism resort (Rebollo and Baidal, 2003). However, these are hard to apply in modern mass tourism destinations, and even more so to historical or current destinations at earlier stages of development.

In addition, statistical techniques used to measure the impacts of large numbers of tourists are quite crude. Several measures used to compare concentrations of tourists at different destinations, for example the Tourist Intensity Rate, Tourist Penetration Rate and Tourism Density Ratio (Harrison, 1992a, p. 12; Dumbrovská and Fialová, 2014; McElroy, 2015), provide only a rough indication of impact. Their use is neither standardized nor widespread; they rarely distinguish between high and low seasons; and they cannot take into account the many historical and cultural factors that affect the social and economic impacts of tourism on destinations.

In the end, numbers are not the only factor: large numbers of tourists may be absorbed relatively comfortably, as in world cities with a long history of receiving visitors, whereas even a few visitors to destinations unaccustomed to tourists, *especially* those who want to understand local culture and meet local people, may be highly disruptive. We have thus refrained from polarizing small- and large-scale tourism, recognizing that 'sustainability' (in so far as it can be defined) is not automatically in an inverse relationship to scale: 'large' does not automatically mean 'ugly', any more than 'small' is necessarily 'beautiful' (Harrison, 2012).

About This Book: A Preview

In this book, we have endeavoured to balance theoretical perspectives, historical approaches and case studies of specific regions. Following this introduction, the six chapters in Section 2 present a range of theoretical approaches, in which several established scholars of mass tourism build on their previous work. Aramberri (Chapter 2) and Butcher (Chapter 3) focus on general features of mass tourism, with the former arguing that while tourism's global growth has been uneven, it has benefited countries and regions with a comparative advantage in tourism resources. For Aramberri, mass tourism is here to stay and, rather than continually arguing about its impacts, the priority is to understand what is going on. By contrast, Butcher challenges the moral superiority often claimed by advocates of 'alternative' or 'ethical' forms of tourism, and argues mass tourism should be celebrated not only for the economic benefits it brings, but also for its role in democratizing travel and contributing to the self-fulfilment, empowerment and 'moral autonomy' of increasing numbers of travellers.

Bianchi (Chapter 4) adopts a Marxist political economy approach. He situates mass tourism, especially in the Southern Mediterranean, in a global economic system dominated by neoliberal free market ideology, where the dynamics of capital accumulation and forces of production and reproduction – the 'logics' of capitalism – are reflected in the differential articulation of international tourism corporations with national and local capitals and played out through political–institutional arrangements in different societies. By contrast, Savelli and Manella (Chapter 5) summarize the emergence of Italian tourism and discuss how social scientists have analysed tourism's impacts, its technologies and forms of organization, new and (especially) urban destinations, and the regionalization of tourism and its governance that have accompanied globalization. They note that perceptions and practices of travel and tourism have changed: tourism may be a form of nomadism, an escape from self, a base for constructing new identities, a mode of self-fulfilment, and an indicator of well-being.

The last two chapters in Section 2 focus more on the issue of sustainability. Weaver (Chapter 6) suggests 'mass' and 'alternative' tourism can merge in forms of enlightened or sustainable mass tourism, where economies of scale and competitively driven innovation are combined with more sensitive and ethical features of alternative tourism, and he uses evidence of best practice from protected national parks and urban beach resorts to demonstrate that mass tourism can indeed incorporate elements of sustainability. By contrast, Holden (Chapter 7) notes the problematic reciprocal relationship of tourism with climate change, the impacts of mass tourism on ecosystems and natural habitats, and the need for governments and planners to counter tourism's negative consequences and promote pro-environment behaviour among tourists. Perhaps less of an optimist than Weaver, he asks how far such individual or collective tourist behaviour will occur voluntarily.

The authors of the five chapters in Section 3 take a historical approach. Heeley's (Chapter 8) very personal account of tourism development in Britain depicts how public and private capital – local authorities, the aristocracy and 'local grandees', railway companies and an assortment of private entrepreneurs – variously combined to bring about 19th-century urban tourism development. By contrast, Sharpley (Chapter 9) traces the process of 'Butlinization', showing how the operational principles of Billy Butlin, a key player in providing tourism for the masses in British holiday camps, are reflected in more recent forms of all-inclusive tourism, including cruise holidays and all-inclusive beach resorts in the Caribbean. Farr (Chapter 10), too, looks at British tourism, and his wide-ranging account of post-1970s' British seaside resorts, and their depiction in film and print, reveals a narrative and perception of 'declinism' only partly confirmed by the realities. Some resorts clearly survived, with or without efforts to re-brand and re-present them as 'heritage', while others did decline, despite (sometimes half-hearted) political attempts to prioritize British seaside tourism.

Bricker (Chapter 11) and Duval (Chapter 12) shift attention away from the UK. The former reviews mass tourism in protected areas managed by the National Park Service (NPS) in the USA, outlining the emergent role and administration of the NPS, and the extent and economic importance of large-scale tourism (especially through the operation of concessions). She details the continued challenges faced by the NPS in pursuing its twin (and sometimes conflicting) goals of enabling people to enjoy the parks and preserving landscapes in their natural state. By contrast and more generally, Duval addresses the links between transport and tourism, highlighting the continued tension between consumer and commercial interests, between the desire for national sovereignty and free movement, especially in air travel, and the role of international regulation, designed to control and regularize such movement.

In Section 4, there are seven chapters on mass tourism in specific destinations. Ryan (Chapter 13) examines mass tourism in the context of China, by far the most important emerging global destination and a much-valued source of tourists, and the role of government policy in promoting tourism, including mass domestic tourism, as a form of economic development. Cohen (Chapter 14) examines Chinese and Russian visitors in Thailand, noting their different characteristics as sightseers and vacationers, respectively, the destabilizing effect of mass tourism on the image of destinations, and the impact of changing economic conditions on the volume of inbound tourism.

Attention then shifts to Bulgaria and then the Mediterranean, with chapters by Ivanov (Chapter 15), Andrews (Chapter 16) and Jeffrey and Bleasdale (Chapter 17). All discuss destinations where tourism is both seasonal and highly concentrated, and all acknowledge mass tourism has brought considerable economic benefits to resident populations. However, their overall assessments are quite nuanced. For Ivanov, although mass tourism has had negative consequences in Bulgaria, and both the private sector and government could do more to make Bulgarian tourism more sustainable, he considers mass tourism to have been largely beneficial, not only economically but also socially, culturally and environmentally. Andrews is more circumspect: she acknowledges tourism's economic advantages for Mallorca but notes, too, the disruptive impact of in-migration, the environmental degradation, water shortages, and resident and tourist 'enclaves'. That said, she reflects that local authorities are attempting to redress these issues and that mass tourism in the island is likely to continue. A very different situation is found in Tunisia where, as Jeffrey and Bleasdale note, tourism has been promoted as a tool for economic and social development, with an explicit Westernizing ethos (reportedly appreciated by the Tunisian middle class). However, Western tourism arrivals have plummeted since terrorist attacks and while the authors offer five possible solutions, including targeting tourists from the Middle East and emerging countries, notably China and Russia, any success will ultimately depend on the re-establishment of stability, which is likely to take time.

Finally, there are two chapters on island destinations. In the first, the late Greg Ashworth and John Tunbridge (Chapter 18) describe Malta's efforts to move from mass beach tourism to heritage tourism. However, the promotion seems to have been half-hearted, the 'heritage' market fickle and diverse, often blurring into beach tourism, and the messages of heritage attractions contested, while supplementary but necessary cultural attractions (e.g. gastronomy, shopping and entertainment) are unavailable. In Chapter 19, Wilkinson argues that, while employment benefits might be considerable, the financial returns from Caribbean cruise ship tourism are exaggerated, especially when one considers leakages from imports, the cost of new piers for ever-larger ships, and the use of private islands. His assessment is unequivocal: mass cruise tourism is badly managed by destinations ignorant of its complexity or its costs. It is a mess.

Clearly, our coverage of the global aspects of mass tourism is somewhat uneven and the assessments of mass tourism by contributors to this book are varied. This is no surprise: they approach the issues from different theoretical perspectives and focus on different destinations at different times. Despite the alleged standardization said to characterize mass tourism, it is full of complexities and one size does not fit all. It is hoped, though, that together we have started a wider and more balanced debate about the nature and role of mass tourism in the 'small world' we inhabit, some of us as participants in a continued 'academic romance' (Lodge, 1984) and every one of us as global citizens in a world of increasing interdependence, where 'development' is a continuing issue for us all.

References

Agarwal, S. and Shaw, G. (eds) (2007) *Managing Coastal Tourism Resorts: A Global Perspective*. Channel View Publications, Clevedon.

Amin, L. (2016, 4 March) Lesbos: a Greek island in limbo over tourism, refugees and its future. *The Guardian*.

Andrews, H. (2011) *The British on Holiday: Charter Tourism, Identity and Consumption*. Channel View Publications, Bristol.

Aramberri, J. (2010) *Modern Mass Tourism*. Emerald, Bingley.

Arlt, W.G. and Li, X. (2016) China's outbound tourism: history, current development, and outlook. In: Li, X. (Robert). (ed.) *Chinese Outbound Tourism 2.0*. Apple Academic Press, Waretown, New Jersey, pp. 3–19.

Ashworth, G. and Tunbridge, J. (1990) *The Tourist-Historic City*. Belhaven Press, London.

Baranowski, S. (2005) Radical nationalism in an international context: strength through joy and the paradoxes of Nazi tourism. In: Walton, J. (ed.) *Histories of Tourism: Re-presentation, Identity and Conflict*. Channel View Publications, Clevedon, UK, pp. 125–143.

Becken, S. (2013) A review of tourism and climate change as an evolving knowledge domain. *Tourism Management Perspectives* 6, 53–62. DOI: 10.1016/j.tmp.2012.11.006.

Berger, R. (2015) *European City Tourism Study 2015*. Berger, Munich.

Berghoff, H., Korte, B., Schneider, R. and Harvie, C. (eds) (2002) *The Making of Modern Tourism:The Cultural History of the British Experience, 1600–2000*. Palgrave, Houndmills, Basingstoke.

Bianchi, R. (2015) Towards a new political economy of global tourism revisited. In: Sharpley, R. and Telfer, D. (eds) *Tourism Development: Concepts and Issues*, 2nd edn. Channel View Publications, Bristol, pp. 287–331.

Black, J. (1992) *The British Abroad: The Grand Tour in the Eighteenth Century*. Sutton Publishing, Stroud, UK.

Blanchar, C. (2015) Barcelona Mayor introduces one-year ban on new tourist accommodation. Available at: http://elpais.com/elpais/2015/07/02/inenglish/1435828213_655223.html (accessed 11 May 2015).

Boissevain, J. (ed.) (1996) *Coping with Tourists: European Reaction to Mass Tourism*. Berghahn Books, Oxford.

Borsay, P. and Walton, J.K. (eds) (2011) *Resorts and Ports: European Seaside Towns Since 1700*. Channel View Publications, Bristol, UK.

Boyer, M. (2007) *Le Tourisme de Masse*. L'Harmattan, Paris.

Bramwell, B. (ed.) (2004) *Coastal Mass Tourism: Diversification and Sustainable Development in Southern Europe*. Channel View Publications, Clevedon.

Butcher, J. (2003) *The Moralisation of Tourism: Sun, Sand … and Saving of the World*. Routledge, London.

Butler, R.W. (1980) The concept of a tourist area cycle of evolution: implications for management of resources. *The Canadian Geographer/Le Géographe canadien* 24(1), 5–12. DOI: 10.1111/j.1541-0064.1980.tb00970.x.

Butler, R. (1992) Alternative tourism: the thin end of the wedge. In: Smith, V. and Eadington, W. (eds) *Tourism Alternatives: Potentials and Problems in the Development of Tourism*. University of Pennsylvania Press, Philadelphia, pp. 31–46.

Butler, R. (ed.) (2006a) *The Tourism Area Life Cycle, Vol.1: Applications and Modifications*. Channel View Publications, Bristol.

Butler, R. (ed.) (2006b) *The Tourism Area Life Cycle, Vol. 2: Conceptual and Theoretical Issues*. Channel View Publications, Bristol.

Buzard, J. (1993) *The Beaten Track: European Tourism, Literature and the Ways to 'Culture', 1800–1918*. Clarendon Press, Oxford.

Casson, L. (1974) *Travel in the Ancient World*. John Hopkins University Press, Baltimore, Maryland.

Chen, G. (2015) Typology, tourist. In: Jafari, J. and Ziao, H. (eds) *Encyclopedia of Tourism*. Springer, Basel, Switzerland.

Cohen, E. (1972) Toward a sociology of international tourism. *Social Research* 39(1), 164–182.

Cohen, E. (1979) A phenomenology of tourist experiences. *Sociology* 13(2), 179–201. DOI: 10.1177/003803857901300203.

Cohen, E. (1992) Alternative tourism: a critique. In: Singh, T.V., Fish, M., Smith, V. and Richter, L. (eds) *Tourism Environment: Nature, Culture, Economy*. Inter-India Publications, New Delhi, pp. 269–278.

Cole, S. (2008) *Tourism, Culture and Development: Hopes, Dreams and Realities in East Indonesia*. Channel View Publications, Clevedon.

Corbin, A. (1994) *The Lure of the Sea*. Polity Press, Cambridge.

Croall, J. (1995) *Preserve or Destroy: Tourism and the Environment*. Calouste Gulbenkian Foundation, London.

Dann, G. and Parrinello, G.L. (2009) Setting the scene. In: Dann, G. and Parrinello, G. (eds) *The Sociology of Tourism: European Origins and Developments*. Emerald, Bingley, pp. 1–64.

de Kadt, E. (ed.) (1979) *Tourism: Passport to Development?* Oxford University Press, Oxford.

Dedieu, F. and Mathieu, B. (2015) Le Tourisme Peut-il Sauver La France? Available at: http://lexpansion.lexpress. fr/actualite-economique/le-tourisme-peut-il-sauver-la-france_1665209.html (accessed 7 April 2015).

Department for Culture, Media and Sport (DCMS) (2015) *Backing the Tourism Sector: A Five Point Plan*. DCMS, London.

Dumbrovská, V. and Fialová, D. (2014) Tourist intensity in capital cities in central Europe: comparative analysis of tourism in Prague, Vienna and Budapest. *Czech Journal of Tourism* 3(1), 5–26. DOI: 10.2478/cjot-2014-0001.

Dwyer, L., Forsyth, P. and Dwyer, W. (2010) *Tourism Economics and Policy*. Channel View Publications, Bristol.

eTurboNews (2015) Dubai rapidly approaching goal of becoming medical tourism hub. Available at: www.eturbonews.com (accessed 7 September 2015).

Ghimire, K.B. (2001) *The Native Tourist: Mass Tourism within Developing Countries*. Earthscan, London.

Gilbert, E. (1954) *Brighton: Old Ocean's Bauble*. Methuen, London.

Guangrui, Z. and Lew, A. (2003) Introduction: China's tourism boom. In: Lew, A., Vu, L., Ap, J. and Guangrui, Z. (eds) *Tourism in China*. Haworth Press, New York, pp. 3–11.

Guide, T.C. (2016) 2015 statistics. Available at: https://www.travelchinaguide.com/tourism/2015statistics/ (accessed 10 March 2017).

Harrison, D. (2015) Development theory and tourism in developing countries: what has theory ever done for us? *International Journal of Asia Pacific Studies* 11(1), 53–82.

Harrison, D. (2017) Looking East but learning from the West? mass tourism and emerging nations. *Asian Journal of Tourism Research* 1(2), 1–36.

Hibbert, C. (1987) *The Grand Tour*. Methuen, London.

Hannam, K. (2009) The end of tourism? Nomadology and the mobilities paradigm. In: Tribe, J. (ed.) *Philosophical Issues in Tourism*. Bristol, Channel View Publications, pp. 101–113.

Harrison, D. (1992a) International tourism and the less developed countries: the background. In: Harrison, D. (ed.) *Tourism and the Less Developed Countries*. Belhaven, Chichester, pp. 1–18.

Harrison, D. (1992b) Tourism to less developed countries: the social consequences. In: Harrison, D. (ed.) *Tourism and the Less Developed Countries*. Belhaven, Chichester, pp. 19–34.
Harrison, D. (1996) Sustainability and tourism: reflections from a muddy pool. In: Briguglio, L., Archer, B., Jafari, J. and Wall, G. (eds) *Sustainable Tourism in Islands and Small States: Issues and Policies*. Pinter, London, pp. 69–.89.
Harrison, D. (2001) Less developed countries and tourism: the overall pattern. In: Harrison, D. (ed.) *Tourism and the Less Developed World*. CAB International, Wallingford, pp. 1–22.
Harrison, D. (2012) Tourism: is small beautiful? In: Singh, T.V. (ed.) *Critical Debates in Tourism*. Channel View Publications, Bristol, pp. 80–84.
Harrison, D. (2014) Tourism and development: from development theory to globalisation. In: Lew, A., Hall, C.M. and Williams, A.M. (eds) *The Wiley Blackwell Companion to Tourism*. Wiley, Chichester, UK, pp. 143–154.
Holden, A. (2007) *Environment and Tourism*, 2nd edn. Taylor and Francis, London.
Hudson, R. (ed.) (1993) *The Grand Tour, 1592–1796*. The Folio Society, London.
Jafari, J. (1989) Sociocultural dimensions of tourism: an English language literature review. In: Bystrzanowski, J. (ed.) *Tourism as a Factor on Social Change: A Sociocultural Study*. European Co-ordination Centre for Research and Documentation in Social Studies, Vienna, pp. 17–60.
Jakle, J.A. (1985) *The Tourist: Travel in Twentieth Century North America*. University of Nebraska Press, Lincoln.
Jones, T. (2016, 19 February) Doesn't tourism's Faustian pact need a review? Ask Italy. *The Guardian*.
Keeley, G. (2014) Barcelona fights back against the Britons behaving badly. *The Times* 3.
Krippendorf, J. (1987) *The Holiday Makers: Understanding the Impact of Leisure and Travel*. Heinemann Professional Publishing, Oxford.
Lanfant, M.-L. (1992) International tourism reconsidered: the principle of the alternative. In: Smith, V.L. and Eadington, W.R. (eds) *Tourism Alternatives: Potentials and Problems in the Development of Tourism*. University of Pennsylvania Press, Philadelphia, pp. 88–112.
Lencek, L. and Bosker, G. (1999) *The Beach: The History of Paradise on Earth*. Penguin, Harmondsworth.
Lodge, D. (1984) *Small World: An Academic Romance*. Secker and Warburg, London.
Lofgren, O. (1999) *On Holiday: A History of Vacationing*. University of California Press, Berkeley, California.
MacCannell, D. (1999) *The Tourist: A New Theory of the Leisure Class*, 3rd edn. University of California Press, Los Angeles.
Macnaught, T.J. (1982) Mass tourism and the dilemmas of modernization in Pacific island communities. *Annals of Tourism Research* 9(3), 359–381. DOI: 10.1016/0160-7383(82)90019-6.
McElroy, J. (2015) Sustainable island tourism. In: Chambers, D. and Rakic, T. (eds) *Tourism Research Frontiers: Beyond the Boundaries of Knowledge*. Emerald, Bingley, pp. 123–140.
McGibbon, J. (2000) *The Business of Alpine Tourism in a Globalising World*. Vetterling Druck, Rosenheim.
Mitchell, J. and Mann, S. (2010) *Tourism and Poverty Reduction: Pathways to Prosperity*. Earthscan, London.
Mowforth, M. and Munt, I. (2016) *Tourism and Sustainability: Development, Globalisation and New Tourism in the Third World*, 4th edn. Routledge, Abingdon.
Obrador Pons, P., Crang, M. and Travlou, P. (eds) (2009) *Cultures of Mass Tourism: Doing the Mediterranean in the Age of Banal Mobilities*. Ashgate, Farnham.
Ousby, I. (1990) *The Englishman's England: Taste, Travel and the Rise of Tourism*. Cambridge University Press, Cambridge.
O'Reilly, K. (2000) *The British on the Costa del Sol: Transnational Identities and Local Communities*. Routledge, London.
Pearce, P.L. (2005) *Tourist Behaviour: Themes and Conceptual Schemes*. Channel View Publications, Clevedon.
Pechlaner, H. and Nordhorn, C. (2016) Perspectives of hospitality. *Presentation at ITB (Internationale Tourismus-Börse) Berlin*, 10 March 2016.
Picard, M. (1996) *Bali: Cultural Tourism and Touristic Culture*. Archipelago Press, Singapore.
Pimlott, J. (1947) *The Englishman's Holiday: A Social History*. Harvester Press, Hassocks.
Plomer, W. (ed.) (1992) *Kilvert's Diary, 1870–1879*. Cresset Press, London.
Rebollo, J.F.V. and Baidal, J.A.I. (2003) Measuring sustainability in a mass tourist destination: pressures, perceptions and policy responses in Torrevieja, Spain. *Journal of Sustainable Tourism* 11(2–3), 181–203. DOI: 10.1080/09669580308667202.
Ryan, C. (2003) *Recreational Tourism: Demand and Impacts*. Channel View Publications, Clevedon.

Segreto, L., Manera, C. and Pohl, M. (eds) (2009) *Europe at the Seaside: The Economic History of Mass Tourism in the Mediterranean*. Berghahn Books, New York.
Settis, S. (2016) *If Venice Dies*. New Vessel Press, New York.
Sharpley, R. (2003) Tourism, modernisation and development on the island of Cyprus. *Journal of Sustainable Tourism* 11(2–3), 246–265.
Sharpley, R. (2009) *Tourism Development and the Environment: Beyond Sustainability*. Earthscan, London.
Shields, R. (1991) *Places on the Margin: Alternative Geographies of Modernity*. Routledge, London.
Smith, V. (1992) Boracay, Philippines: a case study in 'alternative' tourism. In: Smith, V.L. and Eadington, W.R. (eds) *Tourism Alternatives: Potentials and Problems in the Development of Tourism*. University of Pennsylvania Press, Philadelphia, pp. 135–157.
Smith, V. (2000) Typology, tourist. In: Jafari, J. (ed.) *Encyclopedia of Tourism*. Routledge, London, p. 608.
Smith, V. (2001) Power and ethnicity in 'paradise': Boracay, Philippines. In: Smith, V. and Brent, M. (eds) *Hosts and Guests Revisited: Tourism Issues of the 21st Century*. Cognizant, New York, pp. 141–152.
Smith, V. and Brent, M. (2001) Introduction. In: Smith, V. and Brent, M. (eds) *Hosts and Guests Revisited: Tourism Issues of the 21st Century*. Cognizant, New York, pp. 1–11.
Smith, V. (ed.) (1977) *Hosts and Guests: The Anthropology of Tourism*. University of Pennsylvania Press, Philadelphia.
Soane, J. (1993) *Fashionable Resort Regions: Their Evolution and Transformation*. CAB International, Wallingford.
Telfer, D. (2015) The evolution of development theory and tourism. In: Sharpley, R. and Telfer, D. (eds) *Tourism and Development: Concepts and Issues*, 2nd edn. Channel View Publications, Bristol, pp. 31–73.
Tjolle, V. (2014) Venice revolts against tourism. TravelMole. Available at: http://travelmole.com/news_feature. php?news_id=2012843 (accessed 16 September 2016).
Towner, J. (1996) *An Historical Geography of Recreation and Tourism in the Western World*. Wiley, Chichester.
Travis, J. (1993) *The Rise of the Devon Seaside Resorts, 1750–1900*. Exeter University Press, Exeter.
Turner, L. and Ash, J. (1975) *The Golden Hordes: International Tourism and the Pleasure Periphery*. Constable, London.
United Nations World Tourism Organization (UNWTO) (2013) *Domestic Tourism in Asia and the Pacific*. UNWTO, Madrid.
United Nations World Tourism Organization (UNWTO) (2015) *World Tourism Barometer*. 13. UNWTO, Madrid.
Urbain, J.-D. (2003) *At the Beach*. University of Minnesota Press, Minneapolis.
van Tubergen, A. and van der Linden, S. (2002) A brief history of spa therapy. *Annals of the Rheumatic Diseases* 61(3), 273–275. DOI: 10.1136/ard.61.3.273.
Waldren, J. (1996) *Insiders and Outsiders: Paradise and Reality in Mallorca*. Berghahn Books, Oxford.
Walton, J.K. (1978) *The Blackpool Landlady: A Social History*. Manchester University Press, Manchester.
Walton, J.K. (2000) *The British Seaside: Holidays and Resorts in the Twentieth Century*. Manchester University Press, Manchester.
Walton, J.K. (2005) Introduction.In: Walton, J.K. (ed.) *Histories of Tourism: Representation, Identity and Conflict*. Channel View Publications, Clevedon, pp. 1–18.
Weaver, D. (2006) The 'plantation' variant of the TALC in the small-island Caribbean. In: Butler, R. (ed.) *The Tourism Area Life Cycle, Vol. 1: Applications and Modifications*. Channel View Publications, Clevedon, pp. 81–184.
Weaver, D. (2012) Towards sustainable mass tourism: paradigm shift or paradigm nudge? In: Singh, T.V. (ed.) *Critical Debates in Tourism*. Channel View Publications, Bristol, pp. 28–34.
Weaver, D. (2013) Asymmetrical dialectics of sustainable tourism: towards enlightened mass tourism. *Journal of Travel Research* 53(2), 131–140.
Weber, M. (1949) Shils, E.A., Finch, H.A., Press, Free. and York, New. (eds) *The Methodology of the Social Sciences*. Free Press, New York.

20

Tourists, Mobilities and Paradigms

Introduction: Kuhn and Paradigms

Published more than five decades ago, Kuhn's *'The Structure of Scientific Revolutions'* (Kuhn, 1962) is a 20th century classic. Its central thesis, on changes in scientific thinking, has been the subject of much discussion and numerous revisions, many by Kuhn himself (Kuhn, 1970a, b, c, 1974, 1987, 2000, 2012). Moreover, while he was not the only philosopher of science to question 'the received view' of scientific theories (Suppe, 1977, pp. 4–5), his work on paradigms and paradigm change has been especially important, reportedly changing the prevailing image of science (Hacking, 2012).

When first discussing 'normal science', Kuhn (1962) portrayed it as passing through four stages: first, a more or less random collection of facts; second, a pre-paradigm stage, characterized by competing schools of thought; third, a period when one school's paradigm becomes pre-eminent and, finally, normal science, when an agreed world view and standard methods of problem-solving comprise a paradigm into which newcomers to the science are socialized. A period of consensus and stability follows: the dominant paradigm is unquestioned, problem-solving more efficient, and progress in accumulating scientific knowledge – as defined by the scientific community - is made. Eventually, however, disagreements emerge over the problems to be solved; new terms emerge, old ones are redefined, and there is 'misunderstanding' among different schools of thought. A period of crisis ensues, alternative paradigms emerge which are incommensurable with the dominant paradigm and, through argument or (more) conversion, there is a revolutionary shift in (some) scientists' world views. Eventually, 'after the last holdouts have died, the whole profession will again be practising under a single, but now different, paradigm' (Kuhn, 2012).

Problems with Paradigms

As indicated in the following pages, Kuhn's work stimulated immediate debate and prompted him to make substantial revisions in his position. However, the issues continue to centre on how far paradigms are characterized by mutual incommensurability, whether they change through conversion to new ways of thinking or because previous theories have been disproved or falsified (Toulmin, 1970; Bird, 2013) and, for present purposes, how far the social sciences are characterized by paradigms and paradigm change. This debate is especially relevant to tourism studies, both theoretically and in matters of empirical research for, as shown below, in recent years several tourism scholars have advanced strong claims that a mobilities perspective is, in effect, a new and more appropriate paradigm for the study of tourism, especially in non-Western societies.

The overall purpose of this paper is to assess the validity of such claims. However, first it is necessary to examine Kuhn's changing views on the nature of paradigms and then to summarize arguments as to how far paradigms exist in the social sciences. This provides the context for asking how far what is claimed to be the 'New Mobilities paradigm' is indeed a paradigm in any meaningful Kuhnian sense, rather than a useful perspective that is quite commensurable with different theoretical and empirical approaches to tourism and other forms of travel.

At the outset, it should be noted that all disputants, including Popper and Kuhn, who differ in many other respects, accept that natural sciences *are* sciences (Kuhn, 1970c, pp. 3,6; Popper, 1970). However, soon after *The Structure of Scientific Revolutions* was published, contradictions in Kuhn's concept of paradigm became evident (Masterman, 1970). Consequently (Kuhn, 1970b, 1974, pp. 459–482; Hacking, 2012, pp. xvii–xxv), he redefined 'paradigm' as both a disciplinary matrix and exemplar. The former is the global or world view of an independent scientific community and includes symbolic generalizations, shared commitments to 'particular models', and shared values (Kuhn, 1970a, pp. 182–185). The fourth component of the disciplinary matrix is the exemplar, the second and narrower sense in which Kuhn initially used the term paradigm. This is a template for 'concrete problem-solutions', accepted by the scientific group, taught to students of science, and is the guide to their research activities (Kuhn, 1970a, 1974, pp. 173–208, pp. 186, 459–482, 463).

Paradigms, then, are disciplinary matrices and, more locally, exemplars, and in both senses they are the property of the appropriate scientific community, which functions 'as a producer and validator of sound knowledge' (Kuhn, 1974, p. 463). As Kuhn puts it: 'A paradigm is what the members of a scientific community, and they alone, share' (1974, p. 460). The community is defined by its disciplinary matrix, by its world view, underpinned by commonly held commitments and values, and by its dominant exemplar, which is the community's accepted way of solving problems.

Another issue is how much dissent over problem-solving approaches among adherents to an exemplar can be accommodated before different problems and the need for different solutions are perceived. After all, even within normal science, with its shared commitments and value consensus, agreement is never total. Unfortunately, neither Kuhn nor his supporters resolved this matter. Perhaps, as Hacking notes, new disciplines are to some extent mutually incomprehensible (Hacking, 2012, p. xxxiii), but that simply fudges the question: at what *point* does incommensurability across paradigms (both as world views and exemplars) occur?

However, Kuhn does suggest how incommensurability might be approached. In particular, he links paradigm change to the acquisition of a new language. As competing paradigms emerge, and old terms take on new meanings, some adherents of the previously accepted paradigm familiarize themselves with the new language, slipping into it 'without a decision having been made' (Kuhn, 2012). But bilingualism is not enough. The emergent paradigm will be 'native' only to the newly initiated, for whom it is, so to speak, a first language. By contrast, those who once accepted the earlier paradigm – even those who are bilingual – remain outsiders, 'foreigners in a foreign environment' (Kuhn, 2012), until they experience conversion and make the 'gestalt switch' which, for Kuhn, is 'at the heart of the revolutionary process' (2012, p. 203).

In effect, Kuhn abandoned the notion of paradigm, especially as a disciplinary matrix, and focused on incommensurability, which he defines as 'a sort of untranslatability, localized to one or another area in which two lexical taxonomies differ' (Kuhn, 2000). Every scientific community has a shared lexicon, which validates its standards and activities and, simultaneously, maintains 'its isolation from practitioners of other specialities' (Kuhn, 2000). Knowledge across scientific communities, with different lexical taxonomies, is *more or less* commensurable, but the *extent* incommensurability occurs depends on the emergence of new forms of normal science, and overlap in scientific lexicons decreases *as scientific communities become more specialized.* The greater the specialization, the more incommensurability is encountered.

Scientific communities exist at different levels. For Kuhn, the most global is 'the community of all natural scientists' (2012, p. 176), though it could be argued that, above this, there is an even wider community, comprised of scientists of all backgrounds who follow science as 'a vocation' (Weber, 1948, pp. 129–156). The next level is made up of 'the main scientific professional groups' (physicists, chemists, astronomers, and so on), followed by such major subgroups as organic chemists and radio astronomers (Kuhn, 2012). However, it is only *below* this level that normal science is found; here, in the world of specialist conferences, peer-reviewed papers, citations and interest-based networks, are 'the producers and validators of scientific knowledge' (Kuhn, 2012), and it is into these communities, the arbiters of scientific progress (Kuhn, 2012), that newcomers are socialized and converts welcomed.

Paradigms and the Social Sciences

Kuhn was reportedly horrified at how social scientists 'misappropriated' and 'mangled' his theory of paradigms (Walker, 2010). He recognized commonalities in the development of natural sciences and other fields of endeavour, 'none necessarily unique to science', but considered it was their '*conjunction*' which set the scientific activity apart' (Kuhn, 2012, My emphasis). More particularly, he believed social and natural sciences differed in several key respects (Kuhn, 2012). The former, categorized by him as 'proto-sciences' (1970b, p. 244), are comprised of competing schools, lack concrete achievements and a single, dominant paradigm (either worldview or exemplar), and have not built on the work of 'classics' of the field. In addition, social scientists must continually re-interpret inherently unstable socio-political systems (Kuhn, 2000), and seek validation of their choice of subject matter and their findings from *outside* their communities, rather than from their peers who, in normal science, are the sole validators of progress (Kuhn, 2012).

Despite such objections, many social scientists persisted – and persist – in seeking paradigms, especially in the sense of a disciplinary matrix (even though it was a concept Kuhn largely abandoned). Van den Berghe, disillusioned with social theory, suggested the 'paradigm of evolutionary biology' should be bridged 'with several of the major lines of thinking in the social sciences, especially ... economics, exchange theory, game theory, behaviorism, and, at the 'macro' level, Marxism and ... even some brands of functionalism' (van den Berghe, 1978, p. 45), thus incorporating almost the entire range of social sciences as a single paradigm.

Slightly less comprehensively, Ritzer saw paradigms as 'subsuming theories, methods, exemplar and image of the subject matter of sociology' (1978, p. 2) which he considered to be dominated by 'social facts, social definition, and social behaviour', derived respectively from Durkheim, Weber and Skinner (Ritzer, 1978, p. 2). He also suggested that sociobiology, structuralism and environmentalism were *en route* to 'paradigmatic status' (1978, p. 3), and advocated 'paradigmatic integration' (1978, p. 10), for which he was heavily criticized by Eckberg and Hill (1979).

Many other suggestions followed. Hammersley lists positivivsm, constructivism and critical theory (Hammersley, 1992, p. 133); McGregor and Murnane characterize the human and social sciences as defined by the post-positivist 'paradigm' (McGregor and Murnane, 2010, p. 419), and Cresswell, defining paradigms variously as philosophical assumptions, epistemologies and ontologies, or research methodologies, and noting methods themselves may be qualitative, quantitative or mixed, adds pragmatism and advocacy to post-positivism and constructivism, seeing all as 'alternative knowledge claims (Cresswell, 2003). Punch, too, opts for positivism and constructivism as 'the main paradigm positions' (2013, pp. 16–17). By contrast, after rightly noting that quantitative and qualitative methods are ways of obtaining data, and not paradigms, Mertens produces yet another list, preferring dialectical pluralism, pragmatism and 'transformative' paradigms, the last of which portrays research, the outcome of negotiations of researcher and researched, as necessarily oriented to social justice and community empowerment (Mertens, 2007, Mertens, 2012). Similarly, Shannon-Baker discusses four 'paradigmatic perspectives', notably pragmatism, transformative-emancipation, dialectics and critical realism, that together comprise a variety of tool-kits available to the fledgling researcher (Shannon-Baker, 2015).

It is noteworthy that incommensurability, as a distinguishing feature of paradigms, figures little in such discussions. Indeed, many commentators advocate combining paradigms in research (Mackenzie and Knipe, 2006, p. 200; Punch, 2013) and mixed methods has itself been accorded the status of a separate paradigm, enabling researchers to combine different research techniques (Morgan, 2007, p. 73; Denscombe, 2008; Yvonne Feilzer and Feilzer, 2010).

While most contenders for paradigmatic status might be described as methodologies (i.e. systems of approaching research) or distinctive research techniques, there are exceptions. Some have isolated a key, pre-defined aspect of social life and developed a conceptual framework around it, as when Rojek and Urry focus on the mobility of people and cultures (Rojek and Urry, 1997, p. 11) and when Urry, more comprehensively, argues that 'mobilities rather than societies should be at the heart of a reconstituted sociology' (2000, p. 210). His work on mobilities, and that of several others, led in the mid-2000s to a coalescence of interests around this theme, the first issue of the journal *Mobilities*, and somewhat tentative claims in two important editorials that there was a 'new mobilities paradigm' (NMP), though in neither case was the term 'paradigm' actually defined (Hannam *et al.*, 2006; Sheller and Urry, 2006).

As put forward in these editorials, the NMP is a cross- and interdisciplinary perspective where the emphasis is on

> complex networks by which hosts, guests, buildings, objects and machines are contingently brought together to produce certain performances in certain places at certain times (Sheller and Urry, 2006).

The focus is no longer on (allegedly) static social structures but on dynamic social processes, inter-linking structures and systems, negotiated cross-border networks, connectivities and exclusions, on movements of bodies and images, even of places. Mobilities is a lens, opening up the vista of 'a distinctive social science that is productive of different theories, methods, questions and solutions' (Urry, 2007). Drawing on a variety of theoretical insights, ideologies and methods, on 'paradigmatic fragments' from 'multiple archives', a mobilities approach highlights 'various kinds of movement of people, or ideas, or information or objects. And in so doing this paradigm brings to the fore theories, methods and exemplars of research that have been mostly subterranean, out of sight' (Urry, 2007).

For some, at least, mobilities approaches are potentially able to revolutionize social science. The 'new paradigm' is nothing less than a reformulation of sociology as the study of the 'social as mobility', where 'the social is an open system of mobilities (and immobilities) or networks' (Mavrič and Urry, 2009).

Is the NMP a paradigm? Urry once considered Kuhn's concept of paradigm to be inappropriate to the social sciences (Urry, 1973, pp. 464–466, 1995, pp. 41–42). Later, though, he suggested the NMP was 'derived from Kuhn's exemplary analysis of normal science, scientific exemplars and what constitutes scientific revolution' (2007, p. 18), a position repeated in 2016, with the distinction 'between a general disciplinary matrix, which we term here 'paradigm', as opposed to specific models of good research which we refer to as 'exemplars' (Sheller and Urry, 2016, p. 11). Now, though, the assertion is buttressed by evidence of the increased importance of mobilities approaches: in the increased number of citations; in numerous peer-reviewed papers in journals, book chapters and commentaries and, third, in the emergence of several active mobilities-oriented academic and other institutions in Europe, the USA and Australia (Sheller and Urry, 2016, pp. 14–15).

However, while Sheller and Urry deduce from such evidence that the NMP is, indeed, a paradigm, authors of another important recent commentary on mobilities, in the journal founded to further work from this perspective, are more circumspect, preferring instead to refer to it as a 'field' of study:

> After 10 years, the vibrant and interdisciplinary set of dialogues that characterize the field has largely rendered moot the question of whether there really was a mobilities turn or paradigm shift. More interesting, therefore, are questions of how links and connections, histories and futures, become intertwined in 'state-of-the-art' mobilities research (Faulconbridge and Hui, 2016, p. 2).

The nature of mobilities contributions may indeed be more interesting than the labels used to define any shift towards this perspective, and it is not intended here to minimize (or review) the work of mobility theorists and researchers over the last decade. Indeed, the significance of their achievements is noted in the two recent reviews just cited. However, the NMP is quite evidently *not* a paradigm in the Kuhnian sense *as defined in this paper*. It is no 'exemplar', rigorously developed and guarded by an exclusive community of scientists committed to agreed ways of puzzle-solving, insulated from the outside world, participating in peer-based networks of normal science, and possessing a distinctive scientific lexicon or language which is *incommensurable* to outsiders. In fact, it is the very *opposite*, incorporating a wide range of disciplines and theoretical orientations, including Simmel and complexity theory, constructed by intellectually mobile scholars, who 'parasitically' draw on 'emancipatory' social movements for support and who, in turn, contribute their insights to civil society (Urry, 2000).

To summarize: among the perspectives discussed in this section, paradigmatic status has been claimed, *inter alia*, for the social sciences, sociobiology, theoretical presuppositions, worldviews, ways of looking at the world, positivism, post-positivism, critical realism, critical theory, pragmatism, mixed methods, constructivism, interpretivism, dialectical pluralism, and transformative, advocacy and metaphysical approaches, and the NMP. However, if the argument in the first section of this paper is accepted, whatever their respective merits, none can be considered paradigms in a Kuhnian sense.

What emerges, instead, is a motley collection of worldviews, methodologies, even entire constellations of social sciences, that simply confirms Kuhn's portrayal of social sciences as comprised of competing schools, lacking concrete achievements or a dominant paradigm, which have advanced but little from their founders (Heyl, 1975, p. 66; Keith, 2006, p. 12). As Bergman notes, 'there are as many paradigms as there are authors who feel the need to distinguish a meta, grand, and middle-range theoretical approach from alternatives' (Bergman, 2010, p. 173). Similarly, Eckberg and Hill, who list the fundamentally different views of twelve prominent sociologists, all claiming to present 'Kuhnian paradigms' (Eckberg and Hill, 1979), suggest paradigm hunts succeed only 'by corrupting Kuhn's model of social science' (Eckberg and Hill, 1979), and argue that while disciplinary matrices exist in sociology, as in all disciplines, 'a Kuhnian paradigm approach is inapplicable, for no truly extended research can take place at this level' (1979, p. 934).

Arguably, sociology and other social sciences are different from, perhaps less 'scientific' than, the natural sciences. However, if that were so, Kuhn's framework, developed to look at changes in scientific thinking, 'to analyse the paradigmatic status of the discipline' (Eckberg and Hill, 1979), becomes redundant. Alternatively, if social sciences *are* like the natural sciences, we should be able to find exemplars – but there are none. Instead, there is 'constant arguing, bickering, and debate, but very little agreement' (Eckberg and Hill, 1979), which is further demonstrated in their extensive exchanges with Ritzer (Hill and Eckberg, 1981; Ritzer, 1981).

'Mobilities' in Tourism Studies

Discussions on paradigms in tourism research have focused on much the same methodologies as those in the social sciences generally and have been similarly inconclusive (Dann, 1997). The lack of consensus and the disagreement over basic terms is apparent in a relatively recent, co-authored paper on tourism paradigms, where Tribe first conflates paradigms and ideologies and then cites neoliberalism and sustainability as competing paradigms (Tribe *et al.*, 2015, p. 34). By contrast, Jamal manages to see tourism studies both as a 'young *disciplinary* matrix' and as an 'immature *field*' (Tribe *et al.*, 2015, p. 38, My emphasis), while Dann opposes the conflation of paradigms and ideologies, claims that Tribe (along with most tourism scholars, but not Dann) is ignorant of European social theory and thus suffers from 'Anglo-centricism'. He then archly sidesteps the entire paradigm issue by prioritizing the need for 'a sociolinguistic change in patterns of communication rather than a xenophobic air-brushing of history' (Tribe *et al.*, 2015, p. 41).

Clearly, though, the mobility of people within and across borders is central to studies of travel and tourism and 'the new mobilities paradigm' has been taken up by some tourism scholars as a way of reconceptualizing and re-situating tourism, hoping to move tourism studies from the periphery of mainstream social research to a central role in an interlinked system of immobilities and mobilities (Hannam, 2014Hannam, 2014). Adherents to the mobilities position claim it brings together 'the situated and mobile nature of tourism and allows analysis from many (inter-personal, regional, global) perspectives' (Mavrič and Urry, 2009). This theoretical emancipation, where the emphasis is less on structures *per se* than on 'networks of mobilities and immobilities', is also claimed to facilitate 'new methodological approaches', though those cited – ethnographic methods, observation, cyberethnography, interviews, time diaries, the use of visual and literary sources, digital recording and historical research (Mavrič and Urry, 2009) – are less methodologies than established research techniques, none of which are especially innovative or specific to any single theoretical perspective.

Despite such objections, prominent scholars have applied the NMP in tourism studies, and Erik and Scott Cohen, in particular, make important claims for its relevance in addressing tourist phenomena (Cohen and Cohen, 2012a, 2015a, b). Their contributions to the debate are especially worthy of attention for three reasons. First, they provide a clearer, more concentrated and yet more comprehensive argument in favour of applying the NMP to tourism, developed over several years, than any found elsewhere in the tourism literature. Second, both authors come with impeccable credentials as tourism scholars. Indeed, Erik Cohen, in particular, has consistently published on a wide range of topics since the early 1970s and is justifiably considered one of the founders of tourism studies as a field of enquiry (Cohen, 2013). Third, the combination of these two factors – the quality of the intervention and the pedigree of the authors – should ensure their contributions are fully recognized and, where necessary, subjected to the most careful of criticism, and it is for such intellectual – and not personal – reasons that the importance of their writing is recognized in this paper.

Cohen and Cohen make three claims: first, as applied by them, the NMP is a paradigm. Second, there is a dominant 'discourse of authenticity' in tourism studies and, third, NMP is especially appropriate in analysing the development of tourism in 'emerging' societies. However, it is argued here that the first claim is conceptually problematic, the second is historically inaccurate, and the third demonstrably wrong.

Defining a 'paradigm' for tourism studies

The argument in this section, and the rest of this paper, focuses especially on Cohen and Cohen and the claim that their version of the NMP (which may not be recognized by other mobilities theorists) is more appropriate to tourism studies than other perspectives.

From the outset, NMP's paradigmatic status seems highly dubious: it is referred to variously as a 'paradigm' (2012a, p. 2180), a 'model' (2012a, p. 2181), a 'theoretical approach', a 'theory', and an 'innovative insight' (2012a, p. 2185) which, though, cannot 'offer a set of basic (predictive) propositions, which could be evaluated in empirical research' (Cohen and Cohen, 2012a), a stance which rather rules out the NMP as an exemplar. Later, paradigm change is said to involve neither a fundamental shift in epistemologies, nor an empirical shift in public policy; rather, it is 'the substitution of one theoretical approach by another' (2015b, p. 159). And then, acknowledging a lack of consensus among mobilities theorists, they resort to calling NMP 'a nascent paradigm' (2015b, pp. 161, 162).

Despite such conceptual ambiguity, it is alleged that by prioritizing movement over structure the NMP 'has the qualities of Kuhn's (Kuhn, 1962, 1970a, b) formulation of a paradigm shift on the theoretical level' (Cohen and Cohen, 2015b). However, such a claim can be sustained only by ignoring Kuhn's reflections on his own work, especially the role of the community of scientists, language lexicons and incommensurability, which for him are the defining features of paradigms.

The inadequacy of the 'discourse of authenticity'

The second major claim by Cohen and Cohen is that pre-mobilities Western studies of tourism rely on the dominant 'discourse of authenticity', which emanated from MacCannell's argument 'that moderns seek authenticity outside modernity' (Cohen and Cohen, 2012a). Such an approach is deemed ethnocentric, Western-oriented and 'culture bound' (2012a, p. 2180) because 'tourists from the emerging regions, particularly Asians and Africans, are not looking for authenticity as modern Westerners *allegedly* did', and so 'current theories are inapplicable to tourism from the emerging regions' (Cohen and Cohen, 2015b, My emphasis).

Following Winter (2009, p. 23), Cohen and Cohen later expand their critique to include 'the main theoretical approaches in modernist studies of tourism', asserting that all have been 'explicitly formulated for the study of 'Western tourism' (2015a, p. 11), an indictment that apparently includes all perspectives ever applied to tourism, from grand theory, through middle-range theory, to interactionism (Harrison, 2001, pp. 69–77), though they do baulk at supporting the call for 'alternative social sciences in non-Western societies' (Cohen and Cohen, 2015b).

The concept of authenticity as a motivating factor has undoubtedly featured prominently in tourism studies (Wang, 1999, 2000; Cohen, 2007) and in heritage studies (Park, 2014), and Cohen and Cohen have themselves contributed to this literature in their discussion of 'hot' and 'cool' authentication (Cohen and Cohen, 2012b). However, it is misleading to claim 'the discourse of authenticity' is the dominant characteristic of 'Modernist tourist theory' (though maintaining it is 'allegedly' sought by Western tourists leaves some room for manoeuvre).

First, even in their own work, neither Erik nor Scott Cohen have *prioritized* the quest for authenticity as a key motivator of Western tourists. Indeed, the former has long been a critic of MacCannell, asserting 'it is senseless to talk about *the* typical tourist' (Cohen, 1979, p. 21), and authenticity *as a motivating factor* features but little in his impressive and extensive published output (http://sociology.huji.ac.il/sites/default/files/sociology/files/cohen_cv.pdf, accessed 26 June 2017; Harrison, 2007, p. 73). Moreover, as Erik Cohen himself reflects: 'I hardly followed a theoretical paradigm, which would direct the line of my questioning. My interest, particularly in tourism, was on issues which were new and unexplored' (Cohen, 2013, p. 107).

By contrast, Scott Cohen has focused more on mobilities as a concept, but in his work, too, the 'discourse of authenticity' is largely absent, and when discussed it is a function of a more general 'need to escape'. Indeed, he acknowledges that motivations are *multidimensional*, need to be contextualised, and are changing over time' (Cohen, 2010, My emphasis).

Second, claiming that Western tourism studies have been dominated by the discourse of authenticity ignores the bulk of academic writing on tourism. This includes psychologists with a special interest in motivation which, for them, is largely accessed through empirical research, especially attitude surveys (Pearce, 1982, pp. 48–67, Pearce, 2005; Ross, 1994, pp. 51–53). They rarely ascribe any importance to authenticity as a motivating factor, and while Pearce did note the distinction between 'real' or 'staged' activities (Pearce, 1982), he later concluded authenticity was 'weakening as an academic concept' (2005, pp. 141–142).

The 'discourse of authenticity' is also absent from the literature on tourism economics (Dwyer and Forsyth, 2006; Dwyer *et al.*, 2010), from accounts of the historical development of European tourism (Walton, 2005; Borsay and Walton, 2011), and as a motivating factor in anthropological writing on tourism. Indeed, Bruner considered the concept of authenticity 'a red herring' (Bruner, 1991, p. 241), and Selwyn remarked that 'authenticity' is one of 'a number of reductive terms (along with 'gaze,' 'escape,' and 'consumption') ... made to carry more analytical weight than they can bear' (Selwyn, 2007, p. 50). Finally, 'the discourse of authenticity', as a reflection of tourist *motivation*, is virtually absent in the extensive literature on tourism as a form of development (Harrison, 2001; Sharpley and Telfer, 2015).

The new mobilities paradigm and tourism in emerging countries

It is suggested, then, that the argument that 'the discourse of authenticity' is the dominant perspective in Western approaches to tourism cannot be sustained. Such dominance is absent in the distinguished body of work of some of its strongest advocates, and in the various social science disciplines, including psychology, that contribute to the field of tourism studies.

The third claim of Cohen and Cohen, related to the second, is that while NMP has its limitations, most notably its tendency towards positivism (2015a, p. 15) and its implicit association with 'neoliberal, individualistic tendencies' (2015b, p. 163) (assertions likely to be rejected by many other mobilities theorists), when used in conjunction with other concepts it is less ethnocentric and more universally appropriate than 'the discourse of authenticity' and other (unspecified) Western approaches to the study of tourism in 'emerging' countries, including in East and South-east Asia.

First, it should be noted that appeals for more emphasis on non-Western tourism are not new (Towner, 1995; Winter *et al.*, 2009; Chang, 2015; King, 2015a) and that, with notable exceptions (Ghimire, 2001; Singh, 2009; Winter *et al.*, 2009), general treatments of intra-regional and domestic tourism in Asia have been relatively few. However, this is less so for international tourism in East and South-east Asia. Over more than three decades, Erik Cohen has examined tourism's impacts in Thailand (Cohen, 2000, Cohen, 2001), and many other Western scholars have researched tourism in South-east Asia, along with an increasing number of scholars from the region (Hitchcock *et al.*, 1993, Hitchcock *et al.*, 2010; Yamashita *et al.*, 1997; Chon, 2000; Hall and Page, 2000; Teo *et al.*, 2001; King, 2015b; Harrison *et al.*, 2017a, b, c, d. There is also a well established body of research on Japanese tourism, though it has not always been recognized (Graburn, 1983; Cooper, 2005, Special issue: Japanese tourism, part 2, 2007; Guichard-Anguis and Moon, 2008).

Second, the interest of Western academics in mass tourism only emerged some time after MacCannell's pioneering work (MacCannell, 1976; Jafari and Aaser, 1988) and there was a similar time lag before the development of academic interest in Japanese tourism (Williams, 2013, pp. 47–48). Delays also occurred in research on Chinese tourism, and were exacerbated by several factors: a lack of language skills among Western scholars; the insistence in many Western universities that staff publish in a few 'A-ranked' journals, published only in English; the earlier under-representation of Asian scholars in Western institutions (though this is changing); and the fact that tourism research funds (especially in South-east and East Asia) are often directed at marketing and policy-related projects, resulting in a decidedly descriptive approach and a failure to theorize the relationships of the emergent Asian 'centres' to their 'peripheries'.

However, as Chinese mass tourism (inbound and outbound) developed, it increasingly attracted academic attention. Some interest emerged in the 1980s (Guangrui, 1989; Mings and Liu, 1989), much more in the 1990s (Oakes, 1998; Chee-Beng *et al.*, 2001) and since the early 2000s there has been an explosion of publications, in both English and Chinese (Kong and Cheung, 2009; Xie and Li, 2009; Hsu *et al.*, 2010; Huang, 2011; Tsang and Hsu, 2011; Arlt, 2013; Ryan and Huang, 2013; Xiang, 2013; Andreu *et al.*, 2014, p. 346; Harrison, 2016, p. 23; Huang and Chen, 2016; Li, 2016; Dai *et al.*, 2017, p. 148). Indeed, Zhang, Lan, Qi and Wu report on a survey of 16,024 academic papers on tourism articles published in China from 2003 to 2012 (Zhang *et al.*, 2017). Such a body of scholarship, drawing on the wide range of theoretical perspectives, methodologies and research techniques available in 'Western' social science, belies suggestions that Chinese tourism has been neglected or that 'Western' social science is inappropriate to the study of tourism in non-Western societies. Rather, it demonstrates considerable and increasing levels of cooperation among ethnic Chinese and Western tourism academics, often arising from student–supervisor collaboration, evidenced in the increasing number of papers published in Chinese- and English-language peer reviewed journals (Law and Cheung, 2008; Xiao *et al.*, 2010; Xie, 2013; Tse, 2015; Zhong *et al.*, 2015), including many published in *Tourism Management* and (less so) *Annals of Tourism Research* (Andreu *et al.*, 2014).

Undoubtedly, much comparative work remains to be done; new approaches to tourism, in Western or non-Western societies, are needed; detailed comparison of the motivation, behaviour and perceptions of tourists from different national, ethnic or cultural backgrounds would be valuable (Chang, 2015), as would studies of cross-cultural interaction involving Chinese and non-Chinese (as residents and tourists). However, such topics are matters for empirical research and require no paradigm change.

Third, if (as argued in this paper) a defining feature of paradigms is the existence of distinctive exemplars, pieces of research uniquely informed and structured by the paradigm from which they have emerged, such links should be evident in applications of the NMP to tourism phenomena. However, while Cohen and Cohen list several 'important areas for problem-oriented work', including social justice, terrorism and environmental sustainability (2012a, pp. 2177–2186), no link is established with the NMP. Indeed, as King notes, 'rather than the concept of mobility, the issues listed can be understood in terms of the very straightforward notions of encounter and interaction' (2015b, pp. 2–3).

Another test of the empirical usefulness of the NMP is its relevance to tourism in non-Western societies. For Cohen and Cohen, it is especially useful in analysing the growth of tourism in 'emerging' regions (2015a), and trace formal, commercial tourism back to visiting friends and relatives (VFR), outings and pilgrimage, the growth of new urban middle classes, especially the Chinese, and the growth of outbound tourism, especially from Japan and China. However, this descriptive, historical account owes nothing to the NMP; indeed, Cohen and Cohen soon jettison their conceptual framework and, apart from occasionally introducing new terminology – for example, mobility constellations (2015a, p. 33) – their overview of tourism in emerging regions is fully accessible to the general reader.

In fact, the appropriateness of the NMP to non-Western societies is questionable. As Coles notes, after a decade of writing on mobilities, we still know little, 'in comparative terms, about tourism mobilities beyond Europe, North America and Australasia' (Coles, 2015, p. 65). Could it be, he suggests, that scholars working in or on developing or emerging regions may not consider such topics 'a particularly relevant or pressing research agenda' (2015, p. 65)?

Elsewhere, too, it is evident that productive writing about mobilities need not depend on adherence to a mobilities *paradigm*. Contributors to edited collections purportedly written to demonstrate the value of the NMP as a paradigm show little inclination to exchange one lexicon for another, or adopt another theoretical perspective (Sheller and Urry, 2004; Hannam *et al.*, 2014). Some contributions may be more informed by mobilities theory than others, but in neither collection is there any evidence of exclusivity or incommensurability, a welcome characteristic also found in work of Urry (e.g. Urry, 1995, Urry, 2000, 2007) and in Sheller's excellent texts on the 'consumption' of the Caribbean (Sheller, 2003) and the ways it is being transformed through tourism (Sheller, 2009). A similar assessment applies to an analysis of vintage postcards sent from the Turkish town of Smyrna (Izmir) (Andriotis and Mavrič, 2013), research on slum tourism in India (Diekmann and Hannam, 2012), and Moslem migrant workers in Sharm Elsheikh (Elgammal and Wilbert, 2013). In many such cases, 'mobilities' can simply be replaced by 'movement', with no loss of sense or meaning, which does raise a wider issue: is there any kind of tourism research that does *not* involve an analysis of mobility?

If it is indeed the case that examples such as those cited here are entirely commensurable with alternative stances towards tourism and other forms of travel, two responses are possible. First, there is a tier of scholars, perhaps the majority, who focus on some aspect of mobility in tourism and other forms of travel, or on relationships between mobilities and immobilites, for whom the NMP paradigm is either implicit or intuitive. If so, as indicated above, it is a category of fellow travellers that includes many tourism scholars.

Alternatively, it could be argued there is a smaller and exclusive community of mobilities scholars, members of which are constrained or governed by rules of membership, and with established and *exclusive* patterns of research and problem-solving. However, advocates of NMP assert quite the opposite, claiming it to be characterized by, and benefiting from, a wide range of theories, methodologies and research techniques, which hardly qualify it as a 'disciplinary matrix' even in the earlier and discarded Kuhnian version.

Conclusions

This paper has focused on several key issues. First, Kuhn's concept of paradigm is shown to have been defined by several central features: scientific knowledge is the property of specialist communities of scholars, who share a common language, an agreed perspective on the nature of scientific problems and how they are solved, and whose approach is incommensurable to others – scientists or non-scientists – outside this community.

Second, despite Kuhn's objections, social scientists have attempted to apply his concept to their disciplines, especially in the sense of a disciplinary matrix. A necessarily brief review of some of their efforts shows the concept of 'paradigm' to be differentially defined and hotly contested. It has been applied to a range of worldviews and methodologies, even the entire range of social sciences, and those who advocate the NMP as a distinctive paradigm are but last in a long line of claimants. Faced with such a plethora of positions, it is hardly surprising that some have come to view the notion of paradigm as 'a cultural cliché ... [which] is now almost meaningless' (Gorard, S. (ed), 2004) or an unnecessary diversion from theoretical development and research priorities (Walker, 2010).

Third, it is suggested that arguments that the NMP is especially applicable to tourism studies (whether or not accepted by more established mobilities theorists) are flawed. The depiction of 'paradigm' is ambiguous, the dominance of the 'discourse of authenticity' in the tourism literature has not been demonstrated, and tourism in 'emerging regions' can be described without resort to a new 'paradigm'. Indeed, mounting evidence of scholarly writing on tourism in East Asia, both in Chinese and English, indicates that ethnic Chinese and Western scholars, often in collaboration, have increasingly, profitably and productively directed their attention to aspects of Chinese tourism without recourse to a 'non-Western' or 'Asian' paradigm.

Does it matter whether or not the NMP, especially when applied to tourism studies, can be defined as a paradigm? It is argued here that the tighter and more restricted definition, emphasizing exclusivity of membership and language, special expertise and incommensurability, is the only definition that makes sense. Other definitions serve only to confuse. What matters more, though, is how far adherents to different theoretical positions, operating different analytical frameworks, can nevertheless accommodate alternative viewpoints, and it is suggested here that opportunities for cross-fertilization of mobilities approaches and other positions held by tourism scholars are sufficient to indicate that separate paradigms, in the Kuhnian sense, neither exist nor are necessary. In closing, it is hoped a few examples will suffice.

First, mobilities and immobilities are clearly inter-related. While mobilities theorists direct attention to movement and processes of different types of travellers across time and place, others who compare the movement of holidaymakers, business tourists, economic migrants, refugees or pilgrims, and the various (formal and informal, legal and illegal) agencies involved in their movement, might prefer rather to situate such movement within the socio-economic, cultural and political context of their origins and their destinations. The continued emergence of East-West tourism, for example, can be understood only within this wider context.

Second, the nature of 'place' can change. Societies may be both the origin and destination of travellers, while travellers in different categories may move from place to place: when they first arrive, a place is a destination, but when they leave it becomes a (secondary?) place of origin (as for holidaymakers visiting multiple destinations, or refugees moving across Europe). In like manner, tourist centres (e.g. the 'West') may become, or simultaneously are, peripheries (e.g. for the Chinese).

Third, the mobilities identified by such commentators as Hannam *et al.* (2014) reflect processes of traveller movement. The ways things move and are moved (mobilities), modes of transport, their control and usage (automobilities), and the technologies of modern travel and communication used, albeit for different purposes, by different categories of traveller, all contribute to the underlying processes of travelling, at places of origin, during the journey, and at the place(s) of destination. Seen phenomenologically, social interaction, the role performance of the travellers and other social actors, whether or not by choice, and the (more or less) negotiated outcomes all constitute the social dynamics and the (apparently solid but in fact highly flexible) *core* of the social institutions and structures which characterize places of traveller origin and their destinations. The likelihood that some mobilities theorists may baulk at the reference to 'structure' is recognized, but the point here is that reference to and analysis of structure is *compatible* with an analysis of mobilities.

Finally, in so far as mobilities approaches complement others, and scholars from different perspectives can learn from one another, tourism studies is characterized by commensurability rather than exclusivity. What, then, is in the word 'paradigm'? In the 'field' of tourism studies, with its plethora of approaches and disciplines, it seems to lead to conceptual confusion and little empirical clarity. To give a final example: when moving from the study of 'Western' tourist to that of the 'Asian' or 'Chinese' tourist (all categories which cannot be assumed to be homogeneous), no 'paradigm shift' is necessary; there is, though, a need for more comparative research by social scientists from different national and disciplinary backgrounds, using tools provided them by 'Western' social science – until, that is, someone has invented a 'non-Western' social science. As yet, however, no such alternative is on the horizon.

References

Andreu, R., Claver, E. and Quer, D. (2014) Destination attributes and Chinese outbound tourism to Europe. *Journal of China Tourism Research* 10(3), 275–291.

Andriotis, K. and Mavrič, M. (2013) Postcard mobility: going beyond image and text. *Annals of Tourism Research* 40(1), 18–39.

Arlt, W.G. (2013) The second wave of Chinese outbound tourism. *Tourism Planning & Development* 10(2), 126–133. DOI: 10.1080/21568316.2013.800350.

Bergman, M.M. (2010) On concepts and paradigms in mixed methods research. *Journal of Mixed Methods Research* 4(3), 171–175. DOI: 10.1177/1558689810376950.

Bird, A. (2013) Thomas Kuhn. In: Zalta, E.N. (ed.) *The Stanford Encyclopedia of Philosophy*. Available at: http://plato.stanford.edu/archives/fall2013/entries/thomas-kuhn (accessed 22 October 2015).

Borsay, P. and Walton, J.K. (eds) (2011) *Resorts and Ports: European Seaside Towns Since 1700*. Channel View Publications, Bristol.

Bruner, E.M. (1991) Transformation of self in tourism. *Annals of Tourism Research* 18(2), 238–250. DOI: 10.1016/0160-7383(91)90007-X.

Chang, T.C. (2015) The Asian wave and critical tourism scholarship. *International Journal of Asia Pacific Studies* 11(1), 83–101.

Chee-Beng, T., Cheung, S.C.H. and Hui, Y. (eds) (2001) *Tourism, anthropology and China*. White Lotus, Bangkok.

Chon, K.S. (ed.) (2000) *Tourism in Southeast Asia: A new direction*. The Haworth Hospitality Press, New York.

Cohen, E. (1979) Rethinking the sociology of tourism. *Annals of Tourism Research* 6(1), 18–35. DOI: 10.1016/0160-7383(79)90092-6.

Cohen, E. (2000) *The commercialized crafts of Thailand*. University of Hawai'i Press, Honolulu.

Cohen, E. (2001) *Thai Tourism: Hill Tribes, Islands and Open-ended Prostitution*, 2nd edn. White Lotus, Bangkok.

Cohen, E. (2007) 'Authenticity' in tourism studies: *Aprés La Lutte*. *Tourism Recreation Research* 32(2), 75–82. DOI: 10.1080/02508281.2007.11081279.

Cohen, S.A. (2010) Searching for escape, authenticity and identity: experiences of 'lifestyle travellers'. In: Morgan, M., Lugosi, P. and Ritchie, J.R.B. (eds) *The Tourism and Leisure Experience: Consumer and Management Perspectives*. Channel View Publications, Bristol, pp. 27–42.

Cohen, S.A. (2013) A portrait of Erik Cohen. *Anatolia* 24(1), 104–111. DOI: 10.1080/13032917.2013.785694.

Cohen, E. and Cohen, S.A. (2012a) Current sociological theories and issues in tourism. *Annals of Tourism Research* 39(4), 2177–2202. DOI: 10.1016/j.annals.2012.07.009.

Cohen, E. and Cohen, S.A. (2012b) Authentication: hot and cool. *Annals of Tourism Research* 39(3), 1295–1314. DOI: 10.1016/j.annals.2012.03.004.

Cohen, E. and Cohen, S.A. (2015a) A mobilities approach to tourism from emerging world regions. *Current Issues in Tourism* 18(1), 11–43. DOI: 10.1080/13683500.2014.898617.

Cohen, E. and Cohen, S.A. (2015b) Beyond Eurocentrism in tourism: a paradigm shift to mobilities. *Tourism Recreation Research* 40(2), 157–168. DOI: 10.1080/02508281.2015.1039331.

Coles, T. (2015) Tourism mobilities: still a current issue in tourism? *Current Issues in Tourism* 18(1), 62–67. DOI: 10.1080/13683500.2014.937325.

Cooper, M. (ed). (2005) Special issue: Japanese tourism, part 1. *Tourism Review International* 9.

Cooper, M. (ed.) (2007) Special issue: Japanese tourism, part 2. 11. Tourism Review International.

Cresswell, J.W. (2003) *Research Design: Qualitative, Quantitative and Mixed Methods Approaches*, 2nd edn. Sage Publications, London.

Dai, B., Jiang, Y., Yang, L. and Ma, Y. (2017) China's outbound tourism – stages, policies and choices. *Tourism Management* 58, 253–258. DOI: 10.1016/j.tourman.2016.03.009.
Dann, G.M.S. (1997) Paradigms in tourism research. *Annals of Tourism Research* 24(2), 472–474. DOI: 10.1016/S0160-7383(97)80022-9.
Denscombe, M. (2008) Communities of practice: a research paradigm for the mixed methods approach. *Journal of Mixed Methods Research* 2(3), 270–283.
Diekmann, A. and Hannam, K. (2012) Touristic mobilities in India's slum spaces. *Annals of Tourism Research* 39(3), 1315–1336. DOI: 10.1016/j.annals.2012.02.005.
Dwyer, L. and Forsyth, P. (eds) (2006) *International Handbook on the Economics of Tourism*. Edward Elgar, Cheltenham.
Dwyer, L., Forsyth, P. and Dwyer, W. (2010) *Tourism and economics policy*. Channel View Publications, Bristol.
Eckberg, D.L. and Hill, J.L.H. (1979) The paradigm concept and sociology. *American Sociological Review* 44(6), 925–937.
Elgammal, I. and Wilbert, C. (2013) Tourism mobilities, marriages and 'mimicking': some experiences of domestic migrant tourist workers in Sharm Elsheikh. *Annals of Leisure Research* 16(4), 332–347. DOI: 10.1080/11745398.2013.862899.
Faulconbridge, J. and Hui, A. (2016) Traces of a mobile field: ten years of mobilities research. *Mobilities* 11(1), 1–14. DOI: 10.1080/17450101.2015.1103534.
Feilzer, M.Y. (2010) Doing mixed methods research Pragmatically: implications for the rediscovery of pragmatism as a research paradigm. *Journal of mixed methods research* 4(1), 6–16. DOI: 10.1177/1558689809349691.
Ghimire, K.B. (ed.) (2001) *The Native Tourist: Mass Tourism Within Developing Countries*. Earthscan, London.
Gorard, S. (ed) (2004) Sceptical or clerical? theory as a barrier to the combination of research methods. *Journal of Educational Enquiry* 5, 1–21.
Graburn, N. (1983) *To pray, pay and play: The Cultural structure of Japanese domestic tourism*. Centre des Hautes Etudes Touristiques, Aix-en Provence.
Guangrui, Z. (1989) Ten years of Chinese tourism research: profile and assessment. *Tourism Management* 10(1), 51–62.
Guichard-Anguis, S. and Moon, O. (eds) (2008) *Japanese Tourism and Travel Culture*. Routledge, London.
Hacking, I. (2012) Introductory essay. In: Kuhn, T.S. (ed.) *The Structure of Scientific Revolutions*, 50th Anniversary Edition, with an Introductory Essay by Ian Hacking. The University of Chicago Press, Chicago and London, pp. vii–xxxvii.
Hall, C.M. and Page, S. (2000) *Tourism in South and South-East Asia*. Butterworth-Heinemann, Oxford.
Hammersley, M. (1992) The paradigm wars: reports from the front. *British Journal of Sociology of Education* 13(1), 131–143. DOI: 10.1080/0142569920130110.
Hannam, K.and (2014) Tourism mobilities. In: Lew, A.L., Hall, C.M. and Williams, A.M. (eds) *The Wiley Blackwell Companion to Tourism*. Wiley, Chichester, pp. 78–87.
Hannam, K., Sheller, M. and Urry, J. (2006) Editorial: mobilities, immobilities and moorings. *Mobilities* 1(1), 1–22. DOI: 10.1080/17450100500489189.
Hannam, K., Butler, G. and Paris, C.M. (2014) Developments and key issues in tourism mobilities. *Annals of Tourism Research* 44(1), 171–185. DOI: 10.1016/j.annals.2013.09.010.
Harrison, D. (2001) Tourism and less developed countries: key issues. In: Harrison, D. (ed.) *Tourism and the Less developed World: Issues and Case Studies*. CABI International, Wallingford, pp. 23–46.
Harrison, D. (2007) Developing a framework for analysing tourism phenomena. *Current Issues in Tourism* 10(1), 61–86.
Harrison, D. (2016) Looking East but learning from the West? Mass tourism and emerging nations. *Asian Journal of Tourism Research* 1(2), 1–36. DOI: 10.12982/AJTR.2016.0014.
Harrison, D., Eades, J. and King, V.T. (eds) (2017a) *Tourism in East and Southeast Asia, Volume 1: The Background: Concepts, Region and History*. Routledge, London.
Harrison, D., Eades, J. and King, V.T. (eds) (2017b) *Tourism in East and Southeast Asia, Volume 2: Stakeholders in the Political Economy of Tourism*. Routledge, London.
Harrison, D., Eades, J. and King, V.T. (eds) (2017c) *Tourism in East and Southeast Asia, Volume 3: The Global and the Local: Tourism, Culture and Identity*. Routledge, London.
Harrison, D., Eades, J. and King, V.T. (eds) (2017d) *Tourism in East and Southeast Asia: Volume 4: Tourism and Place: Environments and Contexts*. Routledge, London.

Heyl, J.D. (1975) Paradigms in social science. *Society* 12(5), 61–67. DOI: 10.1007/BF02699921.

Hill, L. and Eckberg, D.L. (1981) Clarifying confusions about paradigms: a reply to Ritzer. *American Sociological Review* 46(2), 248–252. DOI: 10.2307/2094985.

Hitchcock, M., King, V.T. and Parnwell, M. (eds) (1993) *Tourism in Southeast Asia*. Routledge, London and New York.

Hitchcock, M., King, V.T. and Parnwell, M. (eds) (2010) *Heritage Tourism in Southeast Asia*. NIAS Press, Copenhagen.

Hsu, C., Huang, J. and Huang, S. (2010) Tourism and hospitality research in mainland China: Trends from 2000 to 2008. In: Pearce, D.G. and Butler, R.W. (eds) *Tourism Research: A 20-20 Vision*. Goodfellow, Oxford, pp. 147–160.

Huang, S (S). (2011) Tourism as the subject of China's doctoral dissertations. *Annals of Tourism Research* 38(1), 316–319. DOI: 10.1016/j.annals.2010.08.005.

Huang, Songshan (Sam). and Chen, G. (2016) Current state of tourism research in China. *Tourism Management Perspectives* 20(1), 10–18. DOI: 10.1016/j.tmp.2016.06.002.

Jafari, J. and Aaser, D. (1988) Tourism as the subject of doctoral dissertations. *Annals of Tourism Research* 15(3), 407–429. DOI: 10.1016/0160-7383(88)90030-8.

Keith, B. (2006) The nature of sociological knowledge. *The American Sociologist* 37(2), 6–14. DOI: 10.1007/s12108-006-1001-0.

King, V.T. (2015a) Substantive and conceptual issues in tourism research: a personal engagement with Southeast Asia. *International Journal of Asia Pacific Studies* 11(1), 15–31.

King, V.T. (2015b) Current issues in tourism: Thailand and beyond – introduction to special issue. *International Journal of Asia Pacific Studies* 11(1), 1–13.

Kong, H. and Cheung, C. (2009) Hotel development in China: a review of the English language literature. *International Journal of Contemporary Hospitality Management* 21(3), 341–355. DOI: 10.1108/09596110910948332.

Kuhn, T.S. (1962) *The Structure of Scientific Revolutions*. The University of Chicago Press, Chicago and London.

Kuhn, T.S. (1970a) *The Structure of Scientific Revolutions*, 2nd edn. The University of Chicago Press, London and Chicago, pp. 173–208.

Kuhn, T.S. (1970b) Reflections on my critics. In: Lakatos, I. and Musgrave, A. (eds) *Criticism and the Growth of Knowledge*. Cambridge University Press, Cambridge, pp. 231–278.

Kuhn, T.S. (1970c) Logic of discovery or psychology of research? In: Lakatos, I. and Musgrave, A. (eds) *Criticism and the Growth of Knowledge*. Cambridge University Press, Cambridge, pp. 1–24.

Kuhn, T.S. (1974) Second thoughts on paradigms. In: Suppe, F. (ed.) *The structure of scientific theories*. University of Illinois Press, Urbana Illinois.

Kuhn, T.S. (1987) What are scientific revolutions? In: Kruger, L., Dalston, L. and Heidelberger, M. (eds) *The Probabilistic Revolution*. Cambridge University Press, Cambridge, pp. 7–22.

Kuhn, T.S. (2000) *The Road since Structure. In Philosophical Essays, 1970–1993, with an Autobiographical Interview*. University of Chicago Press, Chicago.

Kuhn, T.S. (2012) *(With an introductory essay by Ian Hacking) The Structure of Scientific Revolutions*, 4th edn. The University of Chicago Press, Chicago and London.

Law, R. and Cheung, P. (2008) An analysis of publications in leading tourism journals and its implications on China tourism research. *Journal of China Tourism Research* 4(1), 78–97. DOI: 10.1080/19388160802099840.

Li, X. (2016) *Chinese Outbound Tourism 2.0*. Apple Academic Press, Oakville, Canada.

MacCannell, D. (1976) *The Tourist: A New Theory of the Middle Class*. Macmillan, London.

Mackenzie, N. and Knipe, S. (2006) Research dilemmas: paradigms, methods and methodology. *Issues in Educational Research* 16, 193–205.

Masterman, M. (1970) The nature of a paradigm. In: Lakatos, I. and Musgrave, A. (eds) *Criticism and the Growth of Knowledge*. Cambridge University Press, Cambridge, pp. 59–90.

Mavrič, M. and Urry, J. (2009) Tourism studies and the new mobilities paradigm. In: Jamal, T. and Robinson, M. (eds) *The Sage Handbook of Tourism Studies*. Sage, London, pp. 645–657.

McGregor, S.L.T. and Murnane, J.A. (2010) Paradigm, methodology and method: intellectual integrity in consumer scholarship. *International Journal of Consumer Studies* 34(4), 419–427. DOI: 10.1111/j.1470-6431.2010.00883.x.

Mertens, D.M. (2007) Transformative paradigm: mixed methods and social justice. *Journal of Mixed Methods Research* 1(2), 212–235.

Mertens, D.M. (2012) What comes first? The paradigm or the approach? *Journal of Mixed Methods Research* 6(4), 255–257. DOI: 10.1177/1558689812461574.
Mings, R.C. and Liu, W. (1989) Emerging tourism in China — the case of Xian. *Tourism Management* 10(4), 333–336. DOI: 10.1016/0261-5177(89)90012-5.
Morgan, D.L. (2007) Paradigms lost and pragmatism regained: methodological implications of combining qualitative and quantitative methods. *Journal of Mixed Methods Research* 1(1), 48–76.
Oakes, T. (1998) *Tourism and Modernity in China*. Routledge, London.
Park, H. (2014) *Heritage Tourism*. Routledge, London.
Pearce, P. (1982) *The Social Psychology of Tourist Behaviour*. Pergamon Press, Oxford.
Pearce, P. (2005) *Tourist Behaviour: Themes and Conceptual Schemes*. Channel View Publications, Clevedon.
Popper, K. (1970) Normal science and its dangers. In: Lakatos, I. and Musgrave, A. (eds) *Criticism and the Growth of Knowledge*. Cambridge University Press, Cambridge, pp. 51–58.
Punch, K.F. (2013) *Introduction to Social Research: Quantitative and Qualitative Approaches*. Sage, Los Angeles and London.
Ritzer, G. (1978) Reflections on the paradigmatic status of sociology. *Mid-American Review of Sociology* 3, 1–15.
Ritzer, G. (1981) Paradigm analysis in sociology: clarifying the issues. *American Sociological Review* 46(2), 245–248. DOI: 10.2307/2094984.
Rojek, C. and Urry, J. (1997) Transformations of travel and theory. In: Rojek, C. and Urry, J.. (eds) *Touring Cultures: Transformation of Travel and Theory*. Routledge, London, pp. 1–19.
Ross, G.F. (1994) *The Psychology of Tourism*. Hospitality Press, Melbourne.
Ryan, C. and Huang, S. (eds) (2013) *Tourism in China: Destinations, Planning and Experiences*. Channel View Publications, Bristol.
Selwyn, T. (2007) The political economy of enchantment: formations in the anthropology of tourism. *Suomen Antropologi: Journal of the Finnish Anthropological Society* 32(2), 48–70.
Shannon-Baker, P. (2015) Making paradigms meaningful in mixed methods research. *Journal of Mixed Methods Research* 9(1), 1–16.
Sharpley, R. and Telfer, D. (eds) (2015) *Tourism and Development: Concepts and Issues*, 2nd edn. Channel View Publications, Bristol.
Sheller, M. (2003) *Consuming the Caribbean*. Routledge, London.
Sheller, M. (2009) The new Caribbean complexity: mobility systems, tourism and spatial rescaling. *Singapore Journal of Tropical Geography* 30(2), 189–203. DOI: 10.1111/j.1467-9493.2009.00365.x.
Sheller, M. and Urry, J. (2006) The new mobilities paradigm. *Environment and Planning A: Economy and Space* 38(2), 207–226. DOI: 10.1068/a37268.
Sheller, M. and Urry, J. (2016) Mobilizing the new mobilities paradigm. *Applied Mobilities* 1(1), 10–25. DOI: 10.1080/23800127.2016.1151216.
Sheller, M. and Urry, J. (2004) Places to play, places in play. In: Sheller, M. and Urry, J. (eds) *Tourism Mobilities: Places to Play, Places in Play*. Routledge, London, pp. 1–10.
Singh, S. (2009) *Domestic Tourism in Asia: Diversity and Divergence*. Earthscan, London.
Suppe, F. (1977) Introduction. In: Suppe, F. (ed.) *The Structure of Scientific Theories*. University of Illinois Press, Urbana and Chicago, pp. 3–5.
Teo, P., Chang, T.C. and Ho, K.C. (eds) (2001) *Interconnected Worlds: Tourism in Southeast Asia*. Elsevier, Oxford.
Toulmin, S. (1970) Does the distinction between normal and revolutionary science hold water? In: Lakatos, I. and Musgrave, A. (eds) *Criticism and the Growth of Knowledge*. Cambridge University Press, Cambridge, pp. 39–45.
Towner, J. (1995) What is tourism's history? *Tourism Management* 16(5), 339–343. DOI: 10.1016/0261-5177(95)00032-J.
Tribe, J., Dann, G. and Jamal, T. (2015) Paradigms in tourism research: a trialogue. *Annals of Tourism Research* 40(1), 28–47.
Tsang, N.K.F. and Hsu, C.H.C. (2011) Thirty years of research on tourism and hospitality management in China: a review and analysis of journal publications. *International Journal of Hospitality Management* 30(4), 886–896. DOI: 10.1016/j.ijhm.2011.01.009.
Tse, T.S.M. (2015) A review of Chinese outbound tourism research and the way forward. *Journal of China Tourism Research* 11(1), 1–18. DOI: 10.1080/19388160.2014.984798.

Urry, J. (1973) Thomas S. Kuhn as sociologist of knowledge. *The British Journal of Sociology* 24(4), 462–473. DOI: 10.2307/589735.

Urry, J. (1995) *Consuming Places*. Routledge, London.

Urry, J. (2000) *Sociology Beyond Societies: Mobilities for the Twenty-first Century*. Routledge, London.

Urry, J. (2007) *Mobilities*. Polity, Cambridge.

van den Berghe, P.L. (1978) Bridging the paradigms. *Society* 15(6), 42–49. DOI: 10.1007/BF02697775.

Walker, T.C. (2010) The perils of paradigm mentalities: revisiting Kuhn, Lakatos, and Popper. *Perspectives on Politics* 8(2), 433–451. DOI: 10.1017/S1537592710001180.

Walton, J.K. (2005) *Histories of Tourism: Representation, Identity and Conflict*. Channel View Publications, Clevedon.

Wang, N. (1999) Rethinking authenticity in tourism experience. *Annals of Tourism Research* 26(2), 349–370. DOI: 10.1016/S0160-7383(98)00103-0.

Wang, N. (2000) *Tourism and Modernity: A Sociological Analysis*. Pergamon, Oxford.

Weber, M. (1948) Science as a vocation. In: Gerth, H.H. and Wright Mills, C. (eds) *From Max Weber: Essays in Sociology*. Routledge and Kegan Paul, London.

Williams, D. (2013) 30 years of tourism research: a bibliography of Japanese tourism (1980-2009). *Josai International Bulletin* 21(7), 43–70.

Winter, T. (2009) Asian tourism and the retreat of anglo-western centrism in tourism theory. *Current Issues in Tourism* 12(1), 21–31. DOI: 10.1080/13683500802220695.

Winter, T., Teo, P. and Chang, T.C. (eds) (2009) *Asia on Tour: Exploring the Rise of Asian Tourism*. Routledge, London.

Xiang, Y. (2013) The characteristics of independent Chinese outbound tourists. *Tourism Planning & Development* 10(2), 134–148. DOI: 10.1080/21568316.2013.783740.

Xiao, H., Su, D. and Li, M. (2010) Diffusion of knowledge across linguistic communities: the case of using 'foreign' sources for tourism research in China. *Journal of China Tourism Research* 6(4), 326–342. DOI: 10.1080/19388160.2010.527569.

Xie, P.F. (2013) The 'great wall' in tourism research – a portrait of Geoff Wall. *Anatolia* 24(2), 282–287. DOI: 10.1080/13032917.2013.807390.

Xie, Y. and Li, M. (2009) Development of China's outbound tourism and the characteristics of its tourist flow. *Journal of China Tourism Research* 5(3), 226–242. DOI: 10.1080/19388160903149965.

Yamashita, S., Kadir, H.D. and Eades, J.S. (eds) (1997) *Tourism and Cultural Development in East Asia and Oceania*. Penerbit Universiti Kebangsaan, Malaysia, Bangi.

Zhang, L., Lan, C., Qi, F. and Wu, P. (2017) Development pattern, classification and evaluation of the tourism academic community in China in the last ten years: from the perspective of big data of articles of tourism academic journals. *Tourism Management* 58, 235–244. DOI: 10.1016/j.tourman.2016.03.007.

Zhong, L., Wu, B. and Morrison, A.M. (2015) Research on China's tourism: a 35-year review and authorship analysis. *International Journal of Tourism Research* 17(1), 25–34. DOI: 10.1002/jtr.1962.

21

Anthropologists, Development and Tourism Networks: Encounters and Shadows of a Colonial Past

Introduction

The aim of this paper is to situate and describe some of the problems that emerge when sociologists/social anthropologists from developed societies become involved in 'development' work in developing countries. The paper is in three distinct sections: a brief overview of the debates about anthropologists' roles during colonialism, a discussion of tourism academics' attitudes to applied work and, finally, a more personal reflection of some of the issues arising from this writer's experience in working with various stakeholders on tourism in developing countries.

An Historical Overview: Anthropologists and Colonialists: The Debates

In 1983, at the third decennial conference of the Association of Social Anthropologists, held in Cambridge, there was a section of papers on 'Anthropology in the Eighties' and, as part of this discussion, there was a sub-theme of 'applied anthropology'. The issue was addressed by some of the most notable anthropologists of the time, and tended to centre on the role anthropologists – or at least some of them – had played during the period of British colonialism, most particularly in Africa.

Discussion was clearly heated, and this is reflected in the ensuing publication (Grillo and Rew, 1985). In the book, 11 anthropologists, with highly respectable pedigrees, discuss their role, and that of anthropology, across a wide range of contexts. Most were academics – e.g. from Cambridge, Sussex, London and Toronto – but there was also a specialist consultant and an aid agency official. Several had been involved to a greater or lesser extent with colonial administrations (especially Gulliver), and/or in providing advice to 'native peoples' (Layton in Northern Australia, Cheater in independent Zimbabwe, Whisson in apartheid-governed South Africa, and Strathern in Papua New Guinea). Notably, too, in much of the discussion, but especially in Grillo's introduction, there are periodic references to the attitudes and practices of previous generations of anthropologists, including such notables as Malinowski, Mair and Firth, and to the establishment in 1926 of what was to become the International African Institute, an organization which supported many activities of social anthropologists working in colonial Africa (Grillo, 1985).[1]

In passing, it is no exaggeration to suggest that, irrespective of individual attitudes towards colonial authorities among anthropologists, access to 'their' societies was largely facilitated by the colonial system, albeit sometimes reluctantly. Indeed, Malinowski, the 'father' of the British fieldwork tradition, found himself in the Trobriand Islands *only* because, as a Pole resident in the UK during the First World War, his alternative was internment as an enemy alien!

Why were such apparently historical matters of interest in the mid-1980s? First, after the Second World War (especially with the onset of independence for many former colonial territories), applied anthropology was considered marginal, even (according to some commentators) tarnished by its association with colonialism. It was no longer appropriate to conceive of such societies as 'primitive'. Second, with the increasing popularity of phenomenological sociology in the 1960s and 1970s, and the recognition by social anthropologists that anthropological approaches and techniques were as relevant to urban contexts (and peoples of the 'fourth world') in developed societies as to 'traditional societies', the distinguishing characteristics of social anthropology, in both *locus* and *methods* of study, had become increasingly blurred (Chambers, 2000). In general, proponents of social/cultural anthropology were looking for a new role, a search given added focus during the 1970s, when more anthropology graduates were being produced and (as a result) were seeking employment opportunities. Then, as now, the relevance of academic studies to the 'requirements' of the job market were a major concern.

The 1980s reprise of anthropologists' relationships to colonial governments, in particular, revealed, first, a series of polarized stereotypes of (apparently) different kinds of anthropology, accompanied by numerous but implied value judgements and, second, a categorization of anthropologists' attitudes to (and indeed definitions of) applied anthropology.

An indication of the stereotypes prevailing in the mid-1980s, as discussed by Graburn (1983) and informed by the writer's own experience, is given in Table 21.1, which contrasts perceptions of 'pure' and 'applied' anthropology, here defined as the activities of anthropologists that are non-theoretical and goal-oriented. The adjectives are really self-explanatory, but it is especially noteworthy that those who (allegedly) pursue anthropology for anthropology's sake (whatever that might mean) are contrasted with those of a more practical, empirical bent and, certainly if it is the former who are doing the classifying, not to the latter's advantage.

As Grillo pointed out, such polarization was unjustified even in the 1920s, as well as in the 1980s. There was no reason, for example, to suppose that applied anthropology was less intellectually demanding than 'pure' anthropology, and applied research can produce data and ideas of great relevance to the corpus of disciplinary knowledge, including theory. Nevertheless:

Table 21.1. Pure vs applied anthropology.

Pure anthropology	Applied anthropology
Abstract and theoretical	Practical and empirical
Disinterested	Interested, engaged and action-oriented
Objective	Subjective
Moral	Immoral
Academic	Non-academic
Discipline-based	Client-based (e.g. government, aid agencies, NGOs)
Intellectual	Policy-oriented
Comprehensive	Half-baked
Exponents are wise	Exponents are 'experts'
Paid by the state in *universities* – not polytechnics!	Paid by government in aid agencies or (worse??) by the private sector?

> We must face the fact that an applied/theoretical opposition persists in anthropology and affects both 'sides': one abjuring practice, the other theory. Why does anthropology, more than any other social science, appear to make such heavy weather of this distinction? What, one wonders, would Keynes have made of it? If, then, the meaning of applied anthropology is to be found in its rejection by those in the mainstream of the subject, we have to ask why this rejection has occurred (Graburn, 1983).

The second issue, which really follows from the first, is that it was possible to categorize anthropologists according to where they stood on the pure vs applied debate. Grillo suggests they can be placed in one of three camps: principled rejectionists, monitorists and activists (Grillo, 1985, pp. 28–31).

Principled rejectionists are those who might suggest anthropologists are not qualified to intervene in social matters, that those with whom they would have to deal make the effort dirty and unworthwhile, and/ or that they should resist incorporation into a system they dislike and even despise.

Monitorists are those who assess what is going on and carry out research 'to investigate specific plans, policies or projects and their implementation' (Graburn, 1983), thus pursuing the anthropology *of* development without actually being *in* development, so to speak. Like St. Paul, they are *of* but not *in* the world.

> Finally, activists 'are actively engaged in customer-contractor relationships in which they are called upon to devise and review development policies' (Graburn, 1983). They are at the opposite end of the spectrum to the rejectionists, and (unlike the latter) believe at least in the possibility that their actions can improve the existing situation. Quoting Belshaw, Grillo suggests that such a view is characterized by the belief or hope that 'the application of anthropological knowledge will moderate the bad and enhance the good' (1985, p. 30).

This debate over the role of anthropologists in the colonial period is no mere historical footnote. Since 2001, for instance, the UK government-sponsored research councils have launched several major initiatives that focus on terrorism, including the Economic and Social Research Council's programme entitled 'Domestic Management of Terrorist Attacks'. Initially, the focus was on assessing the risk of terrorist attacks, and managing and monitoring the authorities' response to them (www.publications. parliament.uk/pa/cm200203/cmselect), but studies of terrorism have since multiplied on a massive scale. Most relevant to the present context, in 2008:

> A £1.3 million ESRC [Economic and Social Research Council] and AHRC [Arts and Humanities Research Council] programme, jointly funded by the [UK] Foreign Office on 'Combating Terrorism by Countering Radicalisation,' was pulled after academics claimed it was tantamount to asking researchers to act as spies for British Intelligence (Attwood, 2009).

Similar debates over the militaristic or nationalistic applications of anthropology have occurred in North America. In 1919, Boas angrily suggested that some fellow anthropologists had 'prostituted science by using it as a cover for their activities as spies' (Mcfate, 2005), while Margaret Mead and Ruth Benedict, both prominent American anthropologists, assisted the US government during the Second World War. Later, the infamous Project Camelot, initiated in 1964, sought anthropologists' assistance in defeating protest movements in developing countries (Horowitz, 1967), and a few years later the American Anthropological Association was split by acrimonious debate over the role of anthropologists with fieldwork experience in Thailand who offered their services to Project Agile, a programme of the US Agency for International Development designed to ensure Thailand remained 'in the Free World' (Wakin, 1992). Similarly, Hickey, the author of an ethnography of a Vietnamese village, (unsuccessfully) advised the US government on its conduct of the Vietnam/American War (Mcfate, 2005). The advent of the Cold War gave added impetus to the development of Area Studies in the USA and, once again, many anthropologists were engaged by the state for political purposes, though others, by contrast, were targeted for their involvement in the civil rights movements of the 1940s and 1950s (Price, 1998).

Most recently, since 2007, as part of its *Human Terrain System*, the US army has experimented in attaching anthropologists to combat brigades in Iraq and Afghanistan (Rohde, 2007;

Shweder, 2007). The aim has been to provide 'direct social science support in the form of ethnographic and social research, cultural information research, and social data analysis that can be employed as part of the military decision making process' (Marlowe, 2007). How far this has been achieved is not yet known (though little evidence of success seems to have emerged to date), but the effort has again led to discussion of the anthropologists' role in war zones (Johnson, 2007; Marlowe, 2007).

Unity despite diversity

Clearly, concerns over anthropological involvement with government, especially at times of perceived insecurity, are as relevant today as they were in the 1920s, 1940s or 1980s, and current attitudes among sociologists and anthropologists reflect these earlier viewpoints. Such disagreements are entirely consistent with individual integrity. However, this should not disguise the fact that these debates, even when acrimonious, occur *within* an established scientific discipline. Despite their differences, anthropologists (and by extension other social scientists) occupy a key position in what might be described as the Western intellectual tradition, best characterized by Weber's famous reference to 'Science as a Vocation'. As members of a network of scientists operating in a world characterized by disenchantment, by rationality, they are committed to understanding what they have previously selected as 'culturally significant' by exercising their (value-based) *vocation*, or calling, in the pursuit of 'science', which is organized in special disciplines in the service of self-clarification and knowledge of inter-related facts' (Weber, 1948). And where disagreements *do* emerge, as is inevitable, they are mediated from within the scientific community, by those who, to all intents and purposes, are part of a common culture.

The Pure-applied Division Among Tourism Academics

The extent to which anthropologists are willing to work with, for or against governments continues to exercise concern in the current international climate, and principled rejectionists, monitorists and activists are still distinguishable among academics interested, more widely, in matters pertaining to 'development'. At the same time, it must also be recognized that since 1945, and especially since the 1980s, the development context has also changed. There are new generations of leaders in what were once described as 'new nations'; 'transitional economies' (really a synonym for another set of new nations) have emerged with the disappearance of the 'Third World' and the subsequent merging of the 'First' and the 'Second'; transnational companies and non-governmental organizations (NGOs) are now major players on the world economic stage, and unilateral and multilateral aid has become increasingly important to developing societies. And while in this increasingly globalized context there has perhaps been a further dilution of policy makers' interest in (what they regard as) 'pure' academic subjects (including social anthropology), there has *also* been an increased preference for 'useful' qualifications. What is required, it would seem, is not academic excellence but *expertise*.[2]

More specifically, the divisions discussed in the previous section are evident among academics who focus their attention on tourism, either as a series of inter-related phenomena worthy of academic study or, more generally, as a 'tool' for development. A passing mention of those initially involved in establishing tourism as an academic subject in the 1970s highlights some of the tensions. In 1989, for example, MacCannell, who was influential in establishing tourism as a proper subject of anthropological enquiry (1976), and who is situated firmly at the academic end of the continuum, was to reflect as follows:

> Perhaps this is an appropriate place to note that while I may seem to have overlooked or neglected to report the sources of support for my research on tourism and travel, this is not the case. I have never received any institutional funding for this work. This was certainly not my intention at first.... Now I admit to a certain perverse pleasure in the knowledge that none of this work is on anyone else's balance sheet or ledger.... Nor have I taken any fees for consultation on matters of travel and tourism.... (MacCannell, 1989).

While MacCannell was quite willing to accept institutional funding for other research projects (MacCannell, 1992, pp. xi-xii), and presumably felt obliged to accept his university salary, towards tourism, at least, he apparently adopted a rejectionist position. It is echoed by many others. Sometimes this rejection of any involvement with policy-making goes further, and comes close to opposition to tourism as a system and to tourists as a category. MacCannell was thus led to ask: 'What is an expeditionary force without guns Tourists?' (1989, p. xviii). Similarly, Nash was clear, in the 1970s, that tourism is a form of 'imperialism' (Nash, 1978), and Graburn likens tourism in developing countries to a form of prostitution, penetration and rape (1983).

Not all tourism academics are hostile towards tourism. By definition, all those who participate in tourism *management*, or who are involved in *planning*, at local, national, regional or even international level, might best be described at least as monitorists, and many take a truly active role in promoting tourism, or some types of it. Valene Smith, another major influence in the development of the anthropology of tourism, for example, has long been a tour guide. Contributors to the first edition of her volume *Hosts and Guests* (Smith, 1978) adopt a variety of stances, as do those of de Kadt's similar collection, which soon followed (De kadt, 1979).

It is also the case that some anthropologists of tourism have changed their views over time. By 1996, for example, Nash seems to have rejected the notion that tourism was a form of imperialism, and was arguing instead that while anthropologists should become more involved in applied work, they 'still have a way to go towards scientific maturity' (Nash, 1996, p. 169). However, most sociologists and anthropologists working on tourism write from a perspective which, at best, can be described as monitorist, tending to be decidedly ambivalent to tourism and indicating little respect for the tourists they study. They prefer, instead, to 'side' with destination communities (as if they somehow constitute a homogeneous unit), thus raising, yet again, the question long ago posed by Howard Becker: 'Whose side are we on?' (Becker, 1967; Liebling, 2001).

Such attitudes can be explained in several ways. First, as Kit Jenkins, one of the few to have combined successful academic work with consultancy, points out, this 'great divide' between tourism academics and tourism practitioners arises, in part, because they 'occupy very different work situations, and those academics who do participate in tourism projects tend to be approached only *after* strategies and methods have been agreed' (Jenkins, 1999). However, this is not the only reason. The second is that because tourism academics were mainly educated in liberal higher education establishments in the 1960s and 1970s, some (like MacCannell) have been opposed to capitalism from the outset. This would help explain why consultancy work, especially for the private sector, is often considered to compromise academic independence, and even (rightly or wrongly) to be seen as an out and out betrayal and loss of intellectual integrity (Mowforth and Munt, 1998).

True, there is continued advocacy of stakeholder partnership; environmentalists and economists often contribute to applied and policy-oriented research, and there seems to be little objection to an involvement in ecotourism or community-based tourism (CBT), perhaps because they are considered 'politically correct' (Rowe, 2005). By contrast, though, academics tend to distrust such major development agencies as the UNWTO, the World Bank and the Asian Development Bank, and are quick to voice their disapproval of mass tourism and the role of transnational companies in it, especially in developing countries (Rowe, 2005). In particular, as indicated elsewhere (Harrison, 2004), direct cooperation with the private sector is rare, and those who *do* manage to cross this 'great divide' find that institutions in which they work, especially in the traditional university sector, give them little credit for doing so (Mars, 2004).

Third, even if they *want* to become involved in policy-oriented research, they are often unable to do so. A brief examination, for example, of the list of 139 consultants in the UK's Tourism Society in 2003 reveals that a mere six were based in academic institutions, and this is no indication that the 'lucky' six obtained many (even any) consultancy contracts! Indeed, consultancies are awarded on the basis of prior consultancy experience which, by definition, most academics lack. And the attitude of many consultants and contract-awarding bodies towards academics, justified or not, is hardly encouraging. On being shown a copy of the *Annals of Tourism Research*, for example, one leading British tourism consultant exclaimed: 'Who writes this stuff? And who reads it?' The obvious answer to this partly rhetorical question is that academics write it, because that is what they are paid to do, and academics read it (for the same reason). For their part, consultants and policy makers are *not* paid to read journals, rarely have the time, and usually lack the inclination. Instead, they write or commission reports, which may or may not be peer-reviewed, and which are read by others from a similar background: other consultants, NGO personnel, officials in aid agencies and international institutions, and so on. And the net result? Two largely separate networks – academic and non-academic – and two largely separate bodies of literature, written in different styles, according to different criteria, and for very different audiences.

Finally, it can be argued that the anthropological project is, in many respects, based on a form of intellectual tourism and, at least for some residents in tourist destination communities, anthropologists appear as but another kind of tourist. Faced with such an embarrassment, and with due regard to the low opinion they might already have of tourists, it should occasion no surprise that anthropologists act somewhat defensively to protect their reputation (MacCannell, 1976; Mintz, 1977; Errington and Gewertz, 1989; Crick, 1994; Berno, 1999; Nash, 2001; Lacy and Douglass, 2002)!

The situation I am describing may be slightly different in Australia and New Zealand, where links between academic studies and consultancy work seem closer, simply because of the way higher education (especially applied studies) has emerged. However, in the UK and, I suspect, in other parts of Europe and North America, partly because of the trends I have described, the divide between practitioners and academics, and their respective outputs, continues to be deep and wide.

The Anthropologist in Developing Countries – Examples from Tourism

Despite the kind of hurdles I have been describing, anthropologists/sociologists do sometimes work in developing societies,[3] perhaps as consultants with aid agencies or the private sector, and/or through international links made with universities. In so doing, they move from operating within a largely academic network where, despite differing perceptions towards 'applied' work, widely disputed theoretical positions, less than perfect administrative environments, and debates over the cross-cultural applicability of research methods (Berno, 1996), there is still a wide degree of consensus over the value of 'research' and 'scholarship' and (for those with formal training in social anthropology) the lasting importance of a relatively long exposure to the 'field' of enquiry, as well as the need to submit research outcomes to peer group review. On moving away from this network and entering the 'development' environment, and thus encountering a new range of 'stakeholders', with quite differing characteristics, such a consensus cannot be assumed to continue. Consultants, aid workers and NGO representatives, and politicians all have their own networks, and their own agendas, which may or may not coincide or overlap with that of the academic.

Consultants, for instance, operate in a highly competitive environment. In the UK, they are likely to have graduated from 'redbrick' universities, with qualifications in economics or geography, or in such applied disciplines as urban planning. Once appointed, they may use fairly standard templates, and the terms of reference generally require them to produce speedy, succinct and unreferenced reports and recommendations. Under severe time constraints and deadlines, they often have to juggle time spent on one consultancy with that on others. There is little emphasis on (or time to carry out) primary research. Instead, focus is largely on secondary sources, and there may be unashamed borrowing from official and other documents. Tourism consultants, for instance, rely heavily on *Lonely Planet* or *Rough Guide* publications. And their conclusions, which may indeed be correct, are rarely informed by academic publications on their area of research, or validated according to academic rules of evidence, and they do not have the luxury of numerous caveats and qualifications.

Example: In one international tourism consultancy, funded by the Asian Development Bank, consultants were charged, *inter alia*, with anticipating the impacts of infrastructural developments (such as new or upgraded roads and airports) on local communities in South-east Asia. Most sites were visited for, *at most*, 2 days, and sometimes for as little as a morning or an afternoon, before leaving for the next destination.

By contrast, NGOs in developing countries are likely to be staffed by relatively young and committed expatriates, who often form the basis of a distinct social segment in urban centres, and also constitute an equally distinct market segment in the demand for housing and the use of Western-oriented bars and restaurants. Assisted by members of the local elite, whose employment frequently circulates among NGOs, they are reliant on aid from international donors, and continually have to submit applications for continued funding. They operate within an ethos which is highly budget-conscious and distinctly managerial, but nevertheless lack experience in carrying out research or running businesses.

Example: People employed by NGOs may have a very different perspective on tourism to the visiting consultant (or academic). Indeed, in development circles generally, it has long been fashionable to express contempt for tourists and tourism. Nowadays, however, NGOs are becoming more involved in running their own 'pro-poor' tourism projects, under the umbrella of 'community-based tourism'. At a 4-day conference in Taveuni, Fiji, in July 1998, convened by the New Zealand Overseas Development Agency, international donors and local NGOs involved in ecotourism projects over the previous 5 years throughout the region agreed that, measured in terms of economic sustainability, most were demonstrably *unsuccessful*. A lack of business and marketing skills and experience on the part of NGO personnel, unrealistic expectations by donors as to the length of training in hospitality required for community members, along with an underestimate of the time-frame needed before adequate economic returns would materialize, and a failure to utilize the expertise of the private sector were just some of the factors held responsible for the almost universal failure of ecotourism projects in the region (Tourism Research Consultants, 1999).

Another example: Even successful projects operated by NGOs may not match the performance of the private sector, though it must also be accepted that sometimes benefits from the former's involvement are difficult to measure. In Lao PDR, for instance, by 2008 the Nam Ha Ecotourism Project, which commenced in 1999, had received some US$900,000 from international donors, but the direct *gross* revenue from its operations (primarily training guides and running treks) to the population of about 2000 in eight minority villages over this period was no more than 25% of the income from aid donors. By contrast, *annual* tourist expenditure on the Mekong island of Don Det, in southern Laos, where small locally owned and privately financed guest houses and restaurants are the norm, is estimated to be in the region of US$ 450,000 (Harrison and Schipani, 2007).

It might be expected that Western anthropologists would fit most easily into networks of academics in developing countries, and productive and successful collaboration and friendships do indeed develop among academics across cultures. At the same time, it cannot be *assumed* that there will be a meeting of minds. Even in the West, academics from apparently similar social scientific traditions, for example the USA, France, Germany and the UK, can experience huge difficulties in cross-cultural discussion and research.[4]

In fact, the expression and practice of science is inevitably filtered through historical, structural and cultural contexts in developed and developing societies. Indeed, if the development of an international community of scholars following (Western) science as a vocation (Weber, 1948) is difficult across developed societies apparently characterized by rationality, in those that are developing or transitional, described by Riggs as 'prismatic' societies (1964), possibilities for misunderstanding and disagreement are even more marked. Ideas and institutions related to scientific practice will be differentially incorporated into their political, social and cultural structures, and similarities in institutional *form* may disguise considerable discrepancies in *function*, as indicated in Fig. 21.1.[5]

In some South-east Asian societies, for example, social scientists in universities (like their colleagues) receive what, by Western standards, is a pitifully low salary, amounting to little more than US$30 a month. As a consequence, much sought-after appointments as national experts on international projects come to be their major source of income, and a *per diem* of US$75

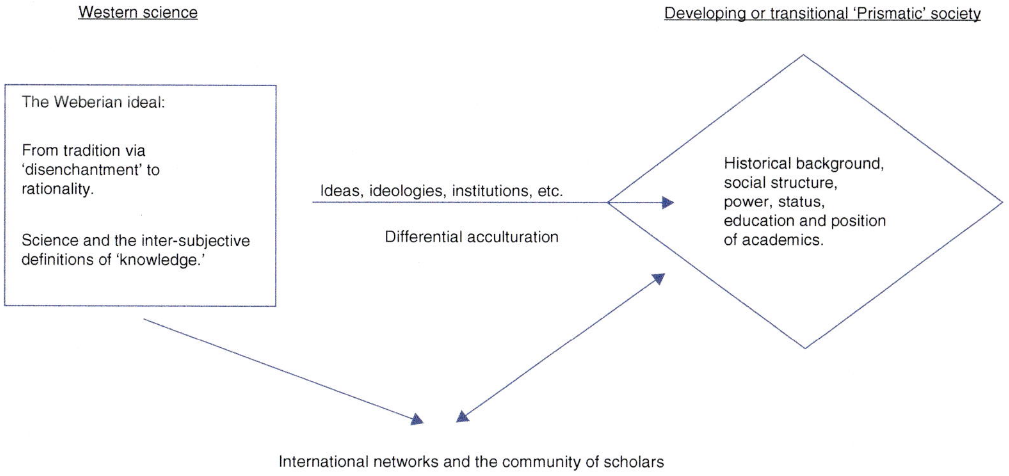

Fig. 21.1. From ideal western 'science' to 'prismatic societies'.

is more than twice their monthly salary. In addition to being financially advantageous, consultancy may attract prestige and lead to the accumulation of considerable cultural and/or political capital as well as a greatly enhanced income, also providing much valued opportunities for overseas travel. The local consultant might *also* be expected (in a modern equivalent of the potlatch) to contract out all or some of the work involved in the project to his/her less-qualified colleagues or students.

Furthermore, depending on the society, the university education of academics might have been quite different from their Western counterparts. In Lao PDR, Vietnam and Cambodia (and countries in the former Soviet bloc) there was considerable training in collecting numerical data but relatively little in more qualitative techniques, *especially* techniques of participant-observation – a problem which can only be compounded where populations have long been subjected to dictatorial rule and may continue to be subjected to surveillance by government agents.

Example: In an EU-funded international project involving cross-cultural cooperation of two European universities with two South-east Asian universities, designed to further understanding of how tourism can benefit the poor, considerable problems arose because of different interpretations towards fieldwork and, more generally, on how the research should be conducted. For the Europeans, the research was an enjoyable, intrinsic and necessary part of their university positions; by contrast, for the South-east Asian partners, although research was also part of their normal university duties, *this* kind of research, funded as it was by the European Commission, was categorized more as a form of consultancy, to be remunerated according to known and commonly used pay scales. It was a perspective also evident in other parts of the project, and drivers of vehicles and local officials who *necessarily* had to accompany the research team similarly expected a *per diem*, which was invariably several times their monthly salary.

Expectations of the *type* of research to be conducted also varied. For the Europeans, time spent in the field was a bonus, to be extended by eking out the *per diem* for as long as possible. It meant the opportunity to get the feel of a place, to observe, to participate, and try to make sense of data that were being collected through more formal research techniques. By contrast, South-east Asian academics attached great importance to carrying out the research quickly, and then returning to other duties. Indeed, there seemed also to be an expectation that, if fieldwork involved traipsing across muddy fields in the monsoon season to isolated villages, it was more appropriate for junior colleagues to go than the senior academic who was actually awarded the contract.

Similar experiences have emerged in a recent tourism project, working with academics in the Lower Mekong Sub-region. These include a reliance on data obtained from interviews with government and other officials (and a corresponding reluctance to involve other informants, either local residents or tourists), and in one case the national expert, an academic, used students to obtain data he was committed to collecting (an approach which admittedly *could* have been useful, but appears not to have been so in this case). In another instance, one of the national experts (selected by his compatriots) had no expertise in carrying out similar research, with somewhat dire consequences.[6]

Finally, as in developed societies, it cannot be assumed that the agendas or expectations of politicians are identical to those of other stakeholders. Clearly, circumstances vary across societies and over time, and generalizations can be misleading. However, in societies where there is a recent history of highly centralized government as in parts of South-east Asia, politicians and government officials, and some senior academics, may hold their positions as a consequence of party influence rather than established expertise. This can have noteworthy implications. First, there may be little understanding of the project requirements and, second, if they involve widespread consultation at community level, politicians and officials (like some researchers) may be unwilling to follow either the spirit or the letter of the project. Practices that characterize top-down development can be addictive. Third, where positions may have been obtained not through party affiliation but by purchase, a client/patron system may have emerged, where those who purchase a senior position must then sell junior positions in order to recoup their own costs.

Conclusions

The underlying aim of this paper was, first, to indicate that anthropologists have long held contrasting viewpoints about their relationships with, and the value of their research to governing authorities, and it has been suggested that debates preoccupying anthropologists during the colonial period have continued to the present, and still influence the way they work (and are *perceived* to work) with governments and other stakeholders involved in 'development'. Indeed, it is still possible to variously categorize the position of anthropologists towards governments in developing societies as principled rejectionists, monitorists or activists.

Second, it has been shown, more specifically, that such attitudes continue to divide anthropologists and other social scientists working on the role of tourism as a tool for development. Whereas some quite willingly promote tourism, or forms of tourism considered to be sustainable, others prefer a more neutral monitorist position, while yet others actively disparage tourism and, in some cases, the capitalist context in which it thrives.

Finally, the above divisions within networks of social anthropologists and tourism social scientists coexist and are thrown into relief by an even more complex series of overlapping networks when academics work on applied projects in developing countries. Expatriate and local personnel in NGOs and aid agencies, consultants and government officials, may all have markedly different agendas from visiting academics. Indeed, even local academics, formally a part of what might broadly be defined as the Western scientific tradition, reflect the values, status systems and expectations of their own 'prismatic' societies. In such cases, there are further impediments to the marriage of true minds, which here, as always, remains an ideal!

If the account provided here is at all valid, and accords with the experience of researchers elsewhere, it carries several implications which merit further discussion, both in 'academic' terms and for more practical reasons.

1. When academic exchanges across international boundaries are arranged, even within the European context, it cannot be assumed that the intellectual baggage of all participants is going to be identical. It is not. Ideas and intellectual practices can undoubtedly be transferred and shared, but all carry with them the cultural soil in which they prospered, and sometimes spend much time defending, and this has to be recognized.
2. Time must be taken – and allowed for by donor agencies – to enable detailed and sustained discussion of the principles underlying the project, and to build up relationships

among academics. The benefits from immediate, one-off associations are likely to be limited, whereas forms of cooperation developed over the years will inevitably be more productive.

3. Similarly, all participants – but perhaps especially those in donor agencies – need to be aware of the cultural contexts in which projects are being undertaken, and differences which may exist in underlying approaches. Even among academics, 'doing research' may have quite different meanings across cultures.
4. Anthropologists, in particular, note the importance of time spent in the field, as do most sociologists with a phenomenological approach. And – as Robert Chambers has long shown – much of this time should be spent *listening* to local people and learning from them (Chambers, 1994). This applies at all levels, and should include the major stakeholders discussed in this paper and *also* 'Tower level' participants, who are most likely to be affected by development programmes, for good or ill.

Other issues may also arise. However, what is necessary, above all, is for anthropologists and other social scientists working in tourism or other sectors, to be *reflexive* of their involvement in development and about their relationships with other stakeholders. Ironically, reflexivity in this context will enable them to theoretically situate their own roles more rigorously *and* to apply their skills and insights more appropriately and efficiently when involved in more practically oriented projects.

Notes

[1] Another example, not mentioned in the Grillo/Rew collection, was the direct involvement of some anthropologists in independence movements. Hilda Kuper, for example, who carried out research among the Swazi, was much associated with King Sobhuza II (Kuper, 1978). She helped the Swazi hierarchy articulate the nature of Swazi tradition (not surprisingly, perhaps, highlighting the importance of the king) when the constitution for independent Swaziland was being formulated. It was finally implemented in 1972, replacing a constitution previously imposed by the British in 1968.

[2] Notably, though, what is *not* usually requested is anthropological expertise! Indeed, the perceived association (correct or otherwise) of anthropology with colonialism led to its withdrawal from many universities in developing countries or (at best) its incorporation as but one aspect of sociology. More generally, in the UK, for instance, the tension between 'applied' and 'pure' disciplinary activity is reflected at numerous levels. Government policy is to increase the number of university students (thus inevitably increasing teaching commitments) but also to emphasize the importance for academics to obtain grants and publish in peer-reviewed journals. This prompts starker divisions between academics who teach and those who do research in the same institutions, as well as highlighting and exacerbating divisions between 'old' and 'new' universities.

[3] In many Western universities, the distinction between anthropology and sociology is now extremely blurred, though separate departments often continue to exist. By contrast, in many developing societies, especially those with a colonial history, anthropology has been rejected – mistakenly or otherwise – as too closely associated with colonialism.

[4] Petty nationalisms and linguistic ignorance frequently emerge, for instance, in discussions over the *language* to be used in academic debate and discussion.

[5] This perception of Western science, derived from Weber and portrayed in Fig. 21.1, and earlier, is not universally accepted but, for Weber, the validity of scientific knowledge and its related 'truths' ultimately rest in the inter-subjective nature of the scientific *community*. 'This itself reflects the evaluative ideas which 'dominate the investigator and his age', and the community is the arbiter of the value and significance of research (1949, p. 84). Anthropology, then, does not *have* to follow the precedents of, say, empirical social anthropology, though this clearly continues to predominate in both developed and developing societies (cf. Bernard, 2006). In so far as the views of postmodernists who consider qualitative research techniques as 'the telling of tales', as stories cobbled together for various purposes (Lincoln and Denzin, 2000) are *shared* by other social scientists, they too will be 'doing science' (though in developing societies, more misunderstandings and tensions than those described in the following pages are likely to occur).

[6] It is recognized that this is a subjective account, written from personal experience. National consultants may have very different perspectives of their international counterparts who, as Aziz points out (Aziz, 2003), may be quite ignorant of local contexts, and may also lack the time (or the inclination) to familiarize themselves with them. Not surprisingly, this, too, can lead to considerable suspicion or scepticism among local residents, including local academics.

References

Attwood, R. (2009) Study of terrorism steps up to new level. Times Higher Education. Available at: www.timeshighereducation.co.uk/ story.asp?storyCode=209437

Aziz, H. (2003) From the consultant's couch. *Tourism in Focus* 46(Spring), 7.

Becker, H.S. (1967) Whose side are we on? *Social Problems* 14(3), 239–247. DOI: 10.2307/799147.

Bernard, H.R. (2006) *Research Methods in Anthropology: Qualitative and Quantitative Approaches*, 4th edn. AltaMira Press, Oxford.

Berno, T. (1996) Cross-Cultural research methods: content or context? A Cook Islands example. In: Butler, R. and Hinch, T. (eds) *Tourism and Indigenous Peoples*. Thomson Business Press, London, pp. 376–395.

Berno, T. (1999) When a guest is a guest: Cook Islanders view tourism. *Annals of Tourism Research* 26(3), 656–675.

Chambers, R. (1994) The origins and practice of participatory rural appraisal. *World Development* 22(7), 953–969. DOI: 10.1016/0305-750X(94)90141-4.

Chambers, E. (2000) Applied ethnography. In: Denzin, N.K. and Lincoln, Y.S. (eds) *Handbook of Qualitative Research*. Sage, London, pp. 851–869.

Crick, M. (1994) *Resplendent Sites, Discordant Voices: Sri Lankans and International Tourism*. Harwood, Switzerland.

De kadt, E. (1979) *Tourism: Passport to Development*. Oxford University Press (for the World Bank), New York.

Errington, F. and Gewertz, D. (1989) Tourism and anthropology in a post-modern world. *Oceania* 60(1), 37–54. DOI: 10.1002/j.1834-4461.1989.tb00350.x.

Graburn, N.H.H. (1983) Tourism and prostitution. *Annals of Tourism Research* 10(3), 437–443. DOI: 10.1016/0160-7383(83)90068-3.

Grillo, R. (1985) Applied anthropology in the 1980s: retrospect and beyond. In: Grillo, R. and Rew, A. (eds) *Social Anthropology and Development Policy: ASA Monographs 23*. Tavistock, London, pp. 1–36.

Grillo, R. and Rew, A. (eds) (1985) *Social Anthropology and Development Policy: ASA Monographs, Vol. 23.*Tavistock, London.

Harrison, D. (2004) Working with the tourism industry: a case study from Fiji. *Social Responsibility* 1(1), 1–39.

Harrison, D. and Schipani, S. (2007) Lao tourism and poverty alleviation: community-based tourism and the private sector. *Current Issues in Tourism* 10(2–3), 194–230. DOI: 10.2167/cit310.0.

Horowitz, I.L. (1967) *The Rise and Fall of Project Camelot: Studies in the Relationship between Social Science and Practical Politics*. M.I.T. Press, Cambridge, MA.

Jenkins, C. (1999) Tourism Academics and practitioners: bridging the great divide. In: Pearce, D.G. and Butler, R.W. (eds) *Contemporary Issues in Tourism Development*. Routledge, London, pp. 52–64.

Johnson, E.M. (2007) The primate diaries: notes on science. Politics and culture from a primate in the human zoo: anthropology goes to war. Available at: http://primatediaries.blogspot.com/2007/10/anthropology-goes-to-war (accessed 17 October 2009).

Kuper, H. (1978) *Sobhuza II: Ngwenyama and King of Swaziland*. Holmes and Meier, New York.

Lacy, J.A. and Douglass, W.A. (2002) Beyond authenticity: the meanings and uses of cultural tourism. *Tourist Studies* 2(1), 5–21.

Liebling, A. (2001) Whose side are we on? Theory, practice and allegiances in prisons research. *British Journal of Criminology* 41(3), 472–484. DOI: 10.1093/bjc/41.3.472.

Lincoln, Y.S. and Denzin, N.K. (2000) The seventh moment: out of the past. In: Denzin, N.K. and Lincoln, Y.S. (eds) *Handbook of Qualitative Research*. Sage, London, pp. 1047–1065.

MacCannell, D. (1976) *The Tourist: A New Theory of the Leisure Class*. Macmillan, London.

MacCannell, D. (1989) *The Tourist: A New Theory of the Leisure Class*, 2nd edn. Macmillan, London.

MacCannell, D. (1992) *Empty Meeting Grounds*. Routledge, London and New York.

Marlowe, A. (2007) Anthropology goes to war: there are some things the army needs in Afghanistan, but more academics are not at the top of the list. *The Weekly Standard*. 13(11). Available at: http://www.weeklystandard.com/Utilities/printer_preview.asp

Mars, G. (2004) Refocusing with applied anthropology. *Anthropology Today* 20(1), 1–2. DOI: 10.1111/j.0268-540X.2004.00244.x.

Mcfate, M. (2005) Anthropology and counterinsurgency: the strange story of their curious relationship. Military Review. Available at: www.army.mil/professionalwriting/volumes/volume3/August_2005/7_05_2_pf.html (accessed 13 May 2007).

Mintz, S. (1977) Infant, victim and tourist: the anthropologist in the field. *John Hopkins Magazine* 27, 54–60.

Mowforth, M. and Munt, I. (1998) *Tourism and Sustainability: New Tourism in the Third World*. Routledge, London.

Nash, D. (1978) Tourism as a form of imperialism. In: Smith, V.L. (ed.) *Hosts and Guests: The Anthropology of Tourism*. Blackwell, Oxford, pp. 33–47.

Nash, D. (1996) *The Anthropology of Tourism*. Pergamon Press, Oxford.

Nash, D. (2001) On travelers, ethnographers and tourists. *Annals of Tourism Research* 28(2), 493–496. DOI: 10.1016/S0160-7383(00)00053-0.

Price, D.H. (1998) Cold War anthropology: collaborators and victims of the national security state. *Identities* 4(3–4), 389–430. DOI: 10.1080/1070289X.1998.9962596.

Riggs, F.W. (1964) *Administration in Developing Countries: The Theory of Prismatic Society*. Houghton Mifflin, Boston, MA.

Rohde, D. (2007) Army enlists anthropology in war zones. New York Times. Available at: http://nytimes.com/gst/fullpage.html (accessed 30 October 2007).

Rowe, D. (2005) Some critical reflections on research and consultancy in cultural tourism planning. *Tourism Culture & Communication* 5(3), 127–137. DOI: 10.3727/109830405774545071.

Shweder, R.A. (2007) A true culture war. The New York Times. Available at: http://www.nytimes.com/2007/10/27/opinion/ 27shweder.html (accessed 30 October 2007).

Smith, V.L. (ed.) (1978) *Hosts and Guests: The Anthropology of Tourism*. Blackwell, Oxford.

Tourism Research Consultants (1999) *Report on the Pacific Ecotourism Workshop. Taveuni, Fiji Islands, 28–31 July, 1998*. Convened and funded by the New Zealand Overseas Development Agency, Wellington.

Wakin, E. (1992) *Anthropology Goes to War: Professional Ethics and Counterinsurgency in Thailand*. University of Wisconsin Press, Madison.

Weber, M. (1948) Science as a vocation. In: Gerth, H. and Mills, C.W. (eds) *From Max Weber*. Routledge and Kegan Paul, London, pp. 129–156.

22

Looking Ahead

Over the last seven decades, successive advocates of international tourism have confidently asserted that tourism has indeed provided a 'passport to development'. And except for 2001 and the terrorist attacks on the USA, or the outbreak of SARS in 2003, international tourism has long seemed to be the most reliable and accessible form of development. With the apparently standard proviso 'as far as one can see', it was confidently predicted the trend would continue and, until early 2020, there was no reason to doubt the evidence of 70 years. For virtually every year over this period, international tourism increased. International tourist trips in 2019 were calculated to be 1.4 billion (UNWTO, 2019, p. 1) and as late as 25 January 2020, the UNWTO was suggesting that growth in international tourism would continue to outpace the global economy (www.unwto.org, accessed 20 January 2020):

> The UNWTO Secretary-General Zurab Pololikash stressed that 'in these times of uncertainty and volatility, tourism remains a reliable economic sector'. Against the backdrop of recently downgraded global economic perspectives, international trade tensions, social unrest and geopolitical uncertainty, 'our sector keeps outpacing the world economy and calling upon us to not only grow but to grow better', he added (UNWTO, 2020a; www.unwto.org, 20 January 2020).

Then the unthinkable happened. The 'reliable economic sector' was no longer reliable. In January 2020, the virus COVID-19 started to concern medical professionals and, after a period of denial in China, established itself as a major threat to global prosperity, rapidly expanding from East and South-east Asia to the West. On 30 January the World Health Organization declared the outbreak to be a Public Health Emergency of International Concern, and on 11 March upgraded it to a pandemic. By 21 September, 31,296,553 cases had been reported, resulting in 965,857 deaths worldwide. Most cases were in the USA, India, Brazil, Russia, Peru, Columbia, Mexico, South Africa and Spain, while Peru, Brazil, Chile, Spain, Mexico, the USA, the UK and Italy recorded the mostdeaths as a proportion of their populations. Altogether, 188 countries have been affected, the global economy shattered, everyday economic and social activities have been severely curtailed, and country after country has been sent into lockdown (www.worldometers.info/coronavirus, accessed 21 September 2020). The social impacts from lockdown have been dire: entire populations restricted to their homes for weeks, sometimes months, loved ones separated, to the extent that people have tragically been prevented visiting sick or dying relatives, attending funerals or weddings, and carrying out the most basic of economic activities.

Air travel – so vital to international trade – was effectively brought to a catastrophic halt, from which some companies are unlikely to recover, and the leisure and hospitality sector, so often in the past the leader of recovery, closed down, as did the many attractions so popular among tourists. With offices, pubs and restaurants closed, once vibrant city centres have come to resemble ghost towns and, because hospitality relies most on close human contact – hitherto its greatest asset – it will be among the last to recover.

Cruising, which had been undergoing a boom time, has also been catastrophically impacted by the pandemic. The problem first emerged on Carnival's *The Diamond Princess*, and more than 700 passengers were infected with the coronavirus, leading to 13 deaths. In February the ship was quarantined by Japan off Yokohama.

> Its sister ship the Grand Princess was later held off Oakland, California, with 78 passengers testing positive for the virus, while other ships, such as Holland America Line's Westerdam, were forced to bounce between ports as country after country refused to let potentially infected ships reach land (Hancock, 2020).

Such problems impacted unfavourably on the cruise sector. Despite the fact that only 82 people are reported to have died from coronavirus on cruise ships (Hancock, 2020), they now have an unenviable reputation, from which it will take a long time to recover. The situation is made worse by the fact that – to cut costs – the ships are mainly registered in the Bahamas and Panama, and so cruise companies cannot claim government assistance. Meanwhile, 338 ships have been docked, but costs still have to be met. Carnival, the world's largest cruise company, for instance, has to pay US$1 billion a month simply to maintain its fleet. In addition, as at 7 June 2020, 60,000 crew members were stranded on board ships, waiting to see if they could be repatriated to their home countries.

> The industry has ceased to function altogether. For it, coronavirus has been the perfect storm. It has gone from being an industry worth $46bn (£37 bn) a year, with 26 million passengers per annum, to an almost total standstill overnight (Bloom, 2020).

At the time of writing, cruise companies are not planning to resume schedules until later in the year. P and O cruises, for example, has written off the summer season and will resume operations in October, with a huge advertising campaign and heavily discounted prices.

Clearly, the economic costs have been stark. In the USA, for instance, nearly 39 million Americans reportedly lost their jobs in 9 weeks (Rushe, 2020) and, as at 7 August, 2020, 5,042,905. coronovirus cases were reported in the USA, leading to 162,990 deaths. In addition, it has been estimated that 54 million people in the USA will need assistance from food banks, food stamps and other forms of aid (Lakhani, 2020). In the UK, the number claiming unemployment benefit was nearly 2.1 million (Partington and Partridge, 2020) and the UK economy shrank by a record 20.4% (Narwan, 2020). The Office for Budget Responsibility (OBR) warned that the coronavirus could cause a 35% fall in Gross Domestic Product (GDP) if lockdown were to last for 3 months, and GDP could crash by around 13% for 2020 as a whole, while Britain's budget deficit is expected to surge to the highest level since the Second World War, surpassing the damage caused by the 2008 financial crisis (Guardian, 2020). Elsewhere in Europe, Germany was plunged into recession, and in the first quarter of 2019 the economies of Italy, France and Spain collapsed, with the GDP in France falling by 5.8% in the first quarter (before the full impact of coronavirus), and that of Italy by 4.7% (Rogers, 2020). Indeed,

> The EU saw the sharpest drop in GDP for more than a decade as coronavirus lockdowns brought Europe's economies to a standstill, new data show. GDP in the eurozone tumbled 3.1% in the first three months of 2020, compared with the same period a year earlier. The EU saw a fall of 2.6% over the same period. There (sic) are the biggest drop since 2009, according to Eurostat (Hurst, 2020).

According to the International Monetary Fund, the world faces the worst recession since the Great Depression of the 1930s. Tourism has been hit especially badly: according to the UNWTO, potential impacts of COVID-19 could vary: 850 million to 1.1 million fewer international tourist arrivals; US$910 billion to US$1.2 trillion loss in export revenues from tourism and 100 to 120 million direct tourism jobs at risk (UNWTO, 2020b).

The range of possible impacts is an indication of the degree of uncertainty is estimated that Britain's economy 'is now smaller than it was at its peak in February' (Narwan, 2020).

The crisis, then, has been global in extent and impact, and has affected both domestic and international tourism. At the time of writing (September, 2020) many Western countries, having tentatively attempted to reduce the lockdown and kickstart their economies, while trying to regulate social contact in the hope of avoiding or minimising spikes, are now faced with the prospect of a second wave of the pandemic. In such cases, there is the possibility that local lockdowns might be extended to entire countries. Meanwhile, in Southeast Asia, there is optimism that the pandemic is under control, but it is gaining momentum in such developing regions as Africa, Latin America and South Asia. There, perhaps, the worst is yet to come. Certainly, in mid-September, 2020, the future for the world economy, and tourism, is far from certain. A headline in *The Sunday Times*, 'Hospitality firms face wipeout' (Kelly and Treanor, 2020), augers an uncertain economic future not just for the UK but for the tourism sector across Europe and North America.

Taking the picture as a whole, and provided the pandemic is controlled, Western economies will slowly recover, but depression is inevitable. The impact on tourism will depend to some extent on previous reliance on the sector, and the ability and willingness of people to resume earlier travel plans. Small island states, for example, in which the share of international tourism revenues as a percentage of total exports is relatively high, are especially vulnerable. The UNWTO estimates that in 19 such states the share of tourism is between 50% and 90%, with a further 17 showing an excess of 30% (UNWTO, 2020b). And those who rely on tourism will have most to lose if the virus takes hold. In fact, so far, island states in the Caribbean and the South Pacific have shown considerable resilience, with relatively few deaths attributed to COVID-19 (https://www.ecdc.europa.eu/en/geographical-distribution-2019-ncov-cases, accessed 3 June).

Most analyses of tourism focus on three aspects of the economy, namely travel, accommodation and attractions, which together contextualize both domestic and international tourism, and they provide the foundation of the following discussion. First, tourism and travel are intimately linked: while travel can be for purposes other than tourism, tourism always involves travel from one place to another. At its most rigorous, under lockdown virtually *all* travel was banned, and many millions of people worldwide were confined (for at least 3 months) to their homes.

Domestic and international tourism were effectively curtailed when lockdown was imposed. Holiday plans amounted to nothing as airline companies, along with airports and ground-handling companies, suddenly found themselves without passengers. Nearly one hundred per cent of their operations were grounded. Some airlines went bankrupt, and in common with other industries, staff were laid off. In the UK, British Airways cut up to 12,000 positions from its workforce of 42,000 and has plans for further drastic reductions in staff, while EasyJet laid off 4000 of its cabin crew for 2 months. Elsewhere, Qantas put 20,000 on leave, American Airlines persuaded 700 pilots to take early retirement (Leggett, 2020), and airlines in Norway, Italy, Finland and Sweden have all been in receipt of emergency assistance from government (Kammel, 2020).

The hotel sector was similarly affected. In the USA, for instance, the lockdown led to 70% of hotel rooms being empty, with the loss of four million jobs and in excess of $21 million revenue. Such major hotel chains as Hilton, Marriott, Hyatt and Accor reported dramatic falls in business, and the CEO of Hilton suggested 'It could take several years to return to the hotel demand levels we experienced in 2019' (Touryalai, 2020).

Whereas major hotel chains have the wherewithal to weather major shocks to their business, small and medium-sized businesses are more vulnerable. In the UK, for instance, which in mid-June was still operating a two-metre rule in social distance, deemed impractical in the hospitality sector, it has been suggested that one in ten pubs and hotels face collapse within weeks and need urgent support (Gye, 2020).

After the Lockdown

In August 2020, as the lockdown was tentatively eased and travel restrictions slowly lifted, the tourism sector showed some signs of recovery.

Invariably, domestic tourism – which is several times greater than international tourism – was to lead the way (Anon, 2020a). According to the OECD, it accounts for 75% of the total tourism economy in OECD countries, and while it has been heavily affected by restrictions on the movement of people, is likely to recover more quickly than international travel once such measures are eased (OECD (Organisation for Economic Co-operation and Development), 2020). In the USA, for instance, Memorial Day weekend (25 May) 'brought the biggest day for air travel since the coronavirus took hold' (Puckett, 2020), and in May 2020 the demand for domestic tourism was clearly evident in China, when 115 million took a tourist trip on May Day (Anon, 2020b; Mariano, 2020a), the UK (Robinson, 2020) and New Zealand (Mariano, 2020b). Travel to adjacent countries follows domestic tourism, to be joined later by long haul travel. Early in June 2020, most countries allowed entry only to exceptional cases (Fitzgerald, 2020), though some traditional holiday destinations, for example Portugal, Spain and Greece, had eased restrictions and signalled they were open for business. However, despite the reluctance of some member states, the EU called on most of its 27 states to re-open borders within the bloc on 15 June.

At the time of writing, efforts were being made to recommence international air travel. On 15 June, ten aircraft of EasyJets's 300-strong fleet were in operation, mainly on domestic service, but also routes to Switzerland, France, Italy and Portugal (Topham, 2020a, 2 June), and it announced that in August it would be putting on 75% of its routes, though at a third of its normal peak capacity (Topham, 2020b). However, the holiday experience is likely to be very different from before the pandemic. Passenger loads will have to fall dramatically to facilitate social distancing which, in turn, could be reflected in higher fares (Topham, 2020a, 2 June) Elsewhere, Delta Air Lines is to add flights to ensure that none are at more than 60% capacity (www.reuters.com/article/us-health-coronavirus, accessed 2 June) and other airlines follow similar policies. Furthermore, the need for social distancing will continue at destinations, inevitably reducing the personal contact so valued in the hospitality sector.

In order to stimulate tourist demand, some governments are encouraging outbound tourism by agreeing to air bridges linking them with similar destinations, thus negating the need to subject passengers to quarantine (Paton and Wright, 2020). However, arrangements can change quickly, leading to the reimposition of restrictions and, at least in the short term, some quarantine regulations will continue; health checks will abound at airports and on aircraft, and the need for social distancing means that passenger numbers will be substantially reduced, possibly resulting in higher airfares.

More specifically, the future of air travel in the short term, will depend on a number of factors:

1. Government restrictions. These continue to be a major factor. All member countries of the European Union have introduced controls on air travel, and some (e.g. Denmark and France) have restrictions that will last until beyond September 2020 (Vokrinek, 2020).
2. The cost of travel. Because airlines will have to reduce carrying capacity to allow for social distancing, they will either have to increase the price of travel, thus depressing demand even further, or suffer a loss of revenue.
3. The need for social distancing will affect every stage of the holiday, commencing with arrival at the airport and boarding procedures. Thermal cameras and sanitation booths at major airports may be introduced, cleaning and sanitation of restrooms will be increased, deep cleaning of aircraft and luggage will add time to the trip, and masks will be worn by passengers and flight attendants. In addition, passenger movement during the flight will be strictly controlled, and there will be no queues for toilets.
4. Passenger demand. This is likely to vary considerably. Some evidence exists that there will be a 'bounce back' following the removal of travel restrictions, but this will vary according to the nature of the vacation and the potential travellers. Indeed, more pessimistic projections suggest the pace of recovery could average 5 years (Vokřinek, 2020). Younger travellers may lead the way, whereas older travellers might prefer to wait, or opt for travel to more familiar (and closer) destinations. By contrast, according to some authorities, business travel will take longer to revive (Vokrinek, 2020). Demand for long-haul destinations may take even longer

to recover, and airline sources suggest that it will take at least 2 or 3 years for demand to return to pre-pandemic levels.

Furthermore, there is evidence that demand will be inconsistent. Langham Hospitality Group reports that some of their properties in China were

> recording strong peaks during major holidays with occupancy rates up to 70% in city hotels and reaching near full capacity in resorts, but then quickly falling to as low as 20% within a few days (Vokrinek, 2020).

Indeed, such uncertainties in demand could mean 5–15% of hotels do not re-open at all (Vokrinek, 2020). What is perhaps more certain is that the global collapse of tourism will drive many small and medium enterprises to bankruptcy, and leave those that remain in a more precarious position than they were before the pandemic.

1. There will also be job losses in aviation because of the coronavirus pandemic. In the UK it is calculated that at least 70,000 jobs in the wider aviation industry – including engineering, catering and duty free shopping – were at risk before the end of the summer (Partington and Partridge, 2020).
2. The journey: Deep cleaning of aircraft and luggage will add time to the trip, and masks will be worn by passengers and flight attendants.
3. Once at the destination, too, social distance will still be required to protect hotel staff and other guests, and guests are likely to wear masks when not in their rooms. Staff, too, will have to wear masks, and public spaces will be cleaned regularly and social contact strictly controlled. This means that hotels will 'initially have to operate with heavily reduced services and amenities, more like 'bedroom factories' (Vokrinek, 2020), and it remains to be seen whether or not guests will accept such changes.

Attractions

The third factor that helps form the holiday experience is attractions, which range from the buildings, places or people that tourists travel to experience, as well as major sports events – and most have become unavailable during the lockdown. Because of the virus, most tourist attractions around the world, such as museums, amusement parks and sports venues, have been closed. In the USA, for instance, Disney closed Walt Disney World in Orlando, its Disneyland resort in Los Angeles, and Disney's theme parks in Tokyo, Hong Kong and Shanghai. Universal Studios theme parks in Orlando and Los Angeles announced they were closing temporarily, and St. Patrick's Day parades in Boston, Philadelphia and New Orleans were cancelled.

While domestic travel has started to revive, foreign visitors – who provide much of the income from tourism – have been largely restricted from travelling. In Europe, where it seems the pandemic is gradually coming under control, in mid-August 2020 there was a very gradual opening of some attractions, albeit with restrictions to ensure social distance. However, not all governments are enthusiastic about a return to tourism. The Dutch government is not alone in putting the safety of its residents ahead of the tourism sector:

> In line with government measures to reduce the spread of the coronavirus, all non-essential travel is discouraged. We're proud of our beautiful city, but this is a time to keep yourself and others safe by staying at home and avoiding crowded places (https://www.iamsterdam.com/en/plan-your-trip/practical-info/corona-virus, accessed 8 June).

Where coronavirus is still a threat, attractions remain closed, and those that *are* open are operating in a very insecure environment. In China, which seems to have controlled the pandemic sooner than other countries, attractions have opened, only for some – for example, Shanghai Haichang Ocean Park – to close again when a second wave of infections was feared.

Clearly, the situation in Asia is problematic. However, its experience with attractions provides lessons for Europe and the USA. It is likely that attractions will have to adapt to a 'new normal'. This includes several of the safety measures taken in Chinese theme parks. For instance, temperature checks, extra cleaning, social distancing measures, face masks and visitor caps. We have also seen that the restrictions on high-risk attractions might take longer to lift. For example, cinemas and other indoor leisure venues (Mitchell, 2020).

Developing Countries – and Beyond

Whereas much of the tourism from the West, or indeed from China and other far eastern countries, might come under the heading of mass tourism (Harrison and Sharpley, 2017), some tourism to developing countries has followed a different pattern, in that the country itself, and those who provide hospitality, are part of the attraction. Here, relatively small groups of tourists may seek – often in smaller scale and less luxurious accommodation – to experience a way of life that in some senses is different from their own. Visits by Western tourists to Lao PDR, Cambodia, or to other parts of South-east Asia, for example, fall into this category. Often advertised as forms of ecotourism, or ethnic tourism, and marginal to mainstream tourism literature (Weaver and Lawton, 2007), such ventures are less capital intensive and more likely to operate with smaller profit margins. As a consequence, like small and medium enterprises elsewhere, many will find it difficult to survive the lockdown, especially as international tourists from the West, at least initially, are likely to be wary of an early return to South-east Asia. Alternatively, there are also smaller, niche enterprises catering for wealthier travellers – for example, those wishing to go on African safaris or to the Polar region – where profit margins are higher, and whose tourists would not be put off by higher air-fares.

By contrast, what might be described as 'island tourism' involves small, tropical island states whose tourism attractions, often of the 'sun, sea and sand' variety, appeals specifically to more developed countries within the region. The islands of the South Pacific, for instance, might be regarded as the leisure periphery for Australians and New Zealanders, while tourism in the Caribbean is geared more to the USA (Harrison, 2001; Harrison and Pratt, 2015). Where such dependence exists, the fortunes of island states in a post-COVID-19 world are inextricably linked to those of their more developed partners. Certainly, at the time of writing, tourism in such small island states has more or less ceased, leaving such peripheries totally exposed.

It has been noted earlier that increases in tourism can, in certain circumstances, lead to some parts of the population adopting patterns of tourist behaviour, wearing sun glasses and generally mimicking tourists, and it is also thought that tourism has brought increased independence to certain sections of the population, most notably women and young people. Such changes have often led to a decline in the authority of traditional elders, who have been side-lined by the processes of modernity brought by tourism and commoditization. However, with lockdown and the catastrophic reduction in tourism it brought, such changes may have been halted, or even reversed, and this new situation offers social scientists the opportunity to study an unexpected phenomenon: in brief, what happens when 'development' is brought to an abrupt halt? What was the effect of a loss of tourists and the income that comes with them? Have young people returned to their villages, reversing their former migration, and reverted to more traditional forms of occupation, for example subsistence production? Has women's independence been curtailed with the loss of income that has occurred? In what ways have the young and women compensated for the loss of income from tourism? Have 'traditional' (usually male) authorities reasserted themselves? Have modern and traditional forms of production been able to operate harmoniously together, so that when one fails, society members can fall back on the other? Has the family played a role in such circumstances? How has the crisis in world tourism affected the relationship of metropolitan, tourist-sending countries (including China) with such tourist-receiving countries as Thailand?

Such questions focus on processes of social and economic change, but in periods of crisis rather than periods of growth. Noticeably, too, most are also relevant to tourist destinations in developed societies. Blackpool's problems are not unique.

Conclusion

Themes I discussed in the first chapter included the role of the private sector and the state, the role of China, and the development of 'tradition'. All of these have been very much present from March to August 2020, and their influence will continue long beyond this date. Clearly, though, throughout the crisis the state everywhere has played a central role and, moreover, it has been forced to act (more or less) in conjunction with other states and the United Nations (in the form of the World Health Organization). It is no exaggeration to say that the extent of its role is unprecedented. In Britain, a Conservative government – not strongly predisposed to state intervention – provided millions of workers with a furlough scheme that paid 80% of their salary, which was to last until August, when it was reduced to 50% (Parsley *et al.*, 2020), while in the USA at least 36 million Americans filed for unemployment benefits (Rushe *et al.*, 2020) and similar schemes were implemented elsewhere.

The role of China in world tourism, too, has been crucial. The pandemic first emerged there, late in 2019, and government exercised its role to the full, shutting down entire cities and banning all but the most necessary movement away from home. At a stroke, tourism in China's neighbours and further afield went into immediate decline. And it is tourism *from* China which (it is hoped) will kick-start international tourism again.

The role tradition has played during the pandemic has not been discussed in this afterword but, for the social scientist, questions concerning its role, and the process of commoditization, are crucial. For so long we have discussed the structural impacts of tourism's growth. What happens when the process is reversed?

Finally, some commentators have seen in the pandemic's effects the hope of a greener tourism, a new phoenix rising from the ashes of the old, outdated mass tourism model. In this respect, there is some support from changes in the aviation industry:

> 'The political movement is now' to address the climate risks posed by the aviation industry, analysts, insiders and campaigners say, as governments across the world weigh up bailouts for airlines grounded by the coronavirus pandemic.
>
> Rescue packages need to come with green strings, such as reduced carbon footprints and frequent flyer levies, they warn, or the sector will return to the path that has made it the fastest rising source of climate-wrecking carbon emissions over the past decade (Watts, 2020, 17 May).

Whether or not the pandemic will make it easier to introduce green energy, however, is open to question, and other commentators are more doubtful (Sweney, 2020, 27 May). However, renewed tourism will allow airlines to pension off older and less fuel efficient aircraft. More generally, there is a pressure group that recommends a deliberate policy of degrowth, to encourage more environmentally and socially friendly international tourism, and which sees the present time as appropriate to implement its policies, including restrictions on the quantity of mass transport, equitable allocation of space on transport, disincentives for short-term travel, use of communication technologies for online interaction, and subsidies for small and medium-sized businesses (Fletcher *et al.*, 2020). However, while the policies proposed are comprehensive, there is little evidence that they carry much traction among the big hitters of the international tourism industry, governments anxious to reinstate such profitable sources of foreign exchange, or even members of the general population (whose response to a lessening of travel regulations might well be to take a holiday).

Looking ahead, then, we may consider two possible future scenarios. First, efforts to develop a vaccine against COVID-19 are unsuccessful, and the hospitality sector in both developed and developing countries fails to recover the momentum it enjoyed until early 2020. In such circumstances, international tourism will continue to struggle, against the pandemic and *fears* of the pandemic, to reinstate the successes and tourist numbers of the past. Tourism will grow but little, airlines will continue to retrench, and many small and medium-sized tourism enterprises will struggle to survive, with high rates of failure, especially, perhaps, in developing countries. The hospitality sector, until now a major source of economic growth and employment, will be substantially reduced and the progress of international tourism, hitherto assumed to be unstoppable, will be arrested, major barriers to international travel would remain, and recession in the global economy would continue. And this should not be minimised. By August 2020, the world was in a fully-fledged recession, the extent of which far exceeded any which had previously occurred.

Frankly, though, this scenario is unlikely. The race to produce a vaccine against COVID-19 has been unprecedented, and already there are some signs of success. Much more is now known of the virus, and with knowledge comes the ability to counter it, either with a vaccine or with other measures to reduce its impact. If this is indeed a more accurate reading of the current situation (and I claim no expertise in this area), the short term impacts of the pandemic, difficult though they have been for both the tourism 'industry' and economies more generally, will be overcome, and by 2022 international tourism will again continue its upward trend, albeit from a much smaller base than seemed likely at the end of 2019.

Will it be any different from what would have happened without the intervention of the pandemic? I doubt it. There will be clarion calls among tourism's critics for (yet another) 'new tourism', and residents of over-popular destination areas (e.g. Barcelona and Venice) may (again) attempt to regulate (but not ban) tourism numbers to make their lives more tolerable. Reducing, or even banning, cruise ships would be a start. And it is just *possible* that Western and Chinese tourists will reject mass tourism and embrace smaller-scale alternatives (a move which would create problems of its own). Possible – but unlikely. Instead, outbound tourists from China will lead the way out of COVID-19 – induced recession and destinations in South-east Asia and Europe will gratefully welcome their return, which will initially be tentative and will, in all likelihood, take two or three virus-free years to occur. But: if the coronavirus persists, and no vaccine is developed for its control, the story could be very different. If the tourist industry cannot be revived, the global economy will be condemned to remain in decline, unemployment will remain high in both tourist-sending and tourist-receiving societies, and social and political unrest will be a real possibility.

And the study of tourism? Whether or not the crisis is relatively short-lived, social scientists should take the opportunity offered by changes in international tourism to study the social and cultural impacts of the new situation, influencing as it does so many millions of travellers and members of host societies, in both developed and developing societies. For them, such a situation has not arisen before. COVID-19 is indeed an ill wind, but it also presents unexpected opportunities for research that should not be passed over.

References

Anon (2020a) Coronovirus (COVID-19) recovery hope for hotels: exploring the domestic market. Available at: https://www.siteminder.com/marketing/hotel-digital-maarketing/coronovirus-covid-19-recovery-hotels-domestic-tourism

Anon (2020b) How China is restoring its tourism industry after three months in lockdown. Available at: https://www.express.co.uk/news/world/1278910/china/coronavirus (accessed 15 June).

Bloom, J. (2020) Will we ever take cruise holidays again? BBC news. Available at: https://www.bbc.co.uk/news/business-52182509 (accessed 6 August 2020).

Fitzgerald, R. (2020) Coronavirus travel update: which countries are open and which are closed? Available at: https://www.wanderlust.co.uk/content/coronavirus-travel (accessed 1 June).

Fletcher, R., Mas, I.M., Blazquez-Salom, M. and Blanco-Romero, A. (2020) Tourism, Degrowth and the COVID-19 crisis. Available at: https://politicalecologynetwork.org/2020/03/24 (accessed 13 August 2020).

Guardian (2020) What will coronavirus mean for the British economy? Available at: www.theguardian.com (accessed 13 August 2020).

Gye, H. (2020) Coronavirus latest: One in ten pubs and hotels face collapse within weeks and 'need urgent support'. inews. Available at: https://inews.co.uk/news/coronavirus-uk-latest-pubs-hotels-face-collapse-urgent-support-430602(accessed 6 August 2020).

Hancock, A. (2020) Coronavirus: Is this the end of the line for cruise ships? Financial Times, 7 June.

Harrison, D. (2001) Tourism in small islands and microstates. *Tourism Recreation Research* 26(3), 9–14.

Harrison, D. and Pratt, S. (2015) Tourism in Pacific island countries: current issues and future challenges. In: Pratt, S. and Harrison, D. (eds) *Tourism in Pacific Islands*. Routledge, London, pp. 3–21.

Harrison, D. and Sharpley, R. (2017) Introduction: mass tourism in a small world. In: Harrison, D. and Sharpley, R. (eds) *Mass Tourism in a Small World*. CAB International, Wallingford, pp. 1–14.

Hurst, L. (2020) EU saw the sharpest GDP fall in more than a decade during COVID-19 lockdowns. euronews. Available at: https://www.euronews.com/2020/06/09/eu-saw-sharpest-gdp-fall-in-more-than-a-decade-during-covid-19-lockdowns (accessed 6 August 2020).

Kammel, B. (2020) Desperate airlines turn to European governments for support. *Bloomberg Businessweek*.

Kelly, L. and Treanor, J. (2020) Pubs and restaurants face coronavirus wipeout.. The Sunday times. Available at: https://www.thetimes.co.uk/article/pubs-and-restaurants-face-coronavirus-wipeout-d7brdzgll (accessed 21 September 2020).

Lakhani, N. (2020) People are going to go hungry: pandemic effects could leave 54m Americans without food. *The Guardian*.

Leggett, T. (2020) Coronavirus: how will airlines get flying again? BBC news. Available at: http://www.bbc.co.uk/news/business

Mariano, K. (2020a) Travelling after lockdown: what can we learn from China? Available at: https://www.traveldailymedia.com/travelling-after-lockdown-what-can-we-learn-from-china (accessed 27 May 2020).

Mariano, K. (2020b) New Zealand promotes domestic tourism and makes the world green with envy. Available at: https://www.traveldailymedia.com/new-zealand-domestic-tourism

Mitchell, B. (2020) Coronavirus, Asian attractions and the new normal. Available at: https://blooloop.com/features/asian-attractions-coronavirus

Narwan, G. (2020) Coronavirus: UK economy shrinks by record 20.4%. *The Times*.

OECD (Organisation for Economic Co-operation and Development) (2020) Tourism policy responses to the coronavirus (COVID-19). Available at: https://www.oecd.org/coronavirus/policy-responses/tourism-policy-responses-to-the-coronavirus-covid-19-6466aa20/

Parsley, D., Gye, H. and Newlands, C. (2020) Coronavirus: hospitality sector to be protected as government begin to reduce furlough salary scheme. Available at: https://inews.co.uk/news/business/coronavirus (accessed 7 June 2020).

Partington, R. and Partridge, J. (2020) Airline job losses could be on scale of 1980s, report warns. The Guardian. Available at: https://www.the guardian.com/business/2020/jun/10/airline (accessed 10 June).

Paton, G. and Wright, O. (2020) 'Air bridges' plan revives hope of summer flights. *The Times*.

Puckett, J. (2020) Memorial day weekend data shows a rise in domestic travel. Available at: https://www.cntraveler.com/story/memorial-day-weekend (accessed 29 May 2020).

Robinson, A. (2020) UK travel: staycations set for a boom after lockdown as Britons crave domestic holidays. Available at: https://www.express.co.uk/travel/articles/1275398

Rogers, J. (2020) Profits of Doom: Coronavirus plunges Europe further into recession as Germany's economy collapses after Italy, France and Spain. The Sun, 15 May.

Rushe, D. (2020) US unemployment rises by 2.4m despite easing of coronavirus lockdowns. The Guardian.

Rushe, D., Aratani, L. and Holpuch, A. (2020) 36m Americans now unemployed as another 3m file for benefits. Available at: https://www.theguardian.com/business/2020/may/14 (accessed 13 June 2020).

Sweney, M. (2020, 27 May) Investment in global energy to drop by $400bn. *The Guardian*.

Topham, G. (2020a, 2 June) EasyJet to resume flights across most European routes by August. *The Guardian*.

Topham, G. (2020b) Lots of distancing, no long-distance: airlines cautiously return to the sky. Available at: https://www.theguardian.com/business/2020/jun/13/airlines-easyjet-return-skies-social-distancing (accessed 14 June).

Touryalai, H. (2020) Hotels are getting crushed by coronavirus: how will they recover? Available at: www.forbes.com/sites/halahtouryalai/2020/05/13 (accessed 8th June).

UNWTO (2019). *World Tourism Barometer* 17(1), 1.

UNWTO (2020a) International tourism growth continues to outpace the global economy. Available at: www.unwto.org

UNWTO (2020b) *UNWTO World Tourism Barometer, May 2020: Special Focus on the Impact of COVID-19*. UNWTO, Madrid.

Vokrinek, B. (2020) COVID-19 impacts on European hospitality. Available at: www.cushmanwakefield.com/en/united-kingdom/insights/covid-19-impacts-european -hospitality

Watts, J. (2020, 17 May) Is the Covid-19 crisis the catalyst for greening the world's airlines? *The Guardian*.

Weaver, D.B. and Lawton, L.J. (2007) Twenty years on: the state of contemporary ecotourism research. *Tourism Management* 28(5), 1168–1179. DOI: 10.1016/j.tourman.2007.03.004.

Index

Note: The locators in bold and italics represents the tables and figures respectively.